FlexPDE and Finite Element Method Applications in Thermal Energy Storage and Cavities

Khashayar Hosseinzadeh
Sharif University of Technology, Iran

Mehdi Mahboobtosi
Babol Noshirvani University of Technology, Iran

Davood Domairry Ganji
Babol Noshirvani University of Technology, Iran

A volume in the Advances in Chemical and Materials Engineering (ACME) Book Series

IGI Global
PUBLISHER of TIMELY KNOWLEDGE

Published in the United States of America by
IGI Global
Engineering Science Reference (an imprint of IGI Global)
701 E. Chocolate Avenue
Hershey PA, USA 17033
Tel: 717-533-8845
Fax: 717-533-8661
E-mail: cust@igi-global.com
Web site: http://www.igi-global.com

Library of Congress Cataloging-in-Publication Data

CIP PENDING
ISBN:979-8-3693-0932-2
eISBN:979-8-3693-0933-9

This book is published in the IGI Global book series Advances in Chemical and Materials Engineering (ACME) (ISSN: 2327-5448; eISSN: 2327-5456)

British Cataloguing in Publication Data
A Cataloguing in Publication record for this book is available from the British Library.

For electronic access to this publication, please contact: eresources@igi-global.com.

Advances in Chemical and Materials Engineering (ACME) Book Series

J. Paulo Davim
University of Aveiro, Portugal

ISSN:2327-5448
EISSN:2327-5456

Mission

The cross disciplinary approach of chemical and materials engineering is rapidly growing as it applies to the study of educational, scientific and industrial research activities by solving complex chemical problems using computational techniques and statistical methods.

The **Advances in Chemical and Materials Engineering (ACME) Book Series** provides research on the recent advances throughout computational and statistical methods of analysis and modeling. This series brings together collaboration between chemists, engineers, statisticians, and computer scientists and offers a wealth of knowledge and useful tools to academics, practitioners, and professionals through high quality publications.

Coverage

- Polymers Engineering
- Nanomaterials
- Fracture Mechanics
- Heat Treatments
- Chemical engineering
- Multifuncional and Smart Materials
- Fatigue and Creep
- Toxicity of Materials
- Wear of Materials
- Metallic Alloys

IGI Global is currently accepting manuscripts for publication within this series. To submit a proposal for a volume in this series, please contact our Acquisition Editors at acquisitions@igi-global.com or visit: https://www.igi-global.com/publish/.

The Advances in Chemical and Materials Engineering (ACME) Book Series (ISSN 2327-5448) is published by IGI Global, 701 E. Chocolate Avenue, Hershey, PA 17033-1240, USA, www.igi-global.com. This series is composed of titles available for purchase individually; each title is edited to be contextually exclusive from any other title within the series. For pricing and ordering information please visit https://www.igi-global.com/book-series/advances-chemical-materials-engineering/73687. Postmaster: Send all address changes to above address.

Titles in this Series

For a list of additional titles in this series, please visit:
www.igi-global.com/book-series/advances-chemical-materials-engineering/73687

Building a Low-Carbon Society Through Applied Environmental Materials Science
Takaomi Kobayashi (Nagaoka University of Technology, apan)
Engineering Science Reference • © 2024 • 300pp • H/C (ISBN: 9798369300039) • US $255.00

Advancements in Powder Metallurgy Processing, Applications, and Properties
Shashanka Rajendrachari (Bartin University, Turkey) and Rayappa Shrinivas Mahale (Jain College of Engineering and Research, ndia)
Engineering Science Reference • © 2024 • 300pp • H/C (ISBN: 9781668493854) • US $275.00

Innovative Multifunctional Nanomaterial for Photocatalysis, Sensing, and Imaging
Azad Kumar (Siddharth University, India)
Engineering Science Reference • © 2023 • 283pp • H/C (ISBN: 9781668487433) • US $270.00

Innovations in Materials Chemistry, Physics, and Engineering Research
Eugene de Silva (Virginia Research Institute, USA) and Pramudi Abeydeera (Polymeric Ltd., USA)
Engineering Science Reference • © 2023 • 446pp • H/C (ISBN: 9781668468302) • US $270.00

Handbook of Research on Advanced Functional Materials for Orthopedic Applications
R. Ranjith (SNS College of Technology, India) and J. Paulo Davim (University of Aveiro, Portugal)
Engineering Science Reference • © 2023 • 371pp • H/C (ISBN: 9781668474129) • US $325.00

Sustainable Utilization of Nanoparticles and Nanofluids in Engineering Applications
Sampath Boopathi (Mechanical Engineering, Muthayammal Engineering College, India) and J. Paulo Davim (University of Aveiro, Portugal)
Engineering Science Reference • © 2023 • 372pp • H/C (ISBN: 9781668491355) • US $270.00

Modeling, Characterization, and Processing of Smart Materials
Ajay Kumar (Mechanical Engineering Department, School of Engineering and Technology, JECRC University, India) Parveen Kumar (Mechanical Engineering Department, Rawal Institute of Engineering and Technology, India) Ashish Kumar Srivastava (G.L. Bajaj Institute of Technology and Management, India) and Vikas Goyat (Mechanical Engineering Department, SRM Institute of Science and Technology, India)
Engineering Science Reference • © 2023 • 370pp • H/C (ISBN: 9781668492246) • US $270.00

IGI Global
PUBLISHER of TIMELY KNOWLEDGE

701 East Chocolate Avenue, Hershey, PA 17033, USA
Tel: 717-533-8845 x100 • Fax: 717-533-8661
E-Mail: cust@igi-global.com • www.igi-global.com

Table of Contents

Preface

Welcome to *FlexPDE and Finite Element Method Applications in Thermal Energy Storage and Cavities*. This book is the culmination of extensive exploration into the intricate domains of cavities, phase change materials (PCM), and their profound implications in the realm of thermal energy storage systems (TESS). Authored by Khashayar Hosseinzadeh, Mehdi Mahboobtosi, and Davood Domairry Ganji, this compilation seeks to unravel the complexities surrounding these crucial aspects through the lens of FLEXPDE software, harnessing the power of the Finite Element Method (FEM).

At the heart of our inquiry lies the fervent exploration of problems and phenomena associated with cavities and PCM. Leveraging the robust capabilities of FEM via FLEXPDE software, we delve into the multifaceted world of heat transfer, a fundamental phenomenon resonating across numerous disciplines. The profound significance of heat transfer in diverse domains, including building systems, electronic equipment, and solar collectors, propels our exploration towards uncovering methodologies aimed at enhancing heat transfer efficiencies.

Our endeavor in crafting this book stems from a comprehensive dedication to meeting the diverse needs of our readership. By meticulously dissecting these complex problems, we strive to equip readers with an in-depth understanding while providing the necessary tools through meticulously crafted codes. Moreover, our exploration extends beyond the immediate purview of cavities and PCM, encompassing a holistic review of TESS, addressing one of the pressing concerns in today's world—energy storage systems.

As editors, it has been our unwavering commitment to curate a resource that not only elucidates the intricacies of these phenomena but also serves as a guiding beacon for researchers, engineers, and enthusiasts navigating the realms of thermal energy storage and heat transfer. We extend our gratitude to the contributors, whose expertise and dedication have enriched the depth and breadth of this work.

May this book serve as a catalyst for new insights, innovative solutions, and continued advancements in the realm of thermal energy storage, cavities, and the intricate world of phase change materials.

CHAPTER OVERVIEW

Chapter 1: Introduction to FlexPDE

Our journey begins with an exploration of FlexPDE, a formidable tool adept at solving linear and nonlinear partial differential equations via the finite element method (FEM). Delving into the workings of user-written scripts, this chapter unveils the process of defining geometries, equations, boundary conditions, and parameters within FlexPDE. An in-depth investigation into various script components, meshing techniques, handling nonlinear problems, and the diverse problem types solved using FEM sets the stage for our exploration.

Chapter 2: Basic Idea of Finite Element Method

Building on the foundation laid by FlexPDE, this chapter delves into the intricate realm of the Finite Element Method (FEM). Emphasizing its prowess in numerically solving an array of physics and engineering problems, including heat transfer, FEM's discretization, element types, governing equations, and mesh generation techniques are meticulously studied. This chapter serves as a comprehensive guide to understanding and applying FEM in diverse problem domains.

Chapter 3: Definition of Nanofluids

Nanofluids, heralded as potential alternatives to traditional fluids, take center stage in this chapter. We explore the intricate world of these revolutionary fluids, dissecting their formulation by dispersing nanoparticles in a base fluid. The exploration extends to hybrid and ternary hybrid nanofluids, presenting a detailed exposition of their thermophysical properties and applications. The chapter culminates in a comprehensive investigation into the application of ternary hybrid nanofluids, underlining their wide-ranging potential.

Chapter 4: Introduction to Cavity

Venturing into the realm of cavities, this chapter offers a concise yet comprehensive overview. Exploring natural convection as a key aspect, the chapter delves into the challenges posed by low heat transfer rates between fluids and walls within cavities. Strategies to enhance heat transfer, such as leveraging complex geometries, porous media, and magnetic fields, are meticulously examined. Additionally, the chapter provides an extensive review of cavity-related formulas in the presence of nanofluids.

Chapter 5: The Application of FlexPDE in Cavity Problems

The quest to enhance heat transfer takes precedence in this chapter, where the synergy between nanofluids and porous mediums is explored. Unveiling the significant enhancement in heat transfer achieved by combining these two methods, the chapter navigates through diverse examples, including wavy porous enclosures, octagonal porous mediums, trapezoidal porous cavities, and snowflake porous enclosures. The analysis and optimization of hybrid nanofluids further underscore the chapter's focus on innovative solutions.

Chapter 6: Introduction to Phase Change Materials

Addressing the criticality of energy storage, this chapter presents phase change materials (PCM) as an efficient and sought-after method. Detailing their role in latent heat storage systems and their transformative impact on thermal energy storage, the chapter delves into the economic and environmental significance of PCM. The chapter underscores how these materials store and release thermal energy post-phase shift, aligning technological advancements with environmental sustainability.

Chapter 7: The Application of FlexPDE in Latent Heat Thermal Energy Storage Problems

Highlighting the pressing need for effective energy management strategies, this chapter navigates the landscape of thermal energy storage systems (TESS). Delving into the diverse facets of latent heat storage, the chapter emphasizes the pivotal role played by PCM. Tackling challenges related to PCM's thermal conductivity, the chapter explores solutions involving fins within PCM structures and the utilization of nanofluids. Additionally, it elucidates diverse configurations of fins to enhance PCM performance.

Chapter 8: The Application of FlexPDE in Triplex Latent Heat Thermal Energy Storage Problems

This chapter ventures into the domain of triplex-tube heat exchangers, shedding light on the burgeoning interest in Triple Concentric Tube Heat Exchangers (TCTHX). Offering a comparative view to shell-and-tube heat exchangers, the chapter navigates through the architecture and functionality of TCTHX. Specifically, it explores the enhanced heat transfer between Heat Transfer Fluid (HTF) and PCM facilitated by this unique configuration of three tubes.

Conclusion: Final Remarks

Culminating our exploration, this chapter serves as a compendium of knowledge and practical application. It offers a guide to leveraging FlexPDE for problem-solving, featuring sample codes related to cavities and phase change materials. Consolidating the comprehensive studies undertaken in earlier chapters, this section complements the theoretical discourse with practical implementation in the realm of holes and PCM within FlexPDE software.

IN SUMMARY

As we draw the curtains on this compendium, *FlexPDE and Finite Element Method Applications in Thermal Energy Storage and Cavities*, we reflect on a journey that traversed the intricate landscapes of cavities, phase change materials (PCM), and the formidable tool that is FlexPDE. Authored by Khashayar Hosseinzadeh, Mehdi Mahboobtosi, and Davood Domairry Ganji, this book stands as a testament to our collective pursuit of unraveling complexities and advancing knowledge in thermal sciences.

Through the meticulous exploration of FlexPDE's capabilities, we unveiled its prowess in solving linear and nonlinear partial differential equations, offering readers a comprehensive understanding of its functionalities. The chapters on the Finite Element Method (FEM) elucidated its transformative role in addressing diverse physics and engineering problems, especially within the realm of heat transfer.

The discourse on nanofluids illuminated a promising frontier in heat transmission, showcasing the potential of these innovative fluids in augmenting thermal conductivity. Meanwhile, the exploration of cavities shed light on strategies to ameliorate heat transfer rates, paving the way for improved efficiencies in diverse applications.

The chapters dedicated to the application of FlexPDE in cavity problems and latent heat thermal energy storage delved into practical implementations, showcasing the amalgamation of theory and application. These chapters offered insights into optimizing heat transfer through innovative methodologies like hybrid nanofluids, porous mediums, and unique heat exchanger configurations.

The significance of phase change materials in energy storage systems emerged as a recurring theme, emphasizing their pivotal role in addressing the pressing global demand for efficient and sustainable energy solutions. The integration of FlexPDE in solving PCM-related problems further underscored the practical utility of the software in addressing real-world challenges.

We culminate our exploration with a profound appreciation for the contributors whose expertise and dedication have enriched this compendium. Their insights, coupled with the tireless efforts of the authors, have rendered this book a valuable resource for researchers, engineers, and enthusiasts navigating the complexities of thermal energy storage, cavities, and heat transfer phenomena.

As editors, our aspiration was to craft a compendium that not only elucidates the intricacies of these phenomena but also serves as a guiding beacon for future endeavors in this field. We hope that the knowledge encapsulated within these pages acts as a catalyst, inspiring innovative solutions, fostering deeper understanding, and propelling advancements in the realms of thermal sciences.

In closing, we extend our deepest gratitude to all who contributed to this endeavor and to the readers whose pursuit of knowledge fuels the continued evolution of our understanding in this dynamic domain.

Warmest regards,

Khashayar Hosseinzadeh
Sharif University of Technology, Iran

Mehdi Mahboobtosi
Babol Noshirvani University of Technology, Iran

Davood Domairry Ganji
Babol Noshirvani University of Technology, Iran

Chapter 1
Introduction to FlexPDE

ABSTRACT

FlexPDE solves partial differential equations that are linear or nonlinear using the finite element method. A user-written script is used by FlexPDE to solve the equation system. The user can enter the problem's geometry, equations, boundary conditions, and other parameters using this script. Topics connected to FlexPDE, such as various script components, meshing, boundary conditions, and solving non-linear problems, have been examined in this chapter. FlexPDE uses finite element analysis to solve linear or non-linear partial differential equations. Systems may represent static boundary value, time-dependent initial/boundary value, or eigenvalue problems.

INTRODUCTION

FlexPDE uses finite element analysis to solve linear or non-linear partial differential equations. Systems may represent static boundary value, time-dependent initial/boundary value, or eigenvalue problems. FlexPDE treats partial differential equation systems in general rather than solving specific equations for a specified application. It simplifies defining application-specific equations, domains, and boundary conditions. This description generates a problem-specific finite element solution procedure. Thus, FlexPDE may create a numerical solution to a wide range of applications without any prior understanding of them. FlexPDE carries out the operations required to convert a description of a system of partial differential equations into a finite element model, solve the system, and present graphical and tabular output of the results from a script created by the user. Additionally, FlexPDE is a "problem solving environment." It carries out all of the tasks required to solve partial differential equation systems, including an editor for writing scripts, a mesh generator for creating finite element meshes, a solver for finding solutions using finite elements, and a graphics system for visualising the results. Without leaving the FlexPDE application environment, the user can update the script, run the problem and view the results, then repeatedly re-edit and re-run. There is no pre-established problem domain or equation list for FlexPDE. The user has complete discretion over which partial differential equations to utilise. The scripting language used by FlexPDE is a "natural" language. It enables the user to communicate the geometry of his problem domain and the mathematics of his partial differential equations system in a manner comparable to that of a coworker.

DOI: 10.4018/979-8-3693-0932-2.ch001

1. FlexPDE ABILITY

FlexPDE is capable of solving systems of first or second order partial differential equations in one, two, or three-dimensional Cartesian geometry, in one-dimensional spherical or cylindrical geometry, or in two-dimensional axisymmetric geometry. (The PDE can handle other geometries if the appropriate terms are used.). FlexPDE can also be used to handle eigenvalue issues when the system is steady-state or time-dependent. One issue can have both time-dependent and steady-state equations. The number of simultaneous equations that can be solved depends on the hardware restrictions of the computer running FlexPDE. Both linear and nonlinear equations are possible. (In nonlinear systems, FlexPDE automatically employs a modified Newton-Raphson iteration procedure.). It is possible to define any number of zones with various material qualities. The continuity of the modelled variables across material interfaces is assumed. Derivative jump conditions follow from the PDE system's assertion. (Discontinuous variables can be handled by CONTACT boundary conditions.). Due to its potential for being very user-friendly, FlexPDE is suggested for usage in educational settings. FlexPDE, however, is not a toy. It can be applied successfully to highly challenging tasks by making full use of its power.

2. HOW FlexPDE WORKS

FlexPDE is a fully integrated PDE solver that combines a number of internal resources to offer a whole system for addressing problems:

1. Full text editing capabilities and a graphical domain preview are offered via a script editing tool with syntax highlighting.
2. In order to produce symbolic Galerkin equations, a symbolic equation analyzer performs spatial differentiation, expands defined parameters and equations, and symbolically applies integration by parts to decrease second order terms. this equation are then symbolically differentiated to get the Jacobian coupling matrix.
3. A mesh generation facility constructs a triangular or tetrahedral finite element mesh over a two or three-dimensional problem domain. In two dimensions, an arbitrary domain is filled with an unstructured triangular mesh. In three-dimensional problems, an arbitrary two-dimensional domain is extruded into a third dimension and cut by arbitrary dividing surfaces. The resulting three-dimensional figure is filled with an unstructured tetrahedral mesh.
4. For steady-state, time-dependent, or eigenvalue issues, a Finite Element numerical analysis tool determines a suitable solution scheme, with different approaches for linear and nonlinear systems. The basis of the finite element might be linear, quadratic, or cubic.
5. An adaptive mesh refinement process evaluates the mesh's suitability and refines the mesh wherever there is a significant amount of inaccuracy. The mesh refinement and solution are iterated by the system until a user-defined error tolerance is reached.
6. A dynamic timestep control mechanism monitors the solution's temporal curvature and modifies the time integration step as necessary to preserve accuracy.
7. A graphical output facility takes any algebraic function from the solution and produces contour, surface, vector, or elevation plots.

8. Text reports may be generated using a data export tool in a variety of formats, including basic tables, complete finite element mesh data, CDF, VTK, or TecPlot compatible files.

3. USES OF FlexPDE

PDEs describe most of physics and engineering. Scripted solvers like FlexPDE can be used in almost any engineering or research field.

1. FlexPDE can model studies, forecast results, and evaluate consequences for researchers in many domains. Parameter variations and dependencies can be programmed.
2. FlexPDE is a tool that engineers can use to perform conceptual analysis, feasibility studies, and design optimization studies. There is no need for a different tool for each effect because the same program can be used to represent every component of a design.
3. Special-purpose applications that require finite element modeling of partial differential equation systems can use FlexPDE.
4. FlexPDE can teach physics and engineering. One software program can evaluate all systems of interest in a discipline.

Students can interactively alter terms and domains and examine the equations

4. PRINCIPLES

The mathematical expression of a conservation principle, such as the conservation of energy, momentum, or mass, usually takes the form of partial differential equations. Due to the nature of partial differential equations, which deal with continuous functions, a derivative is the end consequence of noting differences at an infinitesimal scale. For instance, it is believed that the temperature distribution in a material varies smoothly from one extreme to the next, causing the values to converge at "zero" separation when we examine the differences between surrounding spots in greater detail. On the other hand, computers use discrete integers, of which only a small number can be stored or processed in a finite amount of time, to perform arithmetic operations. An infinite number of values cannot be analyzed by a computer. So how can we apply a computer to actually solve a problem? The finite element method is one of many methods that have been developed to approximate the behavior of real systems using computers, first in structural mechanics, and later in other fields. Part of its success comes from the fact that it approaches the analysis in the framework of integrals over small patches of the total domain, thus enforcing aggregate correctness even in the presence of microsimulation. The underlying premise is that each local patch of a solution can, at some scale, be well represented by a low order polynomial, regardless of the shape of a solution over the full domain of a problem. This strongly resembles the well-known Taylor series expansion, which uses a few polynomial terms to describe a function's local behavior. For instance, in a two-dimensional heat flow problem, we suppose that if the domain is partitioned into numerous triangular patches, each patch's temperature can be accurately represented by, say, paraboloidal surfaces. By joining the patches, we create a Harlequin surface that, for the solution value but potentially not for its derivatives, complies with the differential limiting assumption of continuity. The sample points at vertices or other locations are referred to as the "nodes" of the mesh, and the patchwork of triangles is known as the computation's "mesh". The procedure is comparable in three dimensions, employing a tetrahedral subdivision of the domain. How do we determine the shape of the approximating patches?

1. Give each vertex of the domain's triangular or tetrahedral subdivision a sample value. Then, many triangles (tetrahedra) share each vertex value.
2. In the partial differential equation, substitute the approximation functions.
3. Integrate the result over the triangles surrounding each vertex after multiplying the result by an importance-weighting function.
4. Find the vertex values that minimize each integral's error by solving.

The continuous PDE problem is effectively transformed into a discrete minimization problem on the vertex values by using this technique, which is also referred to as the "weighted residual" method. This form of the equation is typically referred to as a "weak form" since it does not precisely enforce the PDE at every point in the domain but is right in an integral sense with respect to the domain's triangle division. For various interpolation systems, different sample locations and sample values are used. In FlexPDE, we either employ cubic interpolation (with values at the vertices and two points along each side) or quadratic interpolation (with sample values at the vertices and midpoints of the triangle cells). There are other configurations that are feasible, leading to different "flavors" of finite element methods.

5. SCRIPT IN FlexPDE

A legible text file serves as a problem description script. The file's content is divided into various sections, each of which has a header. The main categories are (Liu & Liu, 2018):

- TITLE: a label that is descriptive for output.
- SELECT: user controls that modify FlexPDE's default behavior.
- VARIABLES: The dependent variables are identified here.
- DEFINITIONS: There are defined useful parameters, relationships, or functions.
- EQUATIONS: There is a partial differential equation for each variable.
- BOUNDARIES: By traversing the domain's edge and connecting line or arc segments to bind the figure, the geometry is described.
- MONITORS and PLOTS: desired graphical output is listed, including any combination of CONTOUR, SURFACE, ELEVATION or VECTOR plots.
- END: finish the script.

6. EULERIAN AND LAGRANGIAN EQUATIONS

The Equations section is assumed to present equations in the Eulerian (Laboratory) frame. Lagrangian Equations is an optional label for the Equations section, in which case FlexPDE will not correct the equations for motion. The user is then required to supply equations suitable for the moving nodes. To make it clear that the equations are appropriate for the laboratory reference frame, use the section label Eulerian Equations. The accepted interpretation is this one.

7. MESHING

7.1 Moving Meshes

During a problem run, FlexPDE enables techniques for shifting the computation mesh and domain borders. Simple enhancements of the current script language serve as the tools for providing this feature.

The motion of the boundaries or mesh has no impact on how ordinary equations are specified. It is expected that equations are given in Eulerian (Laboratory) form. The equations are symbolically corrected for motion using FlexPDE. This strategy yields an Arbitrary Lagrange/Eulerian (ALE) model, where the user can select the mesh velocities:

- A Lagrangian model is produced when the mesh velocity is locked to a fluid velocity. (In circumstances of erratic motion, FlexPDE is not advised since it lacks a mechanism for reconnecting twisted meshes.)
- By specifying a mesh velocity that is distinct from the fluid velocity, the bounding surfaces or fluid's bulk motion can still deform while the mesh integrity is preserved.
- An Eulerian model, which was the default in earlier iterations of FlexPDE, is produced if mesh motion is not specified.g

7.2 Mesh Balancing

Diffusing the coordinates or the mesh velocities is a straightforward way to evenly distribute the computation mesh inside a moving domain boundary. Take our basic example problem, for instance, and modify it to model a sphere of fluctuating size Rm=0.5+0.25×cos(t).

- Diffusing Mesh Coordinates

We specify X and Y surrogate coordinates as follows:

Variables

Phi
Xm = MOVE(x)
Ym = MOVE(y)

We will utilize simple diffusion equations for the mesh coordinates to evenly disperse the positions inside, assuming that boundary conditions will be what determines the actual motions:

Div(Grad(Xm)) = 0
Div(Grad(Ym)) = 0

Using the time derivative of R and geometric principles, we can apply the boundary velocities directly to the mesh coordinates on the blob surface:

VELOCITY(Xm) = -0.25*sin(t)*x/r
VELOCITY(Ym) = -0.25*sin(t)*y/r
** Diffusing Mesh Velocities

As an alternative, we can define mesh velocity variables and substitute X and Y coordinates as follows:

Variables

Phi
Xm = MOVE(x)
Ym = MOVE(y)
Um
Vm

Simply put, the velocity relations are the EQUATIONS for the mesh coordinates:

dt(Xm)=Um
dt(Ym)=Vm

We will apply a diffusion equation to the mesh velocities to evenly disperse the velocities inside:

div(grad(Um))=0
div(grad(Vm))=0

The blob's mesh velocity boundary conditions are as follows:

VALUE(Um) = -0.25*sin(t)*x/r
VALUE(Vm) = -0.25*sin(t)*y/r

We must also apply the hard equivalency of velocity to the mesh coordinates on the blob boundary since the finite element equations used at the boundary nodes are averages over the cells.

VELOCITY(Xm) = Um
VELOCITY(Ym) = Vm

7.3 Controlling Mesh Density

The cell density of the mesh produced by FlexPDE can be adjusted using a variety of ways.

Adaptive Mesh Refinement

The resulting mesh's cell density will be determined by the distance between each point in the bounding segments. In the area around a very small boundary segment, there will be a cluster of little cells.

Maximum Density

The global command
SELECTNGRID = < number>

limits the largest possible cell size. Subject to reduced size constraints from other criteria, the mesh will be constructed with about NGRID cells in the biggest dimension and corresponding size in the smaller dimension.

Explicit Density Control

The parameters MESH_SPACING and MESH_DENSITY can be used to modify the original mesh's cell density. The maximum cell dimension is controlled by MESH_SPACING, and the minimum number of cells per unit distance is controlled by MESH_DENSITY, which is its inverse. The smallest of the several conflicting effects that influence cell size is accepted as the size of a cell. Therefore, a big spacing request is essentially ignored, and the MESH_SPACING and MESH_DENSITY settings only take effect if they are the smallest of the competing influences. The syntax of defined parameters or boundary conditions can be utilized with the MESH_SPACING and MESH_DENSITY controls. These controls may be used as specified parameters and appear in the DEFINITONS section or they may be redefined in later regional redefinition sections. The controls in this application define the volume or area mesh density over a region or the entire domain. The controls MESH_SPACING and MESH_DENSITY can be used with the boundary condition syntax to control the cell density along border segments and may appear anywhere a boundary condition statement may appear. In this application, the controls define the cell spacing on the surface or border curve. The MESH_SPACING or MESH_DENSITY control's value could depend on the spatial coordinate.

Adaptive Mesh Refinement

After creating the initial mesh, FlexPDE will continue to calculate the solution error and adjust the mesh as necessary to achieve the desired precision. To improve the mesh in time-dependent problems where the variables change rapidly, an adaptive refinement procedure will also be used to the starting values of the variables. Unlike the cells produced by this adaptive refining process, which may be undone, cells produced by the original explicit density restrictions are irreversible

8. SOLVING NONLINEAR PROBLEMS

FlexPDE automatically detects nonlinearity in an issue and adjusts its approach. FlexPDE employs a modified Newton-Raphson iteration process as its solution approach. This "descent" method seeks to descend down an energy functional's gradient until the functional's minimal energy is reached (i.e., the gradient reaches zero). The method converges quadratically (the relative error is squared on each iteration) if the functional is roughly quadratic, as it is in straightforward diffusion problems. FlexPDE's built-in default method typically works well enough to find a solution without user input. However, in situations involving severe nonlinearities, it could be required for the user to direct FlexPDE toward a workable solution. Several methods can be applied to aid in the problem-solving process.

9. TIME-DEPENDENT PROBLEMS

The default behavior in nonlinear time-dependent problems is to perform a single Newton step at each timestep, with the understanding that any nonlinearities will be detected by the timestep controller, and that timestep adjustments will ensure an accurate evolution of the system from the provided initial conditions. This mode does not update the derivatives of the solution with respect to the variables, which are computed only once at the start of the timestep.

10. STEADY-STATE PROBLEMS

Nonlinear steady-state problems present a little more challenging set of circumstances. We cannot assure that the system will have a singular solution, and even if it does, we cannot assure that FlexPDE will be able to locate it.

- Start with a Good Initial Value: Finding a solution will be greatly aided by providing a starting value that is close to the ideal solution. Make sure the initial value and the boundary conditions match up especially carefully. If it doesn't, the trial solution might be stimulated by significant excursions, creating problems with the solution.
- Gradually Activate the Nonlinear Terms Using STAGES: FlexPDE's staging feature can be used to gradually raise the nonlinear terms' strength. Allow FlexPDE to start with a linear or nearly linear system and then find a solution that complies with the boundary criteria. Then, utilize this solution as a springboard for a system that is severely nonlinear. You can sneak up on a solution to very bad situations by using staging wisely.
- To stabilize solutions, use artificial diffusion: when an abrupt transition is modeled with a low mesh cell density. In signal processing, a "window function" can be used to smooth the signal. The high frequency components of the signal are eliminated by this, which functions as a low-pass filter. The diffusion operator Div(grad(u)) is a low-pass filter that can be used in partial differential equations to calm oscillations in the solution. In a nutshell, by setting eps=D2/pi2 in steady state or eps=2*D*c/pi in time dependence (where c is the signal propagation velocity), you can utilize a term eps*Div(Grad(u)) in a PDE to smooth oscillations of spatial extent D. In order to maintain dimensional consistency with the other variables in the equation, the term should likewise be scaled as necessary. By using such a term, the spatial frequency components of the solution are only restricted to those that the finite element mesh can sufficiently resolve.
- Use CHANGELIM to Manage Changes: On each Newton-Raphson step, the selector CHANGELIM restricts nodal value changes. As in a one-dimensional Newton iteration, firing down the gradient

will try to step an immense distance to the next trial solution if the trial solution is near a local functional maximum. FlexPDE uses a backtracking approach to discover the appropriate residual reduction step size and limits nodal changes to CHANGELIM times the variable's average value. If the initial value (or any intermediate trial solution) is far from the genuine solution, the default value for CHANGELIM, 0.5, may cause wild excursions from which FlexPDE cannot recover. To compel FlexPDE to converge, reduce CHANGELIM to 0.1 or even 0.01. CHANGELIM=0.01 can converge quickly given an acceptable beginning value. CHANGELIM multiplies the RMS average value, not each local value, therefore its effect fades when a solution is found, ensuring quadratic ultimate convergence.

- Watch Out for Negative Values: FlexPDE approximates with piecewise polynomials. In early stages, rapid solution variation over a single cell will cause severe undershoot. Don't assume a positive variable value. Rewrite your equations in terms of the variable's logarithm if negative values invalidate them. The suggested value of your variable will remain positive even if the logarithm goes negative.
- Problem Recast in a Time-Dependent Form: Steady-state problems are infinite-time limits of time-dependent problems. Rewrite your PDEs using a temporal derivative term to decrease divergence from the steady-state PDE solution. Div(K*GRAD(U))=DT(U) is a good model. A negative divergence suggests a local solution maximum and drives the value down. "Time" is a fake variable like "iteration count" in the steady-state N-R iteration, but the time-dependent formulation allows the timestep controller to drive solution evolution.

11. PROBLEM DESCRIPTOR REFERENCE

The FlexPDE system is script-driven. It reads information from a text file known as a "problem descriptor" or "script" about the equations, domain, auxiliary definitions, and requests for graphical output. You can generate the problem description file using FlexPDE's editor feature or any other ASCII text editor. Only when a word processor has an optional "pure text" output option, where formatting codes have been removed from the file, can it be used. Problem descriptors use an easy-to-learn natural language developed by Robert G. Nelson for use in Lawrence Livermore National Lab's PDS2 system and later in Macsyma, Inc.'s PDEase2. Dr. Gunnar Backstrom's "Simple Fields of Physics by Finite Element Analysis" describes the language. A variety of extensions have been introduced to FlexPDE over time to increase its processing power. This article provides a description of the language as it is used in FlexPDE at the moment. Although FlexPDE scripting language is somewhat comparable to computer programming languages, it is more natural and focused on the description of PDE systems. The majority of engineers, scientists, and college students at the intermediate level who have taken at least one course on partial differential equations may soon become fluent in the language and start creating their own problems. One may think of the FlexPDE issue description language as a shortened version of the Finite Element modeling language. The descriptor's statements give FlexPDE the details it needs to put together a numerical procedure to address the issue. It's critical to realize that FlexPDE problem descriptions don't use a procedural language. The user explains how the system's many parts interact with one another. As would be done in a procedural programming language like C or FORTRAN, he does not specify the steps that must be taken in order to produce the solution. FlexPDE determines the steps required to locate the answer based on the relationships between the pieces of the problem. Numerous presumptions are made by FlexPDE regarding the components of the problem description. If a variable is listed in the VARIABLES section, for instance, it is presumed that:

1. The variable has values across the problem domain and is either a scalar or vector field,
2. Finite element interpolation between the nodes of a computing mesh will be used to approximate it.
3. Over the domain, the variable's values are continuous.
4. The behavior of the variable will be described by a partial differential equation.

It is presumed that the named quantity is what is described if a definition is found in the DEFINITIONS section.

1. the PDE system's auxiliary component
2. over the domain, it can be discontinuous.
3. is not need to obey any PDE.

12. PROBLEM DESCRIPTOR STRUCTURE

As explained in the previous sections, a FlexPDE code consists of parts. Here we will examine this problem in more detail. Each section begins with a proper name and is followed by one or more definitions of the issue (Pustovalova & Karyakin, 2017; Ward, n.d.). Following are acceptable section names:

- TITLE: -defines the problem title
- SELECT: -sets various controls and options
- COORDINATES: -the coordinate system's definition
- VARIABLES: -names of the problem variables
- DEFINITIONS: -the definition of auxiliary quantities and parameters
- MATERIALS: -specifies collections of variables that can be used in a name-based fashion
- BOUNDARY CONDITIONS: sets of boundary conditions that can be applied by name are defined.
- INITIAL VALUES: -sets initial values of variables
- EQUATIONS: -The partial differential equation system is defined.
- CONSTRAINTS: -specifying integral optional constraints
- EXTRUSION: -three-dimensionally expands the domain -describes the projected 3D or 2D space
- BOUNDARIES: -describes the projected 3D or 2D space
- RESOLVE: -supplements mesh refinement control if desired.
- FRONT: -supplements mesh refinement control for moving fronts if necessary.
- TIME: -sets forth the temporal domain
- MONITORS: -selects interim graphic display
- PLOTS: -chooses the ultimate graphic presentation
- HISTORIES: -selects time-summary displays
- END: -identifies when the descriptor ends

The following points should be noted:

- With the exception of the requirement that all files contain a BOUNDARIES section and an END section, the amount of sections utilized in a given problem descriptor can vary.
- Although there is considerable freedom in the arrangement of these sections, it is advised that the user follow the above-mentioned sequencing.
- SELECT and DEFINITIONS may be used more than once.
- Descriptors are dynamically processed from top to bottom, hence forward references are not allowed. Variables and other defined names may be referred to in definitions as long as they were previously defined in the same section or one that came before it.

13. BOUNDARY CONDITIONS

Any partial differential equation system's collection of boundary conditions, which by themselves give the solution its singularity, is a fundamental element. The integration constants that appear in integral calculus are equivalent to the boundary conditions. When is any constant, we say $\int x^2 dx = \frac{1}{3}x^3 + C$. No matter what the value of *C* is, if we differentiate the right hand side, we recover the integrand. In a similar manner, we must integrate twice in order to solve the equation $\frac{\partial^2 u}{\partial x^2} = 0$. First integration yields $\frac{\partial u}{\partial x} + C_1$, whereas second yields C_1x+C_2. The boundary conditions of the problem statement must provide these integration constants. This illustration demonstrates that there are the same number of integration constants as there are nested levels. variations inside the PDE. These constants can be provided, in the general case, by values at each end of an interval, values and derivatives at one end, etc. The most typical application is to supply a value or a derivative at each end of the domain interval. One condition is provided at each end of any coordinate integration path in two or three dimensions when a value or derivative condition is applied throughout the full bounding curve or surface.

14. NATURAL BOUNDARY CONDITIONS

The notion of a flux boundary condition is generalized by the NATURAL boundary condition. It is actually the outward flux of the diffusing quantity in diffusion equations. It is the surface load in stress equations. It may be less obvious in other formulae. To lower the order of second derivative terms in the system equations, FlexPDE employs integration by parts. Application of this technique over a two-dimensional calculation cell yields an interior area integral term and a border line integral term. Forming the same integral in two adjacent computing cells gives the same boundary integral at their interface, except that the direction of integration is opposite in the two cells. If the integrals are combined together to generate the total integral, the shared boundary integrals cancel.

Integration by parts produces the result for the phrase dx(f), where f is an expression with additional derivative components.

$$\text{Integral(dx(f)*dV) = Integral(f*c*dS),} \tag{1}$$

where the differential surface element is denoted by dS and the x-component of the outward surface-normal unit vector is denoted by c. (The appropriate unit-vector component is used in place of c for the Y- and Z- derivative terms, respectively)

Integration by parts produces the result for the phrase dxx(f), where f represents a scalar quantity.

$$\text{Integral(dxx(f)*dV) = Integral(dx(f)*c*dS),} \tag{2}$$

where dS is the differential surface element and c stands for the x-component of the outward surface-normal unit vector. (When handling Y- and Z-derivative terms, the appropriate unit-vector component is used in place of c.)

Integration by parts is equivalent to the divergence theorem when applied to the expression DIV(F), where F signifies a vector quantity comprising additional derivative components.

$$\text{Integral(DIV(F)dV) = Integral(F . n dS),} \tag{3}$$

where the differential surface element is called dS and the outward surface-normal unit vector is called n.

Integration by parts is analogous to the curl theorem when applied to the expression CURL(F), where F denotes a vector quantity containing additional derivative components.

$$\text{Integral(CURL(F) dV) = Integral(n x F dS),} \tag{4}$$

where, once more, n represents the outward surface-normal unit vector and dS the differential surface element.

These integrations are carried out in three dimensions using FlexPDE, together with the volume and surface elements that are appropriate for the geometry. The volume cell in 2D Cartesian geometry is one unit in the Z direction; in 2D cylindrical geometry, it is r*dr*dtheta.

The treatment of outer boundary conditions and internal material interface behavior in FlexPDE is based on this method.

- At internal cell interfaces, it is assumed that all boundary integral terms disappear.
- Unless a NATURAL border condition statement provides an independent evaluation of the boundary integrand, all boundary integral terms are presumed to disappear at internal and external boundaries.

This treatment has a number of consequences:

In equations involving divergence, such as DIV(kF)=0,

- Across interior material interfaces, the quantity (kF.n) will be continuous.
- The value of (kF.n) on the border is specified by the NATURAL boundary condition.

I. Energy will be conserved across material discontinuities if (kF) is heat flux (kF=-k Grad(T)), and the NATURAL border condition determines outward heat flow.
II. The material interface conditions suggested by Maxwell's equations will be met if (kF) is electric displacement (D=-eps Grad(V)) or magnetic induction (B=Curl(A)), and in the electric case, the NATURAL boundary condition will specify the surface charge density.

In equations involving curl, such as CURL(kF)=0,

- Across interior material interfaces, the amount (k n x F) will be continuous.
- The value of (k n x F) on the border is specified by the NATURAL boundary condition.
- If (kF) is magnetic (H=(1/mu) Curl(A)) or electric (E=-Grad(V)), Maxwell's equations will satisfy the material interface conditions, and the magnetic case's NATURAL boundary condition will specify the surface current density.

It should be noted that these requirements need not be accomplished by directly writing the equations with the DIV or CURL operators. The same methodology will be applied to any legitimate differential equivalent in the problem's coordinate system. Also take note of the close connection between the PDE and the NATURAL boundary condition.

- A differential operator's argument becomes the subject of integration by parts and creates a matching component of the NATURAL boundary condition if that argument also contains a differential operator.
- The accompanying NATURAL boundary condition must be multiplied by the same factor as the PDE if the PDE is multiplied in any way.
- When transferred to the left side of the equation, the related PDE terms' signs must match those of the NATURAL boundary conditions.
- FlexPDE is given the integrand of the surface integral produced by the integration by parts, which is otherwise considered to be zero, by the NATURAL boundary condition statement.

REFERENCES

Liu, Z., & Liu, Z. (2018). Numerical Solution of PDEs. *Multiphysics in Porous Materials*, 79-89.

Pustovalova, O., & Karyakin, M. (2017). Usage FlexPDE Package in the Courses of Mathematical Modeling and Mechanics. *Conference Proceedings. New Perspectives in Science Education 2017.*

Ward, W. (n.d.). *Propellant Preheat System Optimization*. Academic Press.

Chapter 2
Basic Idea of Finite Element Method

ABSTRACT

The finite element method (FEM) is covered in this chapter. For the numerical solution of different physics and engineering problems, including heat transport issues, the finite element method is a particularly potent tool. An analytical method-like mathematical function cannot be obtained using FEM. This chapter provides a thorough study of several FEM-related topics, including relationships and governing equations, discretization of equations, two- and three-dimensional isoparametric elements, mesh generation, etc.

INTRODUCTION

FEM is a very powerful numerical method for solving various physics and engineering problems (Reddy, 2021b). Different problems in engineering and physics such as heat transfer, fluid flow, structure analysis and mass transfer can be solved by using the finite element method. Finding answers to many problems with analytical or semi-analytical methods is impossible or very difficult due to complications in geometry, material properties and loading. In the analytical method, a mathematical function is provided to express the response of the system, which can be used to calculate the response of the system at any point. This method requires solving the differential equation governing the system, which cannot find an exact answer due to the complexities in loading, geometry and material properties. In semi-analytical methods, the response of the system is assumed as a series, and these series must satisfy the minimum boundary conditions. In the FEM as a numerical method, the differential equation governing the system is tried to be converted into a set of algebraic equations, by solving them simultaneously, the response of the system will be calculated (Dhatt et al., 2012). Using the finite element method, the approximate values of the response in a series of discrete points of the continuous system are obtained. In this method, a continuous system is divided into small units called finite elements, each of which is connected with adjacent elements at points called nodes. Dividing a large and continuous system into small systems is called discretization. In the finite element method, instead of analyzing the whole system, we write the relationships for each small component, and then the relationships of all components are assembled

DOI: 10.4018/979-8-3693-0932-2.ch002

for the whole system (Zienkiewicz et al., 2005). The general concept of this method is that we divide the entire domain or the same environment that we want to investigate (for example, fluid, structure) into a series of small subsections, which are called finite elements or components. These elements are connected in a series of points called nodes. In other words, the finite element method is a method that divides a continuous system into discrete systems and the displacement field is considered as a piecewise continuous function (David Müzel et al., 2020; Nikishkov, 2004; Sabat & Kundu, 2020).

1. STEPS OF FINITE ELEMENT ANALYSIS

Using the finite element method in a problem requires doing a series of specific steps accurately. In general, the main steps of solving by the finite element method are as follows:

Step 1. Discretization of the Desired Problem and Selection of Element Type

The first step in the FEM is to divide the desired solution area into finite elements. In this step, we have to divide the object into smaller components by choosing the appropriate element with the specified number of nodes. The selection of the element should be such that it can show the real physical behavior of the problem. Choosing the type of element, the number of its nodes, the number of degrees of freedom of each node, and the number of elements considered for the whole body are among the first steps of a finite element analysis. In a finite element analysis, the most time is spent on elementing and meshing because the higher the quality of the mesh, the higher the accuracy of the obtained results.

Step 2. Selection of Interpolation Functions

The quantity that we want to be calculated and specified after the FEM is applied to the problem is selected in this section. For example, in problems related to fluids, the variable field is velocity, and in problems related to mechanics of solids, the variable field is displacement. To interpolate field variables on the element, interpolation functions are used. It is worth noting that the polynomial degree depends on the number of nodes assigned to the element. Other quantities that we can calculate in the problem are related to the field variable using relationships, and its values are calculated in the nodes and then in the whole object.

Step 3. Find Element Properties

The matrix equation for the finite element must be created that relates the nodal values of the unknown function to other parameters. For this purpose, different approaches can be used, the most convenient of which are the Galerkin method and the variational approach.

Step 4. Assembling Element Equations

We know that our problem is defined for the whole object, not for each element, so we have to assemble the equations. To obtain the global system of equations for the entire solution region, we must assemble all the element equations.

Step 5. Apply Boundary Conditions

Any type of boundary condition may be applied on the boundaries of the object or the desired area, but it should be noted that two types of boundary conditions cannot be applied on the same boundary. For example, in the mechanics of solids, in order to obtain the correct answers, the effect of the applied conditions on the object must be entered in the global stiffness matrix and the general force vector.

Step 6. Solving the System of Universal Equations

By applying the boundary conditions, we now have a system of equations with which the unknown nodal values of the field variable can be calculated. The universal equation system is usually symmetric. Direct and iterative methods can be used to solve the global equation system.

Step 7. Calculation of Other Results

In many problems we need to calculate additional parameters. By obtaining the field variable at the nodes, other quantities that are related to the field variable can be calculated at the nodes and from there at all points of the object or region. For example, in mechanical problems, stresses and strains are very important in addition to displacements, which are obtained after solving the system of global equations.

If the physical formulation of the issue is a differential equation, the Galerkin method is the most popular finite element formulation method.

There are several ways to convert the problem's physical formulation into its discrete finite element equivalent. The Galerkin method is the most widely used strategy for the problem's finite element formulation if the physical formulation is known as a differential equation. The variational formulation of the finite element equations is typically used if the physical problem can be expressed as the minimization of a functional.

2. GALERKIN METHOD

Let's use a straightforward one-dimensional example to explain how the Galerkin technique can be used to formulate finite elements (Pepper & Heinrich, 2005). Assume we must numerically answer the following differential equation:

$$a\frac{d^2u}{dx^2}+b=0, \quad 0\le x\le 2L \tag{1}$$

with boundary conditions

$$\begin{aligned} &u\Big|_{x=0}=0 \\ &a\frac{du}{dx}\Big|_{x=2L}=R \end{aligned} \tag{2}$$

where u is a hypothetical answer. As seen in Figure 1, we will use two linear one-dimensional finite elements to address the problem

Figure 1. Two linear one-dimensional finite elements with internal function interpolation

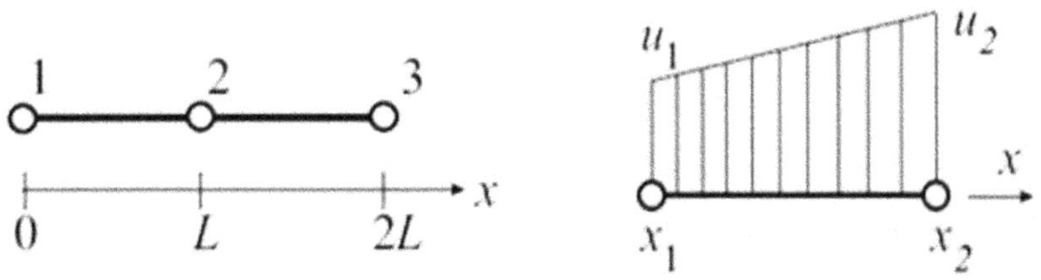

First, take a look at the finite element in Figure1 to the right. There are two nodes in the element, and the function $u(x)$ can be approximated as follows:

$$u = N_1 u_1 + N_2 u_2 = [N]\{u\}$$
$$[N] = [N_1 \quad N_2] \tag{3}$$
$$\{u\} = \{u_1 \quad u_2\}$$

Where N_i are the so called shape functions

$$N_1 = 1 - \frac{x - x_1}{x_2 - x_1}$$
$$N_2 = \frac{x - x_1}{x_2 - x_1} \tag{4}$$

which employ the node values of $u(x)$ for interpolation. The discrete global equation system should be used to find the nodal values u_1 and u_2. When u is substituted into the differential equation using its nodal values and shape functions, the estimated form is as follows:

$$a \frac{d^2}{dx^2}[N]\{u\} + b = \psi \tag{5}$$

where ψ is a residual that is not zero since a function's approximation representation inside a finite element is present. By multiplying terms in the aforementioned equation by shape functions, integrating over the element, and equating to zero, the Galerkin approach achieves residual minimization:

$$\int_{x_1}^{x_2} [N]^T a \frac{d^2}{dx^2}[N]\{u\}dx + \int_{x_1}^{x_2} [N]^T bdx = 0 \tag{6}$$

The differential equation for the finite element has the following discrete form when integration by parts is used:

$$\int_{x_1}^{x_2} \left[\frac{dN}{dx}\right]^T a \left[\frac{dN}{dx}\right] dx\{u\} - \int_{x_1}^{x_2} [N]^T bdx - \begin{Bmatrix} 0 \\ 1 \end{Bmatrix} a \frac{du}{dx}\bigg|_{x=x_2} + \begin{Bmatrix} 1 \\ 0 \end{Bmatrix} a \frac{du}{dx}\bigg|_{x=x_1} = 0 \tag{7}$$

For a finite element, this relation is typically given as:

$$\begin{aligned} &[k]\{u\} = \{f\} \\ &[k] = \int_{x_1}^{x_2} \left[\frac{dN}{dx}\right]^T a \left[\frac{dN}{dx}\right] dx \\ &\{f\} = \int_{x_1}^{x_2} [N]^T bdx + \begin{Bmatrix} 0 \\ 1 \end{Bmatrix} a \frac{du}{dx}\bigg|_{x=x_2} - \begin{Bmatrix} 1 \\ 0 \end{Bmatrix} a \frac{du}{dx}\bigg|_{x=x_1} \end{aligned} \tag{8}$$

$[k]$ and $\{f\}$ are known as the stiffness matrix and load vector, respectively, in solid mechanics. It is simple to determine the stiffness matrices and the load vectors in the case of two finite elements of length L under consideration:

$$\begin{aligned} &[k_1] = [k_2] = \frac{a}{L}\begin{bmatrix} 1 & -1 \\ -1 & 1 \end{bmatrix} \\ &\{f_1\} = \frac{bL}{2}\begin{Bmatrix} 1 \\ 1 \end{Bmatrix}, \quad \{f_2\} = \frac{bL}{2}\begin{Bmatrix} 1 \\ 1 \end{Bmatrix} + \begin{Bmatrix} 0 \\ R \end{Bmatrix} \end{aligned} \tag{9}$$

For the two distinct finite elements, the aforementioned relations offer finite element equations. By assembling the element equations, a global equation system for the domain with two elements and three nodes can be created. In our straightforward example, it is evident that components communicate with one another at the node with global number 2. The complete set of global equations is:

$$\frac{a}{L}\begin{bmatrix} 1 & -1 & 0 \\ -1 & 2 & -1 \\ 0 & -1 & 1 \end{bmatrix}\begin{Bmatrix} u_1 \\ u_2 \\ u_3 \end{Bmatrix} = \frac{bL}{2}\begin{Bmatrix} 1 \\ 2 \\ 1 \end{Bmatrix} + \begin{Bmatrix} 0 \\ 0 \\ R \end{Bmatrix} \tag{10}$$

The global equation system finally looks like this after applying the boundary condition $u(x=0)=0$:

$$\frac{a}{L}\begin{bmatrix}1 & 0 & 0\\ 0 & 2 & -1\\ 0 & -1 & 1\end{bmatrix}\begin{Bmatrix}u_1\\ u_2\\ u_3\end{Bmatrix}=\frac{bL}{2}\begin{Bmatrix}0\\ 2\\ 1\end{Bmatrix}+\begin{Bmatrix}0\\ 0\\ R\end{Bmatrix} \tag{11}$$

Values at nodes u_i are the outcomes of solving a set of linear algebraic equations. The shape functions can be used to determine the value of u at any location inside a finite element. For a=1, b=1, L=1, and R=1, the differential equations finite element solution is displayed in Figure 2.

Figure 2. Comparison of the precise solution with the finite element solution

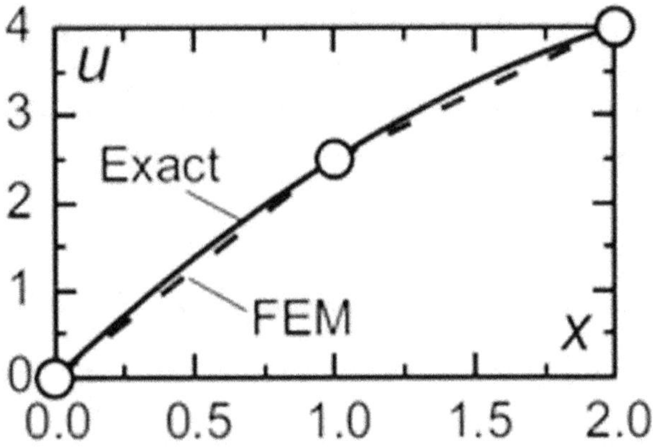

A quadratic function is the exact solution. Piece-wise linearity characterizes the finite element solution when the simplest element is used. By using more basic elements or by using components with more complex shape functions, a more precise finite element solution can be produced. It is important to note that the finite element approach delivers precise values of u at nodes. (just for this particular problem). If a quadratic solution is needed, finite elements with linear shape functions yield precise nodal values. The cubic solution, for example, receives precise nodal values from quadratic elements.

3. VARIATIONAL FORMULATION

The differential equation

$$\begin{aligned}&a\frac{d^2u}{dx^2}+b=0, \quad 0\le x\le 2L\\ &u\Big|_{x=0}=0\\ &a\frac{du}{dx}\Big|_{x=2L}=R\end{aligned} \tag{12}$$

In solid mechanics, with a=EA has the following physical significance. In Fig. 3, a one-dimensional bar with cross-sectional area A and an elastic modulus of E is described as being under tension while being subjected to dispersed load b and concentrated load R at its right end. Such a challenge can be expressed as a potential energy functional Π minimization problem:

$$\Pi = \int_{L} \frac{1}{2} a \left(\frac{du}{dx} \right)^2 dx - \int_{L} budx - Ru\Big|_{x=2L} \tag{13}$$

Figure 3. Tension of the one-dimensional bar subjected to a distributed load and a concentrated load

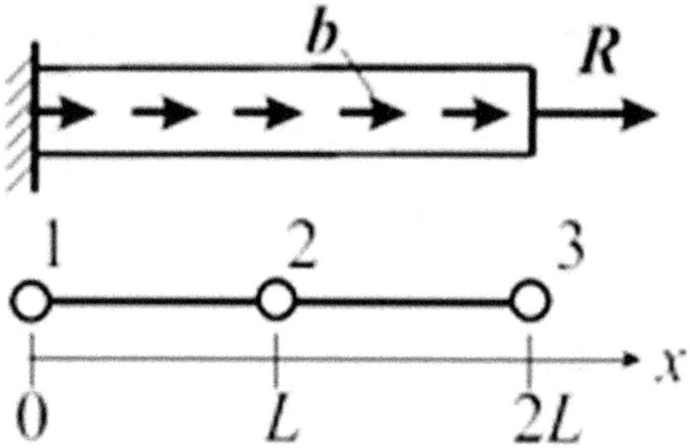

The potential energy for the second finite element can be expressed as follows using the representation of $\{u\}$ with shape functions (1.3) and (1.4):

$$\Pi_e = \int_{x_1}^{x_2} \frac{1}{2} a \{u\}^T \left[\frac{dN}{dx} \right]^T \left[\frac{dN}{dx} \right] \{u\} dx - \int_{x_1}^{x_2} \{u\}^T [N]^T bdx - \{u\}^T \begin{Bmatrix} 0 \\ R \end{Bmatrix} \tag{14}$$

The following is required for the minimum of Π:

$$\delta \Pi = \frac{\partial \Pi}{\partial u_1} \delta u_1 + \ldots + \frac{\partial \Pi}{\partial u_n} \delta u_n = 0 \tag{15}$$

which is equivalent to

$$\frac{\partial \Pi}{\partial u_i} = 0 \,, \quad i = 1 \ldots n \tag{16}$$

It is simple to verify that the finite element equilibrium equation produced by the Galerkin method and the differentiation of Π in relation to u_i are the same:

$$\int_{x_1}^{x_2} \left[\frac{dN}{dx}\right]^T EA \left[\frac{dN}{dx}\right] dx\{u\} - \int_{x_1}^{x_2} \left[N\right]^T b dx - \begin{Bmatrix} 0 \\ R \end{Bmatrix} = 0 \qquad (17)$$

Example: Obtain shape functions for the three-noded, one-dimensional quadratic element. Use the -1≤ξ≤1 local coordinate system.

Figure 4. Example geometry

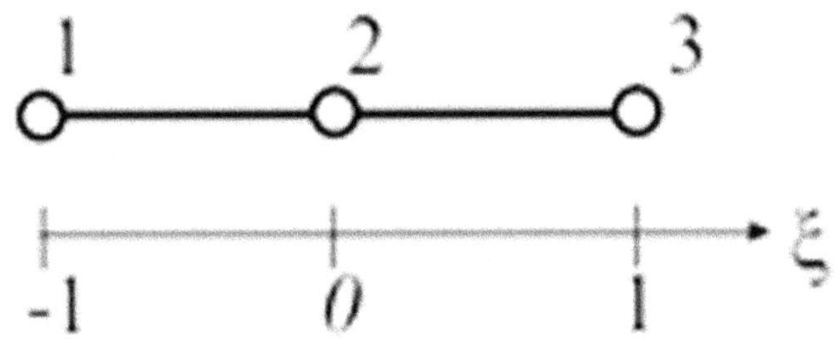

Solution: with shape function, any field contained within an element is displayed as:

$$u(\xi) = \sum N_i u_i \;, \quad i = 1,2,3 \qquad (18)$$

The approximation function should equal its nodal value at nodes:

$$\begin{aligned} u(-1) &= u_1 \\ u(0) &= u_2 \\ u(1) &= u_3 \end{aligned} \qquad (19)$$

The shape functions for the element can be quadratic polynomials because it has three nodes. (with three coefficients). You may write the shape function N_1 as follows:

$$N_1 = \alpha_1 + \alpha_2 \xi + \alpha_3 \xi^2 \qquad (20)$$

The following system of equations defines the unknown coefficients α_i:

$$\begin{aligned} N_1(-1) &= \alpha_1 - \alpha_2 + \alpha_3 = 1 \\ N_1(0) &= \alpha_1 = 0 \\ N_1(1) &= \alpha_1 + \alpha_2 + \alpha_3 = 0 \end{aligned} \tag{21}$$

The answer is $\alpha_1 = 0, \alpha_2 = -\frac{1}{2}, \alpha_3 = \frac{1}{2}$. As a result, the shape function N_1 equals:

$$N_1 = -\frac{1}{2}\xi(1-\xi) \tag{22}$$

Similar results can be obtained to show that N_2 and N_3 shape functions are equal to:

$$\begin{aligned} N_2 &= 1 - \xi^2 \\ N_3 &= \frac{1}{2}\xi(1+\xi) \end{aligned} \tag{23}$$

4. FINITE ELEMENT EQUATIONS FOR HEAT TRANSFER

Problem Statement

Take an isotropic body with temperature-dependent heat transport as an example (Hutton, 2004; Pepper & Heinrich, 2005). The following is how a fundamental heat transfer equation appears:

$$-\left(\frac{\partial q_x}{\partial x} + \frac{\partial q_y}{\partial y} + \frac{\partial q_z}{\partial z}\right) + Q = \rho c \frac{\partial T}{\partial t} \tag{24}$$

Here, the elements of heat flow via the unit area are q_x, q_y, and q_z; The rate of internal heat generation per unit volume is $Q=Q(x,y,z,t)$; Material density is measured by ρ, heat capacity by c, temperature by T, and time by t. The elements of heat flow can be expressed as follows using Fourier's law:

$$\begin{aligned} q_x &= -k\frac{\partial T}{\partial x} \\ q_y &= -k\frac{\partial T}{\partial y} \\ q_z &= -k\frac{\partial T}{\partial z} \end{aligned} \tag{25}$$

where k is the media's thermal conductivity coefficient. The fundamental heat transfer equation is as follows when Fourier's relations are substituted:

$$\frac{\partial}{\partial x}\left(k\frac{\partial T}{\partial x}\right)+\frac{\partial}{\partial y}\left(k\frac{\partial T}{\partial y}\right)+\frac{\partial}{\partial z}\left(k\frac{\partial T}{\partial z}\right)+Q=\rho c\frac{\partial T}{\partial t} \tag{36}$$

The following sorts of boundary conditions are presumptively possible:

1. Specified temperature

$$T_s = T\left(x, y, z, t\right) \text{on } S_1$$

2. Specified heat flow

$$q_x n_x + q_y n_y + q_z n_z = -q_S \text{ on } S_2$$

3. Convection boundary conditions

$$q_x n_x + q_y n_y + q_z n_z = h(T_S - T_e)\text{on } S_3$$

where T_s is the unknown surface temperature, h is the convection coefficient, and T_e is the known environmental temperature.

4. Radiation

$$q_x n_x + q_y n_y + q_z n_z = \sigma\varepsilon T_S^4 - \alpha q_r \text{ on } S_4$$

where q_r is the incoming heat flow per unit surface area, ε is the surface emission coefficient, α is the surface absorption coefficient, and σ is the Stefan-Boltzmann constant. A temperature field for a body at time t=0 must be specified for transient problems:

$$T\left(x, y, z, 0\right) = T_0\left(x, y, z\right)$$

5. FINITE ELEMENT DISCRETIZATION OF HEAT TRANSFER EQUATIONS

Finite elements within a domain V are joined at nodes. All relationships for a finite element will be written. Using connectivity information, global equations for the domain can be constructed from finite element equations. Functions of shape N_i are employed for temperature interpolation inside a finite element:

$$T = [N]\{T\}$$
$$[N] = [N_1 \quad N_2 \quad \ldots] \tag{27}$$
$$\{T\} = \{T_1 \quad T_2 \quad \ldots\}$$

The following interpolation relation for temperature gradients is obtained by differentiating the temperature interpolation equation:

$$\begin{Bmatrix} \frac{\partial T}{\partial x} \\ \frac{\partial T}{\partial y} \\ \frac{\partial T}{\partial z} \end{Bmatrix} = \begin{bmatrix} \frac{\partial N_1}{\partial x} & \frac{\partial N_2}{\partial x} & \ldots \\ \frac{\partial N_1}{\partial y} & \frac{\partial N_2}{\partial y} & \ldots \\ \frac{\partial N_1}{\partial z} & \frac{\partial N_2}{\partial z} & \ldots \end{bmatrix} \{T\} = [B]\{T\} \tag{28}$$

In this case, $\{T\}$ is a vector of node temperatures, $[N]$ is a matrix of shape functions, and $[B]$ is a matrix for interpolating temperature gradients. The fundamental heat transfer equation can be rewritten in the following way using the Galerkin method:

$$\int_V \left(\frac{\partial q_x}{\partial x} + \frac{\partial q_y}{\partial y} + \frac{\partial q_z}{\partial z} - Q + \rho c \frac{\partial T}{\partial t} \right) N_i dV = 0 \tag{29}$$

The following relations are reached after applying the divergence theorem to the first three terms:

$$\int_V \rho c \frac{\partial T}{\partial t} N_i dV - \int_V \begin{bmatrix} \frac{\partial N_i}{\partial x} & \frac{\partial N_i}{\partial y} & \frac{\partial N_i}{\partial z} \end{bmatrix} \{q\} dV = \int_V Q N_i dV - \int_S^V \{q\}^T \{n\} N_i dS$$
$$\{q\}^T = [q_x \quad q_y \quad q_z] \tag{30}$$
$$\{n\}^T = [n_x \quad n_y \quad n_z]$$

where n is an external normal to the body's surface. The discretized equations are as follows when boundary conditions are added to the aforementioned equation:

$$\int_V \rho c \frac{\partial T}{\partial t} N_i dV - \int_V \begin{bmatrix} \frac{\partial N_i}{\partial x} & \frac{\partial N_i}{\partial y} & \frac{\partial N_i}{\partial z} \end{bmatrix} \{q\} dV = \int_V Q N_i dV$$
$$- \int_S^V \{q\}^T \{n\} N_i dS + \int_{S_1}^V \{q\}^T \{n\} N_i dS + \int_{S_2} q_S N_i dS - \int_{S_3}^{S_1} h(T - T_e) N_i dS - \int_{S_4} \left(\sigma \varepsilon T_S^4 - \alpha q_r \right) N_i dS \tag{31}$$

Noteworthy is the fact that

$$\{q\} = -k[B]\{T\} \tag{32}$$

For situations involving heat transfer, discretized finite element equations have the following finite form:

$$\begin{aligned}
&[C]\{T\} + ([K_c] + [K_h] + [K_r])\{T\} = \{R_T\} + \{R_Q\} + \{R_q\} + \{R_h\} + \{R_r\} \\
&[C] = \int_V \rho c [N]^T [N] dV \\
&[K_c] = \int_V k [B]^T [B] dV \\
&[K_h] = \int_{S_3} h [N]^T [N] dS \\
&[K_r]\{T\} = \int_{S_4} \sigma \varepsilon T^4 [N] dS \\
&\{R_T\} = -\int_{S_1} \{q\}^T \{n\} [N]^T dS \\
&\{R_Q\} = \int_V Q [N]^T dS \\
&\{R_q\} = -\int_{S_2} q_S [N]^T dS \\
&\{R_h\} = -\int_{S_3} h T_e [N]^T dS \\
&\{R_r\} = -\int_{S4} \alpha q_r [N]^T dS
\end{aligned} \tag{33}$$

6. DIFFERENT TYPE PROBLEMS

The general equation mentioned above can be used to derive equations for several types of problems:

- Stationary linear problem

$$([K_C] + [K_h])\{T\} = \{R_Q\} + \{R_q\} + \{R_h\} \tag{34}$$

- Stationary nonlinear problem

$$([K_C] + [K_h] + [K_r])\{T\} = \{R_Q(T)\} + \{R_q(T)\} + \{R_h(T)\} + \{R_T(T)\} \tag{35}$$

- Transient linear problem

$$[C]\{T(t)\}+([K_C]+[K_h(t)])\{T(t)\}=\{R_Q(t)\}+\{R_q(t)\}+\{R_h(t)\} \tag{36}$$

- Transient nonlinear problem

$$\begin{aligned}&[C(T)]\{T\}+([K_C](T)+[K_h(T,t)]+[K_r](T))\{T\}\\&=\{R_Q(T,t)\}+\{R_q(T,t)\}+\{R_h(T,t)\}+\{R_r(T,t)\}\end{aligned} \tag{37}$$

7. TWO-DIMENSIONAL TRIANGULAR ELEMENT

The first finite element proposed for continuous problems was the triangular finite element. It can serve as an introduction to other aspects due of its simplicity. Fig. 5 depicts a triangular finite element in the coordinate system *xy*. The element has three nodes, hence the displacements *u* and *v* are approximated linearly:

$$\begin{aligned}u(x,y)&=N_1u_1+N_2u_2+N_3u_3\\v(x,y)&=N_1v_1+N_2v_2+N_3v_3\\N_i&=\alpha_i+\beta_ix+\gamma_iy\end{aligned} \tag{38}$$

Shape functions $N_i(x,y)$ can be figured out from the equation system below:

$$u(x_i,y_i)=u_i\ ,\quad i=1,2,3 \tag{39}$$

Figure 5. The simplest two-dimensional element is the triangular finite element

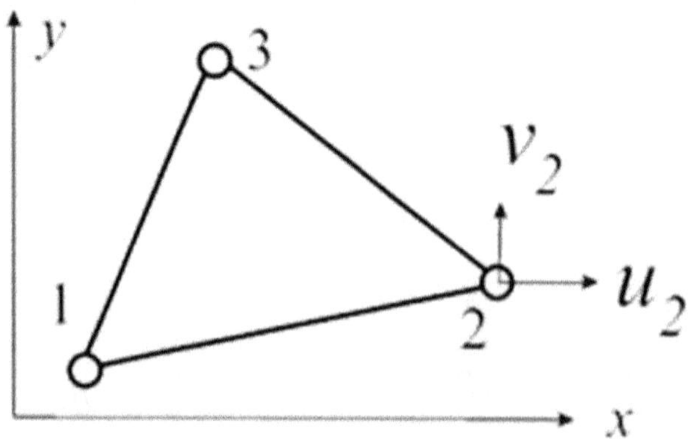

The following shape functions can be used to represent the triangle element:

$$
\begin{aligned}
N_i &= \frac{1}{2\Delta}\left(a_i + b_i x + c_i y\right) \\
a_i &= x_{i+1}y_{i+2} - x_{i+2}y_{i+1} \\
b_i &= y_{i+1} - y_{i+2} \\
c_i &= x_{i+2} - x_{i+1} \\
\Delta &= \frac{1}{2}\left(x_2y_3 + x_3y_1 + x_1y_2 - x_2y_1 - x_3y_2 - x_1y_3\right)
\end{aligned}
\tag{40}
$$

where the element area Δ is. When strains are interpolated using nodal displacements, the matrix [*B*] is equivalent to:

$$
\left[B\right] = \frac{1}{2\Delta}\begin{bmatrix} b_1 & 0 & b_2 & 0 & b_3 & 0 \\ 0 & c_1 & 0 & c_2 & 0 & c_3 \\ c_1 & b_1 & c_2 & b_2 & c_3 & b_3 \end{bmatrix}
\tag{41}
$$

For plane problems, the elasticity matrix [*E*] appears as follows:

$$
\left[E\right] = \begin{bmatrix} \lambda + 2\mu & \lambda & 0 \\ \lambda & \lambda + 2\mu & 0 \\ 0 & 0 & \mu \end{bmatrix}
\tag{42}
$$

Where λ and μ are Lame constants:

$$
\begin{aligned}
\lambda &= \frac{\nu E}{\left(1+\nu\right)\left(1-2\nu\right)} && \text{for plane strain} \\
\lambda &= \frac{\nu E}{1-\nu^2} && \text{for plane stress} \\
\mu &= \frac{E}{2\left(1+\nu\right)}
\end{aligned}
\tag{43}
$$

Here, ν is the Poisson's ratio and *E* is the elasticity modulus. The three-node triangular element's stiffness matrix can be calculated as follows:

$$
\left[k\right] = \int_V \left[B\right]^T \left[E\right]\left[B\right] dV = \left[B\right]^T \left[E\right]\left[B\right]\Delta
\tag{44}
$$

Here, it was considered that none of the matrices, [*B*] or [*E*], is dependent on coordinates. The element was thought to have a constant thickness. The strains and stresses within the triangle element are constant because the matrix [*B*]is constant within the element.

8. TWO-DIMENSIONAL ISOPARAMETRIC ELEMENTS

The parametric specification of both coordinate and displacement functions serves as the foundation for isoparametric finite elements. The element shape is specified using the same shape functions that are used to interpolate the displacement field.

Shape Functions

Figure 6 shows two-dimensional linear and quadratic isoparametric finite elements. Shape functions In local coordinates ξ,η ($-1 \leq \xi,\eta \leq 1$), Ni ar$_e$ defined. For the interpolation of displacements and coordinates, the same shape functions are employed:

$$\begin{aligned} u &= \sum N_i u_i \quad , \quad v = \sum N_i v_i \\ x &= \sum N_i x_i \quad , \quad y = \sum N_i y_i \end{aligned} \tag{45}$$

u_i, v_i are displacement values at the nodes of the finite element; x, y are point coordinates, and x_i, y_i are coordinates of the element nodes, where u, v are displacement components at a point with local coordinates (ξ,η). The following relations have a matrix form:

$$\begin{aligned} \{u\} &= [N]\{q\} \\ \{u\} &= \{u \quad v\} \\ \{q\} &= \{u_1 \quad v_1 \quad u_2 \quad v_2 \quad \ldots\} \end{aligned} \tag{46}$$

$$\begin{aligned} \{x\} &= [N]\{x^e\} \\ \{x\} &= \{x \quad y\} \\ \{x^e\} &= \{x_1 \quad y_1 \quad x_2 \quad y_2 \quad \ldots\} \end{aligned} \tag{47}$$

And the nodal values' interpolation matrix is as follows:

$$[N] = \begin{bmatrix} N_1 & 0 & N_2 & 0 & \ldots \\ 0 & N_1 & 0 & N_2 & \ldots \end{bmatrix} \tag{48}$$

Figure 6. The local coordinate system's representation of linear and quadratic finite elements

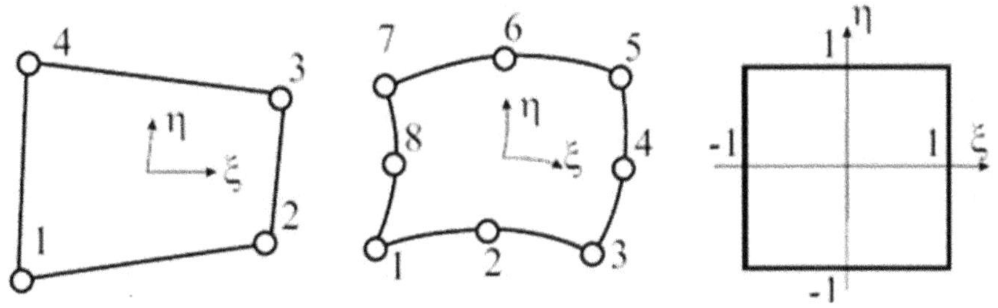

The following formulas provide shape functions for linear and quadratic two-dimensional isoparametric elements (4 nodes):

$$N_i = \frac{1}{4}\left(1+\xi_0\right)\left(1+\eta_0\right) \tag{49}$$

8 node quadratic element:

$$\begin{aligned} N_i &= \frac{1}{4}\left(1+\xi_0\right)\left(1+\eta_0\right) - \frac{1}{4}\left(1-\xi^2\right)\left(1+\eta_0\right) - \frac{1}{4}\left(1+\xi_0\right)\left(1+\eta^2\right) \quad i = 1,3,5,7 \\ N_i &= \frac{1}{2}\left(1+\xi^2\right)\left(1+\eta_0\right), \quad i = 2,6 \\ N_i &= \frac{1}{2}\left(1+\xi_0\right)\left(1+\eta^2\right), \quad i = 4,8 \end{aligned} \tag{50}$$

The following notation is used in the equations above: $\xi_0=\xi\xi_i$, $\eta_0=\eta\eta_i$, where ξ_i,η_i are the values of the local coordinates ξ,η at nodes.

8.1 Strain-Displacement Matrix

The strain vector for a plane problem has three parts:

$$\left\{\varepsilon\right\} = \left\{\varepsilon_x \quad \varepsilon_y \quad \gamma_{xy}\right\} = \left\{\frac{\partial u}{\partial x} \quad \frac{\partial v}{\partial y} \quad \frac{\partial v}{\partial x} + \frac{\partial u}{\partial y}\right\} \tag{51}$$

Using nodal displacements, the strain-displacement matrix is used to calculate strains at any location inside the element and is as follows:

$$
[B] = [B_1 \quad B_2 \quad \dots]
$$

$$
[B_i] = \begin{bmatrix} \dfrac{\partial N_i}{\partial x} & 0 \\ 0 & \dfrac{\partial N_i}{\partial y} \\ \dfrac{\partial N_i}{\partial y} & \dfrac{\partial N_i}{\partial x} \end{bmatrix} \tag{52}
$$

The strain-displacement matrix contains derivatives with regard to the global coordinates x,y, whereas shape functions are defined using the local coordinates ξ,η. The chain rule of partial differentiation can be used to easily translate derivatives from one coordinate system to another:

$$
\begin{Bmatrix} \dfrac{\partial N_i}{\partial \xi} \\ \dfrac{\partial N_i}{\partial \eta} \end{Bmatrix} = \begin{bmatrix} \dfrac{\partial x}{\partial \xi} & \dfrac{\partial y}{\partial \xi} \\ \dfrac{\partial x}{\partial \eta} & \dfrac{\partial y}{\partial \eta} \end{bmatrix} \begin{Bmatrix} \dfrac{\partial N_i}{\partial x} \\ \dfrac{\partial N_i}{\partial y} \end{Bmatrix} = [J] \begin{Bmatrix} \dfrac{\partial N_i}{\partial x} \\ \dfrac{\partial N_i}{\partial y} \end{Bmatrix} \tag{53}
$$

J refers to the Jacobian matrix. The inverse of the Jacobian matrix is used to compute the derivatives with respect to the global coordinates:

$$
\begin{Bmatrix} \dfrac{\partial N_i}{\partial x} \\ \dfrac{\partial N_i}{\partial y} \end{Bmatrix} = [J]^{-1} \begin{Bmatrix} \dfrac{\partial N_i}{\partial \xi} \\ \dfrac{\partial N_i}{\partial \eta} \end{Bmatrix} \tag{54}
$$

Using derivatives of form functions N_i with respect to the local coordinates ξ,η and the global coordinates of element nodes x_i, y_i, the components of the Jacobian matrix are calculated:

$$
\begin{aligned} \frac{\partial x}{\partial \xi} &= \sum \frac{\partial N_i}{\partial \xi} x_i, \quad \frac{\partial x}{\partial \eta} = \sum \frac{\partial N_i}{\partial \eta} x_i \\ \frac{\partial y}{\partial \xi} &= \sum \frac{\partial N_i}{\partial \xi} y_i, \quad \frac{\partial y}{\partial \eta} = \sum \frac{\partial N_i}{\partial \eta} y_i \end{aligned} \tag{55}
$$

Integrals are converted from the global coordinate system to the local coordinate system using the Jacobian matrix $|J|$ determinant:

$$
dV = dxdy = |J| d\xi d\eta \tag{56}
$$

8.2 Element Properties

The following is the calculation for element matrices and vectors:

- stiffness matrix

$$[k] = \int_V [B]^T [E][B] dV \tag{57}$$

- force vector (volume and surface loads)

$$\{P\} = \int_V [N]^T \{p^V\} dV + \int_S [N]^T \{p^S\} dS \tag{58}$$

- thermal vector (fictitious forces to simulate thermal expansion)

$$\{h\} = \int_V [B]^T [E]\{\varepsilon^t\} dV \tag{59}$$

The elasticity matrix E is given by relation (46).

8.3 Integration in Quadrilateral Elements

It is not possible to integrate stiffness matrices and load vectors analytically for the general case of isoparametric elements. Instead, stiffness matrices and load vectors are usually evaluated numerically over quadrilateral regions using Gauss quadrature. The following is the Gauss quadrature formula for the volume integral in the two-dimensional case:

$$I = \int_{-1}^{1}\int_{-1}^{1} f(\xi,\eta) d\xi d\eta = \sum_{i=1}^{n}\sum_{j=1}^{n} f(\xi_i,\eta_i)\omega_i\omega_j \tag{60}$$

where ω_i are the weighting coefficients for the Gauss integration rule and ξ_i,η_i *are* abscissae.

The surface integral is swapped out for line integration along the element side in order to calculate the nodal equivalent of the surface load. The surface load fraction is calculated as follows:

$$\begin{aligned} \{p\} &= \int_S [N]^T \{p^S\} dS = \int_{-1}^{1} [N]^T \{p^S\} \frac{ds}{d\xi} d\xi \\ \frac{ds}{d\xi} &= \sqrt{\left(\frac{dx}{d\xi}\right)^2 + \left(\frac{dy}{d\xi}\right)^2} \end{aligned} \tag{61}$$

Table 1. Abscissae and weights of the Gauss quadrature for $n = 1, 2, 3$

Abscissae and Weights of Gauss Quadrature		
N	ξ_i	ω_i
1	0	2
2	$\pm\frac{1}{\sqrt{3}}$	1
3	0	$\frac{8}{9}$
	$\pm\sqrt{\frac{3}{5}}$	$\frac{5}{9}$

In this example, ξ is a local coordinate along the element side and s is a global coordinate. The nodal equivalent of such load is: if the distributed load is applied to the element side along the normal as shown in Figure 7:

Figure 7. A quadratic element's side was given a normal load distribution

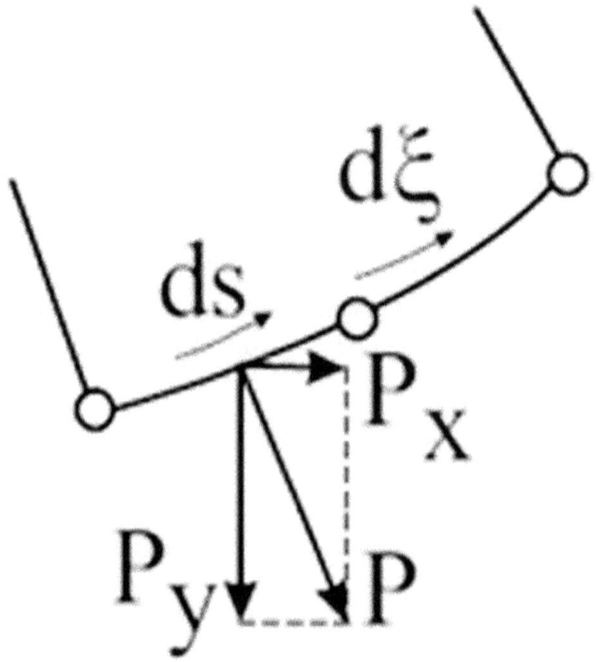

$$\{p\} = \int_S [N]^T p \begin{Bmatrix} \frac{dy}{ds} \\ -\frac{dx}{ds} \end{Bmatrix} \frac{ds}{d\xi} d\xi = \int_{-1}^{1} [N]^T p \begin{Bmatrix} \frac{dy}{ds} \\ -\frac{dx}{ds} \end{Bmatrix} d\xi \quad (62)$$

Example: Determine the nodal equivalents of a distributed load of constant intensity applied to the side of a two-dimensional quadratic element:

Figure 8. Geometry of the problem

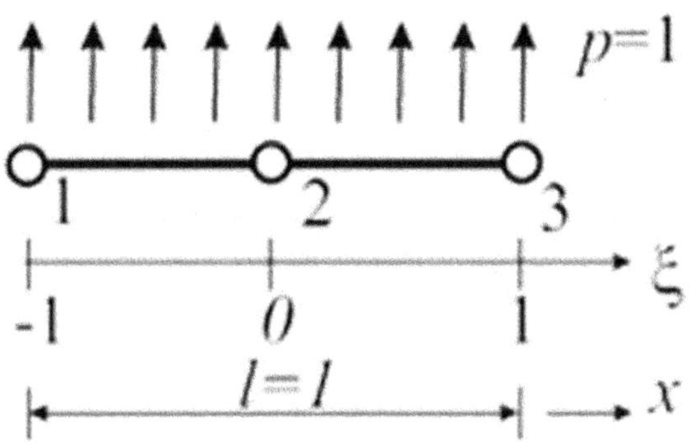

Solution: The dispersed load's nodal equivalent is computed as follows:

$$\{p\} = \int_{-1}^{1} [N]^T p \frac{dx}{d\xi} d\xi \quad (63)$$

or

$$\{p\} = \begin{Bmatrix} p_1 \\ p_2 \\ p_3 \end{Bmatrix} = \int_{-1}^{1} \begin{Bmatrix} N_1 \\ N_2 \\ N_3 \end{Bmatrix} p \frac{dx}{d\xi} d\xi \,, \quad \frac{dx}{d\xi} = \frac{1}{2} \quad (64)$$

The one-dimensional quadratic element's shape functions are:

$$N_1 = -\frac{1}{2}\xi(1-\xi)$$
$$N_2 = 1-\xi^2 \tag{65}$$
$$N_3 = \frac{1}{2}\xi(1+\xi)$$

Integration defines the values of the nodal forces at nodes 1, 2, and 3:

$$p_1 = -\int_{-1}^{1}\frac{1}{2}\xi(1-\xi)\frac{1}{2}d\xi = \frac{1}{6}$$
$$p_2 = \int_{-1}^{1}(1-\xi^2)\frac{1}{2}d\xi = \frac{2}{3} \tag{66}$$
$$p_3 = \int_{-1}^{1}\frac{1}{2}\xi(1+\xi)\frac{1}{2}d\xi = \frac{1}{6}$$

The illustration demonstrates that the nodal equivalents of a distributed load cannot be estimated using a physical technique. It is effective for linear elements but ineffective for more complex ones.

8.4 Calculation of Strains and Stresses

With the use of the displacement differentiation matrix (55) and Cauchy relations (56), strains at each location in an element are calculated:

$$\{\varepsilon\} = [B]\{q\} \tag{67}$$

The Hook's law is used to determine stresses:

$$\{\sigma\} = [E]\{\varepsilon^e\} = [E](\{\varepsilon\} - \{\varepsilon^t\}) \tag{68}$$

where $\{\varepsilon^t\}$ is the free thermal expansion vector:

$$\{\varepsilon^t\} = \{\alpha T \quad \alpha T \quad 0\} \tag{69}$$

The point location where strains and stresses are computed has a considerable impact on their accuracy. The geometric center of the linear element and reduced integration points 2×2 of the quadratic quadrilateral element have the maximum precision for displacement gradients. The best precision for strains and stresses for quadratic elements with 8 nodes is at 2×2 integration points with local coordinates $\xi, \eta = \pm\frac{1}{\sqrt{3}}$. A continuous stress field could be produced with reasonable accuracy by 1) extrapolating stresses from reduced integration points to nodes, and 2) averaging finite element contributions across all nodes in the finite element model. Through the use of quadratic shape functions, later stresses can be extrapolated from nodes. As seen in Fig. 9, a quadratic element in the local coordinate system W, Q has integration points numbered (1)..(4) and corner nodes numbered 1...4. In the quadratic element, we will use linear shape functions for extrapolation. The extrapolation relation can be shown as follows in matrix form:

$$f_i = L_{i(m)} f_{(m)} \tag{70}$$

where known function values at reduced integration locations are represented by $f_{(m)}$; $L_{i(m)}$ is the symmetric extrapolation matrix, and f_i are function values at vertex nodes:

Figure 9. Integration points and vertices for the 8-node isoparametric element are numbered

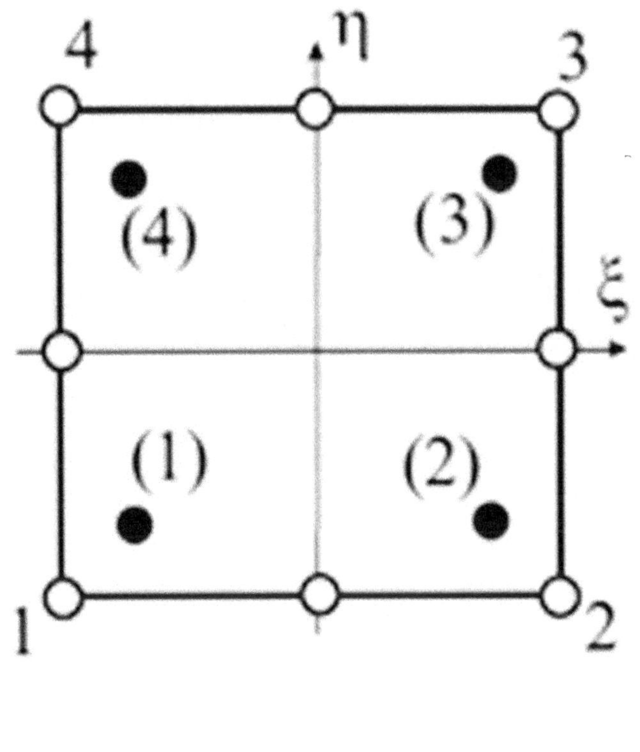

$$L_{i(m)} = \begin{bmatrix} A & B & C & B \\ & A & B & C \\ & & A & B \\ & & & A \end{bmatrix}$$
$$A = 1 + \frac{\sqrt{3}}{2}$$
$$B = -\frac{1}{2} \qquad (71)$$
$$C = 1 + \frac{\sqrt{3}}{2}$$

9. THREE-DIMENSIONAL ISOPARAMETRIC ELEMENTS

9.1 Shape Functions

In Fig. 10, linear 8-node hexahedral (or brick-type) elements and quadratic 20-node three-dimensional isoparametric elements are shown. When something is described as being "isoparametric," it indicates that the geometry and displacement field are parametrically determined and are interpolated using the same functions. Polynomials of the local coordinates ξ,η, and ζ ($-1 \le \xi,\eta,\zeta \le 1$) are utilized as shape functions for interpolation.

Figure 10. Finite elements in three dimensions that are linear and quadratic and how they are represented locally

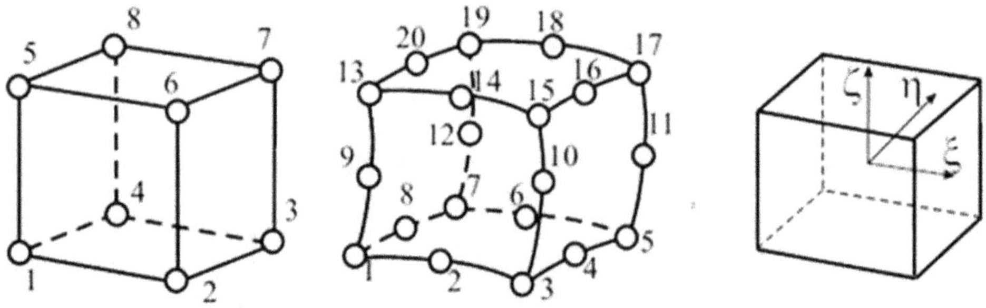

The same shape functions are used to interpolate coordinates and displacements:

$$\{u\} = [N]\{q\}$$
$$\{u\} = \{u \quad v \quad w\} \qquad (72)$$
$$\{q\} = \{u_1 \quad v_1 \quad w_1 \quad u_2 \quad v_2 \quad w_2 \quad \ldots\}$$

$$\{x\} = [N]\{x^e\}$$
$$\{x\} = \{x \quad y \quad z\} \tag{73}$$
$$\{x^e\} = \{x_1 \quad y_1 \quad z_1 \quad x_2 \quad y_2 \quad z_2 \quad \ldots\}$$

Here, x,y,z are point coordinates, and u,v,w are displacements at points with local coordinates ξ,η, and ζ; u_i,v_i,w_i are displacement values at nodes; and x_i,y_i,z_i are coordinates of nodes. The form functions matrix is as follows:

$$[N] = \begin{bmatrix} N_1 & 0 & 0 & N_2 & 0 & 0 & \ldots \\ 0 & N_1 & 0 & 0 & N_2 & 0 & \ldots \\ 0 & 0 & N_1 & 0 & 0 & N_2 & \ldots \end{bmatrix} \tag{74}$$

The linear element's shape functions are equal to:

$$N_i = \frac{1}{8}(1+\xi_0)(1+\eta_0)(1+\varsigma_0)$$
$$\xi_0 = \varsigma\xi_i \ , \ \eta_0 = \eta\eta_i \ , \ \varsigma_0 = \varsigma\varsigma_i \tag{75}$$

The shape functions for the quadratic element with 20 nodes can be expressed as follows:

$$N_i = \frac{1}{8}(1+\xi_0)(1+\eta_0)(1+\varsigma_0)(\xi_0+\eta_0+\varsigma_0-2) \qquad \text{vertices}$$
$$N_i = \frac{1}{4}(1-\xi^2)(1+\eta_0)(1+\varsigma_0) \ , \ i = 2,6,14,18$$
$$N_i = \frac{1}{4}(1-\eta^2)(1+\xi_0)(1+\varsigma_0) \ , \ i = 4,8,16,20 \tag{76}$$
$$N_i = \frac{1}{4}(1-\varsigma^2)(1+\xi_0)(1+\eta_0) \ , \ i = 9,10,11,12$$

ξ_i, η_i and ζ_i in the relationships mentioned above are node local coordinate values for ξ,η, and ζ.

9.2 Strain-Displacement Matrix

Six different strain tensor components are represented by the strain vector $\{\varepsilon\}$:

$$\{\varepsilon\} = \{\varepsilon_x \quad \varepsilon_y \quad \varepsilon_z \quad \gamma_{xy} \quad \gamma_{yz} \quad \gamma_{zx}\} \tag{77}$$

The appearance of the strain-displacement matrix for three-dimensional elements is as follows:

$$[B] = [D][N] = [B_1 \quad B_2 \quad B_3 \quad \ldots] \tag{78}$$

$$[B_i] = \begin{bmatrix} \frac{\partial N_i}{\partial x} & 0 & 0 \\ 0 & \frac{\partial N_i}{\partial y} & 0 \\ 0 & 0 & \frac{\partial N_i}{\partial z} \\ \frac{\partial N_i}{\partial y} & \frac{\partial N_i}{\partial x} & 0 \\ 0 & \frac{\partial N_i}{\partial z} & \frac{\partial N_i}{\partial y} \\ \frac{\partial N_i}{\partial z} & 0 & \frac{\partial N_i}{\partial x} \end{bmatrix} \tag{79}$$

The following steps are taken to get shape function derivatives with regard to global coordinates:

$$\begin{Bmatrix} \frac{\partial N_i}{\partial x} \\ \frac{\partial N_i}{\partial y} \\ \frac{\partial N_i}{\partial z} \end{Bmatrix} = [J]^{-1} \begin{Bmatrix} \frac{\partial N_i}{\partial \xi} \\ \frac{\partial N_i}{\partial \eta} \\ \frac{\partial N_i}{\partial \varsigma} \end{Bmatrix} \tag{80}$$

where it appears that the Jacobian matrix:

$$[J] = \begin{bmatrix} \frac{\partial x}{\partial \xi} & \frac{\partial y}{\partial \xi} & \frac{\partial z}{\partial \xi} \\ \frac{\partial x}{\partial \eta} & \frac{\partial y}{\partial \eta} & \frac{\partial z}{\partial \eta} \\ \frac{\partial x}{\partial \varsigma} & \frac{\partial y}{\partial \varsigma} & \frac{\partial z}{\partial \varsigma} \end{bmatrix} \tag{81}$$

By differentiating displacements defined through form functions and nodal displacement values, it is possible to determine the partial derivatives of x,y,z with respect to ξ,η,ζ:

$$\frac{\partial x}{\partial \xi}=\sum\frac{\partial N_i}{\partial \xi}x_i\ ,\ \frac{\partial x}{\partial \eta}=\sum\frac{\partial N_i}{\partial \eta}x_i\ ,\ \frac{\partial x}{\partial \zeta}=\sum\frac{\partial N_i}{\partial \zeta}x_i$$
$$\frac{\partial y}{\partial \xi}=\sum\frac{\partial N_i}{\partial \xi}y_i\ ,\ \frac{\partial y}{\partial \eta}=\sum\frac{\partial N_i}{\partial \eta}y_i\ ,\ \frac{\partial y}{\partial \zeta}=\sum\frac{\partial N_i}{\partial \zeta}y_i \tag{82}$$
$$\frac{\partial z}{\partial \xi}=\sum\frac{\partial N_i}{\partial \xi}z_i\ ,\ \frac{\partial z}{\partial \eta}=\sum\frac{\partial N_i}{\partial \eta}z_i\ ,\ \frac{\partial z}{\partial \zeta}=\sum\frac{\partial N_i}{\partial \zeta}z_i$$

Using the determinant of the Jacobian matrix, integrals are transformed from the global coordinate system to the local coordinate system:

$$dV = dxdydz = \left|J\right| d\xi d\eta d\varsigma \tag{83}$$

9.3 Element Properties

The form of an element equilibrium equation is as follows:

$$\begin{aligned}&[k]\{q\}=\{f\}\\&\{f\}=\{p\}+\{h\}\end{aligned} \tag{84}$$

Vectors and matrices with elements are:

- stiffness matrix

$$[k]=\int_V [B]^T [E][B]dV \tag{85}$$

- force vector (volume and surface loads)

$$\{P\}=\int_V [N]^T \{p^V\}dV+\int_S [N]^T \{p^S\}dS \tag{86}$$

- thermal vector (forces made up to mimic thermal expansion)

$$\{h\}=\int_V [B]^T [E]\{\varepsilon^t\}dV \tag{87}$$

$[E]$is the elasticity matrix.

$$\left[E\right]=\begin{bmatrix}\lambda+2\mu & \lambda & \lambda & 0 & 0 & 0\\ \lambda & \lambda+2\mu & \lambda & 0 & 0 & 0\\ \lambda & \lambda & \lambda+2\mu & 0 & 0 & 0\\ 0 & 0 & 0 & \mu & 0 & 0\\ 0 & 0 & 0 & 0 & \mu & 0\\ 0 & 0 & 0 & 0 & 0 & \mu\end{bmatrix} \tag{88}$$

where λ and μ are elastic Lame constants that can be represented by the Poisson's ratio ν and the elasticity modulus E:

$$\begin{aligned}\lambda &= \frac{\nu E}{\left(1+\nu\right)\left(1-2\nu\right)}\\ \mu &= \frac{E}{2\left(1+\nu\right)}\end{aligned} \tag{89}$$

9.4 Efficient Computation of the Stiffness Matrix

It takes a lot of zeroed arithmetic operations to calculate the element stiffness matrix by multiplying three matrices. The coefficients of the element stiffness matrix [k]can be written as follows after being multiplied in closed form:

$$\begin{aligned}k_{ii}^{\min} &= \int_V \left[\beta\frac{\partial N_m}{\partial x_i}\frac{\partial N_n}{\partial x_i}+\mu\left(\frac{\partial N_m}{\partial x_{i+1}}\frac{\partial N_m}{\partial x_{i+1}}+\frac{\partial N_m}{\partial x_{i+2}}\frac{\partial N_m}{\partial x_{i+2}}\right)\right]dV\\ k_{ij}^{\min} &= \int_V \left(\lambda\frac{\partial N_m}{\partial x_i}\frac{\partial N_n}{\partial x_j}+\mu\frac{\partial N_m}{\partial x_j}\frac{\partial N_n}{\partial x_i}\right)dV\\ \beta &= \lambda+2\mu\end{aligned} \tag{90}$$

Here, m, n are local node numbers, and i,j are coordinate axes-related indices. (x_1, x_2, x_3). If the coordinate indices rise above 3, the cyclic rule is applied to the previously stated equation.

9.5 Integration of the Stiffness Matrix

In the local coordinate system ξ,η,ζ, the stiffness matrix for three-dimensional isoparametric elements is integrated:

$$\left[k\right]=\int_{-1}^{1}\int_{-1}^{1}\int_{-1}^{1}\left[B\left(\xi,\eta,\varsigma\right)\right]^T\left[E\right]\left[B\left(\xi,\eta,\varsigma\right)\right]\left|J\right|d\xi d\eta d\varsigma \tag{91}$$

The numerical integration process is as follows when the one-dimensional Gauss quadrature rule is applied three times:

$$I = \int_{-1}^{1}\int_{-1}^{1}\int_{-1}^{1} f(\xi,\eta,\varsigma)\,d\xi d\eta d\varsigma = \sum_{i=1}^{n}\sum_{j=1}^{n}\sum_{k=1}^{n} f(\xi_i,\eta_i,\varsigma_k)\,\omega_i\omega_j\omega_k \tag{92}$$

The evaluation of the stiffness matrix for linear elements typically uses integration 2×2×2 and the evaluation of the stiffness matrix for quadratic elements typically uses integration 3×3×3. There is a unique 14-point Gauss-type rule that allows for more precise integration and is useful for three-dimensional quadratic elements.

9.6 Calculation of Strains and Stresses

The assembly procedure is used to put together the global equation system after computing the element matrices and vectors. Displacements at the nodes of the finite element model are provided by the solution of the global equation system. Nodal displacement for each element can be calculated by disassembly.

$$\{\varepsilon\} = [B]\{q\} \tag{93}$$

The Hook's law is used to determine stresses:

$$\{\sigma\} = [E]\{\varepsilon^e\} = [E](\{\varepsilon\} - \{\varepsilon^t\}) \tag{94}$$

where $\{\varepsilon t^j$ represents the free thermal expansion vector:

$$\{\varepsilon^t\} = \{\alpha T \quad \alpha T \quad \alpha T \quad 0 \quad 0 \quad 0\} \tag{95}$$

It should be emphasised that at various points inside finite elements, displacement gradients (and thus strains and stresses) have considerably varying precision. The geometric centre of the linear element and reduced integration points 2×2×2 of the quadratic hexagonal element have the maximum precision for displacement gradients.

9.7 Extrapolation of Strains and Stresses

At 2×2×2 integration points with local coordinates $\xi,\eta,\zeta = \pm\frac{1}{\sqrt{3}}$, displacement derivatives for quadratic elements have the highest degree of precision. It is important to extrapolate result values from 2×2×2 integration points to the vertices of a 20-node element in order to create a continuous field of strains or stresses. (numbering of integration points and vertices is shown in Fig. 11).

Figure 11. Integration points and vertices for the 20 node isoparametric element are numbered

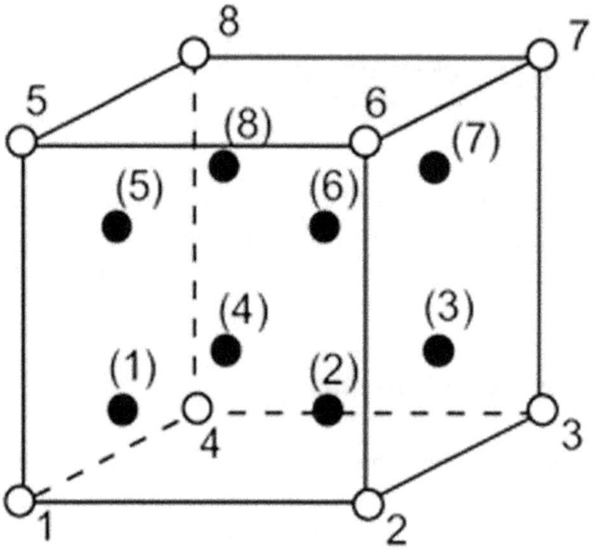

In the local coordinate system ξ,η,ζ, a trilinear extrapolation is employed to calculate the results after 8 integration points:

$$f_i = L_{i(m)} f_{(m)} \tag{96}$$

where known function values at reduced integration locations are represented by $f_{(m)}$; $L_{i(m)}$ is the symmetric extrapolation matrix, and f_i are function values at vertex nodes:

$$L_{i(m)} = \begin{bmatrix} A & B & C & B & B & C & D & C \\ & A & B & C & C & B & C & D \\ & & A & B & D & C & B & C \\ & & & A & C & D & C & B \\ & & & & A & B & C & B \\ & & & & & A & B & C \\ & & & & & & A & B \\ & & & & & & & A \end{bmatrix} \tag{97}$$

$$
\begin{aligned}
A &= \frac{5+\sqrt{3}}{4} \\
B &= -\frac{\sqrt{3}+1}{4} \\
C &= \frac{\sqrt{3}-1}{4} \\
D &= \frac{5-\sqrt{3}}{4}
\end{aligned}
\tag{98}
$$

From integration points, stresses are projected to all elemental nodes. The average of two vertex nodal values can be used to calculate values for midside nodes. Then, for all nodes of the finite element model, contributions from the nearby finite elements are averaged. A continuous field of secondary results specified at model nodes and quadratic variation inside finite elements are produced by averaging. Quadratic shape functions can then be used to interpolate the results to any point within or on the surface of the element.

10. DISCRETE MODEL OF THE PROBLEM

A discrete model of the problem should be supplied in numerical form in order to utilise finite element methods (Huebner et al., 2001; Hutton, 2004; Pepper & Heinrich, 2005). The following is an example of an issue description:

Scalar parameters (such as the number of nodes or elements);
Material properties;
Coordinates of nodal points;
array of connections for finite elements;
arrays of different element materials and element types;
arrays for describing boundary conditions for displacement;
arrays for surface and focused load descriptions;
field of temperature.

Let's record the numerical data for the straightforward issue shown in Figure 12.
Following is a description of the finite element model:

1. Scalar parameters

number of nodes $= 6$
number of elements $= 2$
number of constraints $= 2$
number of loads $= 2$

Figure 12. Problem information

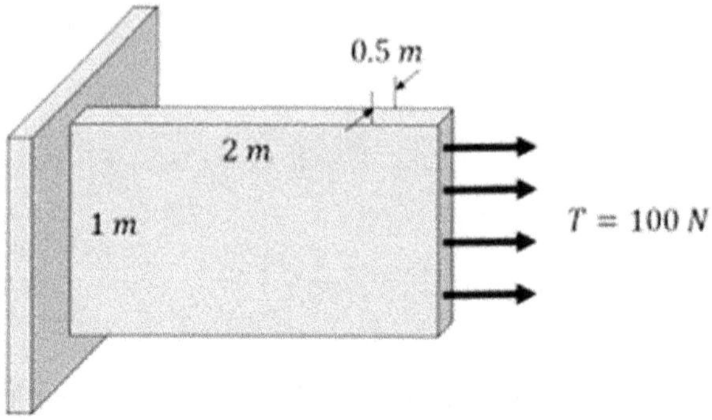

Figure 13. Rectangular element

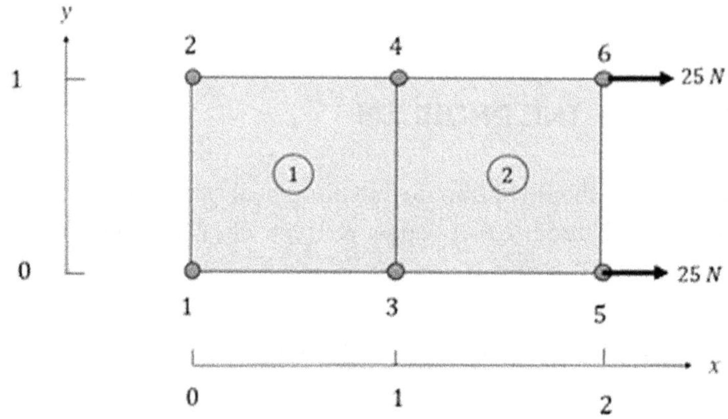

2. Material properties

Elasticity modulus $= 2 \times 10^8$ Mpa
Poisson=0.3

3. Node coordinates (x_1,y_1,x_2,y_2, etc.)
   1) 0 0 2) 0 1 3) 1 0 4) 1 1 5) 2 0 6) 2 1
4. Element connectivity array (counterclockwise direction)
   1) 1 3 4 2 2) 3 5 6 4
5. Constraints (node, direction: x=1; y=2)
   1 1 2 1
6. Nodal forces (node, direction, value)
   5 1 25 6 1 25

The finite element model can be manually programmed for basic examples like the one shown above, however this is impractical for models used in real-world applications. For the purpose of constructing finite element models for complex shapes, numerous automatic mesh generators are employed.

And also for the triangular element we have:

Figure 14. Triangular element

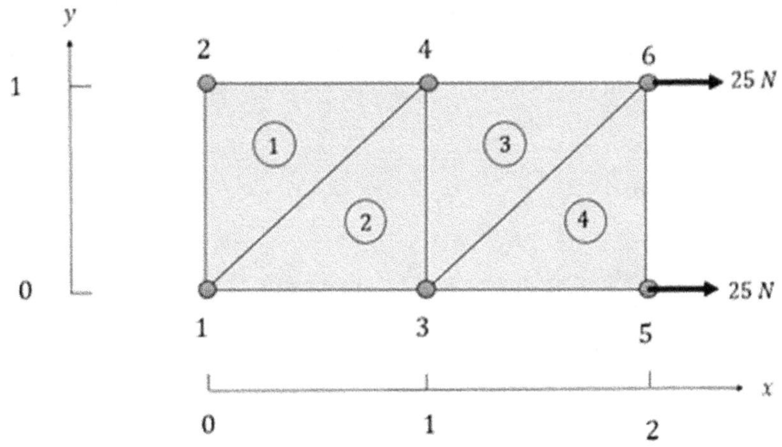

Following is a description of the finite element model:

1. Scalar parameters

number of nodes = 6
number of elements = 4
number of constraints = 2
number of loads = 2

2. Material properties

Elasticity modulus $= 2 \times 10^8$ Mpa
Poisson=0.3

3. Node coordinates (x_1,y_1,x_2,y_2, etc.)
   1) 0 0 2) 0 1 3) 1 0 4) 1 1 5) 2 0 6) 2 1
4. Element connectivity array (counterclockwise direction)
   1) 1 4 2 2) 1 3 4 3) 3 6 4 4) 3 5 6

5. Constraints (node, direction: x=1; y=2)
   1 1 2 1
6. Nodal forces (node, direction, value)
   5 1 25 6 1 25

Example 1

Example: The wall shown in Figure 15 is made of two homogeneous slabs that are in contact with each other. In steady state conditions, the wall temperature is determined by the external surface temperatures θ_1, θ_3 and the interface temperature θ_2. The equilibrium equations of the problem in terms of the mentioned temperatures, if the ambient temperatures θ_0 and θ_4 are known, are as follows:

Figure 15. Example information

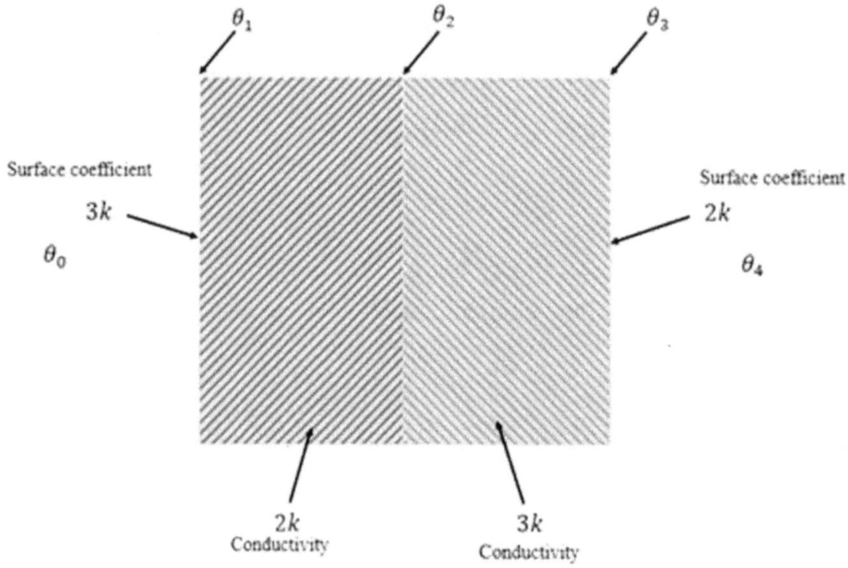

The law of thermal conductivity is $\frac{q}{A} = \bar{k}\Delta\theta$, where q is the total heat flow, A is the surface area, $\Delta\theta$ is the temperature drop in the direction of the heat flow, and $\bar{k}$ is the conductivity or surface coefficient. The state variables in this study are θ_1, θ_2, and θ_3. Using the law of thermal conductivity, the balance equations of the elements are as follows:

For the left surface in surface units: $q_1=3k(\theta_0 - \theta_1)$
For the left slab: $q_2=2k(\theta_1 - \theta_2)$
For the right slab $q_3=3k(\theta_2 - \theta_3)$
For the right surface $q_4=2k(\theta_3 - \theta_4)$

In order to find the governing equations, in terms of state variables, we use the heat flow equilibrium condition, that is $q_1=q_2=q_3=q_4$ we have:

$$3k(\theta_0 - \theta_1) = 2k(\theta_1 - \theta_2) \tag{99}$$

$$2k(\theta_1 - \theta_2) = 3k(\theta_2 - \theta_3)$$

$$3k(\theta_2 - \theta_3) = 2k(\theta_3 - \theta_4)$$

By writing the above equations as a matrix, we will have:

$$\begin{bmatrix} 2k & -2k & 0 \\ -2k & 5k & -3k \\ 0 & -3k & 3k \end{bmatrix} \begin{bmatrix} \theta_1 \\ \theta_2 \\ \theta_3 \end{bmatrix} = \begin{bmatrix} 3k\theta_0 \\ 0 \\ 2k\theta_4 \end{bmatrix} \tag{100}$$

The mentioned equilibrium equations can be derived in a systematic way using the direct stiffness method. By using this technique and applying equilibrium relations for a sample element, we can write:

$$k\begin{bmatrix} 1 & -1 \\ -1 & 1 \end{bmatrix} \begin{bmatrix} \theta_i \\ \theta_j \end{bmatrix} = \begin{bmatrix} q_i \\ q_j \end{bmatrix} \tag{101}$$

where q_i, q_j are the heat flows in the element and θ_i, θ_j are the end temperatures of the element. For the system examined in this example, we have two elements (each slab is an element), so the following relationship is obtained:

$$\begin{bmatrix} 2k & -2k & 0 \\ -2k & 5k & -3k \\ 0 & -3k & 3k \end{bmatrix} \begin{bmatrix} \theta_1 \\ \theta_2 \\ \theta_3 \end{bmatrix} = \begin{bmatrix} 3k(\theta_0 - \theta_1) \\ 0 \\ 2k(\theta_4 - \theta_3) \end{bmatrix} \tag{102}$$

11. MESH GENERATION

11.1 Mesh Generators

For practical analysis, finite element models can include tens of thousands or even hundreds of thousands of degrees of freedom. These meshes cannot be produced manually (Li, 2022). A software tool known as a mesh generator separates the solution domain into several finite element subdomains. Mesh generators come in a variety of forms. We wish to specifically discuss two types for two-dimensional issues: block mesh generators and triangulators.

1. Block mesh generators require some kind of large partitioning at the beginning. The solution domain is divided into a select few relatively small chunks. There should be a uniform form for each block. Typically, mapping approach is used to create the mesh inside the block.
2. Typically, triangulators create irregular meshes inside of arbitrary domains. Mesh generation frequently uses Delaunay triangulation and Voronoi polygons. Later, a triangle mesh can be changed into a quadrilateral element mesh. It is possible to generalise Delaunay triangulation to three-dimensional domains.

11.2 Mapping Technique

Let's say that we wish to create a quadrilateral mesh inside a domain with a curved quadrilateral shape. For this, the mapping method in Fig. 16 may be employed. If a parabola can be used to approximate each side of the curved quadrilateral domain, then the domain resembles an 8-node isoparametric element. The local coordinate system ξ,η maps the domain to a square. After dividing the square in coordinates ξ,η into rectangular pieces, nodal coordinates are converted back to the universal coordinate system of x,y.

Figure 16. Mesh generation with mapping technique

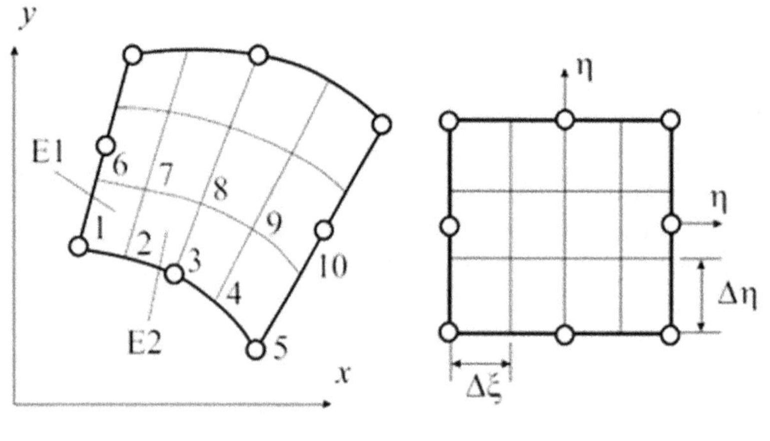

Algorithm of coordinate calculation for node i

n_ξ = number of elements in ξ direction

n_η = number of elements in η direction

$$\text{Row}: R = \frac{i-1}{n\xi+1} + 1$$

$$\text{Column}: C = \text{mod}\left((i-1),(n_\xi+1)\right)+1$$

$$\Delta\xi = \frac{2}{n_\xi}\Delta\eta = \frac{2}{n_\eta}$$

$$\xi = -1 + \Delta\xi(C-1)$$

$$\eta = -1 + \Delta\eta(R-1)$$

$$x = \sum N_k(\xi,\eta)x_k$$

$$y = \sum N_k(\xi,\eta)y_k$$

Connectivities for element *e*

$$\text{Element row: } R = \frac{e-1}{n_\xi}+1$$

$$\text{Element column: } C = \text{mod}\left((c-1),n_\xi\right)+1$$

Connectivities (global node numbers):

$$i_1 = (R-1)(n_\xi+1)+C$$

$$i_2 = i_1 + 1$$

$$i_3 = R(n_\xi+1)+C+1$$

$$i_4 = i_3 - 1$$

To refine (create smaller elements) the mesh near a domain corner, midside nodes can be moved closer to the corner. The midside node should be shifted to the following location if refining is performed on the element side that is parallel to the local axis ξ and the size of the smallest element close to the corner node is Δl:

$$\alpha = \frac{l_m}{l} = \frac{1 + \frac{\Delta l}{l} n_\xi - \frac{2}{n_\xi}}{4(1 - \frac{1}{n_\xi})} \tag{103}$$

Here, n_ξ is the number of elements in the direction of the ξ, l_m is the distance between the corner and midpoint nodes, and l is the length of an element's side.

11.3 Delaunay Triangulation

Take a look at a two-dimensional domain with any shape. Let the domain's inside and exterior points (p_1,p_2,...,p_n) be distinct points. poligon voronoi A region in the plane called V_i is one whose points are closer to node pi than to any other node. As a result, when V_i and V_j are adjacent, the bounds of V_i are the perpendicular bisectors of the lines connecting node p_i to node p_j. In Fig. 17, voronoi polygons are represented by dashed lines. Three nearby polygons share a vertex of a Voronoi polygon. According to T_k, a truangle is formed by joining three points connected to such nearby three polygons. The Delaunay triangulation is the collection of triangles T_k. A quadrilateral element-based mesh can be created using a generated triangle mesh. For this reason, connecting triangles with subsequent quality enhancement is typically used.

Figure 17. Voronoi polygons and delaunay triangulation

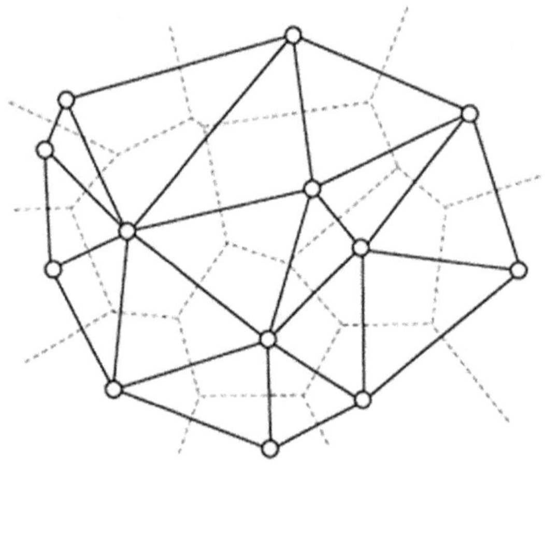

REFERENCES

David Müzel, S., Bonhin, E. P., Guimarães, N. M., & Guidi, E. S. (2020). Application of the finite element method in the analysis of composite materials: A review. *Polymers, 12*(4), 818. doi:10.3390/polym12040818 PMID:32260389

Dhatt, G., Lefrançois, E., & Touzot, G. (2012). *Finite element method.* John Wiley & Sons. doi:10.1002/9781118569764

Huebner, K. H., Dewhirst, D. L., Smith, D. E., & Byrom, T. G. (2001). *The finite element method for engineers.* John Wiley & Sons.

Hutton, D. V. (2004). *Fundamentals of finite element analysis.* The McGraw Hill Companies.

Li, H. (2022). *Graded Finite Element Methods for Elliptic Problems in Nonsmooth Domains* (Vol. 10). Springer Nature. doi:10.1007/978-3-031-05821-9

Nikishkov, G. P. (2004). Introduction to the finite element method. University of Aizu.

Pepper, D. W., & Heinrich, J. C. (2005). *The finite element method: basic concepts and applications.* Taylor & Francis. doi:10.1201/9780203942352

Reddy, J. N. (2021b). Introduction to the finite element method. Cambridge University Press.

Sabat, L., & Kundu, C. K. (2020). History of finite element method: A review. *Recent Developments in Sustainable Infrastructure: Select Proceedings of ICRDSI 2019*, 395-404.

Zienkiewicz, O. C., Taylor, R. L., & Zhu, J. Z. (2005). *The finite element method: Its basis and fundamentals.* Elsevier.

Chapter 3
Definition of Nanofluids

ABSTRACT

Many researchers are interested in nanofluids as an alternative to traditional fluids for increasing heat transmission. Nanofluids are made by dispersing nanoparticles with diameters of around 100 nm in a base fluid. Hybrid and ternary hybrid nanofluids are a novel type of nanofluid that improves the qualities of traditional nanofluids. In this chapter, comprehensive formulas on thermophysical properties (density, viscosity, specific heat capacity, thermal expansion coefficient, thermal conductivity, shape factor, and electrical conductivity) of nanofluids, hybrid nanofluids, and ternary hybrid nanofluids are presented. Considering the wide applications of nanofluids, at the end of the chapter, the application of ternary hybrid nanofluid has been comprehensively investigated.

INTRODUCTION

In his original study, Choi (Hamilton & Crosser, 1962) used nanoparticles to improve the thermal conductivity of several materials (base fluids). Nanomaterial is defined as a mixture of normal fluids and nanoparticles. These nanoparticles may be found in a variety of shapes, including spherical, rod-like, or tubular structures, rectangular blocks, and spheres. In order to improve heat transmission, nanofluids are acknowledged as an alternative to traditional fluids. By dispersing nanoparticles with diameters approximately 100 nm in the base fluid, this class of fluids is produced. As the basis fluid in nanoparticles, substances such engine oil (EO), ethylene glycol (EG), kerosene oil, blood, water, polymer, etc are being used Due to its widespread use in a variety of sciences and businesses. Metallic (Cu, Ag, Si), non-metallic, oxide (CuO, Al_2O_3) nanoparticles have recently attracted lot of interest. The idea of developing a more efficient heat transfer fluid by Hamilton and crosser (1962), was first tested by dispersing micro-sized particles in the base fluid. Interesting work was done to understand the application and thermophysical properties of nanofluids by Devarjan et al. (2018) and Esfahani et al. (2018). Advanced heat transfer fluids known as hybrid nanofluids are made by fusing nanoparticles with conventional base fluids. These very small nanoparticles have a high heat conductivity. They greatly enhance the fluid's capacity for heat transmission when disseminated inside a base fluid. Due to their improved thermal performance, hybrid nanofluids are being studied and used in a variety of industrial and scientific applications. HyNFs is

DOI: 10.4018/979-8-3693-0932-2.ch003

actually made by placing two several nanoparticles, with the same or different size, in a common fluid. Hayat and Nadeem (2017) in their study about the increase of heat transfer by HyNF concluded that even with heat generation, radiation and chemical reaction, the heat transfer of hybrid nanofluid is more than ordinary nanofluid. Thermal analysis for hybrid nanofluid (Al_2O_3–Cu) from behind a cylinder exposed to magnetic field was investigated by Alharbi et al. (2019). In accord with Suresh et al. (2011) the heat conduction characteristics of HyNFs intensify practically linearly with increasing nanoparticle volume concentration. New research on the characteristics of THyNFs has been conducted (Cakmak et al., 2020; Parekh, 2014). Manjunatha et al. (2022) investigated the properties of the THyNF (Tio_2+sio_2+Al_2O_3/H_2O) such as ρ, k and ρCp $_a$nd presented a model for a THyNF to increase the heat transfer pace. in this study, the THyNFs was discovered to have higher thermal conductivity than HyNFs. According to research by Afridi et al. (2019) on the generation of entropy in the flow of HyNF over a flat surface with heat loss, HyNF creates less entropy than regular nanofluid. Ternary hybrid nanoparticles are nanostructures made up of three unique materials or components that provide a flexible platform for integrating varied features. Researchers employ diverse materials, such as metals, semiconductors, or polymers, to develop nanoparticles with increased capabilities, making them useful in domains such as nanotechnology, materials science, and medicine for adapting materials and devices to particular applications. Awan et al. (2023) examines the Ellis hybrid nanofluid flow model with magnetic, Darcy-Forchheimer, and nonlinear thermal radiation effects over the stretched cylinder from a theoretical and computational perspective. The findings of their study demonstrated that greater Darcy-Forchheimer inputs and magnetic parameters result in a reduction in the speed characteristics of mono and hybrid nanofluids. In the world of engineering, heat transfer from a coated surface in a porous medium through which a non-Newtonian hybrid nanofluid flows is of enormous practical significance. The steady free convection flow of a non-Newtonian hybrid nanofluid in a saturated porous media characterized by Forschheimer's extended Darcy law was numerically solved and presented by Sahu et al. (2022). Their research's findings demonstrate that, depending on whether inertia is present or absent, the cross hybrid nanofluid velocity exhibits opposing responses to the Weisenberg number and porosity parameter augmentation. The ternary hybrid nanofluid has the largest surface effect. Furthermore, the ternary hybrid nanofluid has a quicker rate of heat transmission than both hybrid and ordinary nanofluids. Abbas et al. (2022) have studied the hybrid nanofluid flow from a permeable curved surface with nonlinear tension. The findings demonstrate that hybrid nanofluids transmit heat at a slower pace than basic nanofluids. Alharbi et al. (2019). explored the flow of an electroconductive incompressible ternary hybrid nanofluid with heat conduction in a boundary layer including metallic nanoparticles and magnetic induction effects across an extended cylinder. They also underlined that the variety of ternary hybrid NPs improves the thermophysical properties of the base fluid substantially. On an unstable spinning disk, Acharya et al. (2022) investigated the entropy generation of HyNF GO+$Fe3O_4$/$_H2O$ The results show that for increasing magnetic field values, entropy generation increases, but for radiation and nanoparticle concentration, the opposite is true, and the HyNF has lower entropy generation than the normal nanofluid.

1. NANOFLUIDS

A nanofluid is a perfectly uniform and stable suspension of nanoscale solid particles in a base fluid that enhances its thermal and, to a lesser extent, rheological qualities without affecting the physical and chemical properties of the base fluid. Researchers have given greater attention to the improvement in thermal behavior of these fluids compared to their rheological behavior. Fluids such as water, oils, and ethylene glycol are often used in heat transfer and energy transporters in industries. With increased worldwide rivalry in a variety of sectors, as well as the significance of energy in production costs, these businesses are focusing on the development of sophisticated and novel fluids with high thermal indices. Heat transfer via displacement can be enhanced and increased by modifying the shape and characteristics of the fluid, changing the boundary conditions, or raising the fluid's conductivity coefficient. For this growth, many strategies have been recommended. Researchers have found that adding micro or bigger particles to fluids is one way ahead since the conductivity of solids is substantially higher than that of fluids.

2. ADVANTAGES OF NANOFLUIDS

Some of the advantages and potential capabilities of nanofluids are as follows:

2.1 Improved Heat Transfer and Stability

A decrease in the size of a solid's particles, combined with an increase in their number per unit mass, results in an increase in the specific surface area. As a result, the specific surface area of nanometer-sized particles is approximately 1000 times that of micrometer-sized ones. The surface of the particles is effective in heat transmission, and the usage of nanofluid increases the level of heat transfer. The nanoparticles utilized generate a very high surface area for heat transmission, which might be advantageous for nanofluids. When the surface formed for heat transmission in nanoparticles is compared to the surface of ordinary micrometer powders, it is clear that nanoparticles have a remarkable ability to promote heat transfer and create a stable suspension. It should be mentioned that one of the issues with adding micro-sized particles to the base fluid is their quick sedimentation, which may be greatly reduced by shrinking the size to the nano scale.

2.2 Reducing the Power Required for Fluid Pumping

Increasing the rate of displacement heat transfer in common energy-carrying fluids necessitates an increase in fluid velocity, which raises the Reynolds number and, as a consequence, the Nusselt number and, as a result, the displacement heat transfer coefficient or h. This increase in speed within the equipment necessitates an increase in the pump's power consumption. However, if the nanofluid is utilized for heat transfer at a given speed, the increase in heat transfer will be due to an increase in the fluid's thermal conductivity. raising the quantity of heat transfer by two times, for example, involves raising the pump power by around ten times. While adding nanoparticles to the base fluid increases the thermal conductivity of the resultant nanofluid by around 3 times, the same 2 times increase in heat transmission may be obtained without increasing the pump. As a result, another advantage of nanofluids is that they reduce energy costs and pump power usage.

2.3 Reducing the Size of Heat Transfer Systems

Because of the capacity of nanofluid to increase heat transfer, for example, to transmit a specific quantity of heat, the needed heat exchanger will have a lower volume and size when nanofluid is used for heat transfer instead of conventional fluid.

2.4 Reduction in Costs

The use of nanofluid results in significant savings in the investment and operational expenses of industrial units due to the decrease in the power consumption of fluid transfer pumps on the one hand and the reduction in the size and weight of heat transfer equipment on the other.

3. TYPES OF NANOFLUIDS

There are several beneficial combinations of nanoparticles and base fluids. All sorts of nanofluids may be grouped into four groups based on particle type: ceramic, metal, alloy nanofluids, and lastly nanofluids with various forms of carbon or carbon base. Each of these instances is explained in more detail below.

3.1 Ceramic Nanofluids

Ceramic particles were the first materials studied for nanofluids due to their ease of manufacture and chemical stability in solution. Ceramics are divided into three categories.

1- Oxides such as alumina and zirconia (zirconium oxide)
2- non-oxides such as carbides, nitrides, and silicides
3- Composites, which are made up of oxides and non-oxides, each have their own set of properties.

3.2 Metallic Nanofluids

Although less studies have been conducted on nanofluids containing metal nanoparticles compared to oxide nanoparticles, the results are promising. Metal nanoparticles have a higher effective thermal conductivity coefficient compared to oxide nanoparticles.

3.3 Alloy Nanofluids

This type of nanofluid is prepared by mechanical alloy process or inert gas condensation.

3.4 Carbon Based Nanofluid

Polymer nanofluids are regarded as one of the most ideal and strongest alternatives in the field of cooling fluids due to their exceptional qualities, stability, relatively simple fabrication process, and acceptable viscosity. A modest quantity (approximately 1% by volume) of carbon nanotubes in ethylene glycol or oil increases the thermal conductivity of these fluids by 40% and 150%, respectively. The addition of 1% by weight carbon nanotubes more than doubles the thermal conductivity of epoxy, whereas the same amount of carbon fibers enhances thermal conductivity by 40%. As a result, carbon nanotubes offer enormous promise for improving the thermal characteristics of polymers and, as a result, developing composites for heat management applications. Depending on how the graphite sheets are rolled, nanotubes become conductive or semi-conductive; the twist angle of a nanotube, defined as the angle between the axis of its hexagonal pattern and the axis of the tube, determines whether it is conductive or non-conductive.

4. PREPARATION OF NANOFLUIDS

Nanofluids are often prepared using one of two methods:

4.1 One-Step Method

This process involves producing nanoparticles and introducing them into the base fluid in a single step. To make copper nanofluid in one step, for example, copper vapor (made by gas vapor condensation) is directly connected with a base fluid with a low vapor pressure (e.g., ethylene glycol), yielding copper nanofluid with good stable characteristics. can be.

4.2 Two-Step Method

The initial stage in this process is to create nanoparticles in the form of dry powder, which is generally accomplished by condensation with an inert gas. The generated nanoparticles are then disseminated in the fluid in the next step. The primary point in this procedure is the accumulation of nanoparticles as a result of their attachment to one another, which is one of its drawbacks. Ultrasonic equipment is used in this procedure to disperse particles and decrease clumping. One-step and two-step techniques are effective for oxide nanoparticles but less so for metal nanoparticles.

5. CHARACTERISTICS AND PROPERTIES OF NANOFLUIDS

In terms of the characteristics of nanofluids, it should be mentioned that the remarkable and unexpected rise in their thermal conductivity coefficient is the most significant observation made about them. The form and size of the particles, their volume in the solution, and the thermal properties of the base fluid and the nanoparticles are the primary determinants of this increase. Second, the researchers noticed an odd rise in the nanofluid's viscosity relative to the base fluid. To determine the proper pumping power and the displacement heat transfer coefficient, such as Prandtl and Reynolds numbers (viscosity functions), the viscosity of nanofluids must be calculated (Namburu et al., 2007; Xuan & Roetzel, 2000).

5.1 Thermal Conductivity Coefficient

Of all the physical characteristics of nanofluids, the thermal conductivity coefficient is thought to be the most intricate and significant. The performance of heating or cooling fluids can be greatly enhanced by generating a suspension of certain nanoparticles. The following are the primary causes of this:

- Suspended nanoparticles increase the surface area and heat capacity of the fluid
- Increased effective thermal conductivity coefficient is a result of suspended nanoparticles.
- The interaction and collision between particles and fluid is strengthened
- Fluctuation and turbulence of the fluid intensifies
- Transverse temperature gradient is flattened via nanoparticle dispersion.

According to the experimental results of the researchers, the thermal conductivity coefficient of nanofluids is greatly reliant on the thermal conductivity coefficient of nanoparticles, whether metallic or oxide, as well as the volume fraction of nanoparticles suspended in the nanofluid. The thermal conductivity coefficient is based on static models in which nanoparticles are assumed to be stationary in the base fluid, and it is predicted by base conductivity models such as Maxwell, Hamilton-Crosser, and others using the conductivity coefficients of the main phase components and volume fractions, among others. Furthermore, the dynamic models are based on Brownian motion and are applicable to the energy transmitted between nanoparticle collisions (Goharshadi et al., 2013).

5.2 Laboratory Measurement Methods

In a laboratory, there are three main methods for evaluating the thermal conductivity coefficient:

- Transient Hot Wire (THW)
- Steady State Parallel Plate
- Temperature Oscillation

6. STABILITY OF NANOPARTICLES IN NANOFLUIDS

In the size and scale of nano surface atoms, they increase greatly and become important. In other words, the chemical and physical properties of the surface become much more important than the structure of the material. When nano materials are affected by the ambient temperature, the atoms on the surface of the nano particles become less stable than the internal atoms, and this instability of the particle surface is the main reason for the aggregation of the particles. To stabilize the nanoparticles and prevent clumping and agglomeration of the particles that cause The weight and acceleration of the sedimentation, different methods can be used, including stabilizers, changing the pH or creating ultrasonic oscillation.

7. SINGLE NANOPARTICLES

It is crucial to look at equations and relationships pertaining to the thermophysical characteristics of nanofluids.

The relationships related to the single nanoparticle are as follows (Fallah Najafabadi et al., 2022; Hosseinzadeh, Akbari, Faghiri et al, 2023; Hosseinzadeh, Asadi, Mogharrebi et al, 2021; Hosseinzadeh et al., 2022; Hosseinzadeh, Roshani, Attar et al, 2023; Khan et al., 2022; Moradi et al., 2023; Najafabadi et al., 2022; Rostami et al., 2022):

7.1 Viscosity

$$\mu_{nf} = \frac{\mu_f}{\left(1-\phi_1\right)^{2.5}} \tag{1}$$

In the formula above, μf corresponds to the dynamic viscosity of the base fluid and $\phi 1$ represents the nanoparticle volume fraction.

7.2 Density

$$\rho_{nf} = \left(1-\phi_1\right)\rho_f + \phi_1\rho_{s1} \tag{2}$$

The density of the base fluid (ρf), the density of the nanoparticle ($\rho s1$), and the volume fraction of the nanoparticle ($\phi 1$) are represented in the formula above.

7.3 Specific Heat Capacity

$$\left(\rho c_p\right)_{nf} = \left(1-\phi_1\right)\left(\rho c_p\right)_f + \phi_1\left(\rho c_p\right)_{s1} \tag{3}$$

The base fluid is tied to the f index in the formula above, while the nanoparticle is related to the nf index. It should be mentioned that ρcp, or the Specific heat capacity, is taken into account in this section.

7.4 Thermal Expansion Coefficient

$$\left(\rho\beta\right)_{nf} = \left(1-\phi_1\right)\left(\rho\beta\right)_f + \phi_1\left(\rho\beta\right)_{s1} \tag{4}$$

The volume fraction, temperature, and characteristics of the base fluid and nanoparticles all have an impact on the coefficient of thermal expansion (CTE) of nanofluids, which are suspensions of nanoparticles in it. The total thermal expansion coefficient of the nanofluid changes as a result of the distinct thermal expansion coefficients of the nanoparticles relative to the base fluid.

7.5 Latent Heat

$$(\rho L_f)_{nf} = (\rho L_f)_f(1-\phi) \tag{5}$$

By adding nanoparticles to PCM, the thermophysical properties of NEPCM should be used in Eqs. The formula for latent heat for nanofluids is presented as above. L represents latent heat.

7.6 Thermal Conductivity

$$\frac{k_{nf}}{k_f} = \frac{k_{s1} + (s-1)k_f - (s-1)\phi_1(k_f - k_{s1})}{k_{s1} + (s-1)k_f + \phi_1(k_f - k_{s1})} \tag{6}$$

A nanofluid's thermal conductivity refers to its capacity to conduct heat and is affected by the presence of nanoparticles in a base fluid. as nanoparticles are disseminated in a fluid, they greatly increase the thermal conductivity of the nanofluid as compared to the base fluid. This improvement is due to nanoparticles' higher thermal conductivity, which allows them to more effectively transport heat. The thermal conductivity of nanofluids is an important consideration in applications such as cooling systems, where enhanced heat transfer efficiency is sought. To maximize the thermal conductivity of nanofluids for specific engineering applications, researchers investigate several factors such as nanoparticle concentration, size, and material. In equation 6, s indicates the shape factor of nanoparticles, which is very effective in thermal conductivity. f index related to the base fluid. $s1$ index is related to nanoparticle.

7.7 Shape Factor

The geometric properties of the particles, such as their form, aspect ratio, and surface roughness, are referred to as the shape factor of nanoparticles. Certain forms, such Spherical, Hexahedron, Platelets, Blades, Lamina or other intricate geometries, have unique effects on certain characteristics, including heat conductivity. To maximize the efficiency of nanoparticles in a variety of applications, researchers frequently customize them with certain shapes. Table 1 presents the shape factor values for different shapes of nanoparticles.

Table 1. Shape factor (SF) coefficient for various nanoparticles shapes (Hosseinzadeh, Mogharrebi, Asadi et al, 2020)

Kind of Nanoparticles	Shape	Shape Factor
Spherical		s=3
Hexahedron		s=3.7221
Platelets		s=5.7
Blades		s=8.6
Lamina		s=16.1576

7.8 Electrical Conductivity

$$\frac{\sigma_{nf}}{\sigma_f} = \frac{(1+2\phi_1)\sigma_{s1} + (1-2\phi_1)\sigma_f}{(1-\phi_1)\sigma_{s1} + (1+\phi_1)\sigma_f} \tag{7}$$

The capacity of a nanofluid to conduct electricity, which is impacted by the presence of nanoparticles in a base fluid, is referred to as its electrical conductivity. When conductive nanoparticles are disseminated in a fluid, they increase the total electrical conductivity of the nanofluid over the base fluid alone. The conductivity of the nanoparticles, which can be metallic or carbon-based, allows them to promote the flow of electrical current. Nanofluid electrical conductivity is crucial in applications such as electronics, sensors, and energy devices, where better electrical conductivity is desired. In equation 7, index *f* is related to base fluid and $s1$ is related to nanoparticles.

8. HYBRID NANOFLUID

The equation 8 to 15 are presented for hybrid nanoparticles ():

$$\rho_{\mathrm{hnf}}=\rho_{\mathrm{f}}(1-\phi_2)\left[(1-\phi_1)+\phi_1\left(\frac{\rho_{\mathrm{s1}}}{\rho_{\mathrm{f}}}\right)\right]+\phi_2\rho_{\mathrm{s2}} \tag{8}$$

$$(\rho\beta)_{\mathrm{hnf}}=(\rho\beta)_{\mathrm{f}}(1-\phi_2)((1-\phi_1)+\phi_1\frac{(\rho\beta)_{\mathrm{s1}}}{(\rho\beta)_{\mathrm{f}}})+\phi_2(\rho\beta)_{\mathrm{s2}} \tag{9}$$

$$(\rho c_{\mathrm{p}})_{\mathrm{hnf}}=(\rho c_{\mathrm{p}})_{\mathrm{f}}(1-\phi_2)\left[(1-\phi_1)+\phi_1\frac{(\rho c_{\mathrm{p}})_{\mathrm{s1}}}{(\rho c_{\mathrm{p}})_{\mathrm{f}}}\right]+\phi_2(\rho c_{\mathrm{p}})_{\mathrm{s2}} \tag{10}$$

$$(\rho L_f)_{hnf}=(\rho L_f)_f(1-\phi_2)(1-\phi_1) \tag{11}$$

$$\frac{\sigma_{\mathrm{hnf}}}{\sigma_{\mathrm{bf}}}=1+\frac{3\phi\left(\sigma_1\phi_1+\sigma_2\phi_2-\sigma_{bf}\left(\phi_1+\phi_2\right)\right)}{\left((\sigma_1\phi_1+\sigma_2\phi_2+2\phi\sigma_{bf})-\phi\sigma_{bf}((\sigma_1\phi_1+\sigma_2\phi_2)-\sigma_{bf}(\phi_1+\phi_2))\right)} \tag{12}$$

The following formula is also used to calculate the electrical conductivity of hybrid nanofluids:

$$\frac{\sigma_{nf}}{\sigma_f}=\frac{\left(1+2\phi_1\right)\sigma_{s1}+\left(1-2\phi_1\right)\sigma_f}{\left(1-\phi_1\right)\sigma_{s1}+\left(1+\phi_1\right)\sigma_f}\quad \text{nanofluid} \tag{13}$$

$$\frac{\sigma_{hnf}}{\sigma_{nf}}=\frac{\left(1+2\phi_2\right)\sigma_{s2}+\left(1-2\phi_2\right)\sigma_{nf}}{\left(1-\phi_2\right)\sigma_{s2}+\left(1+\phi_2\right)\sigma_{nf}}\quad \text{hybrid nanofluid}$$

In addition, thermal conductivity and viscosity of hybrid nanofluid can be determined as:

$$\frac{k_{hnf}}{k_{bf}} = \frac{k_{s2} + (s\text{-}1)k_{bf} - (s\text{-}1)\phi_2 (k_{bf} - k_{s2})}{k_{s2} + (s\text{-}1)k_{bf} + \phi_2 (k_{bf} - k_{s2})},$$

$$\frac{k_{bf}}{k_f} = \frac{k_{s1} + (s\text{-}1)k_f - (s\text{-}1)\phi_1 (k_f - k_{s1})}{k_{s1} + (s\text{-}1)k_f + \phi_1 (k_f - k_{s1})} \tag{14}$$

$$\mu_{hnf} = \frac{\mu_f}{(1-\phi_1)^{2.5}(1-\phi_2)^{2.5}} \tag{15}$$

In comparison to traditional nanofluids, hybrid nanofluids provide a number of benefits by blending several forms of nanoparticles inside a base fluid. Their customized characteristics, increased stability, and better thermal conductivity enable superior heat transfer performance. When compared to traditional nanofluids, they are a better option for a variety of sectors due to their adaptability, which makes them beneficial in a wide range of applications. In hybrid nanofluids, two nanoparticles are used and placed in the base fluid. In the formulas related to the properties of hybrid nanofluids, this theorem is clearly evident, and two indices, 1 and 2, are used to separate nanoparticles. Also, the *f* index is related to the base fluid. It should be emphasized that the thermal conductivity of hybrid nanofluids is not determined directly. The thermal conductivity of the mono nanofluid should be determined first, followed by the conductivity of the hybrid nanofluid.

9. TERNARY HYBRID NANOFLUID

Ternary hybrid nanofluids are sophisticated nanofluid formulations that incorporate three types of nanoparticles distributed in a base fluid. The substance, size, and form of these nanoparticles might vary. Ternary hybrid nanofluids have improved thermal and electrical characteristics over traditional nanofluids due to synergistic effects caused by the mixing of various nanoparticle kinds. Ternary hybrid nanofluids may be created to have certain thermal conductivity, electrical conductivity, and stability properties by carefully choosing and combining various nanoparticles. Because of their adaptability in adjusting their characteristics, they are well suited for a wide range of applications, including heat exchangers, cavities and enclosure, phase change material, electronic cooling, and renewable energy systems. Ternary hybrid nanofluids are a complex way to maximizing heat transfer and electrical conductivity, offering novel solutions to a wide range of industrial and technical difficulties. The formulas related to the thermophysical properties of ternary hybrid nanofluids are as follows (Hosseinzadeh, Faghiri, Akbari et al, 2023):

$$\mu_{tnf} = \frac{\mu_f}{(1-\phi_1)^{2.5}(1-\phi_2)^{2.5}(1-\phi_3)^{2.5}} \tag{16}$$

$$\rho_{hnf} = (1-\phi_3)\left\{(1-\phi_2)\left[(1-\phi_1)\rho_f + \phi_1\rho_{s1}\right] + \phi_2\,\rho_{s2}\right\}\phi_3\,\rho_{s3} \tag{17}$$

$$\left(\rho c_p\right)_{tnf} = (1-\phi_3)(1-\phi_2)\left[(1-\phi_1)\left(\rho c_p\right)_f + \phi_1\left(\rho c_p\right)_{s1}\right] + \phi_2\left(\rho c_p\right)_{s2} + \phi_3\left(\rho c_p\right)_{s3} \tag{18}$$

$$\left(\rho\beta\right)_{tnf} = (1-\phi_3)(1-\phi_2)\left[(1-\phi_1)\left(\rho\beta\right)_f + \phi_1\left(\rho\beta\right)_{s1}\right] + \phi_2\left(\rho\beta\right)_{s2} + \phi_3\left(\rho\beta\right)_{s3} \tag{19}$$

$$\left(\rho L_f\right)_{tnf} = \left(\rho L_f\right)_f (1-\phi_1)(1-\phi_2)(1-\phi_3) \tag{20}$$

$$\frac{k_{nf}}{k_f} = \frac{k_{s1} + (sf-1)k_f - (sf-1)\phi_1\left(k_f - k_{s1}\right)}{k_{s1} + (sf-1)k_f + \phi_1\left(k_f - k_{s1}\right)} \quad \textit{nanofluid} \tag{21}$$

$$\frac{k_{hnf}}{k_{nf}} = \frac{k_{s2} + (sf-1)k_{nf} - (sf-1)\,\phi_2\left(k_{nf} - k_{s2}\right)}{k_{s2} + (sf-1)k_{nf} + \phi_2\left(k_{nf} - k_{s2}\right)} \quad \textit{hybrid nanofluid}$$

$$\frac{k_{tnf}}{k_{hnf}} = \frac{k_{s3} + (sf-1)k_{hnf} - (sf-1)\phi_3\left(k_{hnf} - k_{s3}\right)}{k_{s3} + (sf-1) + \phi_3\left(k_{hnf} - k_{s3}\right)} \quad \textit{ternary hybrid nanofluid}$$

$$\frac{\sigma_{nf}}{\sigma_f} = \frac{(1+2\phi_1)\sigma_{s1} + (1-2\phi_1)\sigma_f}{(1-\phi_1)\sigma_{s1} + (1+\phi_1)\sigma_f} \quad \text{nanofluid} \tag{22}$$

$$\frac{\sigma_{hnf}}{\sigma_{nf}} = \frac{(1+2\phi_2)\sigma_{s2} + (1-2\phi_2)\sigma_{nf}}{(1-\phi_2)\sigma_{s2} + (1+\phi_2)\sigma_{nf}} \quad \text{hybrid nanofluid}$$

$$\frac{\sigma_{tnf}}{\sigma_{hnf}} = \frac{(1+2\phi_3)\sigma_{s3} + (1-2\phi_3)\sigma_{hnf}}{(1-\phi_3)\sigma_{s3} + (1+\phi_3)\sigma_{hnf}} \quad \text{tri} - \text{hybrid nanofluid}$$

REFERENCES

Abbas, N., Rehman, K. U., Shatanawi, W., & Malik, M. Y. (2022). Numerical study of heat transfer in hybrid nanofluid flow over permeable nonlinear stretching curved surface with thermal slip. *International Communications in Heat and Mass Transfer, 135*, 106107. doi:10.1016/j.icheatmasstransfer.2022.106107

Acharya, N., Maity, S., & Kundu, P. K. (2022). Entropy generation optimization of unsteady radiative hybrid nanofluid flow over a slippery spinning disk. *Proceedings of the Institution of Mechanical Engineers. Part C, Journal of Mechanical Engineering Science, 236*(11), 6007–6024. doi:10.1177/09544062211065384

Afridi, M. I., Alkanhal, T. A., Qasim, M., & Tlili, I. (2019). Entropy generation in Cu-Al2O3-H2O hybrid nanofluid flow over a curved surface with thermal dissipation. *Entropy (Basel, Switzerland), 21*(10), 941. doi:10.3390/e21100941

Alharbi, S. O., Nawaz, M., & Nazir, U. (2019). Thermal analysis for hybrid nanofluid past a cylinder exposed to magnetic field. *AIP Advances, 9*(11), 115022. doi:10.1063/1.5127327

Alipour, N., Jafari, B., & Hosseinzadeh, K. (2023). Optimization of wavy trapezoidal porous cavity containing mixture hybrid nanofluid (water/ethylene glycol Go–Al2O3) by response surface method. *Scientific Reports, 13*(1), 1635. doi:10.103841598-023-28916-2 PMID:36717648

Awan, A. U., Ali, B., Shah, S. A. A., Oreijah, M., Guedri, K., & Eldin, S. M. (2023). Numerical analysis of heat transfer in Ellis hybrid nanofluid flow subject to a stretching cylinder. *Case Studies in Thermal Engineering, 49*, 103222. doi:10.1016/j.csite.2023.103222

Cakmak, N. K., Said, Z., Sundar, L. S., Ali, Z. M., & Tiwari, A. K. (2020). Preparation, characterization, stability, and thermal conductivity of rGO-Fe3O4-TiO2 hybrid nanofluid: An experimental study. *Powder Technology, 372*, 235–245. doi:10.1016/j.powtec.2020.06.012

Devarajan, M., Parasumanna Krishnamurthy, N., Balasubramanian, M., Ramani, B., Wongwises, S., Abd El-Naby, K., & Sathyamurthy, R. (2018). Thermophysical properties of CNT and CNT/Al2O3 hybrid nanofluid. *Micro & Nano Letters, 13*(5), 617–621. doi:10.1049/mnl.2017.0029

Esfahani, N. N., Toghraie, D., & Afrand, M. (2018). A new correlation for predicting the thermal conductivity of ZnO–Ag (50%–50%)/water hybrid nanofluid: An experimental study. *Powder Technology, 323*, 367–373. doi:10.1016/j.powtec.2017.10.025

Fallah Najafabadi, M., Talebi Rostami, H., Hosseinzadeh, K., & Ganji, D. D. (2022). Hydrothermal study of nanofluid flow in channel by RBF method with exponential boundary conditions. *Proceedings of the Institution of Mechanical Engineers, Part E: Journal of Process Mechanical Engineering.*

Goharshadi, E. K., Ahmadzadeh, H., Samiee, S., & Hadadian, M. (2013). *Nanofluids for heat transfer enhancement-a review*. Academic Press.

Hamilton, R. L., & Crosser, O. (1962). Thermal conductivity of heterogeneous two-component systems. *Industrial & Engineering Chemistry Fundamentals, 1*(3), 187–191. doi:10.1021/i160003a005

Hayat, T., & Nadeem, S. (2017). Heat transfer enhancement with Ag–CuO/water hybrid nanofluid. *Results in Physics, 7*, 2317–2324. doi:10.1016/j.rinp.2017.06.034

Hosseinzadeh, K., Akbari, S., Faghiri, S., & Shafii, M. B. (2023). Optimization approach for MoS2-water ethylene glycol mixture nanofluid flow in a wavy enclosure. *International Journal of Thermofluids, 18*, 100337. doi:10.1016/j.ijft.2023.100337

Hosseinzadeh, K., Asadi, A., Mogharrebi, A., Ermia Azari, M., & Ganji, D. (2021). Investigation of mixture fluid suspended by hybrid nanoparticles over vertical cylinder by considering shape factor effect. *Journal of Thermal Analysis and Calorimetry, 143*(2), 1081–1095. doi:10.100710973-020-09347-x

Hosseinzadeh, K., Faghiri, S., Akbari, S., Kermani, J. R., Jafari, B., & Shafii, M. B. (2023). Effect of ternary hybrid nanoparticles (GO-MgO-TiO2) and radiative heat transfer on the solidification process of PCM inside a triplex tube with hollow fins. *International Journal of Thermofluids, 20*, 100443. doi:10.1016/j.ijft.2023.100443

Hosseinzadeh, K., Mardani, M. R., Paikar, M., Hasibi, A., Tavangar, T., Nimafar, M., ... & Shafii, M. B. (2022). Investigation of second grade viscoelastic non-Newtonian nanofluid flow on the curve stretching surface in presence of MHD. *Results in Engineering, 17.*

Hosseinzadeh, K., Mardani, M. R., Salehi, S., Paikar, M., & Ganji, D. D. (2021). Investigation of micropolar hybrid nanofluid (iron oxide–molybdenum disulfide) flow across a sinusoidal cylinder in presence of magnetic field. *International Journal of Applied and Computational Mathematics, 7*(5), 1–17. doi:10.100740819-021-01148-6

Hosseinzadeh, K., Moghaddam, M. E., Asadi, A., Mogharrebi, A. R., & Ganji, D. D. (2020). Effect of internal fins along with hybrid nano-particles on solid process in star shape triplex latent heat thermal energy storage system by numerical simulation. *Renewable Energy, 154*, 497–507. doi:10.1016/j.renene.2020.03.054

Hosseinzadeh, K., Moghaddam, M. E., Asadi, A., Mogharrebi, A. R., Jafari, B., Hasani, M. R., & Ganji, D. D. (2021). Effect of two different fins (longitudinal-tree like) and hybrid nano-particles (MoS2-TiO2) on solidification process in triplex latent heat thermal energy storage system. *Alexandria Engineering Journal, 60*(1), 1967–1979. doi:10.1016/j.aej.2020.12.001

Hosseinzadeh, K., Moghaddam, M. E., Nateghi, S., Shafii, M. B., & Ganji, D. D. (2023). Radiation and convection heat transfer optimization with MHD analysis of a hybrid nanofluid within a wavy porous enclosure. *Journal of Magnetism and Magnetic Materials, 566*, 170328. doi:10.1016/j.jmmm.2022.170328

Hosseinzadeh, K., Mogharrebi, A. R., Asadi, A., Paikar, M., & Ganji, D. D. (2020). Effect of fin and hybrid nano-particles on solid process in hexagonal triplex latent heat thermal energy storage system. *Journal of Molecular Liquids, 300*, 112347. doi:10.1016/j.molliq.2019.112347

Hosseinzadeh, K., Montazer, E., Shafii, M. B., & Ganji, A. R. D. (2021). Solidification enhancement in triplex thermal energy storage system via triplets fins configuration and hybrid nanoparticles. *Journal of Energy Storage, 34*, 102177. doi:10.1016/j.est.2020.102177

Hosseinzadeh, K., Montazer, E., Shafii, M. B., & Ganji, D. D. (2021). Heat transfer hybrid nanofluid (1-Butanol/MoS2–Fe3O4) through a wavy porous cavity and its optimization. *International Journal of Numerical Methods for Heat & Fluid Flow, 31*(5), 1547–1567. doi:10.1108/HFF-07-2020-0442

Hosseinzadeh, K., Roghani, S., Mogharrebi, A. R., Asadi, A., & Ganji, D. D. (2021). Optimization of hybrid nanoparticles with mixture fluid flow in an octagonal porous medium by effect of radiation and magnetic field. *Journal of Thermal Analysis and Calorimetry*, *143*(2), 1413–1424. doi:10.100710973-020-10376-9

Hosseinzadeh, K., Roshani, M., Attar, M. A., Ganji, D. D., & Shafii, M. B. (2023). Heat transfer study and optimization of nanofluid triangular cavity with a pentagonal barrier by finite element approach and RSM. *Heliyon*, *9*(9), e20193. doi:10.1016/j.heliyon.2023.e20193 PMID:37809404

Khan, S. A., Hayat, T., & Alsaedi, A. (2022). Thermal conductivity performance for ternary hybrid nanomaterial subject to entropy generation. *Energy Reports*, *8*, 9997–10005. doi:10.1016/j.egyr.2022.07.149

Manjunatha, S., Puneeth, V., Gireesha, B., & Chamkha, A. (2022). Theoretical study of convective heat transfer in ternary nanofluid flowing past a stretching sheet. *Journal of Applied and Computational Mechanics*, *8*(4), 1279–1286.

Moradi, M. R., Hosseinzadeh, K., Hasibi, A., & Domiri Ganji, D. (2023). Hydrothermal study on nanobioconvective fluid flow over a vertical plate under the effect of magnetic field. *Numerical Heat Transfer Part B*, 1–15. doi:10.1080/10407790.2023.2241632

Najafabadi, M. F., TalebiRostami, H., Hosseinzadeh, K., & Ganji, D. D. (2022). Investigation of nanofluid flow in a vertical channel considering polynomial boundary conditions by Akbari-Ganji's method. *Theoretical & Applied Mechanics Letters*, *12*(4), 100356. doi:10.1016/j.taml.2022.100356

Namburu, P. K., Kulkarni, D. P., Misra, D., & Das, D. K. (2007). Viscosity of copper oxide nanoparticles dispersed in ethylene glycol and water mixture. *Experimental Thermal and Fluid Science*, *32*(2), 397–402. doi:10.1016/j.expthermflusci.2007.05.001

Parekh, K. (2014). Thermo-magnetic properties of ternary polydispersed Mn0. 5Zn0. 5Fe2O4 ferrite magnetic fluid. *Solid State Communications*, *187*, 33–37. doi:10.1016/j.ssc.2014.02.005

Rostami, A. K., Hosseinzadeh, K., & Ganji, D. D. (2022). Hydrothermal analysis of ethylene glycol nanofluid in a porous enclosure with complex snowflake shaped inner wall. *Waves in Random and Complex Media*, *32*(1), 1–18. doi:10.1080/17455030.2020.1758358

Sahu, S. K., Shaw, S., Thatoi, D. N., & Nayak, M. K. (2022). A thermal management of Darcy-Forchheimer SWCNT–MWCNT Cross hybrid nanofluid flow due to vertical stretched cylinder with and without inertia effects. *Waves in Random and Complex Media*, 1–27. doi:10.1080/17455030.2022.2088889

Suresh, S., Venkitaraj, K., & Selvakumar, P. (2011). Synthesis, characterisation of Al2O3-Cu nano composite powder and water based nanofluids. *Advanced Materials Research*, *328-330*, 1560–1567. doi:10.4028/www.scientific.net/AMR.328-330.1560

Xuan, Y., & Roetzel, W. (2000). Conceptions for heat transfer correlation of nanofluids. *International Journal of Heat and Mass Transfer*, *43*(19), 3701–3707. doi:10.1016/S0017-9310(99)00369-5

Zangooee, M. R., Hosseinzadeh, K., & Ganj, D. D. (2022a). Investigation of three-dimensional hybrid nanofluid flow affected by nonuniform MHD over exponential stretching/shrinking plate. *Nonlinear Engineering*, *11*(1), 143–155. doi:10.1515/nleng-2022-0019

Zangooee, M. R., Hosseinzadeh, K., & Ganji, D. D. (2022b). Hydrothermal analysis of hybrid nanofluid flow on a vertical plate by considering slip condition. *Theoretical & Applied Mechanics Letters*, *12*(5), 100357. doi:10.1016/j.taml.2022.100357

Zangooee, M. R., Hosseinzadeh, K., & Ganji, D. D. (2023). Hydrothermal analysis of Ag and CuO hybrid NPs suspended in mixture of water 20%+ EG 80% between two concentric cylinders. *Case Studies in Thermal Engineering*, *50*, 103398. doi:10.1016/j.csite.2023.103398

Chapter 4
Introduction to Cavity

ABSTRACT

In this chapter, a brief explanation about the basics of cavities has been given. Natural convection has been investigated as one of the important problems in cavities. The low heat transfer rate between the fluid and the walls is one of the basic problems of natural convection. This issue leads to the development of strategies to improve the heat transfer rate, such as increasing the thermophysical properties of the working fluids, making cavities with complex geometries, using cavities filled with porous media, and creating a magnetic field. Also, the review of formulas related to cavities in the presence of nanofluids has been fully reviewed in this chapter.

INTRODUCTION

One of the most significant and commonly used topics in engineering and industry is the analysis of natural convection and fluid flow in cavities of various forms. According to a study of studies in this subject, the shape of different configurations such as triangular, square, trapezoidal cavities, and so on impacts heat transmission, fluid flow, and energy loss. Round holes are used to cool circuit boards, heat exchangers, and solar air heaters, for example. Considering the various applications of natural heat transfer in cavities containing fluid in engineering and industry, especially solar collectors, it is very important to investigate this field. Examining the fundamentals and relations governing each problem is one of the basic issues in solving that problem. Many engineering and industrial systems and devices, including solar collectors, solar dryers, thermal energy storage systems, heat exchangers, etc. They can benefit from understanding the design and performance of cavities in different geometries along with nanofluids that have better thermal characteristics than ordinary fluids, as well as porous media and magnetic fields.

DOI: 10.4018/979-8-3693-0932-2.ch004

1. CONVECTION HEAT TRANSFER

The concept of heat transfer is one of the most essential and commonly used topics in engineering applications. There are various modes of heat transfer, with convective heat transfer being one of the most essential. The displacement heat transfer describes the heat transfer between the surface and the moving fluid, whose temperatures are not the same. Understanding the physical causes and determining the calculating methods of displacement heat transfer are the two main goals of investigating this phenomenon. There are two types of convection heat transfer: free (natural) convection and forced convection. Heat transmission is called free convection when the fluid adjacent to the surface is initially stationary, and forced convection when the fluid adjacent to the surface is moving. Displacement heat transfer occurs in two modes, as previously stated: forced and free or natural. When a fluid is pushed to flow over a desired surface by an external force (such as a pump, fan, or wind), this is referred to as forced displacement. Free movement occurs when the fluid travels without any anomalous causes and just by the buoyant force induced by the temperature difference. If the temperature differential between the object's surface and the surrounding air is insufficient to overcome air resistance, set it in motion, and begin free movement, the heat transfer between the object and the air is conduction. In liquids and gases, it is done by displacement method. example of Natural convection includes the Earth's wind currents, heat rising from a flame, and the suction of a heater chimney. Using a pump to circulate hot or cold water or using a fan to circulate hot or cold air are examples of forced convection (Bejan, 2013).

2. CONVECTION HEAT TRANSFER IN THE CAVITY

Convection heat transfer and fluid flow in cavities are of particular interest due to their numerous engineering applications, including solar thermal systems, heat exchangers, building cooling/heating, room ventilation, cooling of electronic devices, drying devices, thermal energy storage systems, fuel cells. Numerous research looked at how different factors affected the rate of heat transfer and fluid flow in cavities. These parameters include the improved thermal properties of unconventional working fluids, the porosity, shape of cavity and the presence (either fully or partially) of a magnetic field. The use of nanoparticles to improve the thermal characteristics of fluids has been thoroughly examined in Chapter 3. Nanofluids, which have better qualities to regular fluids, can help in heat transfer in cavities. The concept of porosity can be defined as the entire volume of the environment that the empty space occupies, expressed as a fraction. The symbol for porosity is e, and its common value for natural environments is less than 0.6. In addition, a porous medium is a substance composed of a solid network joined together by voids or empty pores. The same existing pores are the source of fluid flow. Furthermore, the distribution of these pores is erratic in terms of shape and size.

2.1 Natural Convection Heat Transfer in Cavities

Natural convection is used in many thermal engineering applications to remove heat without the use of external motion. Natural convection benefits are used in a variety of heat transfer applications, including solar collectors, electronic device cooling, heat exchangers, and energy storage tanks, among others. As a result, theoretical, computational, and experimental methodologies were used to examine the phenomenon of natural convective heat transfer and flow dynamics in cavities. The low heat transfer rate between the fluid and the walls is one of the fundamental drawbacks of natural convection. This resulted in the development of heat transfer rate improvement strategies such as enhancing the thermophysical properties of working fluids, constructing cavities with complicated geometries, employing cavities filled with porous media, attaching fins to the wall(s), and/or providing a magnetic field.

2.1.1 Types of Cavity Geometry

Table 1 compiles a number of cavity geometries that have been used in researchers' research.

2.2 Studying the Problems of Cavities in Heat Transfer

2.2.1 Numerical Studies of Natural Convection in Cavities

The advantage of numerical investigation is obvious in the area of natural convection heat transfer and fluid flow inside cavities, particularly for tackling problems that are difficult to solve experimentally. In general, authors have utilized the following conservation of mass, momentum, and energy equations to solve natural convection heat transfer and fluid flow inside cavities. This is done by various means. In this book, FlexPDE software is considered as a tool for solving problems.

2.2.2 Numerical Studies of Mixed Convection Heat Transfer in Cavities

The combination of free and forced convection heat transfer and fluid flow results in mixed convection. As a result, the fluid motion is generated by two mechanisms: buoyancy induced by thermally non-homogeneous boundaries and pressure caused by the movement of one or more cavity walls. As a result, the Richardson number can be used to describe the ratio of buoyancy to shear forces. The combination of these forces can be extremely beneficial in thermal applications and industrial operations such as float glass manufacture, food processing, nuclear reactors, solar ponds, and crystal development. The ratio of Grashof and Reynolds numbers, known as the Richardson number, is the most essential statistic in mixed convection. This number can be associated with mechanisms that are both natural and forced. As a result, for a lid-driven cavity, mixed convection heat transfer is commonly used.

2.2.3 Experimental Studies of Convection Heat Transfer in Cavities

The majority of the available literature on heat transport in cavities is based on numerical research, with experimental studies receiving less attention. It should also be mentioned that natural convection has received more experimental attention than mixed convection heat transfer. Natural convection in cavities is used in solar energy collectors, silicon surfaces, noise modeling, droplet evaporation, cold-formed steel walls, porous media, and air-filled top open cavity. Thus, it is critical to do experimental research to understand the actual behavior of cavities for various convective heat transfer processes. Water was a popular choice for the working fluid, although it has a low heat conductivity. As a result, different fluids, such as nanofluids, have been used.

Table 1. A number of cavity geometries

Cavity Geometry	Figure
Rectangular (Hwang et al., 2007)	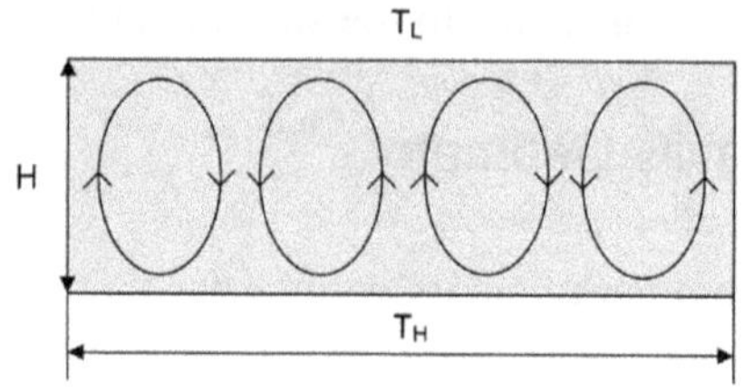
Isosceles Triangular (Rahman et al., 2016)	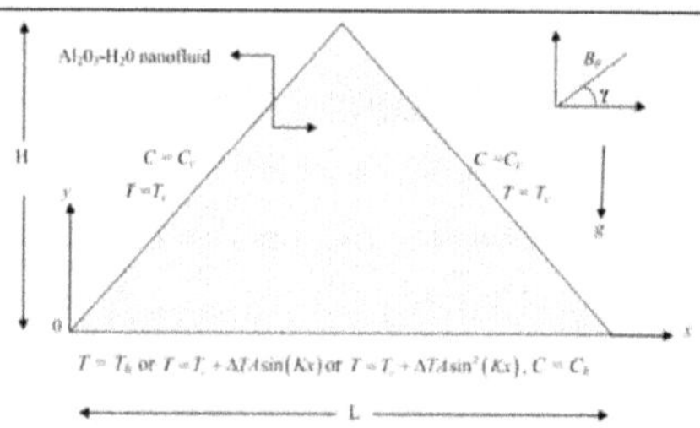
Trapezoidal (Miroshnichenko et al., 2016)	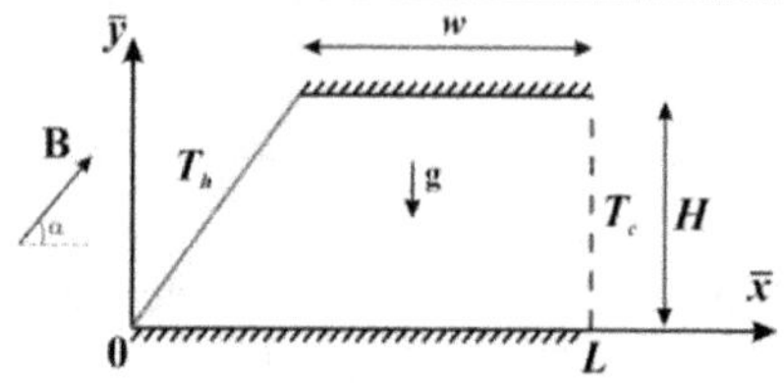
Wavy-walled (Sheremet et al., 2018)	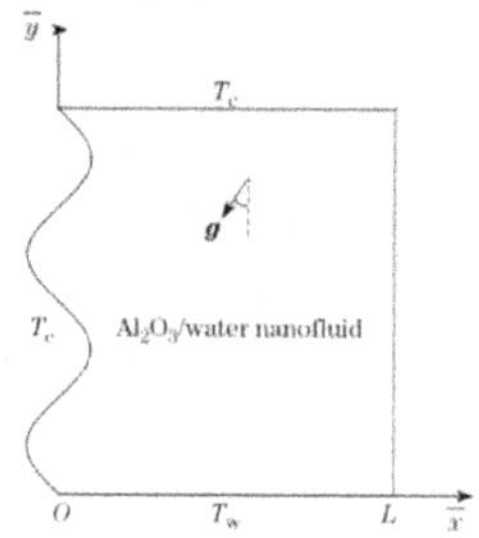
closed elbow-shaped (Ebrahimi et al., 2021)	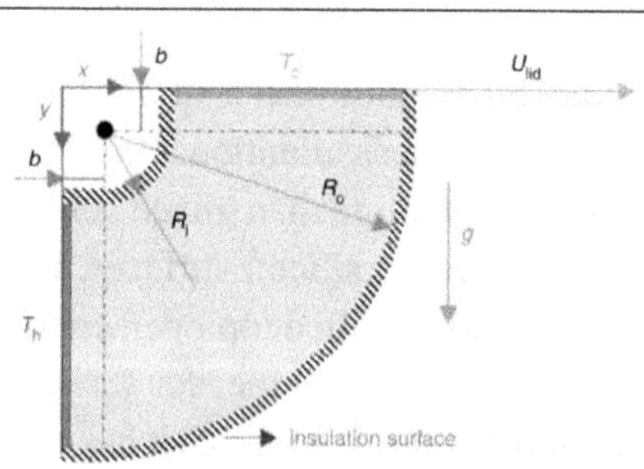

continues on following page

Table 1. Continued

Circular (Sheikholeslami et al., 2019)	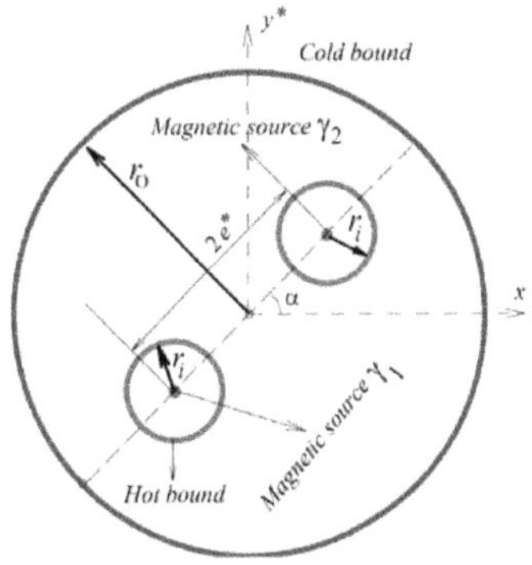
Inclined cube (Heris et al., 2014)	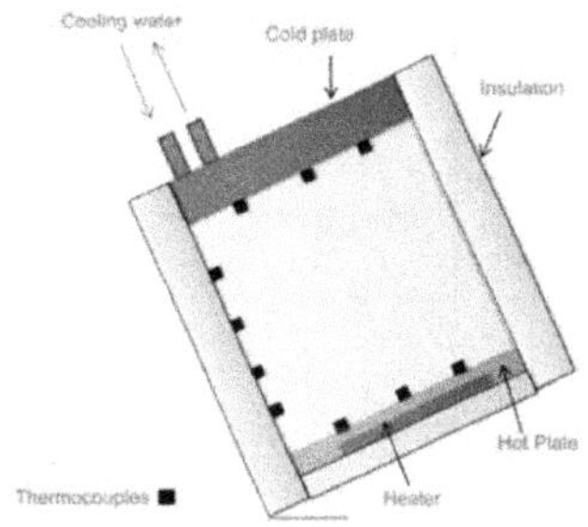
Inclined wavy (Sheremet et al., 2016)	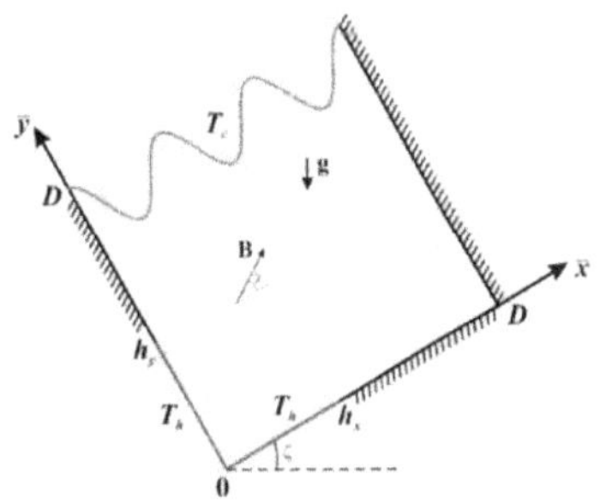
Open wavy (Selimefendigil & Öztop, 2019)	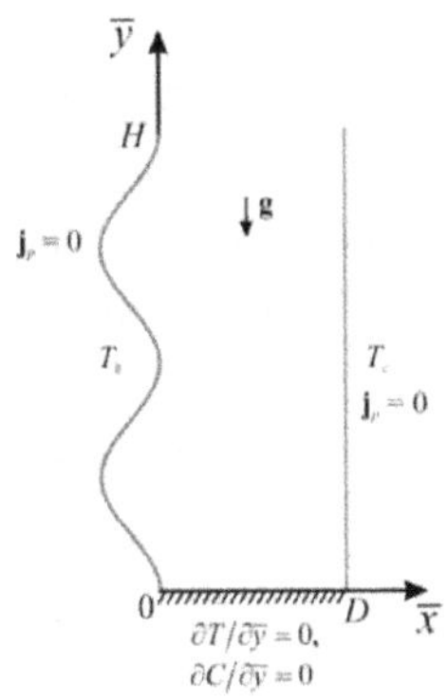

continues on following page

Table 1. Continued

Inclined square cavity (Sathiyamoorthy & Chamkha, 2010)	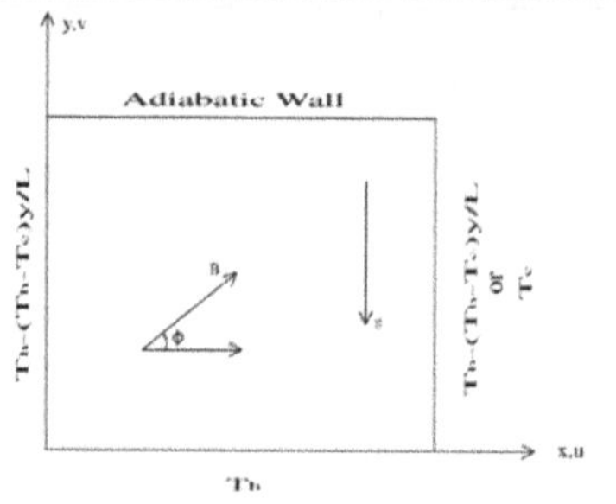
C-Shaped (Chamkha et al., 2016)	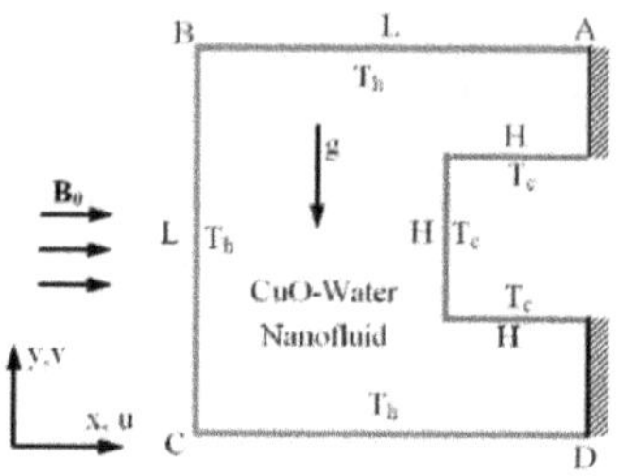
Half-annulus (Hatami et al., 2018)	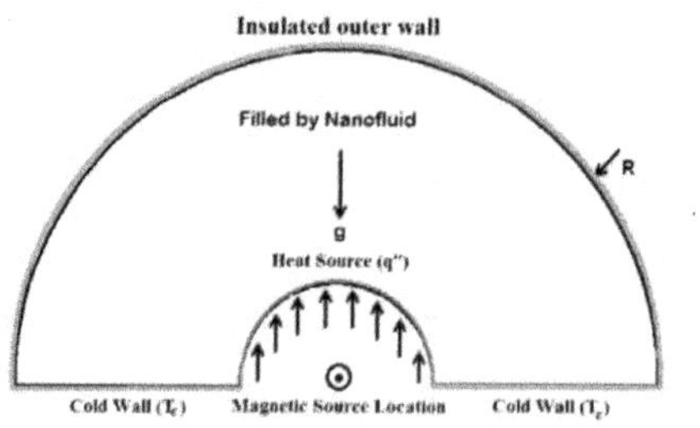

3. GOVERNING EQUATIONS OF CAVITY

3.1. Governing Equations of Cavity Model 1

Continuity equations, momentum equation and energy equation are used in problems related to cavities in the presence of nanofluids. The equations that govern the behavior of the cavity problem in the presence of nanofluids are given in the following section. Equation 1-3 defines continuity and momentum formulas to regulate the fluid flow in the boundary range of the problems in 2D.

$$\frac{\partial u}{\partial x} + \frac{\partial v}{\partial y} = 0 \tag{1}$$

$$\frac{\mu_{nf}}{K}u-\sigma_{hnf}B_0^2\left[v\left(\sin\gamma\right)\left(\cos\gamma\right)-u\left(\sin\gamma\right)^2\right]=-\frac{\partial p}{\partial x} \tag{2}$$

$$\frac{\mu_{nf}}{K}v-\sigma_{hnf}B_0^2\left[u\left(\sin\gamma\right)\left(\cos\gamma\right)-v\left(\cos\gamma\right)^2\right]=-\frac{\partial p}{\partial y}+g\left(\rho\beta\right)_{hnf}\left(T-T_c\right) \tag{3}$$

The equations related to the convective heat of nanofluid under the influence of magnetic field are investigated in the cavity chamber. Lorentz forces, thermal radiation effects and Darcy model are used. The thermal analysis of the cavity problem is done by discretizing the energy equation given below:

$$\begin{aligned}&\frac{1}{\varepsilon}\left(\left(\rho c_p\right)_{hnf}^{-1}\frac{\partial q_r}{\partial y}+\left(u\frac{\partial T}{\partial x}+v\frac{\partial T}{\partial y}\right)\right)=k_{hnf}\left(\frac{\partial^2 T}{\partial x^2}+\frac{\partial^2 T}{\partial y^2}\right)\left(\rho c_p\right)_{hnf}^{-1},\\&q_r=\frac{-4\sigma_e}{3\beta_R}\frac{\partial T^4}{\partial y},\quad T^4\cong 4T_c^3T-3T_c^4\end{aligned} \tag{4}$$

In equations 1- 4, x, y are Horizontal and vertical space coordinates, u and v are the velocity in x and y direction, is dynamic viscosity of nanofluids dealt with by Brinkman model, is Electrical conductivity of nanofluids, is Permeability, is Magnetic field, is acceleration of gravity, is coefficient of thermal expansion, is temperature, is Porosity of the porous medium, is effective heat capacitance, is effective thermal conductivity. Formulas related to nanofluids are comprehensively stated in chapter 3. It should also be noted that the effective heat capacity is in the situation where there is no pressure gradient.

For having a normalized presentation of the involved variables which brings about a more clarified and polished outlook on analysis proceedings, non- dimensionalization prove greatly beneficial. Thus, the corresponding dimensionless parameters and stream function take form as the following:

$$\begin{aligned}&v=-\frac{\partial\psi}{\partial x},u=\frac{\partial\psi}{\partial y},\Delta T=\frac{Lq''}{k_f},\\&\theta=\frac{\left(T-T_c\right)}{\left(\Delta T\right)},\left(X,Y\right)=\frac{\left(x,y\right)}{L},\Psi=\frac{\psi}{\alpha_{nf}}\end{aligned} \tag{5}$$

By some simplification and tweaking, the ultimate expressions for the governing equations emerge as below:

$$\begin{aligned}&\frac{\partial^2\Psi}{\partial X^2}+\frac{\partial^2\Psi}{\partial Y^2}=-\frac{A_6}{A_5}Ha\left[\frac{\partial^2\Psi}{\partial Y^2}\left(\sin^2\gamma\right)+\frac{\partial^2\Psi}{\partial X^2}\left(\cos^2\gamma\right)+2\frac{\partial^2\Psi}{\partial X\partial Y}\left(\sin\gamma\right)\left(\cos\gamma\right)\right]\\&-\frac{A_3A_2}{A_4A_5}\frac{\partial\theta}{\partial X}Ra\end{aligned} \tag{6}$$

$$\varepsilon\left(\left(\frac{\partial^2\theta}{\partial X^2}\right)+\left(1+\frac{4}{3}\left(\frac{k_f}{k_{hnf}}\right)Rd\right)\left(\frac{\partial^2\theta}{\partial Y^2}\right)\right)=-\frac{\partial\theta}{\partial Y}\frac{\partial\Psi}{\partial X}+\frac{\partial\Psi}{\partial Y}\frac{\partial\theta}{\partial X} \quad (7)$$

Also, the constant parameters which included in the final equations above are defined as follows:

$$A_1=\frac{\rho_{hnf}}{\rho_f},\ A_2=\frac{\left(\rho C_p\right)_{hnf}}{\left(\rho C_p\right)_f},A_3=\frac{\left(\rho\beta\right)_{hnf}}{\left(\rho\beta\right)_f},A_4=\frac{k_{hnf}}{k_f},$$
$$A_5=\frac{\mu_{hnf}}{\mu_f},\ A_6=\frac{\sigma_{hnf}}{\sigma_f}, \quad (8)$$
$$Ra=\frac{gK\left(\rho\beta\right)_f L\Delta T}{\mu_f\alpha_f},\ Ha=\frac{\sigma_f KB_0^2}{\mu_f},\ \delta_s=\frac{k_{hnf}}{\left[k_s\left(1-\varepsilon\right)\right]}$$

represents the radiation parameter, is Rayleigh number, is Hartmann number and is Modified thermal conductivity ratio.

Further, the equations 9 and 10 are presenting for local Nusselt number and average Nusselt number:

$$Nu_{loc}=\left(\frac{k_{hnf}}{k_f}\right)\frac{1}{\theta}\left(1+\frac{4}{3}Rd\left(\frac{k_{hnf}}{k_f}\right)\right) \quad (9)$$

$$Nu_{ave}=\frac{1}{2\pi}\int_0^{2\pi}Nu_{loc}dr \quad (10)$$

3.2 Governing Equations of Cavity Model 2

In another style of cavity problems, the cavity is placed under an external magnetic field with a uniform magnitude () in the horizontal direction, which induces the Lorentz force and as a result forced displacement, also a slow and steady flow of nanofluid in the cavity is considered. According to the above assumptions, the conservation equations of mass, momentum and energy for the flow and temperature fields of the mixed nanofluid can be expressed as follows (Kaluri & Basak, 2011):

$$\frac{\partial u}{\partial x}+\frac{\partial v}{\partial y}=0 \quad (11)$$

$$\rho_{nf}(u\frac{\partial u}{\partial x}+v\frac{\partial u}{\partial y})=-\frac{\partial p}{\partial x}+\mu_{nf}(\frac{\partial^2 u}{\partial x^2}+\frac{\partial^2 u}{\partial y^2}), \tag{12}$$

$$\rho_{nf}(u\frac{\partial v}{\partial x}+v\frac{\partial v}{\partial y})=-\frac{\partial p}{\partial y}+\mu_{nf}(\frac{\partial^2 v}{\partial x^2}+\frac{\partial^2 v}{\partial y^2})-\frac{\sigma B_0^{\,2} v}{\rho_{nf}}-g\rho_{nf}\beta_{nf}(T^*-T_C), \tag{13}$$

$$u\frac{\partial T^*}{\partial x}+v\frac{\partial T^*}{\partial y}=\alpha_{nf}(\frac{\partial^2 T^*}{\partial x^2}+\frac{\partial^2 T^*}{\partial y^2})-(\frac{1}{(\rho C_p)_{nf}})\frac{\partial q_r}{\partial y}, \tag{14}$$

where

$$q_r=-(4\frac{\sigma_r}{\chi_r})\frac{\partial T^4}{\partial y}, T^4\cong -3T_C^{\,4}+4T_C^{\,3}T \tag{15}$$

In equations (11)-(14), *u* and *v* denote the horizontal and vertical components of velocity, respectively. Moreover, T^*, *p* and *g* stands for the temperature, the pressure and gravitational acceleration in the opposite direction of *Y*, orderly. The thermophysical features of the nanofluid including the ρ_{nf} density, μ_{nf} viscosity, β_{nf} coefficient of thermal expansion, $(Cp)_{nf}$ heat capacity, and α_{nf} thermal diffusivity It is given in full in Chapter 3.

In order to scrutinize the impacts of crucial variables on the characteristics of the nanofluid flow, the following dimensionless parameters are introduced:

$$X=\frac{x}{L},Y=\frac{y}{L},U=\frac{uL}{\alpha_f},V=\frac{vL}{\alpha_f},T=\frac{T^*-T_C}{T_h-T_C},P=\frac{pL^2}{\rho_f\alpha_f^{\,2}},\vartheta_f=\frac{\mu_f}{\rho_f},\Pr=\frac{\vartheta_f}{\alpha_f},Ra=\frac{\beta_f(T_h-T_C)L^3}{\vartheta_f\alpha_f},$$
$$Ha=B_0L\sqrt{\frac{\sigma}{\rho_{nf}\vartheta_f}},Rd=\frac{4\sigma_rT_C^{\,3}}{\chi_rk_f},P=-\gamma(\frac{\partial U}{\partial X}+\frac{\partial V}{\partial Y}). \tag{16}$$

Here, *Ra*, *Pr*, *Ha* and *Rd* represent the Rayleigh number, Prandtl number, Hartmann number and radiation parameter. *Ha* is a benchmark of Lorentz volumetric force and represents the severity of the magnetic field. Also, the magnitude of γ is equal to 107· which is used for eliminating the pressure from these equations. By substituting these variables to the governing equations, the non-dimensional form of PDEs can be written as follows (Qi et al., 2018):

$$\frac{\partial U}{\partial X}+\frac{\partial V}{\partial Y}=0 \tag{17}$$

$$U\frac{\partial U}{\partial X}+V\frac{\partial U}{\partial Y}=\gamma\frac{\rho_f}{\rho_{nf}}\frac{\partial}{\partial X}(\frac{\partial U}{\partial X}+\frac{\partial V}{\partial Y})+\Pr\frac{\vartheta_{nf}}{\vartheta_f}(\frac{\partial^2 U}{\partial X^2}+\frac{\partial^2 U}{\partial Y^2}), \tag{18}$$

$$\begin{aligned}U\frac{\partial V}{\partial X}+V\frac{\partial V}{\partial Y}=\gamma\frac{\rho_f}{\rho_{nf}}\frac{\partial}{\partial Y}(\frac{\partial U}{\partial X}+\frac{\partial V}{\partial Y})+\Pr\frac{\vartheta_{nf}}{\vartheta_f}(\frac{\partial^2 V}{\partial X^2}+\frac{\partial^2 V}{\partial Y^2})\\ -Ha^2\Pr V+\frac{(1-\phi)\rho_f\beta_f+\phi\rho_p\beta_p}{\rho_{nf}\beta_{nf}}Ra\Pr T,\end{aligned} \tag{19}$$

$$U\frac{\partial T}{\partial X}+V\frac{\partial T}{\partial Y}=\frac{\alpha_{nf}}{\alpha_f}(\frac{\partial^2 T}{\partial X^2}+\frac{\partial^2 T}{\partial Y^2})+\frac{4}{3}Rd\frac{(\rho C_p)_f}{(\rho C_p)_{nf}}\frac{\partial^2 T}{\partial Y^2}, \tag{20}$$

The Nu_{ave} along the inner hot surface is described by the following relation:

$$\begin{aligned}Nu_{Local}&=(1+\frac{4}{3}\frac{k_f}{k_{nf}}Rd)(\frac{k_{nf}}{k_f})\frac{\partial T}{\partial n}\\ Nu_{ave}&=\frac{1}{L_T}\int_0^{L_T}Nu_{Local}\,d\eta,\end{aligned} \tag{21}$$

in which, n indicates the normal direction at the hot wall of the enclosed length (L_T).

REFERENCES

Bejan, A. (2013). *Convection heat transfer*. John Wiley & Sons. doi:10.1002/9781118671627

Chamkha, A., Ismael, M., Kasaeipoor, A., & Armaghani, T. (2016). Entropy generation and natural convection of CuO-water nanofluid in C-shaped cavity under magnetic field. *Entropy (Basel, Switzerland)*, *18*(2), 50. doi:10.3390/e18020050

Ebrahimi, D., Yousefzadeh, S., Akbari, O. A., Montazerifar, F., Rozati, S. A., Nakhjavani, S., & Safaei, M. R. (2021). Mixed convection heat transfer of a nanofluid in a closed elbow-shaped cavity (CESC). *Journal of Thermal Analysis and Calorimetry*, *144*(6), 2295–2316. doi:10.100710973-021-10548-1

Hatami, M., Zhou, J., Geng, J., & Jing, D. J. J. O. M. (2018). Variable magnetic field (VMF) effect on the heat transfer of a half-annulus cavity filled by Fe3O4-water nanofluid under constant heat flux. *Journal of Magnetism and Magnetic Materials*, *451*, 173–182. doi:10.1016/j.jmmm.2017.10.110

Heris, S. Z., Pour, M. B., Mahian, O., & Wongwises, S. (2014). A comparative experimental study on the natural convection heat transfer of different metal oxide nanopowders suspended in turbine oil inside an inclined cavity. *International Journal of Heat and Mass Transfer*, *73*, 231–238. doi:10.1016/j.ijheatmasstransfer.2014.01.071

Hwang, K. S., Lee, J. H., & Jang, S. P. (2007). Buoyancy-driven heat transfer of water-based Al2O3 nanofluids in a rectangular cavity. *International Journal of Heat and Mass Transfer*, *50*(19-20), 4003–4010. doi:10.1016/j.ijheatmasstransfer.2007.01.037

Kaluri, R. S., & Basak, T. (2011). Role of entropy generation on thermal management during natural convection in porous square cavities with distributed heat sources. *Chemical Engineering Science*, *66*(10), 2124–2140. doi:10.1016/j.ces.2011.02.009

Miroshnichenko, I. V., Sheremet, M. A., Oztop, H. F., & Al-Salem, K. (2016). MHD natural convection in a partially open trapezoidal cavity filled with a nanofluid. *International Journal of Mechanical Sciences*, *119*, 294–302. doi:10.1016/j.ijmecsci.2016.11.001

Qi, C., Zhao, N., Cui, X., Chen, T., & Hu, J. (2018). Effects of half spherical bulges on heat transfer characteristics of CPU cooled by TiO2-water nanofluids. *International Journal of Heat and Mass Transfer*, *123*, 320–330. doi:10.1016/j.ijheatmasstransfer.2018.02.086

Rahman, M. M., Alam, M. S., Al-Salti, N., & Eltayeb, I. A. (2016). Hydromagnetic natural convective heat transfer flow in an isosceles triangular cavity filled with nanofluid using two-component nonhomogeneous model. *International Journal of Thermal Sciences*, *107*, 272–288. doi:10.1016/j.ijthermalsci.2016.04.009

Sathiyamoorthy, M., & Chamkha, A. (2010). Effect of magnetic field on natural convection flow in a liquid gallium filled square cavity for linearly heated side wall (s). *International Journal of Thermal Sciences*, *49*(9), 1856–1865. doi:10.1016/j.ijthermalsci.2010.04.014

Selimefendigil, F., & Öztop, H. F. (2019). Fluid-solid interaction of elastic-step type corrugation effects on the mixed convection of nanofluid in a vented cavity with magnetic field. *International Journal of Mechanical Sciences*, *152*, 185–197. doi:10.1016/j.ijmecsci.2018.12.044

Sheikholeslami, M., Mehryan, S. A. M., Shafee, A., & Sheremet, M. A. (2019). Variable magnetic forces impact on magnetizable hybrid nanofluid heat transfer through a circular cavity. *Journal of Molecular Liquids*, *277*, 388–396. doi:10.1016/j.molliq.2018.12.104

Sheremet, M. A., Oztop, H. F., & Pop, I. (2016). MHD natural convection in an inclined wavy cavity with corner heater filled with a nanofluid. *Journal of Magnetism and Magnetic Materials*, *416*, 37–47. doi:10.1016/j.jmmm.2016.04.061

Sheremet, M. A., Trîmbiţaş, R., Groşan, T., & Pop, I. (2018). Natural convection of an alumina-water nanofluid inside an inclined wavy-walled cavity with a non-uniform heating using Tiwari and Das' nanofluid model. *Applied Mathematics and Mechanics*, *39*(10), 1425–1436. doi:10.100710483-018-2377-7

Chapter 5
The Application of FlexPDE in Cavity Problems

ABSTRACT

Ways to improve heat transfer are of great interest due to wide applications. Nanofluid is one of the ways to increase heat transfer due to its higher thermal conductivity than normal fluid. Another way to increase heat transfer is to use a porous medium. The simultaneous use of these two methods causes a significant increase in heat transfer because the nanofluid increases the thermal conductivity and the porous medium of the contact surface. In fact, the porous medium is a material consisting of a solid network that is connected by empty spaces or pores between them, and these empty spaces allow fluid to flow in this network. Topics expressed and analysis of heat transfer and optimization of hybrid nanofluids in wavy porous enclosure, octagonal porous medium, trapezoidal porous cavity, and snowflake porous enclosure are among the examples investigated in this chapter.

INTRODUCTION

Increasing energy consumption causes excessive heat generation. This heat production can cause wear in products or reduce their lifespan. Also, this heat can cause a lot of losses in the economic field. Nowadays, ways to improve heat transfer are of great interest due to its widespread use in systems such as buildings, solar collectors, electronic equipment, etc. As it was said, heat transfer is one of the important issues of energy conversion, which is also very important from an economic point of view. Nanofluid is one of the methods of increasing heat transfer which is used a lot recently. The basis of using nanofluids is because nanofluids have higher thermal conductivity than the base fluid. Another method of increasing heat transfer is the use of porous media. Due to the widespread use of porous media in industry, many researches have been conducted in the field of fluid flow and heat transfer in porous media in recent years. A porous medium is a material consisting of a solid network connected by spaces or empty pores between them, or in other words, any material with a solid frame with interconnected pores is called a porous medium. The empty spaces between the solid network allow fluid to flow in this network. The reason that fluid can flow through the voids is that all the voids in the porous medium, which are solid

DOI: 10.4018/979-8-3693-0932-2.ch005

bodies, are completely saturated with fluid. Optimum building insulation, heat exchangers, cooling, collecting underground energy, improving heat transfer through increasing the surface are examples of the wide application of porous media in the industry. Simultaneous application of these two methods (simultaneous use of porous medium and nanofluid) significantly improves the performance of thermal systems. The reason for the increased performance is that the nanofluid increases the effective thermal conductivity and the porous medium increases the contact surface. Sheremet and Pop (2014) investigated the steady-state natural-state displacement heat transfer in a square-shaped porous chamber with solid walls of limited thickness and conductivity filled by a nanofluid. Also, in this investigation, the mathematical model of nanofluid proposed by Buongiorno (2006) was used. Buongiorno developed a complete model of convective heat transfer in nanofluid by considering the relative velocity of nanoparticle-fluid, also Buongiorno proposed the mechanisms that may cause relative motion between solid and fluid phases in nanofluid, one of the main ones being you can point to Brownian diffusion and thermophoresis. Qi et al. (2017) natural convection heat transfer in a rectangular horizontal chamber filled with Ag-Ga nanofluid in different volume fractions of nanoparticles in different Rayleigh numbers was investigated by using Boltzmann's two-phase lattice model. In another investigation by Qi et al., (2016) the effect of nanoparticle size and base fluid was investigated using the two-phase Boltzmann network on the natural displacement heat transfer of nanofluid. They also investigated the heat transfer and displacement of nanofluid in chambers with different rotation angles. Nield et al. (2010) developed a theory of nanofluid convection with double diffusion in a porous medium. Nield et al. (2011). deals with the analytical investigation of double-diffusion nanofluid convection in a porous medium. In this review, a solution for the natural displacement of a vertical plane with a nanofluid containing a binary base fluid such as salt water is presented. There are different momentum models for fluid flow in a porous chamber. Darcy provided the first description of transport in porous materials. Darcy's law states that the discharge rate of the fluid in the porous medium has a direct relationship with the pressure drop and the viscosity of the fluid in a certain distance. It should also be noted that Darcy's law can only be used for incompressible and isothermal laminar flow. Biswas er al. (2022) investigated the effects of semi-active magnetic fields on the enhanced thermal performance of hybrid nanofluid flow in a diagonal wave porous chamber that is heated from below and cooled from around the wave. Aissa et al. (2023) investigated the free convection of nanofluid flow in a wavy porous chamber under a magnetic field using the Galerkin finite element method. They also investigated the entropy generation inside a wavy porous cavity. Haq et al. (2018) investigated the heat transfer of water-based SWCNT in C-sharp cavity under the influence of magnetic field. The investigations conducted by Haq ct al on the range of volume fraction of nanoparticles (0 to 0.2) show that the volume fraction of nanoparticles has a great effect on temperature distribution. Some of the researches and studies that have been mentioned are perhaps a large part of heat transfer because it includes natural and mixed heat transfer and circulation in closed chambers with different geometries such as cylinders, spheres, triangles, ovals, and squares. they take. More complex geometries include different types of geometric configurations, for example curved, convex, wavy and concave walls. One of the important topics of engineering is the study of fluid flow and natural convection in cavities with different shapes. Research shows that different configuration shapes such as square, triangular, trapezoidal holes, etc. affect heat transfer. For example, one of the uses of the round hole is for cooling circuit boards. Sheremet and Pop (2017) investigated the natural convection heat transfer coefficient and fluid flow inside a porous wavy cavity filled with a nanofluid. The investigated domain is a square cavity with an isothermal corrugated wall on the left side and the other walls are smooth. There is also a constant temperature heat source on the right wall. Mondal and

Mahapatra (2021) investigated the reduction of lost energy in a trapezoidal chamber filled with water nanofluid. Berman and Rao (2021) numerically investigated the effect of aspect ratio on natural convection in a wavy porous cavity. In this investigation, the cavity is vertical and its right side wavy wall is kept constant at low temperature. There is a partial heat source on the left vertical wall that keeps the top and bottom walls adiabatic. Investigation of natural convection in a cavity with a non-Newtonian nanofluid including 2 cylinders inside it under thermal radiation and without thermal radiation was done by Barnon et al. (2019) Also, this study showed that changing the hole angle can be effective in heat transfer. Shulepova et al. (2020) used alumina-water solution as a nanofluid in a square cavity and found that regardless of the Reynolds or Rayleigh numbers, dispersion of nanoparticles in water is desirable for heat transfer. It is noteworthy that buoyancy improves heat transfer in confined spaces. Buoyancy is caused by gravity and density imbalance, which is considered an active force. Free convection flow and temperature gradient in a cavity filled with water and copper oxide as the base fluid were investigated by Ghasemi et al (2009). Seleh et al. (2011) also investigated the nanofluid in a trapezoidal chamber to investigate the buoyancy force in heat transfer.

1. EXAMPLE 1: RADIATION AND CONVECTION HEAT TRANSFER OPTIMIZATION WITH MHD ANALYSIS OF A HYBRID NANOFLUID WITHIN A WAVY POROUS ENCLOSURE

1.1 Problem Definition

For an in-depth demonstration of the respective domain, a representative configuration of the wavy star-like enclosure with a circular cavity and its corresponding boundary conditions are featured by Fig.1 in which hybrid nano-sized particles of Fe_3O_4-Al_2O_3 dispersed in 1-Hexanol are bounded by rigid porous media. The isothermal boundary conditions are set accordingly such that it stimulates thermal gradient profiles inside the domain. As such, low-temperature outer surface and heated inner cavity serve this purpose. This set-forth design along with multiple numerically integrated variables can expound on the unexplored aspects of more complex curved geometries and their underlying convective heat transfer physics in an attempt to accommodate universal applications of this branch.

Equations related to geometry are given in chapter 4. Also, the formulas related to the thermophysical properties of nanofluids are stated in chapter 2. Refer to Table 1 to specify the exact thermo-physical properties of the nanofluid used in this problem. Finally, by using the non-dimensionalization of the boundary conditions, the problem will change as follows:

Along all walls:

$$u = v = 0, \ \Psi = 0, \ \frac{\partial \Psi}{\partial n} = 0 \tag{1}$$

On outer wall:

$$T=T_c, \ \theta=0 \tag{2}$$

On inner wall:

$$T = T_h,\ \theta = 1 \tag{3}$$

Figure 1. Adaptive mesh refinement

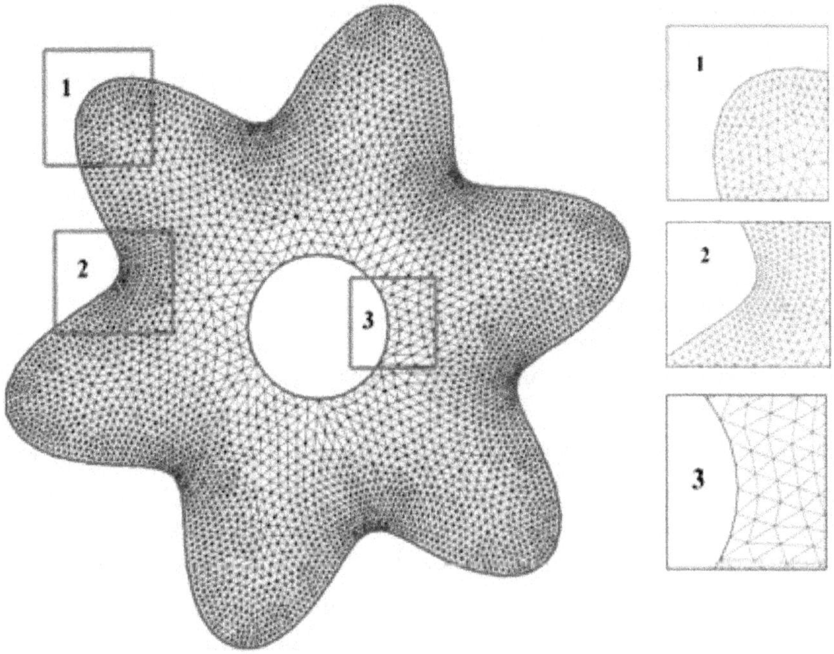

Table 1. Specifications of nanoparticles used in the present research (Hosseinzadeh, Moghaddam, Asadi et al, 2020; Hosseinzadeh, Mogharrebi, Asadi et al, 2020)

Physical Properties	1-Hexanol	Fe_3O_4	Al_2O_3
$\rho\ (\frac{kg}{m^3})$	815	5180	3970
$c_p\ (\frac{J}{kg.^\circ K})$	2384	670	765
$k\ (\frac{W}{m.^\circ K})$	0.193	9.7	25
$\beta \times 10^{-5}(k^{-1})$	55	1.3	1.26

1.2 Numerical Simulation

GFEM has been earning ever-growing utility when it comes to analytical modeling of numerous engineering problems. This is largely due to the strength and flexibility of this method that tends to return exceedingly reliable outcomes, being most likely in a good agreement with experimental results. The computational grid network could be of triangular or multilateral assortments whereby the proportional value of an individual node in the meshwork is defined by a second-order interpolation with typical linear or cubic modes applied more commonly. FlexPDE as an open-source coding compiler and post-processing software package provides a fast-track solving procedure with user-friendly interface. This software also facilitates the grid network generation by introduction of a dynamic meshed domain which is subjected to sustained local refinements or coarsening based on how close the solution is to be converged at the neighboring cells. Moreover, the control volume-based finite element method (CVFEM) which combines finite element with finite volume for superior and more robust computations is employed due to the complex in-effect determinants.

1.3 Impacts of Active Parameter

1.3.1 Meshing Independency and Validation Assessment

In any type of numerical analysis which tries to reflect and bear a resemblance to the real-life problem, the generated grid structure should reach such a quality that further refinement of the meshing does not impact the outcome. Plus, higher degree of quality for the grid network often translates to a more exhaustive procedure as the computational cost grows exponentially with a more refined meshing framework. Thus, an ideal network should find the balance between the cost and accuracy. As demonstrated in Fig. 1, an adaptive and dynamic computational field is exercised to ensure that there is an offset between precision and cost. The density of each section is attributed to its converging criteria and is totally autonomous with regard to the entire converging state of the problem. To further illustrate the credibility of this dynamic network method, the local Nu at the inner cavity under specified conditions for three networks of varied elemental densities ranging from nearly 4000 to 11000 elements is put forth in Fig. 2. In general, the discrepancy in Nu values is roughly detectable except for the grid size of 4000 elements which at some points has gone astray from the pack. Plus, since the other two meshing domains almost parallel with each other in the given plot, the network of 6000 elements is deployed as the main computational field due to its lower cost and faster solving capability. Another vital step that must be taken into account for verification of the veracity of a numerical procedure is to make a thorough comparison between the prospective study and a prior peer-reviewed paper. To address this, Fig. 3 draws out the differentiating and comparability of this study with an investigation by Khanafer et al. (2003) to authenticate the presented data in the ongoing paper. With a glimpse at this plot, there is no appreciable contrast between the two, substantiating the analytical approach taken by this study to resolve the respective numerical problem.

1.3.2 Numerical Analysis

For a wavy porous star-like enclosure with a circular cavity within, this part elaborates on the natural convection, porosity of porous media, radiation strength and magnetic field effectiveness on a nanofluid flow field subjected to isothermal boundary conditions. The resultant figures in the form of streamline contours, isotherm profiles and local Nu plots are presented for each determinant with three specified scales. The scale for Rayleigh number alternate between 10, 50 and 100 while 0.1, 0.3 and 0.5 are the set-points for porosity assessment. The magnetic field strength indicator is associated with Ha number whose values in this analysis starts from 0, meaning there is no any external field, and jumps to 25 and 50 successively. The radiation factor also shifts from 0 to 0.3 and then to 0.6 to have a more clarified understanding of its influence.

Figure 2. Comparison of the local Nusselt number for various grid sizes when Ra=Ha=10, ε=0.4, Sf=3.7

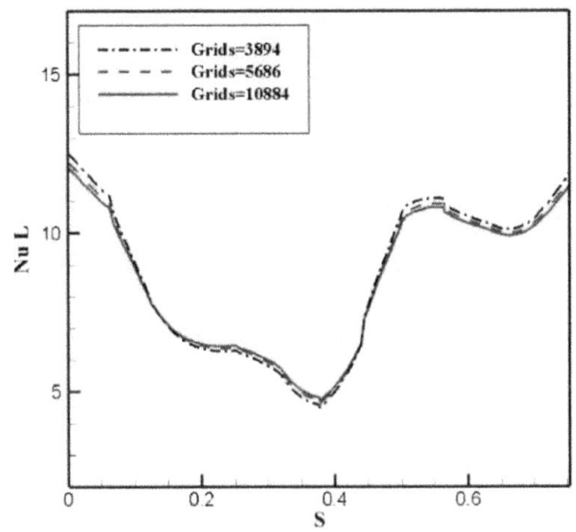

Figure 3. Validation of ongoing study with Khanfar et al. (1996) ϕ=0.01 and Pr=6.2 (Cu-Water)

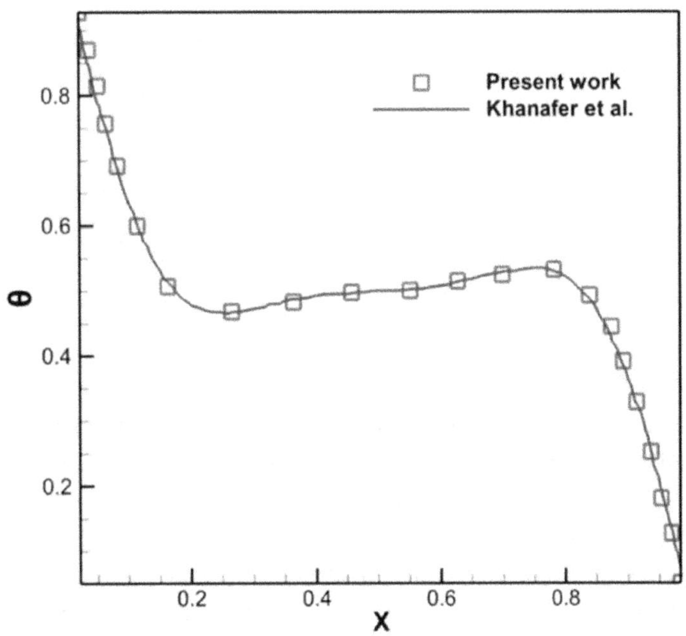

1.3.2.1 Qualification of Porosity Alteration Effectivity

Keeping other variable at their median values, the porosity changes of the porous enclosure give rise to thermal profiles and streamlines of Figure 4. The porosities of 0.1 and 0.6 demonstrably are distinguishable in their heat transfer and flow field performance owing to the more desirable outcome for porosity at 0.1. As evidenced by the respective contours, the penetration of thermal energy from the hot inner cavity to the cold wavy surface at the outer boundary is more profound for lower porosity of 0.1, especially in the upper half.

Figure 4. Streamlines (left) and isotherms (right) for various values of porosity parameters

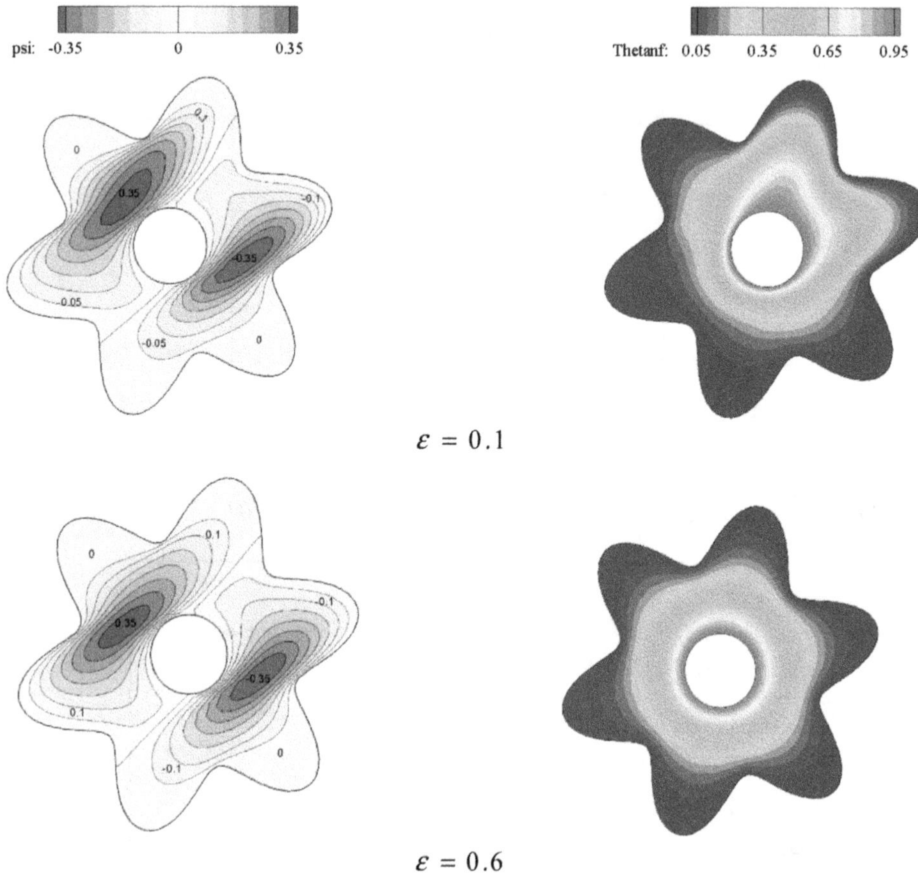

Figure 5 represents local Nu for all proposed porous scenarios in the vicinity of hot inner circular cavity while Ra, Ha and Rd values for all of them are identical. Based on this plot, the local Nu provides some valuable insight with regard to how porosity alteration might play out inside the enclosure. The three lines approximately at two points cross each other, however, the overall trend implies higher values of Nu number along the cavity occurs at lower porosities for the major part of the perimeter except for a marginal share of the boundary. Since Nu signifies the magnitude of convective thermal transfer, higher Nu for lower porosities reflects better natural convection for porous nanofluids with minimal porosities.

Furthermore, another noteworthy perception is related to the difference between investigated porosities. Although, the 0.3 as the porosity registered a more promising result than that of 0.6, porosity set to 0.1 delivered a considerably more robust convection rate.

Figure 5. Local Nusselt number plot for various porosity parameters

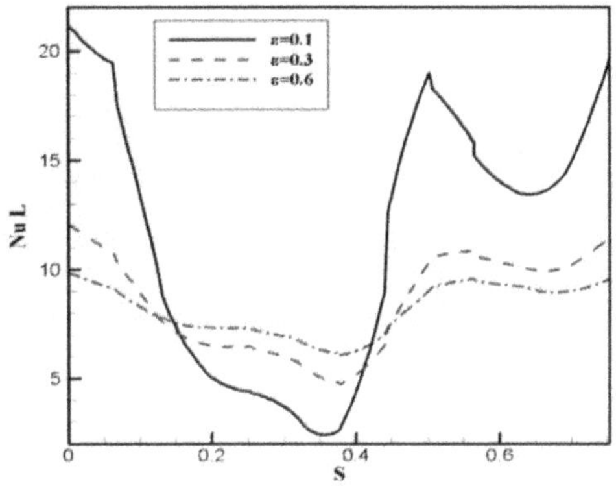

1.3.2.2 Qualification of Rayleigh Number Alteration Effectivity

An established characterization of Buoyancy forces in a Thermo-fluids context is calibrated by a non-dimensionalized number named Rayleigh whereby the strength of natural-convection-induced circulation in a domain is determined. To provide an insight on the individual dominance of this phenomenon over the given problem, the figures for Ra at 10, 50 and 100 are presented by Figure 6 and 7. The most noticeable viewpoint underlines the better thermal profile for higher Ra when 100 is adopted. This superiority highlights the upward movement of free convection currents within the bounded space. Plus, the streamlines contours depict a more cramped flow lines adjacent to the heated cavity. What deduced from the streamlines concerns as to how nonuniform the local flow field runs through the domain with a stratified behavior implying purely conductive heat transfer while the higher Ra corresponds with a more distorted profile. Hence, nonuniformity in streamlines translates to more diffusion and subsequently more improved heat transfer.

This inference proves validated when evaluating Figure 7 in which a plot of local Nu for different conditions related to Ra is displayed. It is evident that average Nu which can be measured by the area under the given local Nu, registers higher numbers for Ra = 100 followed by Ra = 50 and Ra = 10. Furthermore, the two intersections in the plot reveals that there is some section in the domain where natural convection is detrimental for local Nu and as such thermal performance. This tends to occur at the bottom spaces of a system since this part is deprived of the upward movement and is somewhat trapped in the area, suppressing the high heat transfer levels at the upper section.

Figure 6. Streamlines (left) and isotherms (right) for various values of Rayleigh numbers

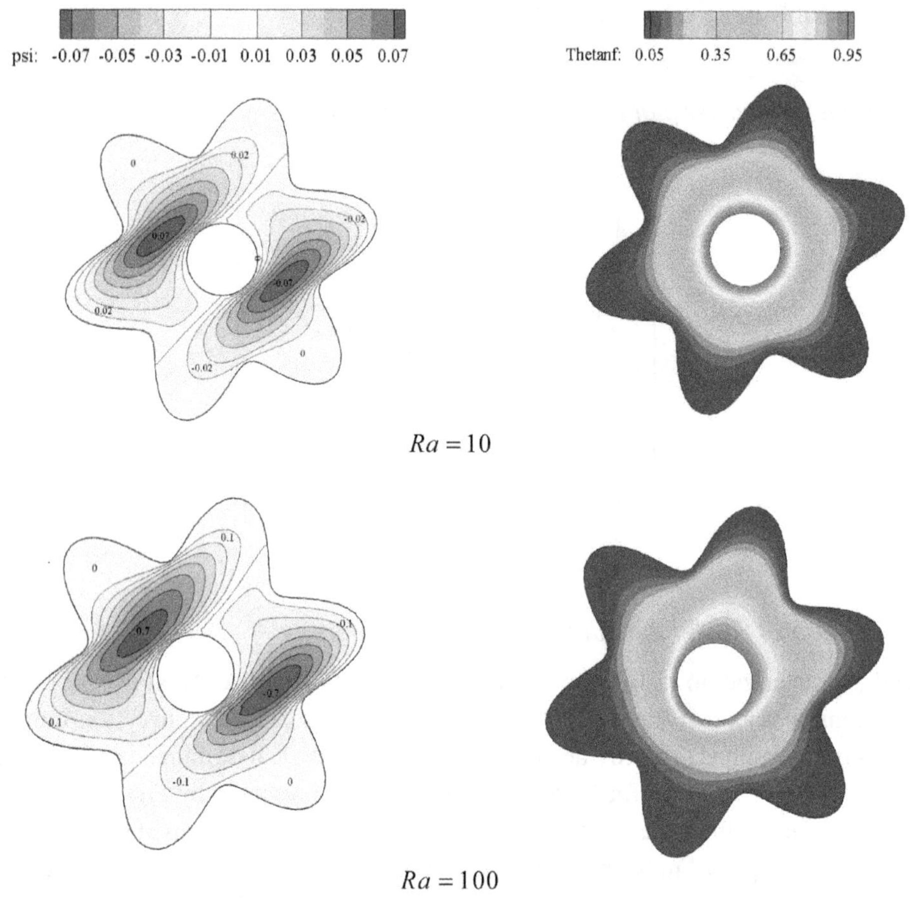

Figure 7. Local Nusselt number plot for various Rayleigh numbers

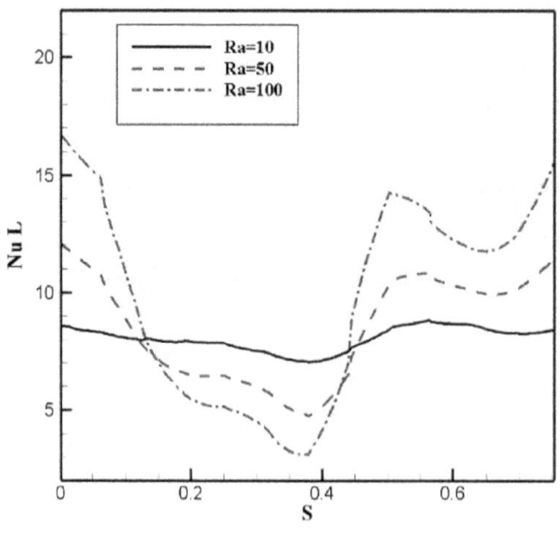

1.3.2.3 Qualification of Hartmann Number Alteration Effectivity

The magnetic field at three specified intensifications is imposed on the nanofluid to experiment with this type of external force and perform a thorough MHD analysis. Hartmann number is arranged to be 0, 25 and 50 as the agreed-upon index of magnetic dominance. Figure 8 delivers the respective data about streamlines and isothermal contours. In absence of magnetic field or at Ha = 0, the nanofluid behaves starkly dissimilar to that of prior scenarios where Ha was set to 25. The isotherms profiles reveal that natural convection induces more powerful vortices in the upward direction when there is no magnetic force deployed compared to all discussed cases. This is substantiated by local Nu plot demonstrated by Figure 9. By disposing of magnetic induction, local Nu undergoes a sizeable growth, topping at roughly 26 whereas introduction of this external factor massively suppresses the natural convection and resultantly thermal efficiency. Another implication for this imposition turns out to be the adversely contained streamlines which are more squeezed and tilted in presence of magnetic field while they spread more stretchily at Ha = 0. All in all, the application of magnetic forces in this context proved futile and unjustified since Lorentz forces stimulates counterflows that offset the Buoyancy-induced vortices, resulting in a less convection and Nu.

Figure 8. Streamlines (left) and isotherms (right) for various values of Hartman numbers

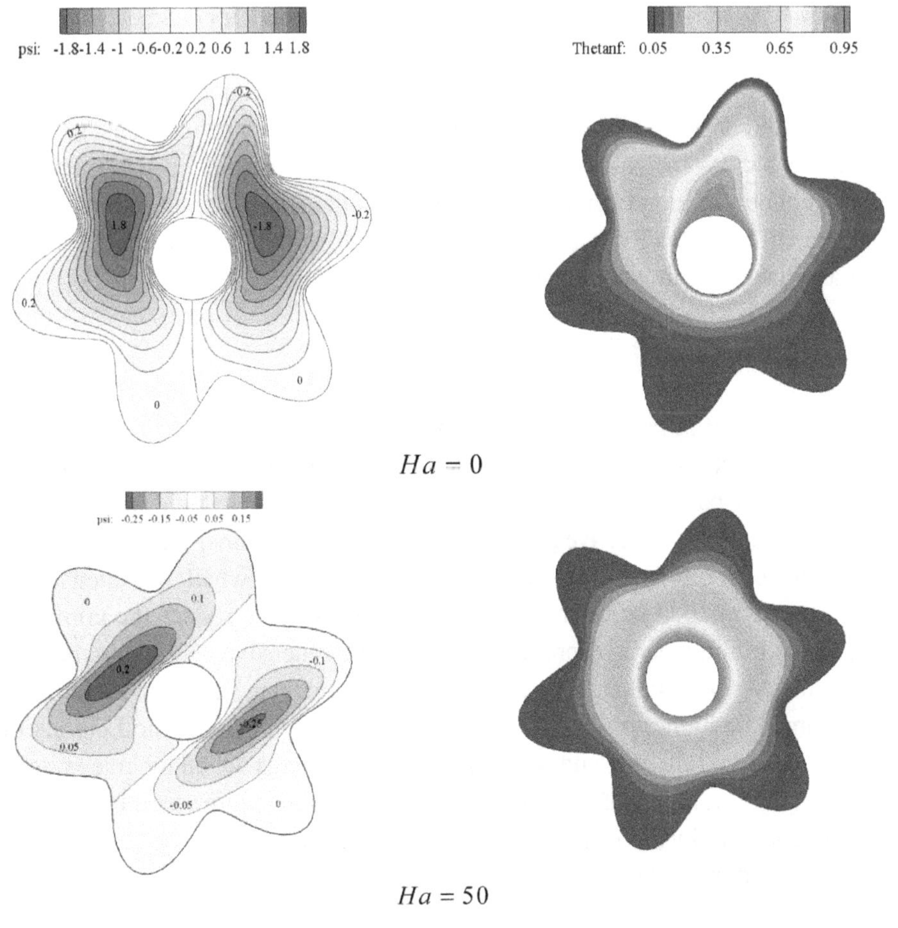

Figure 9. Local Nusselt number plot for various Hartman numbers

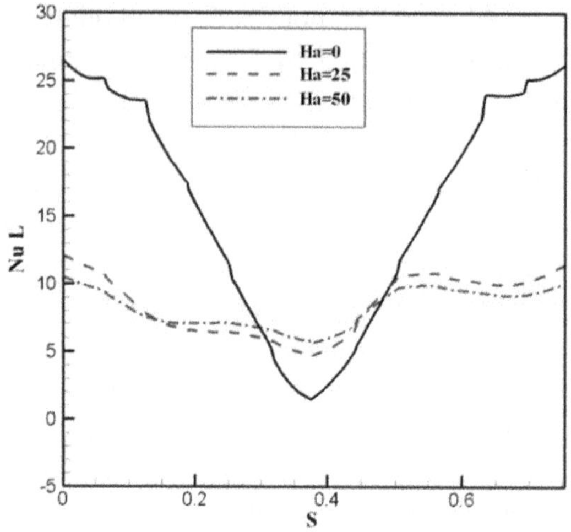

1.3.2.4 Qualification of Radiation Factor Alteration Effectivity

Radiation incorporation in the governing equations facilitates its assessment in the wavy porous enclosure. Rd as a dimensionless form of radiation intensity serves as a benchmark for characterization of this mode of heat transfer phenomenon. Peering over Figure 10, the presented contours of isotherms and streamlines for Rd at 0 which corresponds to non-radiational heat flow and Rd fixed at 0.6 are showcased. There is not any appreçiable difference in the presented profiles as a result of Rd alterations, suggesting that Rd or radiational heat transfer play insignificant roles in the framework that has been numerically evaluated. A somewhat consistent observation from Figure 11 reveals that local Nu or alternatively convective het flux do not correlate with various Rd magnitudes since there is no universal trend that can be singled out from the plot and the Nu is almost independent from this factor.

1.3.3 Optimization and Correlation Development

The large number of influential parameters in this problem necessitates a better assessment that can concisely illustrate the effectivity of eighter variable in individual and combined manners. This study examined four parameters, namely, radiation factor, porosity, magnetic field and natural convection strength by nondimensionalized respective numbers at three specifications. Thus, typically, the required numerical scenarios to cover a fully crossed design would be 81 trials which is somewhat burdensome. To avoid this, the Taguchi method introduces an alternative solution to reduce the numbers of trials by a considerable extent and simultaneously provides a more comprehensive outlook on the given problem.

That is why it gained major applicability in various disciplines and contexts, although its original utility was to accommodate higher quality for goods and products. This method entails three critical steps for fully achievement of fulfilling and cost-effective proceedings which include system, parameter and tolerance designs. By abiding by these rules, the number of trials plummeted from 81 to just 9 runs, hugely reducing the computational cost and burdensome post-processing. The detailed data for final designed trials is presented in Table 2.

Figure 10. Streamlines (left) and isotherms (right) for various values of radiation parameters

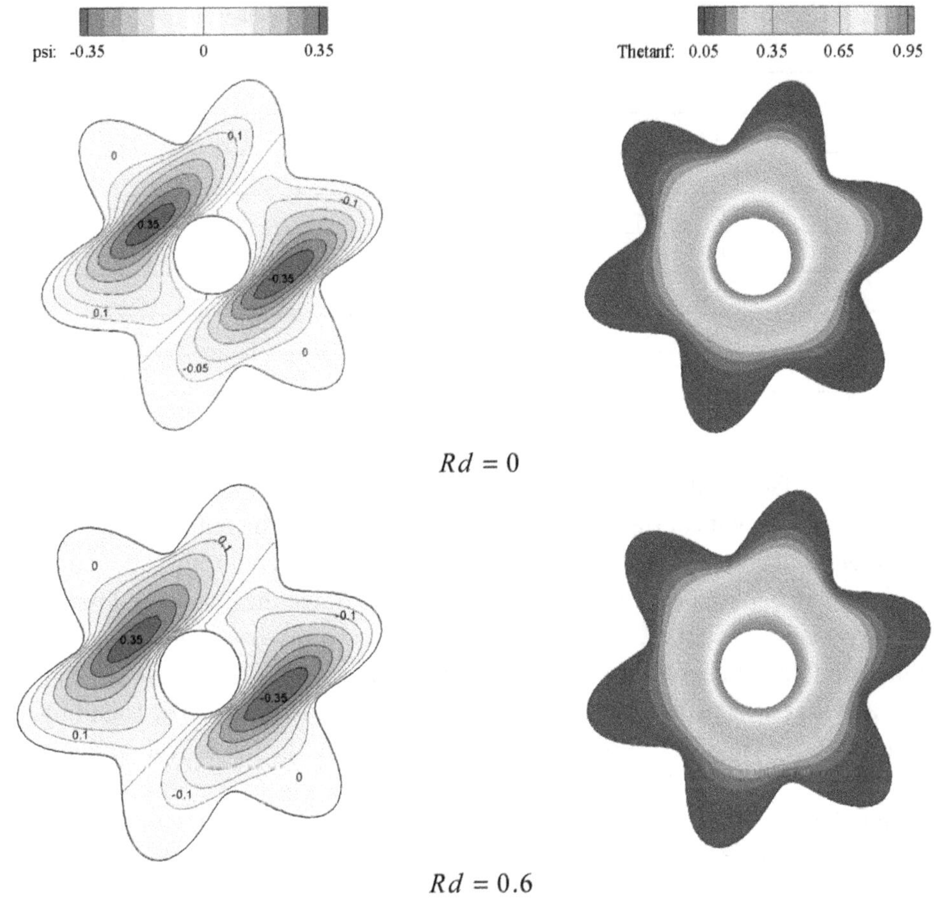

Figure 11. Local Nusselt number plot for various shape factor

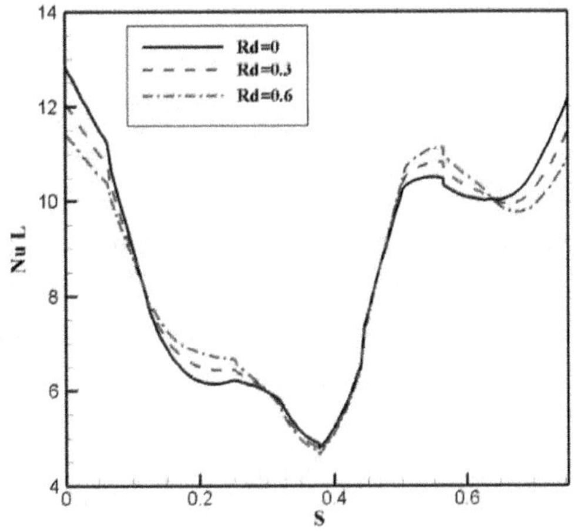

Table 2. Designed tests by Taguchi method

Case Number	Porosity Parameter	Hartman Number	Radiation	Rayleigh Number
1	0.1	25	0.3	50
2	0.1	0	0	10
3	0.3	0	0.3	100
4	0.6	0	0.6	50
5	0.3	50	0	50
6	0.3	25	0.6	10
7	0.1	50	0.6	100
8	0.6	50	0.3	10
9	0.6	25	0	100

RSM or response surface method has been applied widely in multiple optimization problems thanks to its straightforward mathematical frameworks and high accuracy in the output. The final correlations for response variables render prolonged numerical trials unnecessary and brings about a more polished outlook on the interplay between each individual determinant. Since the considered parameters of this problem range over continuous zones, RSM implementation may prove more advantageous. The anticipated correlation from RSM integration with GFEM constitutes a response variable which is the average Nusselt number and four dimensionless control factors of Ha, Rd, Ra and ε. The rounding, divergence and iterations figures may yield some potential errors compared to the numerical output which is dealt with in the following.

The fundamental stage in RSM development contains obtaining a correlation that is progressively getting more matching with the data. To this end, polynomials expressions claim the first priority for their simple interaction and flexibility. In case of inadequate agreement, the order of polynomials tends to get bigger to account for more intricate problems. The following defines a typical second-order polynomial.

$$y = a_0 + \sum_{i=1}^{n} a_i x_i + \sum_{i=1}^{n} a_{ii} x_i^2 + \sum_{i=1}^{n}\sum_{j=1}^{n} a_{ij} x_i x_j \big|_{i<j} + \xi \tag{4}$$

In which hx_i and x_j represent the design variables, a denotes the coefficients while the number of variables and error are determined by n and ξ. Figure 12 demonstrates average Nu in the enclosure by two 3-D plots of two pairs of parameters. The interactive effect of Ra and ε on Nu_{avg} suggests that lower ε or porosity combined with higher Ra yield the best efficiency in Nu_{avg}. However, a similar evaluation between Ha and Rd hints at insignificance of Rd on Nu compared to Ha whose lower values offer a far promising convective performance. Eq. 5 puts forth a comprehensive polynomial correlation for the approximation of average Nu regarding dimensionless control determinants.

$$Nu_{average} = 9.719 - 4.783(\varepsilon) - 0.087(Ha) + 0.044(Ra) + 0.994(Rd) \tag{5}$$

Figure 12. Effects of various operating parameters on average Nusselt number by RSM

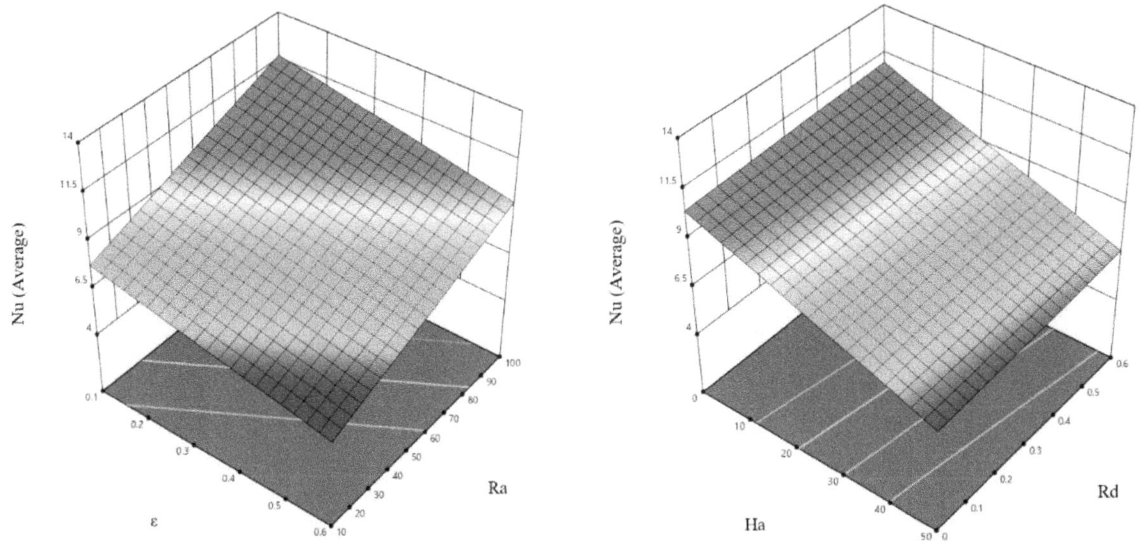

To evaluate how this proposed expression meets accuracy criteria, the analysis of variance (ANOVA) is introduced and applied whose results are tabulated in Table 3. The resultant figures from numerical approach are set side by side with the correlation estimations to weigh the level of agreement or conversely the degree of inconsistency whereby validating the RSM outcome. As evidenced by Table 3, the coefficient of determination (R^2), as a reliable indicator of RSM comparability to the real data, comes to 0.8819. This value fully justifies the application of RSM, validating its correspondence with actual results.

Finally, Figure 13 plots average Nu for the numerical data against the estimated line provided by the polynomial expression in Eq. 5. Considering the number of factors, wide scales for each determinant and the inherent complexity of such problem, the estimation of RSM correlation indicates a reasonable proportionality with actual data. Acquiring and validating the correlation, the optimum state for average Nu can be determined so that the value for each parameter in this universal optimum point are calculated and depicted in Table 4. By plugging in these numbers to the given expression, the optimum for response variable or Nu_{avg} is simply obtainable.

Table 3. The variance analysis (ANOVA)

Source	Sum of Squares	df	Mean Square	F-Value	P-Value	
Model	61.95	4	15.49	1.61	0.3280	significant
A-ε	8.69	1	8.69	0.9035	0.3957	
B-Ha	28.86	1	28.86	3.00	0.1583	
C-Rd	0.5340	1	0.5340	0.0555	0.8253	
D-Ra	23.86	1	23.86	2.48	0.1905	
R^2=0.8947						

Figure 13. Predicted vs actual of the average Nusselt number

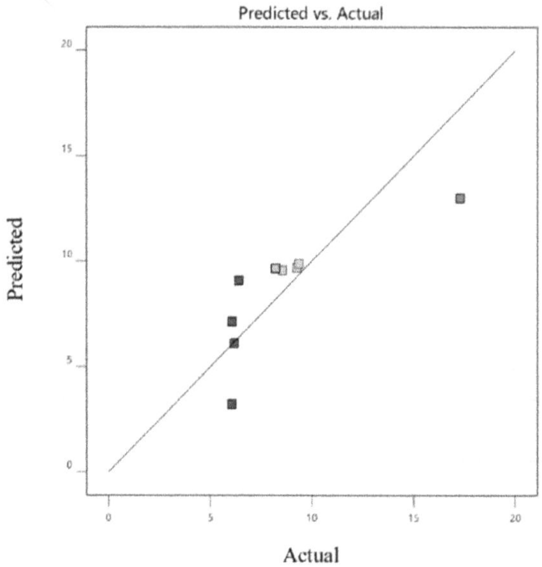

2. EXAMPLE 2: THERMAL ANALYSIS OF SNOWFLAKE POROUS ENCLOSURE CONTAINS Al_2O_3 – ETHYLENE GLYCOL NANOFLUID UNDER THE EFFECT OF A MAGNETIC FIELD BY FEM

2.1 Problem Definition

The geometry and boundary conditions of the present work are shown in Figure 14. The porous cavity is affected by the horizontal magnetic field and is saturated with Al_2O_3-Ethyleneglycol nanofluid. The outer cold surface is a circle with radius $r=r_o$. The inner hot wall is a snowflake. Horizontal magnetic field is employed.

2.1.1 Governing Equations

The equations are developed according to the principles mentioned in Chapter 4. Also, the formulas of thermophysical properties of nanofluids are mentioned in Chapter 2.

Table 4 demonstrate properties of Ethyleneglycol and Al_2O_3 nanoparticles (Sheikholeslami, 2019). Non dimensional quantities are:

Boundary conditions are:

$$\begin{aligned} &\Psi = 0.0 && on\,all\,walls \\ &\theta_{nf} = \theta_s = 0.0 && on\,outer\,wall \\ &\theta_{nf} = \theta_s = 1.0 && on\,inner\,wall \end{aligned} \tag{6}$$

Figure 14. Geometry and boundary conditions

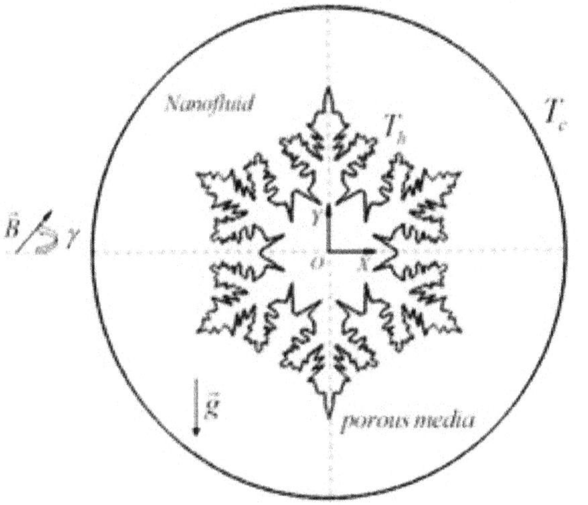

Table 4. Thermo-physical properties of ethylene glycol and Al_2O_3 nanoparticles (Sheikholeslami, 2019)

	$\rho(kg/m^3)$	$k(W/m{\bullet}K)$	$C_p(j/kg{\bullet}K)$	$\beta\times10^5(K^{-1})$	$\sigma(\Omega{\cdot}m)^{-1}$
Ethyleneglycol	1115	0.2463	2400	57	0.75×10^{-3}
Al_2O_3	3970	25	765	55	10^{-10}

3.2 Numerical Simulation

The current simulation has been done in FlexPDE software base on Finite Element Method (FEM). In this software, once the initial mesh is constructed, FlexPDE will continue to estimate the solution error, and will refine the mesh as necessary to meet the target accuracy. In time dependent problems, an adaptive refinement process will also be applied to the initial values of the variables, to refine the mesh where the variables undergo rapid change. Whereas cells created by this adaptive refinement process can later be re-merged, cells created by the initial explicit density controls are permanent, and cannot be un-refined. But, for making sure, a grid independency study has been done According to Figure 15. To be sure about the accuracy of present work comparisons with previous studies are indicated in Figure 16 According to these comparisons the present work is in a very good agreement with previous works.

2.3 Impacts of Active Parameters

First it is worth mentioning that using snowflake for inner wall shows much better performance compared to a circular geometry. To have the same surface area the radius of inner circle must be about $r_i \cong 0.9\,r_o$. Local Nusselt number on outer wall is compared in Figure 17. It is clear that the snowflake geometry caused better heat transfer compared to the circular geometry. Also, by increasing the *Ra* number the Nu_{ave} increases.

Figure 15. Comparison of the local Nusselt number Nu_{local} for different grid sizes at Ra=100, Ha=0, ε=0.9, Nhs=1000, and φ=0.02

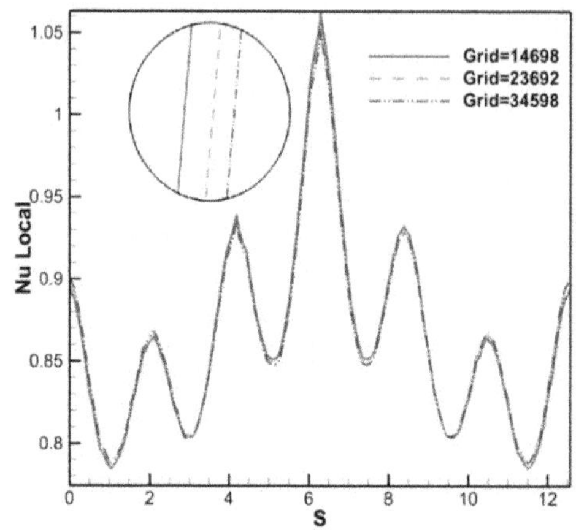

Figure 16. Validation of current work by comparison with previous works (Venter et al., 1996) φ=0.01 and Pr=6.2 (Cu-Water)

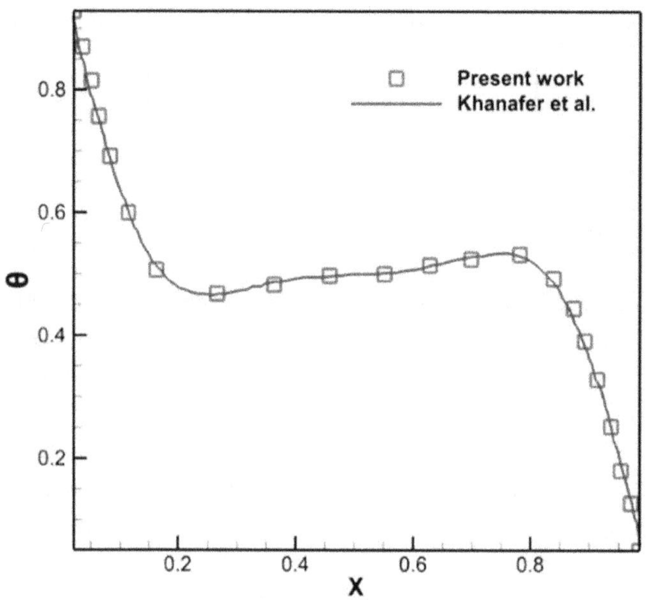

Figure 18 shows the effect of porosity of the porous medium (ε) on the stream function (Ψ), nanofluid dimensionless temperature (θ_{nf}) and solid matrix dimensionless temperature (θ_s). The results show that ε changes do not have a significant effect on Ψ, θ_{nf} and θ_s. The contours Ψ, θ_{nf} and θ_s have almost the same shape and behavior. Fig. 19 shows the effect of the Hartmann number (Ha), which is due to changes in the magnetic field, on Ψ, θ_{nf} and θ_s. Increasing the Hartmann number increases the electric vortices that resist the flow, which reduces the flow velocity (Ψ). However, with increasing Ha, no significant changes occur in isotherm contours (θ_{nf}, θ_s).

Figure 17. Comparing local Nusselt number for snowflake and circular geometries

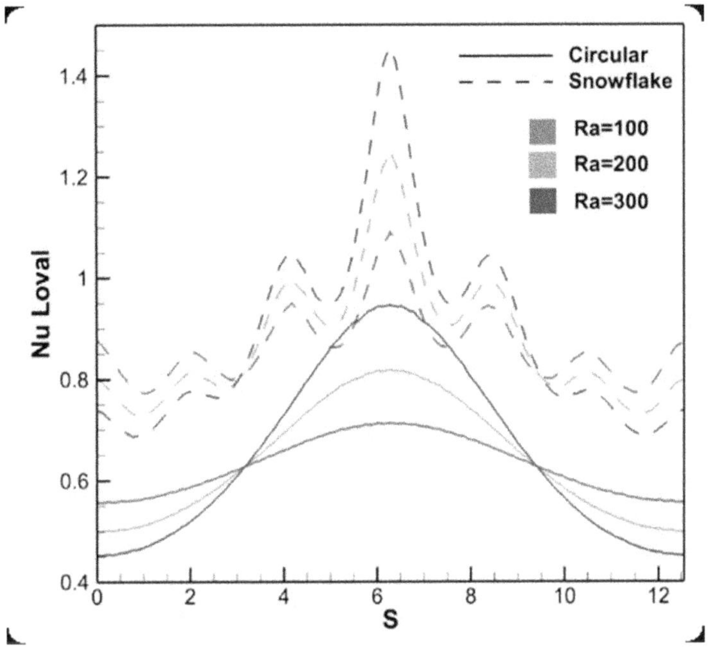

According to Figure 20, which shows the effect of volume fraction (ϕ) of the nanoparticle on Ψ, θ_{nf} and θ_s, with increasing volume fraction of the nanoparticle, the flow velocity (Ψ) decreases, which is due to increased frictional force. By increasing the volume fraction, the isotherm contours (θ_{nf},θ_s) increase, which is due to the improvement of the thermophysical properties of the nanofluid.

The effect of Rayleigh number (Ra) on the contour Ψ is shown in Figure 21. With increasing Ra the buoyancy forces, free convection heat transfer and flow velocity (Ψ) increase.

Figure 22 shows the effect of Rayleigh number (Ra) on thr local Nusselt number (Nuloc). As Ra increases, the buoyancy forces increase with respect to the thermal distribution and strengthens the convective heat transfer. Thus with increasing Rayleigh number (Ra), the local Nusselt number (Nuloc) increases. Fig. 23 shows the effect of Hartmann number (Ha) on the local Nusselt number (Nu_{loc}). With increasing Ha, the electric vortices increase. Electric vortices resist current and reduce the flow velocity (Ψ), the convective heat transfer and local Nusselt number (Nuloc). According to Fig. 24, which shows the effect of volume fraction of the nanoparticle (ϕ) on the local Nusselt number (Nu_{loc}). Increasing ϕ, improves the thermophysical properties of the nanofluid and increases heat transfer and local Nusselt number (Nu_{loc}). Fig. 25 shows the porosity (ε) effect on the local Nusselt number (Nu_{loc}). By increasing porosity, the number of pores inside the cavity increases, which in turn improves the convective heat transfer and relatively improves the local Nusselt number (Nu_{loc}).

Also, by using RSM an equation for the influence of the effective parameters on the average Nusselt number (Nuave) is presented. In Figures 26 to 31, the effect of the parameters Ha, Ra, ϕ and ε on the average Nusselt number (Nuave) is shown. With increasing Ha, the electric vortices increase. This reduces the flow velocity (Ψ), the heat transfer and average Nusselt number (Nuave). Increasing (ϕ) increases heat transfer and average Nusselt number (Nuave). By increasing porosity the convective heat transfer improves and relatively improves the average Nusselt number (Nuave).

Figure 18. Streamlines (Ψ), isotherms for the nanofluid (θn_f) and the solid (θs) at Ra=200, Ha=30, ϕ=0.02

Figure 19. Streamlines (Ψ), isotherms for the nanofluid (θn_{f}) and the solid (θs) at Ra=200, ε=0.5, ϕ=0.02

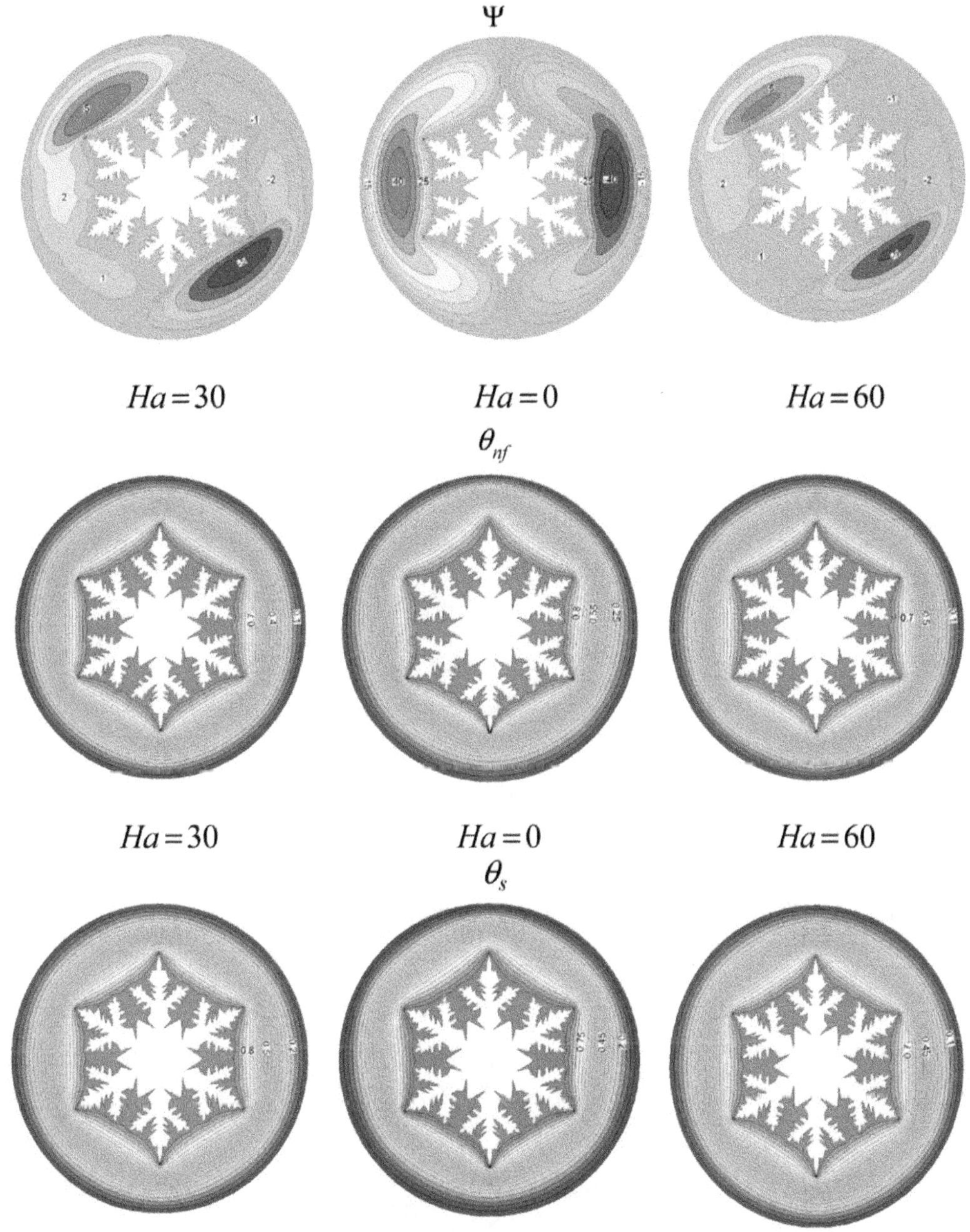

Figure 20. Streamlines (Ψ), isotherms for the nanofluid (θn_{f}) and the solid (θs) at $Ha=0$, $Ra=200$, $\varepsilon=0.5$

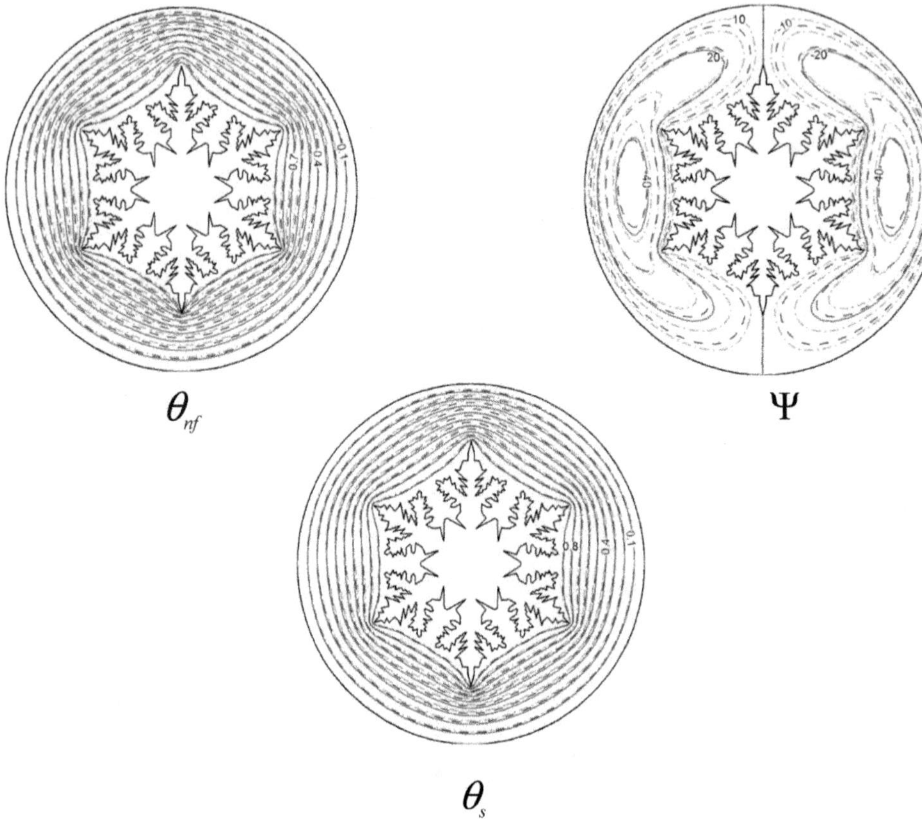

Figure 21. Streamlines (Ψ) at $Ha=0$, $\varepsilon=0.5$, $\phi=0.02$ for $Ra=100$–300

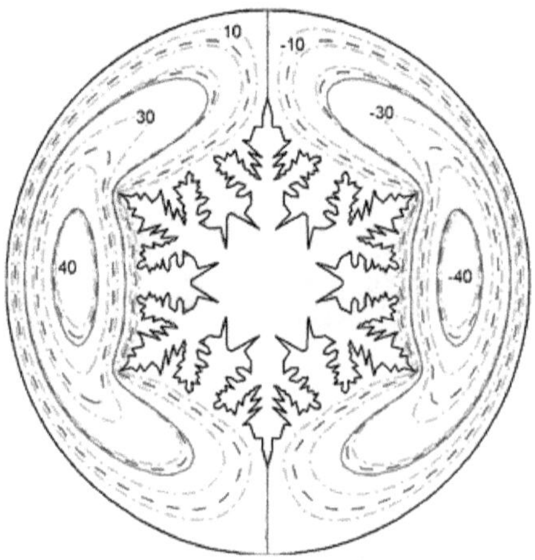

Figure 22. Effect of Ra on Nu_{local} for $\varepsilon=0.5$, $Ha=30$, $\phi=0.02$

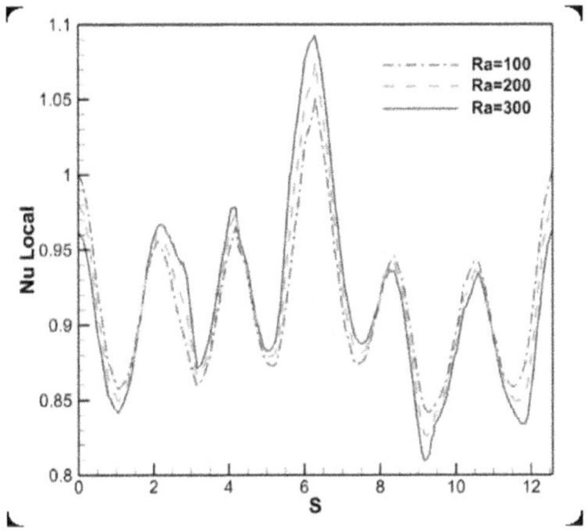

Figure 23. Effect of Ha on Nu_{local} for $\varepsilon=0.5$, $Ra=200$, $\phi=0.02$

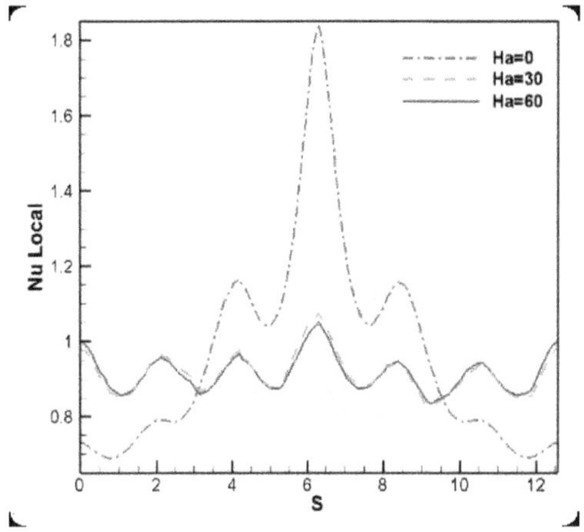

Figure 24. Effect of ϕ on Nul_{ocal} for $\varepsilon=0.5$, $Ha=30$, $Ra=200$

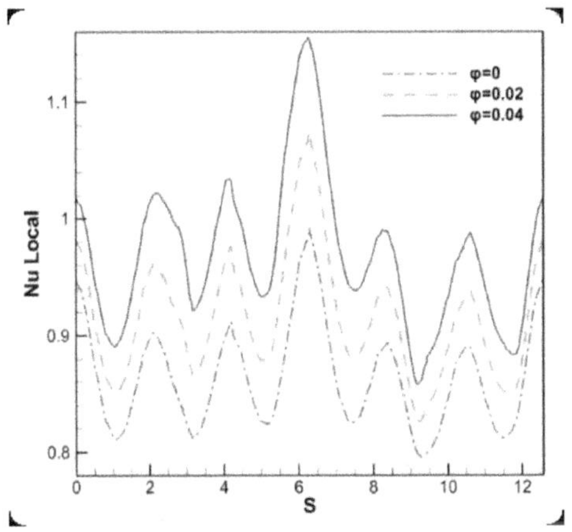

Figure 25. Effect of ε on Nul_{ocal} for $Ha=30$, $Ra=200$, $\phi=0.02$

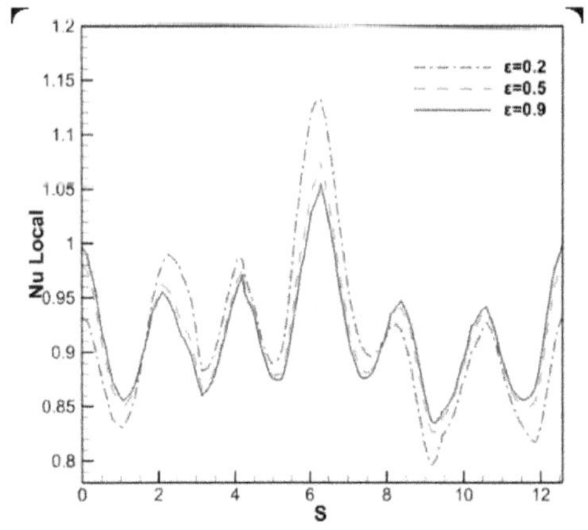

Figure 26. Effect of ε and Ha on Nua_{ve} for $Ra=200$ and $\phi=0.02$

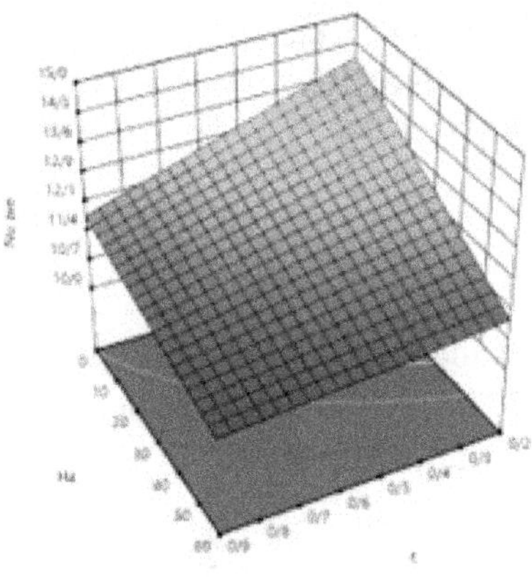

Figure 27. Effect of ε and Ra on Nua_{ve} for $Ha=30$ and $\phi=0.02$

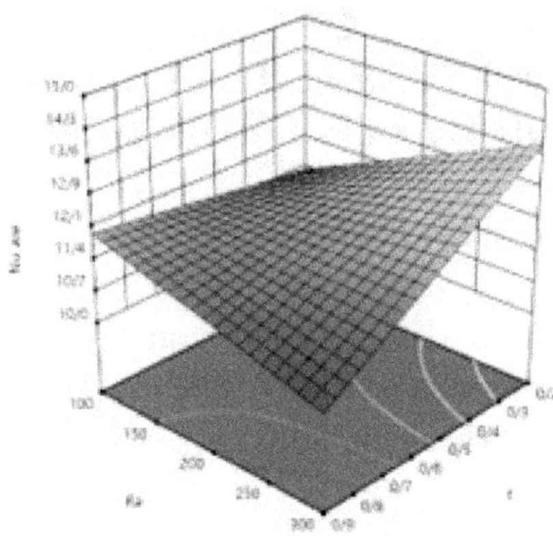

Figure 28. Effect of ε and ϕ on Nuav$_{ef}$or Ra=200 and Ha=30

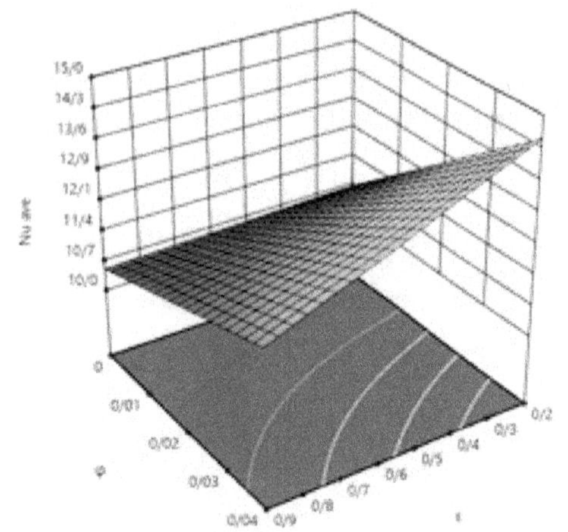

Figure 29. Effect of Ha and Ra on Nu$_{ave}$for ε=0.5 and ϕ=0.02

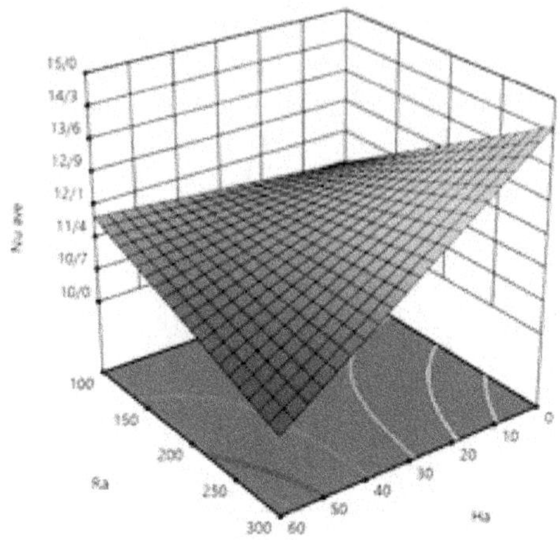

Figure 30. Effect of Ha and ϕ on Nua$_{ve}$for ε=0.5 and Ra=200

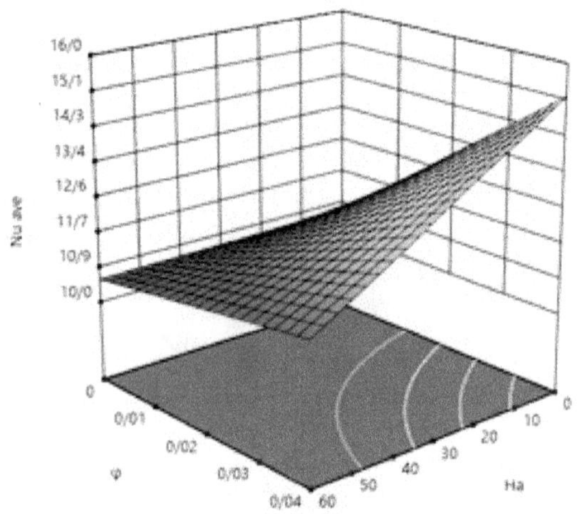

Figure 31. Effect of Ra and ϕ on Nua$_{ve}$for ε=0.5 and Ha=30

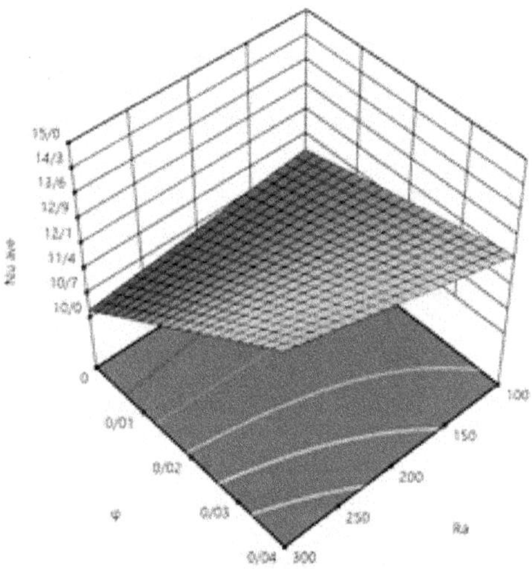

3. EXAMPLE 3: OPTIMIZATION OF HYBRID NANOPARTICLES WITH MIXTURE FLUID FLOW IN AN OCTAGONAL POROUS MEDIUM BY EFFECT OF RADIATION AND MAGNETIC FIELD

3.1 Problem Definition

Figure 32 demonstrates the geometry of the problem with assuming boundary conditions. The cavity enclosure contains water- ethylene glycol (50%-50%) with Mo$S2-_{T}iO2_{h}$ybrid nanoparticles. In addition, Table 5 has been provided for a better understanding of geometry dimensions.

Table 5. Dimensions of the geometry

Parameters	Values
a	180[nm]
b	45[nm]
c	28.57[mm]

Figure 32. Geometry and boundary conditions

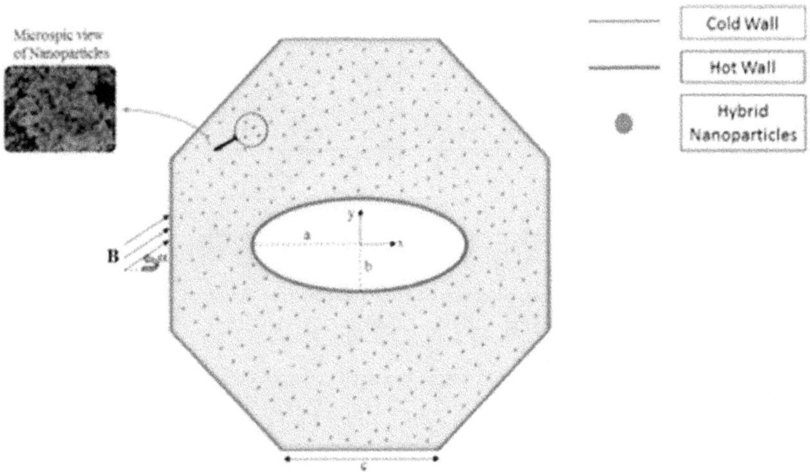

Simulation of nanofluid convective heat under the influence of a magnetic field, in cavity enclosure has been done. By dimensioning the equations according to the relationships of chapter 3, the boundary conditions will change as follows:

$$\begin{aligned}
&\Psi = 0.0 && on\,all\,walls \\
&\theta_{nf} = \theta_s = 0.0 && on\,outer\,wall \\
&\theta_{nf} = \theta_s = 1.0 && on\,inner\,wall
\end{aligned} \tag{7}$$

The thermo-physical properties of nanoparticles and base fluid are stated in Table 6.

3.2 Numerical Simulation

For applying the numerical approach and simulation, the open-source coding software called FlexPDE has been utilized. This software is based on Galerkin's Standard Finite Element Method (GFEM) that uses the second-order interpolation (linear or cubical) between the triangles and quadrilaterals. When the initial meshing has been done, the software continuously the calculation of solution error. Also, it may modify the meshing structure to achieve the desired accuracy. Further, the numerical approach follows the Control Volume-based Finite Element Method (CVFEM) method. CVFEM technique comprises the interaction between finite volume and finite element methods for assessing the complicated geometries.

Table 6. Thermo-physical properties of Ethylene glycol-Water and Hybrid nanoparticles (Hosseinzadeh, Moghaddam, Asadi et al, 2020; Hosseinzadeh, Mogharrebi, Asadi et al, 2020)

Physical Properties	Ethylene Glycol-Water (50%-50%)	MoS_2	TiO_2
$\rho\ (\frac{kg}{m^3})$	1063.8	5060	4250
$c_p\ (\frac{J}{kg.^\circ K})$	3630	397.21	686.2
$k\ (\frac{W}{m.^\circ K})$	0.387	904.4	8.95
$\beta \times 10^{-5}(k^{-1})$	58	2.84	0.9

3.3 Impacts of Active Parameters

3.3.1 Results and Discussion

Figures 33 and 34 exhibit a comparison between the two cases in the same conditions except for the polarization of elliptical. In both cases, the Radiation parameter, the factor of porosity, Rayleigh number, and Hartman number, are 0.5, 0.5, 10, and 10, respectively. According to the Figure 34, though the local Nusselt number in the case of vertical elliptical in some points, is above the case of horizontal elliptical. But given the area below the diagram which represents the average Nusselt number, the horizontal case has a higher average Nusselt number which leads it to a better condition.

Figure 33. Isotherms for horizontal and vertical elliptical

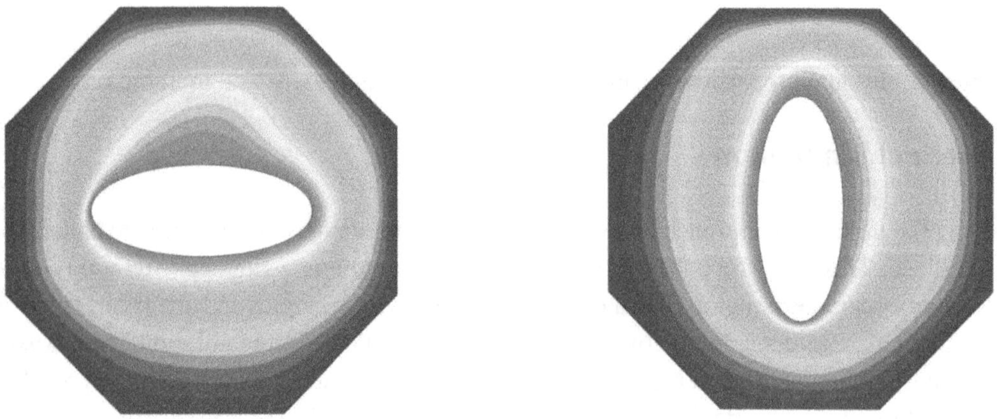

Figure 34. Local Nusselt number plot for horizontal and vertical elliptical

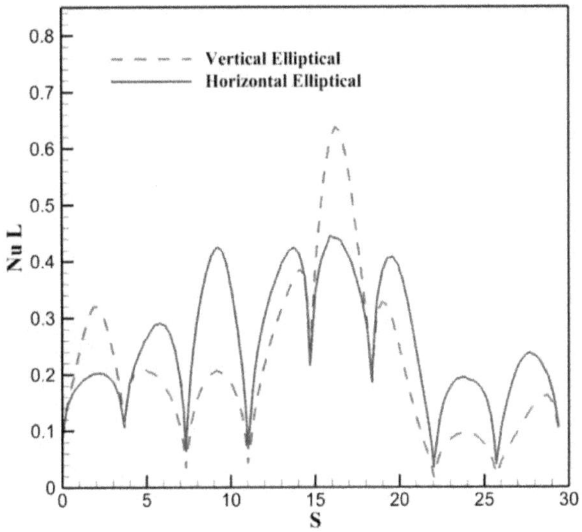

Figure 35 illustrates the contour analysis in ε=0.2 and ε=0.8 for isotherm condition (right) and streamlines (left).

Figure 36 shows the comparison in ε=0.2, ε=0.4 and ε=0.8, as can be seen, the average Nusselt number for ε=0.2 is much higher than the other conditions, which can be deduced that with a decrease in the factor of porosity, and the following increment in conductivity, the values of heat transfer would be enhanced, and in the greatest magnitude it can be represented by NuL=1.*1*

Figures 37 and 38 demonstrate the contour analysis and the influence of Hartman number on convective flow and heat transfer, as it seems, with an augment of Hartman number, due to Lorentz forces and also electrical vortices which resist against the fluid flow, the convective flow, and average Nusselt number, getting declined and consequently heat transfer rate in this condition will be deteriorated.

Figures 39 and 40, also reveals the effect of the Rayleigh number on fluid flow and heat transfer rate. As it portrayed, with the rise of the Rayleigh number due to the enhancement of buoyancy forces convective fluid flow is become improved, and as a result of that, in Figure 39 the average Nusslet number in Ra=100 is the highest values are in relation to the rest. Furthermore, Figures 41 and 42, represent the contours and diagram of the Radiation parameter's effect on convective fluid flow and average Nusselt number. It can be deduced that with increments in the Radiation parameter because of radiation energy losses, the heat transfer rate faces a bit decline.

3.3.2 Optimization of Problem

One of the appropriate methods for reducing the number of tests and determining the effect of each parameter on output characteristics in Studies is test design. In this study, the Taguchi method has been used to design the tests (Taguchi, 1986; Taguchi, 1990). The Taguchi method includes system design, parameter, and tolerance designs. The Porosity parameter, Hartman number, Radiation parameter, and Rayleigh number are the involve variables. Each variable has three levels of change which leads to 81 tests for analyzing their effect. By utilization of the Taguchi method number of tests, have been diminished to nine. Table 7 reveals information about the involved variables in nine proposed tests.

Figure 35. Streamlines (left) and isotherms (right) for various values of porosity parameters

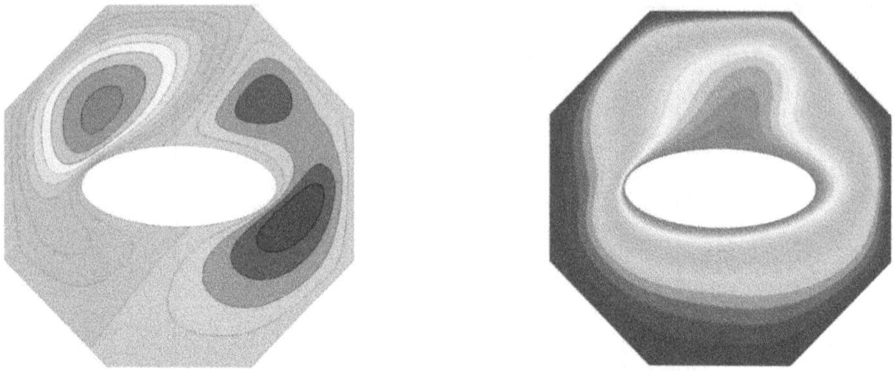

$\varepsilon = 0.2$

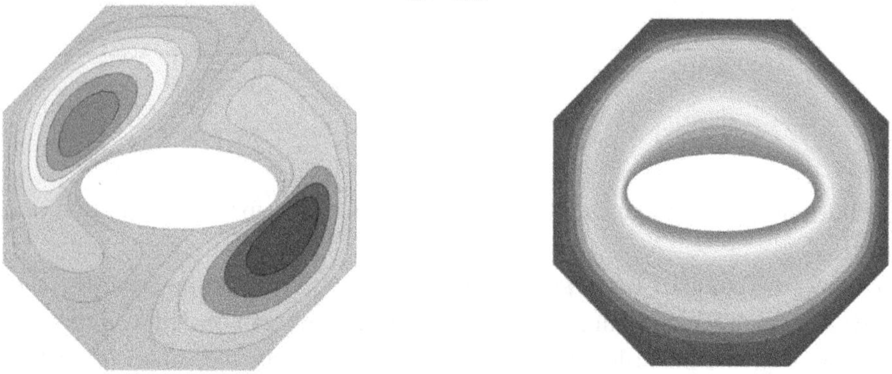

$\varepsilon = 0.8$

Figure 36. Local Nusselt number plot for various prosity parameters

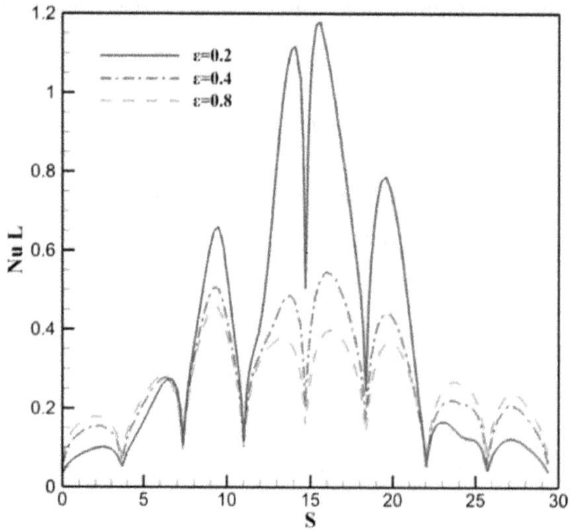

Figure 37. Streamlines (left) and isotherms (right) for various values of Hartman numbers

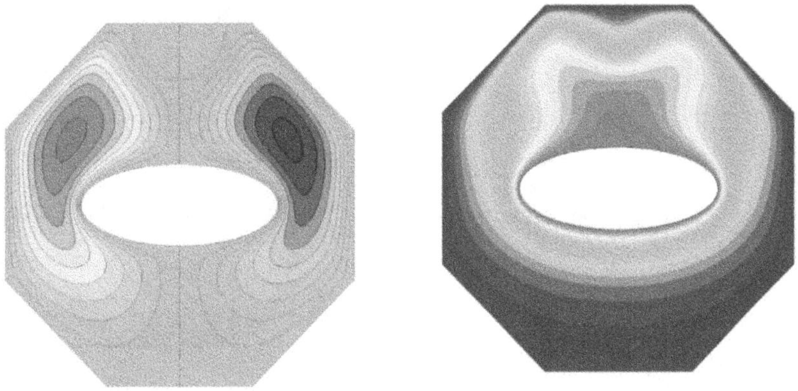

$Ha = 0$

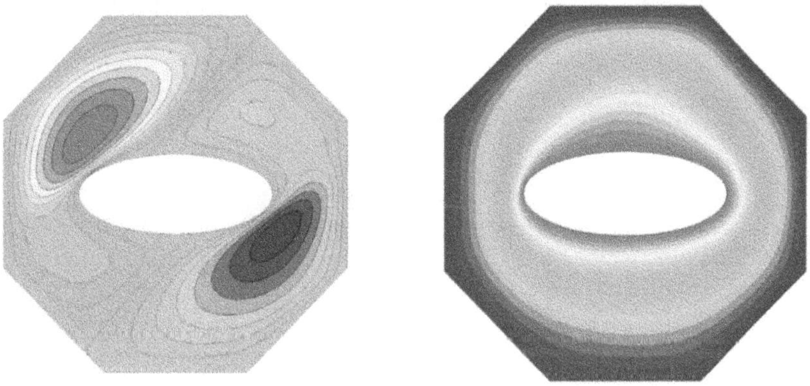

$Ha = 20$

Figure 38. Local Nusselt number plot for various Hartman numbers

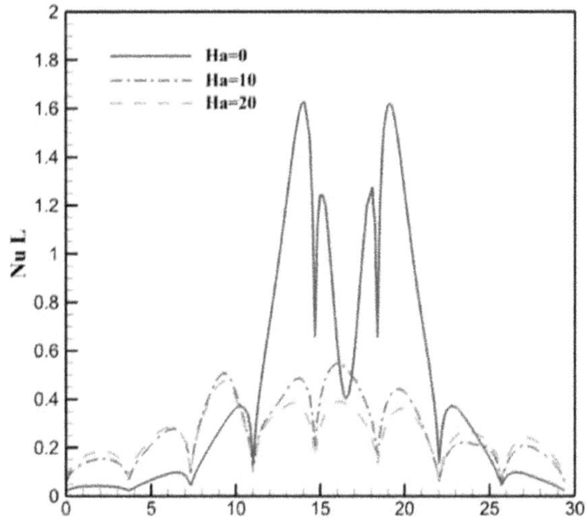

Figure 39. Streamlines (left) and isotherms (right) for various values of Rayleigh numbers

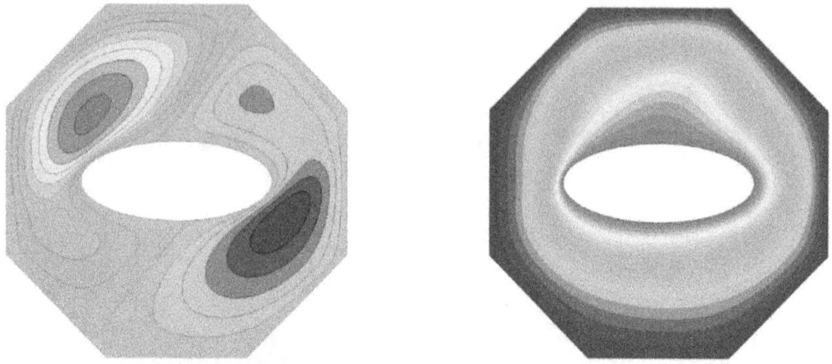

$Ra = 10$

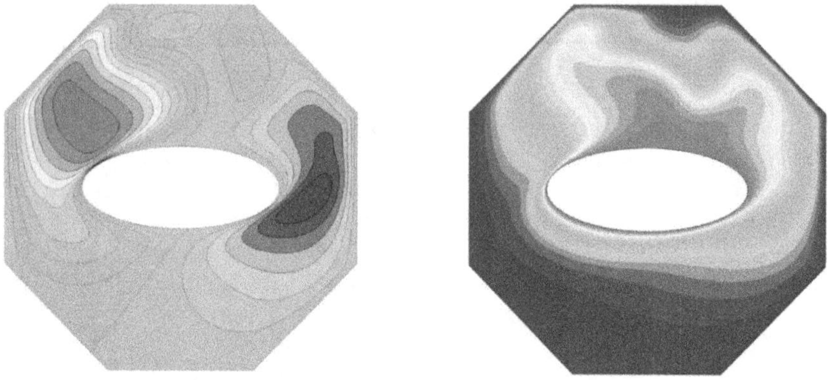

$Ra = 100$

Figure 40. Local Nusselt number plot for various Rayleigh numbers

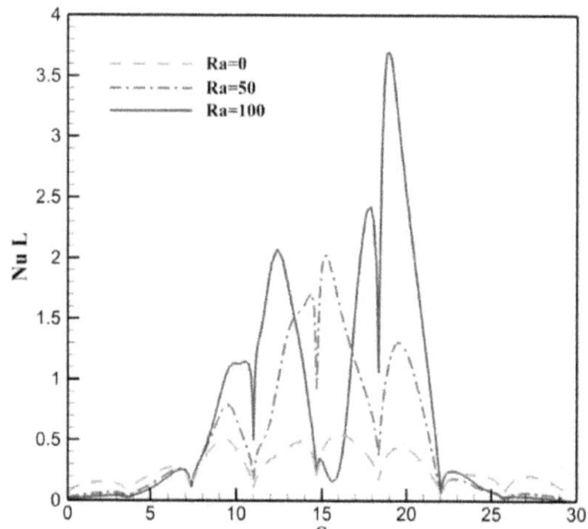

Figure 41. Streamlines (left) and isotherms (right) for various values of Radiation parameters

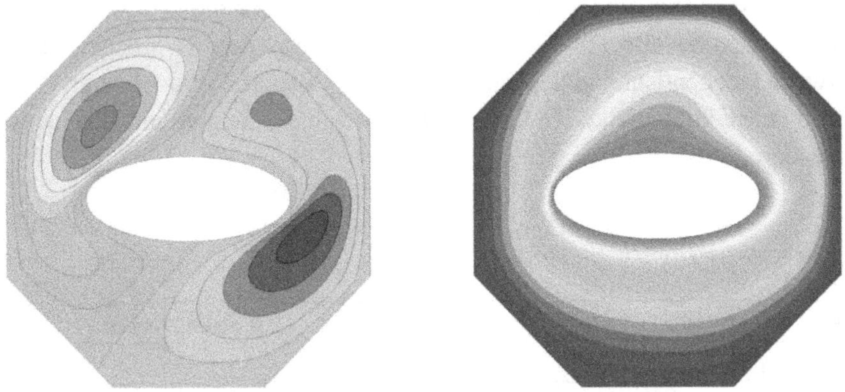

$Rd = 0$

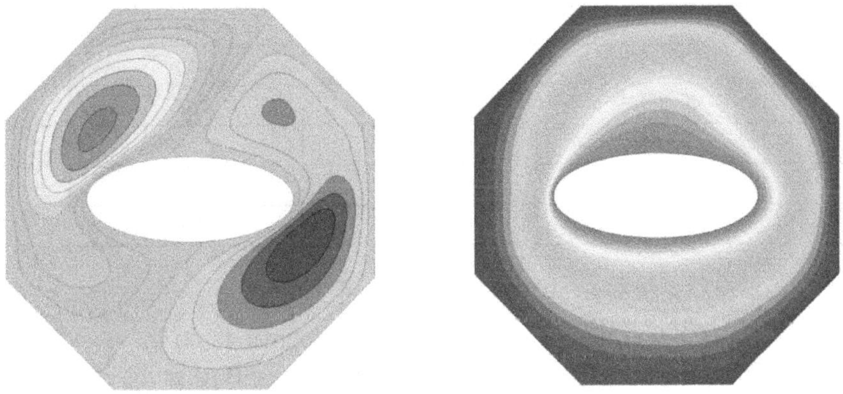

$Rd = 0.8$

Figure 42. Local Nusselt number plot for various radiation parameters

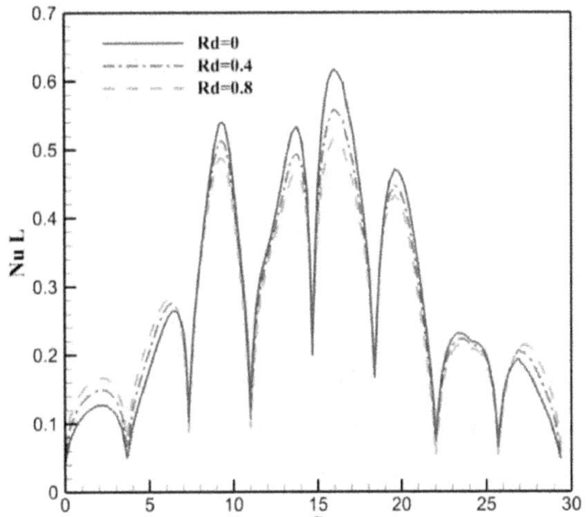

The optimization approach that has been utilized is the Response Surface Methodology (RSM). This method applies a set of mathematical and statistical techniques for the analysis, and modeling of the problems. In order to find the effect of the parameters on the response in a process, RSM is used to create a model between independent and dependent variables. Also, the errors that arise in the numerical applications of this method are related to the rounding of values, the non-convergence and the number of iterations considered.

Table 7. Designed tests by Taguchi method

Case Number	Porosity Parameter	Hartman Number	Rayleigh Number	Radiation Parameter
1	0.8	10	10	0.8
2	0.4	20	10	0.4
3	0.8	20	50	0
4	0.2	0	10	0
5	0.2	10	50	0.4
6	0.2	20	100	0.8
7	0.8	0	100	0.4
8	0.4	0	50	0.8
9	0.4	10	100	0

In RSM, finding a correlation to evaluate the effect of parameters on the objective function is essential. This general relation is usually of the quadratic polynomial type as follows:

$$y = a_0 + \sum_{i=1}^{n} a_i x_i + \sum_{i=1}^{n} a_{ii} x_i^2 + \sum_{i=1}^{n} \sum_{j=1}^{n} a_{ij} x_i x_j + \xi \qquad i \langle j \qquad (8)$$

In above equation, y, a_0, a_i, a_j, a_{ij} are predicted response, constant coefficient, linear coefficient, quadratic coefficient, interaction coefficient, respectively. Also, x_i and x_j are design parameters and ξ illustrate the error. Further, in the present study in order to maximize the average Nusselt number, the effects of Porosity parameter, Hartman number, Radiation parameter, and Rayleigh number have been employed in that correlation. Additionally, the variance analysis (ANOVA) has been applied for the statistical emphasis of involved parameters examination in the regression of prediction.

3.3.3 Analysis of Variance

Table. 8 is provided for the variance analysis (ANOVA) which represents the scale of the discrepancy of predicted and actual outcomes. The R-Squared value is 0.9411 which is considerable and shows a good agreement between predicted and actual values. Moreover, figure. 43 bodes these proper overlaps.

Table 8. The variance analysis (ANOVA)

Source	Sum of Squares	df	Mean Square	F-Value	P-Value	
Model	1428.04	4	357.01	16.21	0.0097	significant
A-ε	98.23	1	98.23	4.46	0.1022	
B-Ha	445.95	1	445.95	20.25	0.0108	
C-Ra	883.79	1	883.79	40.14	0.0032	
D-Rd	0.0641	1	0.0641	0.0029	0.9596	
R^2=0.9411						

Figure 43. Predicted vs. actual of the average Nusselt number

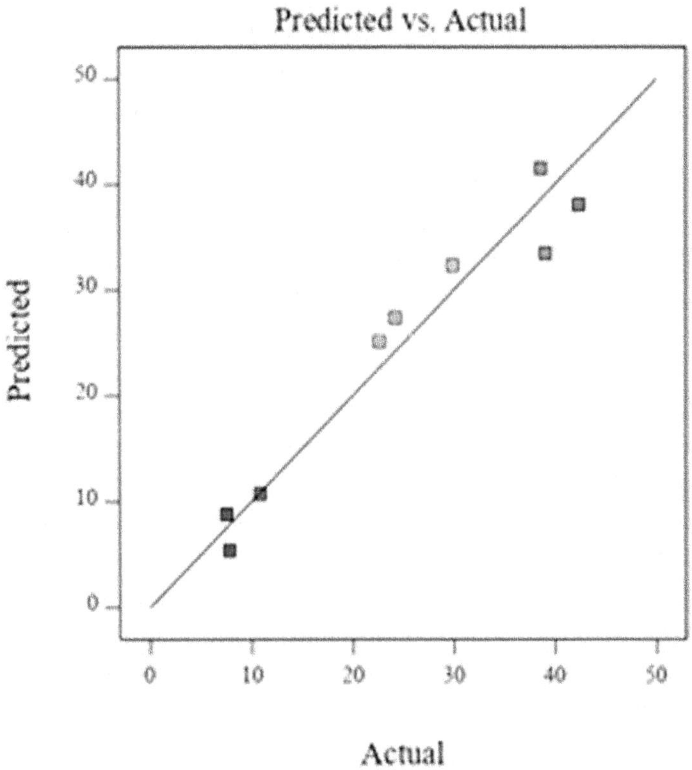

3.3.4 Correlation

Effects of various operating parameters on average Nusselt number by RSM are illustrated in Figure 44. Also, the correlation for achieving maximum Nusselt number is being presented in equation 9:

$$Nu_{average} = 25.118 - 13244(\varepsilon) - 0.862(Ha) + 0.269(Ra) + 0.258(Rd) \tag{9}$$

Figure 44. Contribution of each parameter on the average Nusselt number by the percentage

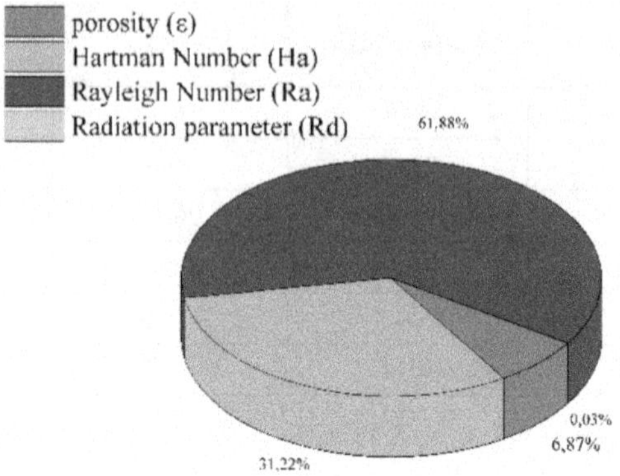

The optimum values of the parameters that include in this correlation can be seen in Table. 9. By applying these values to equation 9 the maximum of average Nusselt number can be calculated. In addition, Figure 45 displays the contribution of each parameter in influencing the average Nusselt number by the percentage.

4. EXAMPLE 4: HEAT TRANSFER HYBRID NANOFLUID (1-Butanol /MoS2 – Fe3O4) THROUGH A WAVY POROUS CAVITY AND ITS OPTIMIZATION

4.1 Problem Definition

Figure 46 schematically displays the complete geometry of the problem with assuming boundary conditions. The cavity contains 1-Butanol based hybrid nanofluid including $MoS_2-Fe_3O_4$ hybrid nanoparticles.

The present investigation carries a new idea to meet the solicitation of numerical simulation of nanofluid convective heat transfer in a corrugated porous enclosure in the existence of a magnetic field. Lorentz forces, the influences of Hartmann number, Rayleigh number, nanoparticle shape factor, the porosity of the porous medium thermal radiation, and the Darcy model are being applied.

Continuity, momentum and energy equations are expressed according to chapter 4. Finally, dimensionless changes the boundary conditions as follows:

$$\begin{aligned} &\Psi = 0.0 && on\ all\ walls \\ &\theta = 0.0 && on\ outer\ wall \\ &\theta = 1.0 && on\ inner\ wall \end{aligned} \tag{10}$$

Additionally, Table 10 indicates the properties of nanoparticles utilized in the current study. Also, the formulas related to nanofluids are fully presented in Chapter 3.

Figure 45. Effects of various operating parameters on average Nusselt number by RSM

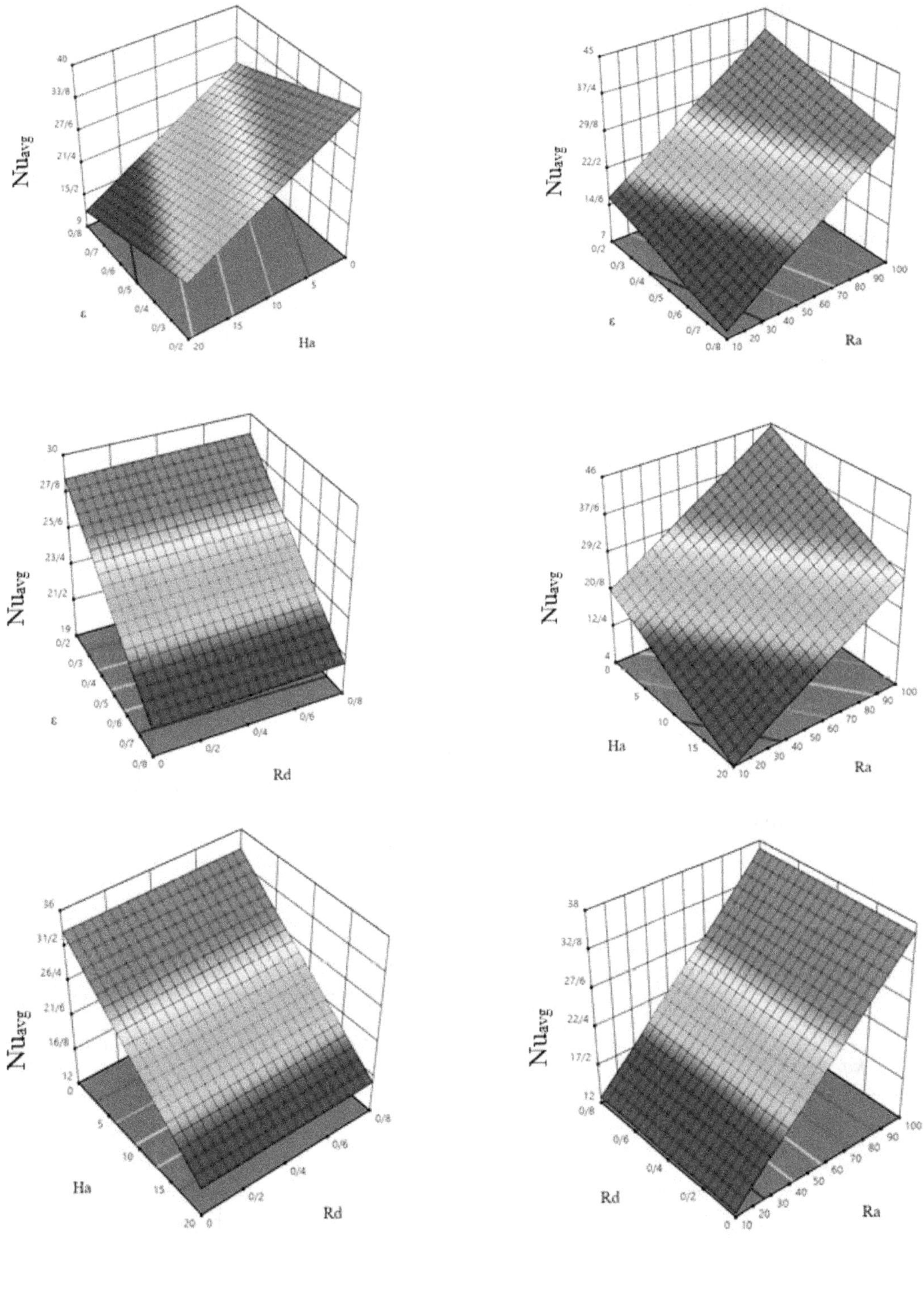

Table 9. Outcome of the optimization procedure by RSM

Optimum Value	Parameter
0.287	ε
3.67	Ha
98.26	Ra
0.75	Rd

Figure 46. Geometry and boundary conditions

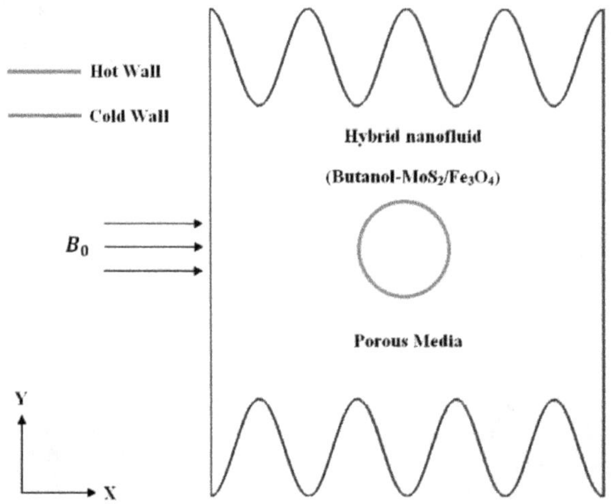

Table 10. Specifications of nanoparticles used in the present research (Hosseinzadeh, Moghaddam, Asadi et al, 2020; Hosseinzadeh, Mogharrebi, Asadi et al, 2020)

Physical Properties	1-Butanol	MoS_2	Fe_3O_4
$\rho\ (\frac{kg}{m^3})$	809	5060	5180
$c_p\ (\frac{J}{kg.^\circ K})$	2338	397.21	670
$k\ (\frac{W}{m.^\circ K})$	0.148	904.4	9.7
$\beta \times 10^{-5} (k^{-1})$	58	2.84	1.3

4.2 Numerical Simulation

The governing Equations for flow are solved by employing Galerkin's Standard Finite Element Method (GFEM) utilizing boundary conditions and the second-order interpolation (linear or cubical) between the triangles and quadrilaterals. In this analysis, the open-source coding software called FlexPDE is used for the simulation. Further, the numerical approach follows the Control Volume-based Finite Element Method (CVFEM) technique. CVFEM technique comprises the interaction between finite volume and finite element techniques for evaluating the complicated geometries. Once the initial meshing has been finished, the software continuously calculated the solution residual. Also, it may modify the meshing structure to achieve the desired accuracy.

4.3 Impacts of Active Parameters

4.3.1 Grid Independence Study and Validation Results

To guarantee that the simulation outputs are autonomous on the grid size of the computational domain, grid independence analysis was done. Figure 47 displays the unstructured and adaptive meshing which has been applied in the current study. It clearly shows that the quality and density of meshing have been enhanced in the edge points. In the present study, three meshes i.e. Mesh 1 with roughly 4700 grids, Mesh 2 with roughly 9000 grids, and Mesh 3 with roughly 14500 grids were tested. The local Nusselt number has been calculated for each gird size, respect to the exterior surface of the trapezoid. As shown, there is a bit different in the three grids for this value. Figure 48 indicates that the results obtained from Mesh 3 slightly deviate from the results obtained by the other two meshes whereas the results predicted by the Meshes 1 and 2 almost overlap with each other. Therefore, for reducing the CPU operation time, Mesh 2 has been chosen to carry out the simulations numerically.

Figure 47. Adaptive mesh refinement

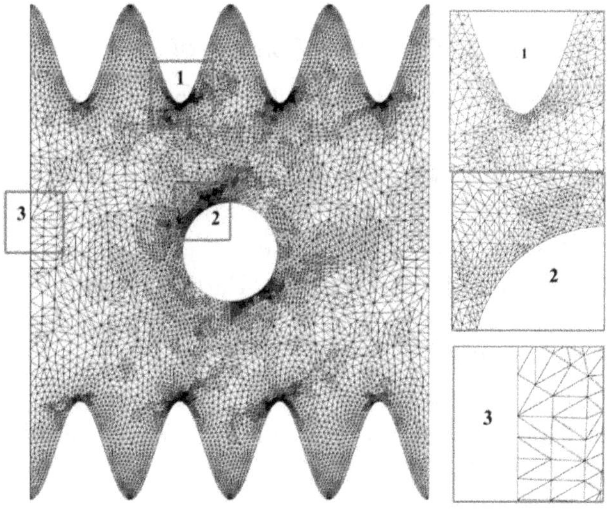

Figure 48. Comparison of the local Nusselt number for various grid sizes when Ra=Ha=10, ε=0.4, Sf=3.7

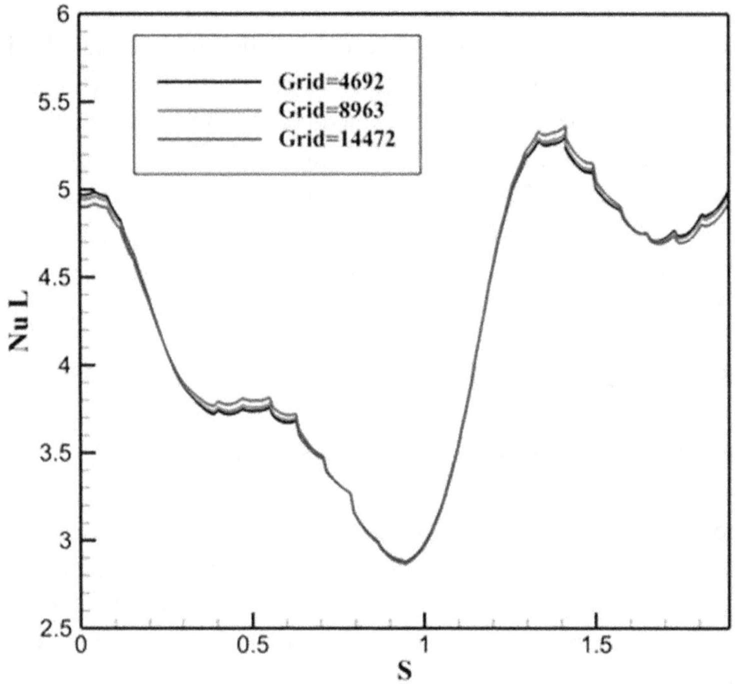

4.3.2 Heat Transfer

This section gives numerical outcomes for the Natural convection in a corrugated enclosure saturated with the hybrid nanofluid porous medium. The streamlines, isotherms, and heat transfer performance with various amounts of the Rayleigh number (Ra=10, 50, and 100), the Hartmann number (Ha=0, 50, and 100), the radiation parameter or nanoparticle shape factor (spherical: 3, hexahedron: 3.7, lamina: 16.1), and porosity of the medium ($0.2 \leq \varepsilon \leq 0.8$) are investigated.

4.3.2.1 Performance Benchmark by Change of Porosity of the Porous Medium

Figure 49 shows the effect of different porosity on isotherms and streamlines owing to heat transfer at Ra=50, Ha=50, and Sf=3.7. As it can be seen, the isotherms are bunched along the heated surface. The fluid which is heated by the inner surface moves to the isothermal left side surface, where it expanded in the direction of the upper left corner between the adiabatic left sidewall and the upper horizontal cold wavy wall. The observation is the same for all three cases.

According to Figure 50, for a given Rayleigh number, Hartmann number, and shape factor, the heat transfer rate decreases from the surface to the fluid by increasing in the porosity of the porous medium. It means as the porosity decreases, the intensity of convection enhances inside the porous enclosure. Figure 50 clearly shows the local Nusselt number for ε=0.2 is much higher than the other conditions, which can be deduced that with a decrease in the factor of porosity, and the following increment in conductivity, the values of heat transfer would be enhanced.

Figure 49. Streamlines (left) and isotherms (right) for various values of porosity parameters

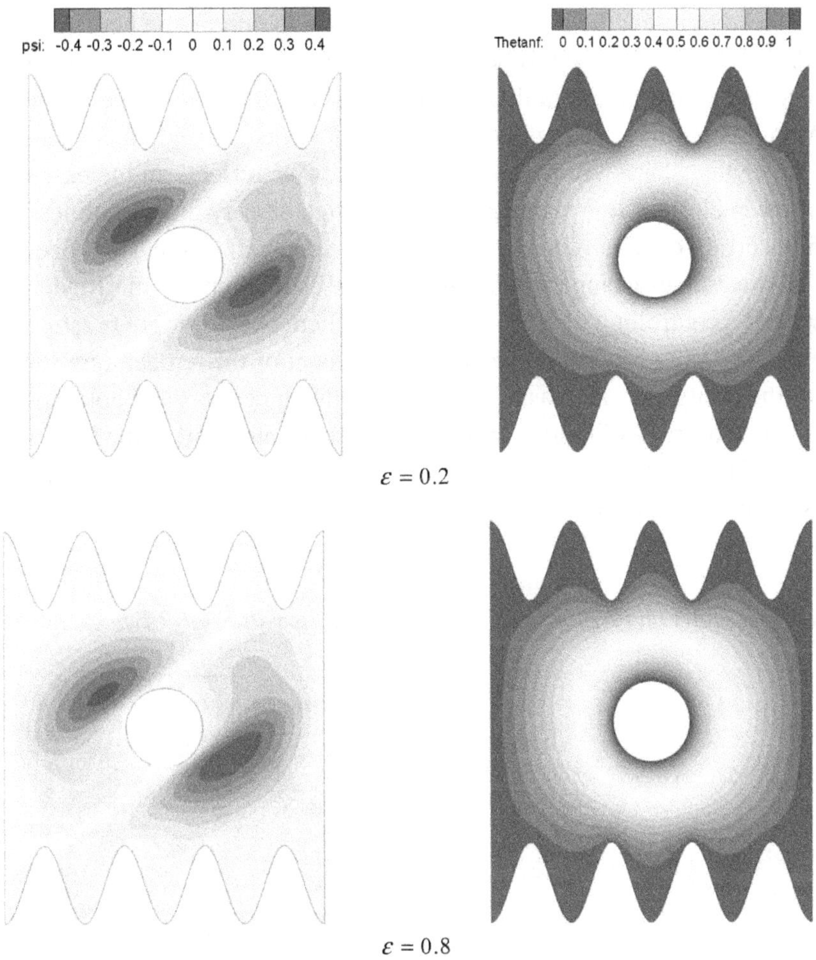

Figure 50. Local Nusselt number plot for various porosity parameters

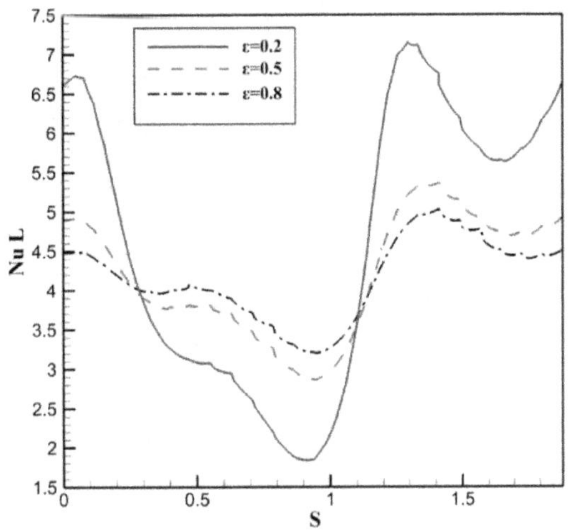

4.3.2.2 Performance Benchmark by Rayleigh Number

Figures 51 and 52 clearly present the influences of the Rayleigh number on heat transfer and fluid flow at Ɛ=0.5, Ha=50, and Sf=3.7. It ought to be noticed that Rayleigh number characterizes an impact of the buoyancy force and enhancement in this factor prompts an increase of the convective flow within the enclosure. Along these lines, for high estimations of *Ra* one can find an enhancement in ψ_{max} and modification of streamline shape. An expansion in *Ra* likewise reflects an increase of nanofluid flow in the top part of the enclosure. It is significant that the high density of streamlines is close to the hot inner wall. Such distortion of streamlines close to the heat source is increasingly perceptible for *Ra* = 100. Variations of streamlines change isotherms. For low estimations of the Rayleigh number (*Ra* = 10) heat conduction is a controlling heat transfer system where isotherms are practically corresponding to the hot inner wall. An expansion in *Ra* causes development of the temperature stratification core in the central part of the cavity with increasingly heating of the upper part and noteworthy cooling of the down section of the enclosure. As Figure 52 portrayed, with the augment of the Rayleigh number owing to the improvement of buoyancy forces, convective fluid flow is become improved. As a result of that, the average Nusselt number in Ra=100 is the highest values are in relation to the rest.

Figure 51. Streamlines (left) and isotherms (right) for various values of Rayleigh numbers

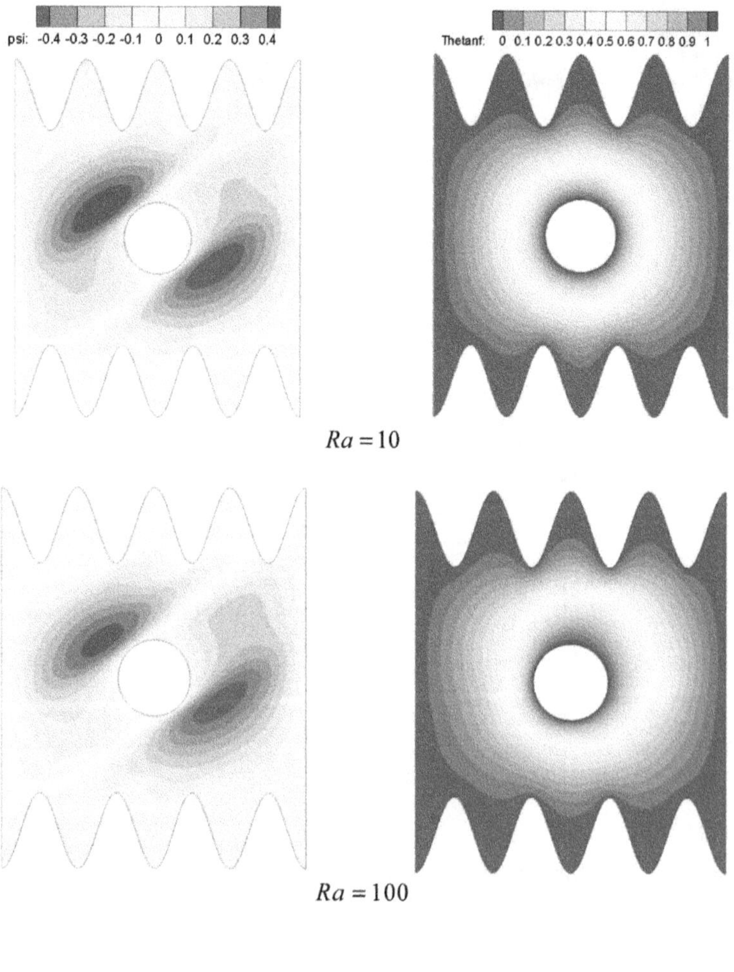

$Ra = 10$

$Ra = 100$

Figure 52. Local Nusselt number plot for various Rayleigh numbers

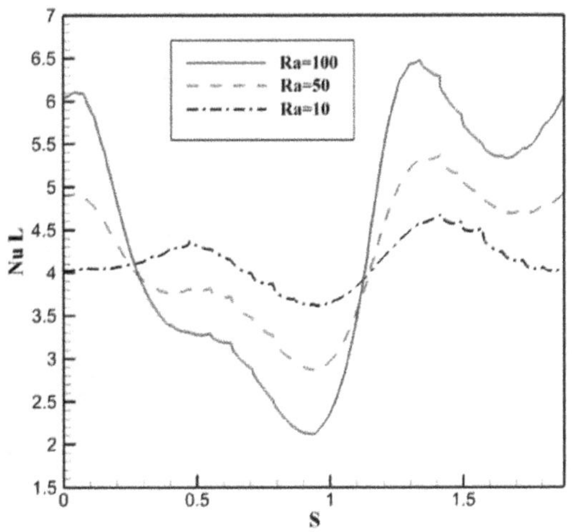

4.3.2.3 Performance Benchmark by Hartmann Number

The influence of the magnetic field parameter on the distributions of isotherms and streamlines are shown in Figure 53 for fixed amounts of Ɛ=0.5, Ra=50, and Sf=3.7. Previous investigations demonstrated that magnetic field impacts cause in decrease of convective heat transfer in enclosures. Without a magnetic field, two vortices are created inside the enclosure which are practically close to isothermal rigid vertical surfaces with progressively heating of the left part and huge cooling of the right part of the enclosure. The major changes in the flow field are detected once the magnetic field is imposed. At the highest Hartmann number estimation, the vortexes turned and placed at the corners of the cavity. Isotherms become denser on the upper part of the enclosure in the absence of a magnetic field. Isotherms intensity's distribution becomes uniform when the magnetic field strength enhances from Ha = 0 to Ha = 100. Magnetic field effect on the different Nusselt number is demonstrated in Figure 54. As it seems, with an augment of Hartman number, due to Lorentz forces and electrical vortices which resist against the fluid flow, the convective flow and local Nusselt number are getting deteriorated.

4.3.2.4 Performance Benchmark by Shape Factor

Figure 55 displays the influence of shape factor on isotherms and streamlines contours. Thermal boundary layer thickness improves with an increment of shape factor. With an increase of the shape factor, the effect of this parameter on streamlines becomes no significant and the distortion of isotherms becomes less than before. Results determine that convective mode gets more powerful with the enhancement of shape factor. In addition, conduction becomes more effective by adding source terms of the magnetic field. Figure 56 displays that the shape factor has a direct relationship with the Nusselt number.

Figure 53. Streamlines (left) and isotherms (right) for various values of Hartman numbers

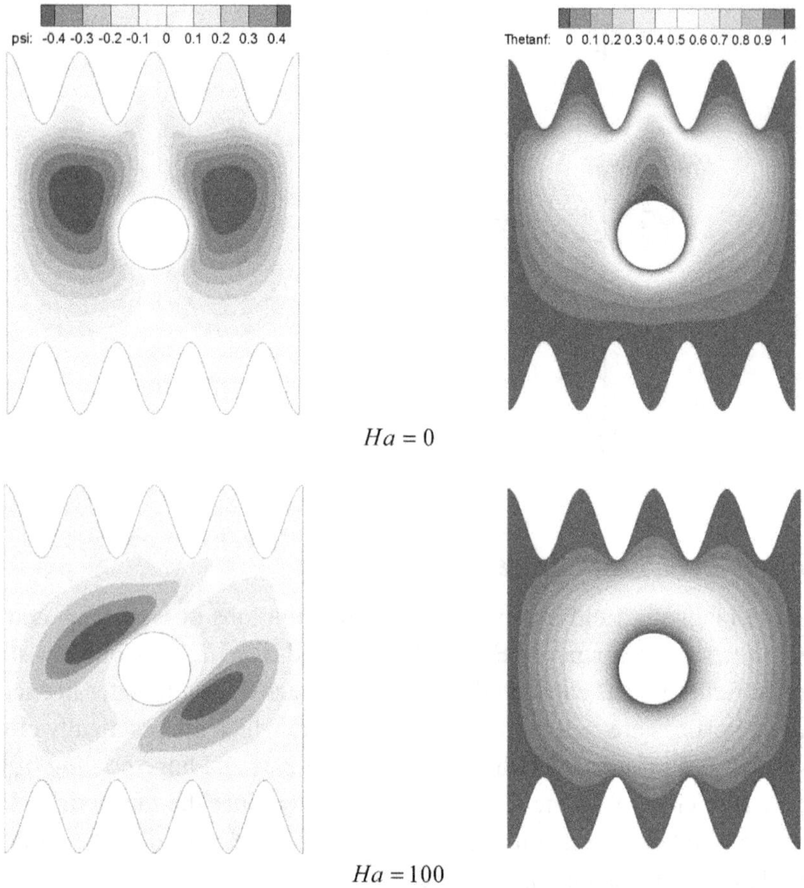

Figure 54. Local Nusselt number plot for various Hartman numbers

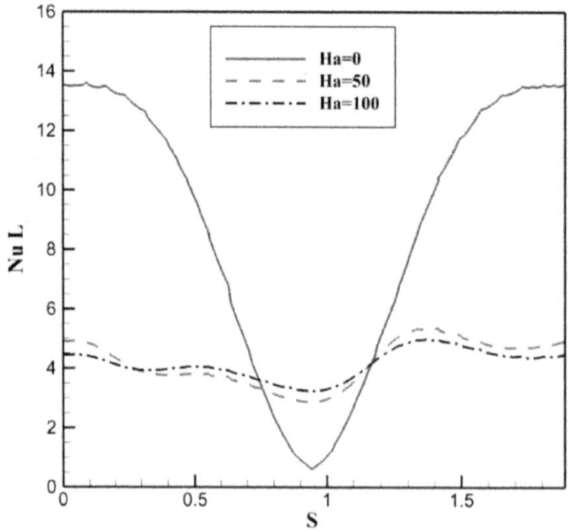

Figure 55. Streamlines (left) and isotherms (right) for various values of Radiation parameters

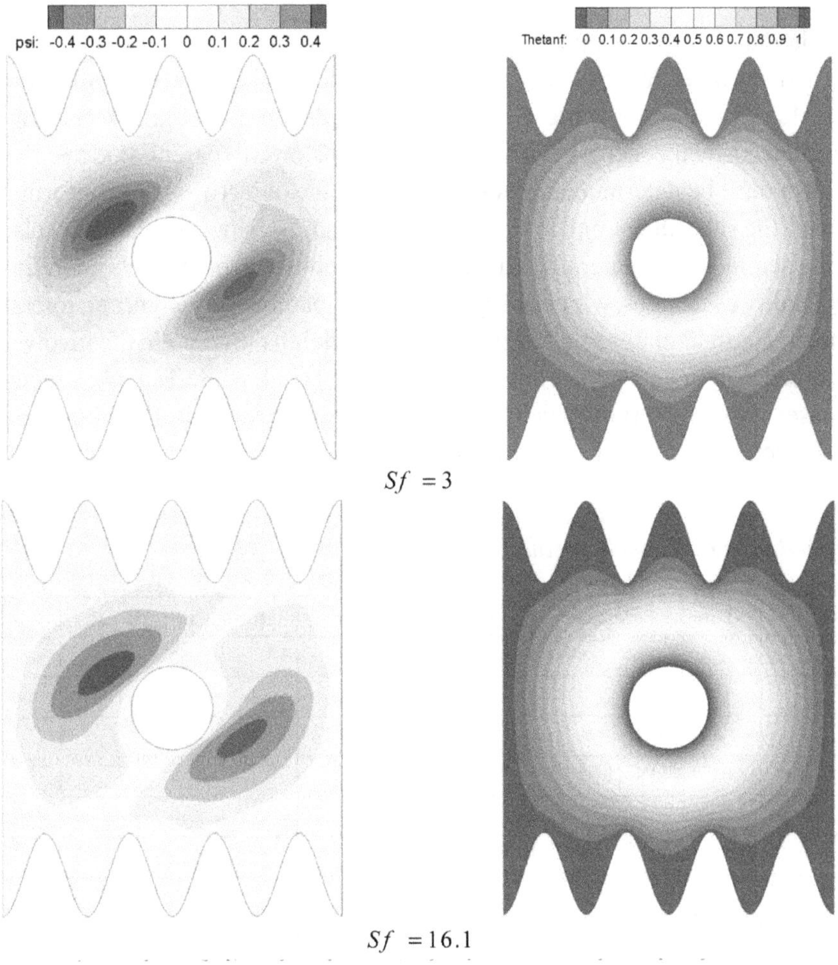

Figure 56. Local Nusselt number plot for various shape factor

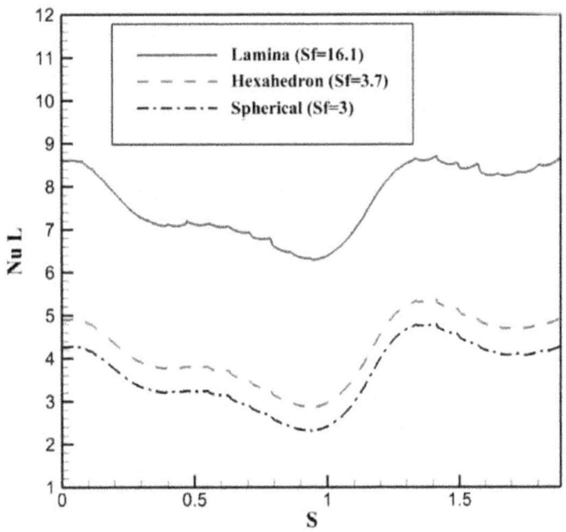

4.3.3 Optimization and Develop an Average Nusselt Number Correlation

The Taguchi method has a broad scope of uses in engineering and industrial issues. Hence, it befitted a topic of interest in many investigations and is a regularly selected technique for optimizing design parameters. This methodology was initially prescribed to improve the quality of products using the application of engineering and statistical concepts. This method depends on two essential ideas; the first is that the quality misfortunes must be determined as deviations from the objectives, not conformance to arbitrary specifications, and the subsequent one is that to acquire high framework quality levels financially, the quality must be structured into the item. Taguchi strategy proposes a three levels procedure to acquire the desired product quality by designs, which are system design, parameter design, and tolerance design. The Porosity parameter, Hartman number, shape factor, and Rayleigh number are the involve variables. Four factors each at three levels need $3^4 = 81$ runs in a full factorial trial, while Taguchi's factorial experiment technique decreases it to only 9 runs which is an incredible improvement. Based on the Taguchi design concept, the L_9 orthogonal array is selected for the simulations as reported in Table 11.

Table 11. Designed tests by Taguchi method

Case Number	Porosity Parameter	Hartman Number	Rayleigh NUMBER	Shape Factor
1	0.8	100	50	3
2	0.5	0	50	16.1
3	0.2	100	100	16.1
4	0.8	0	100	3.7
5	0.5	100	10	3.7
6	0.2	0	10	3
7	0.8	50	10	16.1
8	0.5	50	100	3
9	0.2	50	50	3.7

The optimization approach that has been utilized is the Response Surface Methodology (RSM). This approach applies a series of mathematical techniques and statistical approaches for the analysis, and modeling of the problems. The application of RSM to design optimization is planned for diminishing the cost of expensive analysis methods (e.g., CFD analysis or finite element method) and their related numerical noise. Venter et al. (1996) have examined the benefits of utilizing RSM for design optimization applications. So as to discover the impact of the variables on the response in a procedure, RSM is used to create a model between controllable input parameters and the response variables. Also, the errors that arise in the numerical applications of this method are related to the rounding of values, the non-convergence, and the number of iterations considered.

The initial phase in RSM is to find an appropriate approximation almost near to the true relationship. The most widely recognized structures are low-order polynomials (first or second-order). A second-order model can expressively develop the optimization procedure once a first-order model is not practical because of the interaction between surface curvatures and variables. A second-order model is generally characterized as:

$$y = a_0 + \sum_{i=1}^{n} a_i x_i + \sum_{i=1}^{n} a_{ii} x_i^2 + \sum_{i=1}^{n}\sum_{j=1}^{n} a_{ij} x_i x_j \,|_{i<j} + \xi \tag{11}$$

where x_i and x_j are the design variables, *a* is the tuning factor and *n* is the number of factors. Additionally, ξ illustrate the error. Furthermore, in order to maximize the average Nusselt number, the influences of porosity parameter, Hartman number, shape factor, and Rayleigh number have been employed in that correlation.

Figure 57 deliberates the relations between the nondimensional factors and their impact on the Nu_{ave}. It shows that Nu_{ave} is directly proportional to *Ra* and *Sf* as well as it has an inverse connection with *Ha* and ε. Results determine that convection type gets more powerful with an increase of shape factor. Conduction becomes more effective by loading source terms of the magnetic field. Equation (12) explains the relationship of the nondimensional factors with the average Nusselt number.

$$Nu_{average} = 7.614 - 2.988(\varepsilon) - 0.075(Ha) + 0.0302(Ra) + 0.664(Sf) \tag{12}$$

The obtained models were analyzed statistically by analysis of variance (ANOVA) for investigating whether the mathematical models can satisfactorily describe the suggested responses.

Figure 57. Effects of various operating parameters on average Nusselt number by RSM

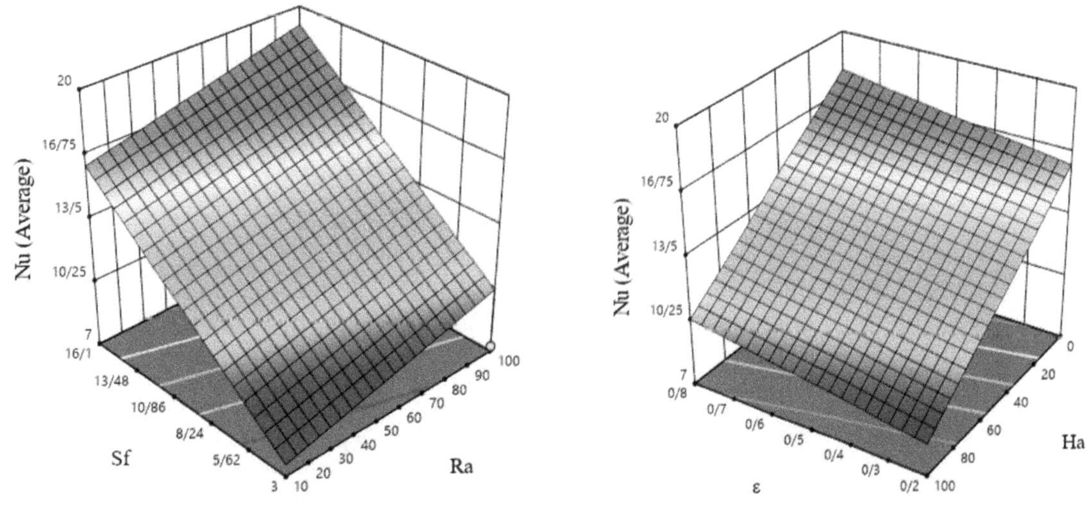

Table 12 is provided for the variance analysis (ANOVA) which represents the scale of the discrepancy of predicted and actual outcomes. The coefficient of determination (R^2) value is 0.8819 which is considerable and shows a good agreement between predicted and actual values. In Figure 58, the numerical data obtained earlier are plotted against the related estimations of the average Nusselt number dependent on the correlated equation (Eq. 12).

Table 12. The variance analysis (ANOVA)

Source	Sum of Squares	df	Mean Square	F-Value	P-Value	
Model	228.78	4	57.20	7.47	0.0386	significant
A-ε	4.82	1	4.82	0.6297	0.4719	
B-Ha	85.96	1	85.96	11.22	0.0286	
C-Ra	11.17	1	11.17	1.46	0.2938	
D-Sf	126.83	1	126.83	16.55	0.0152	
R^2=0.8819						

These outcomes determine how the current correlation enhances the earlier equations by considering the effect of Rayleigh number, Hartmann number, shape factor, and porosity of the porous medium on the heat transfer rate. Note that the solid line represents the good agreement between the numerical outputs and the expectations for each case.

The optimum values of the parameters that include in this correlation can be seen in Table 13. By applying these values to equation 12 the maximum of average Nusselt number can be calculated. In addition, Figure 59 displays the contribution of each parameter in influencing the average Nusselt number by the percentage.

Figure 58. Predicted vs. actual of the average Nusselt number

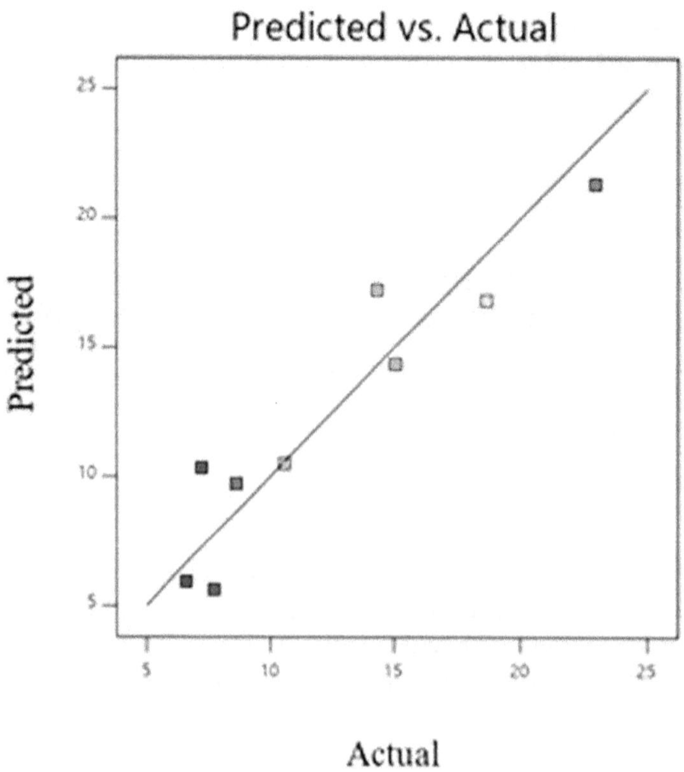

Figure 59. Contribution of each parameter on the average Nusselt number by the percentage

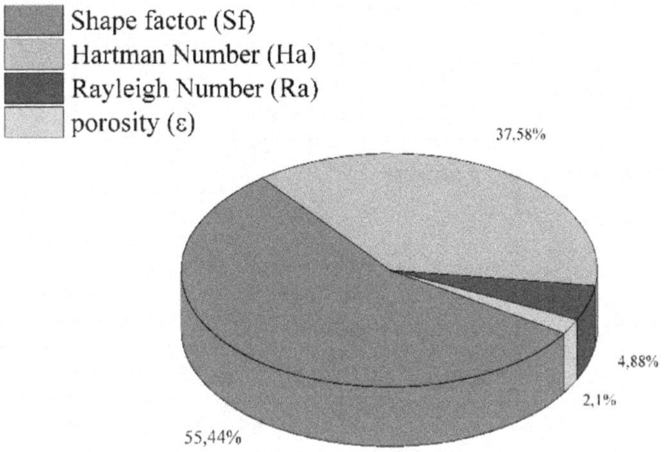

Table 13. Outcome of the optimization procedure by RSM

Parameter	Optimum Value
ε	0.641
Ha	3.384
Ra	99.888
Sf	16.1

5. EXAMPLE 5: OPTIMIZATION OF WAVY TRAPEZOIDAL POROUS CAVITY CONTAINING MIXTURE HYBRID NANOFLUID (WATER/ ETHYLENE GLYCOL Go-Al_2O_3) BY RESPONSE SURFACE METHOD

5.1 Problem Definition

The schematic of the geometry of the problem under consideration with respect to the boundary conditions is shown in Figure 60. The selected geometry is a trapezoidal cavity with a wavy top wall and an inner cylindrical heated boundary that has specific dimensions. The cavity under study contains water-based ethylene glycol Go_Al_2O_3 hybrid nanofluid. In this study, the hybrid nanofluid is assumed incompressible and newtonian. The flow inside the two-dimensional cavity becomes laminar. Nanoparticles and base fluid are considered in thermal equilibrium, And the volume concentration for nanoparticles is 0.05 or 5%. Darcy's law has been used to model the porous medium (Alkanhal et al., 2019). This research has proposed a new solution for numerical simulation of hybrid nanofluid convection heat transfer inside a wavy trapezoidal cavity with the effect of the magnetic field.

The thermophysical properties of selected nanofluids are shown in Table 14.

According to the equations stated in Chapter 4 and dimensionlessization, finally the dimensionless boundary conditions will be as follows:

$$\begin{array}{ll} \Psi = 0.0 & on\ all\ walls \\ \theta = 0.0 & on\ outer\ wall \\ \theta = 1.0 & on\ inner\ wall\ \ and\ outer\ wavy\ wall \end{array} \tag{13}$$

Figure 60. Geometry and boundary conditions

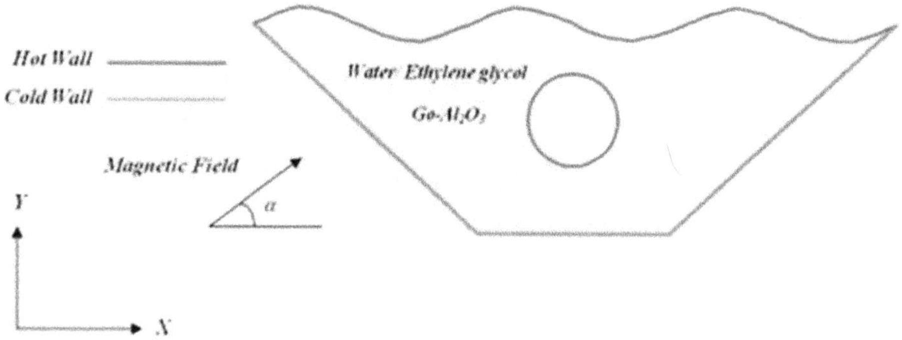

Table 14. Details of nanoparticles utilized in the present study

Physical Properties	Water/Ethylene Glycol	GO	Al_2O_3
$\rho\ (\frac{kg}{m^3})$	1063.8	1800	3970
$c_p\ (\frac{J}{kg.^\circ K})$	3630	717	765
$k\ (\frac{W}{m.^\circ K})$	0.387	5000	25
$\beta \times 10^{-5}(k^{-1})$	58	1.6	1.26

5.2 Numerical Simulation

The standard Galerkin FEM (GFEM) is used to solve the governing equations of the flow. This method provides reliable results due to its high strength and flexibility. The simulations of this research have been done using coding in FlexPDE software, which is an open-source software. Once the initial meshing is finished, the software continuously calculates the remaining solution. It may also change the mesh structure to achieve acceptable accuracy. The software also facilitates the creation of mesh networks by introducing a dynamic mesh domain that is continuously refined or coarsened depending on how close the solution is to convergence in neighboring cells. Moreover, the control volume-based finite element

(CVFEM) method is a powerful numerical method that combines finite volume and finite element methods to discretize complex geometries with a unique formulation. Therefore, the variables are easily analyzed in terms of flux, power and resources. Figure 61 depicts the algorithm created to understand the numerical solution technique better.

Figure 61. The flowchart of the computational algorithm, b) the flowchart of the numerical procedure

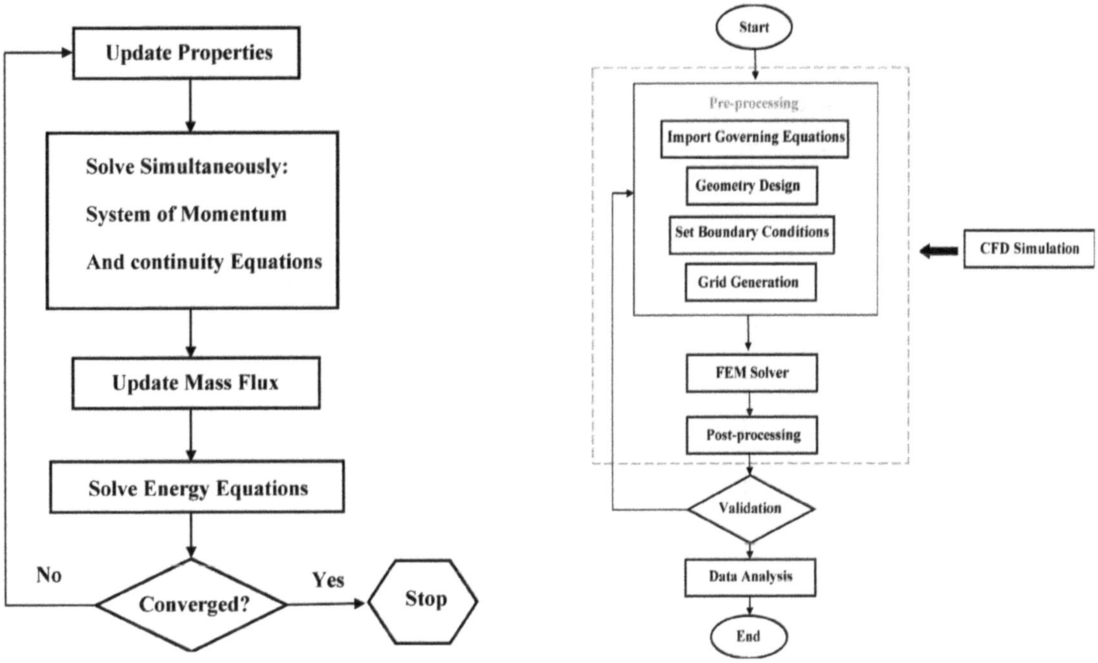

5.3 Impacts of Active Parameters

5.3.1 Meshing Independency and Validation Assessment

Any numerical analysis that attempts to reflect and resemble the real-life problem should generate a grid structure of sufficient quality so that further meshing refinement does not affect the results. Additionally, as the computational cost increases exponentially with a more sophisticated meshing framework, higher quality for the grid network frequently translates to a more laborious process. As a result, the perfect network should strike a balance between price and accuracy. Figure 62 shows this. In order to guarantee that there is a balance between precision and cost, a flexible and dynamic computational field is used. Concerning the problem's overall converging state, each section's density is entirely independent and is attributed to its converging criteria. In order to further demonstrate the validity of this dynamic network method, the local Nu at the inner cavity for three networks with various elemental densities ranging from roughly 3000 to 12,000 elements is presented in Figure 63.

Figure 62. Adaptive mesh refinement

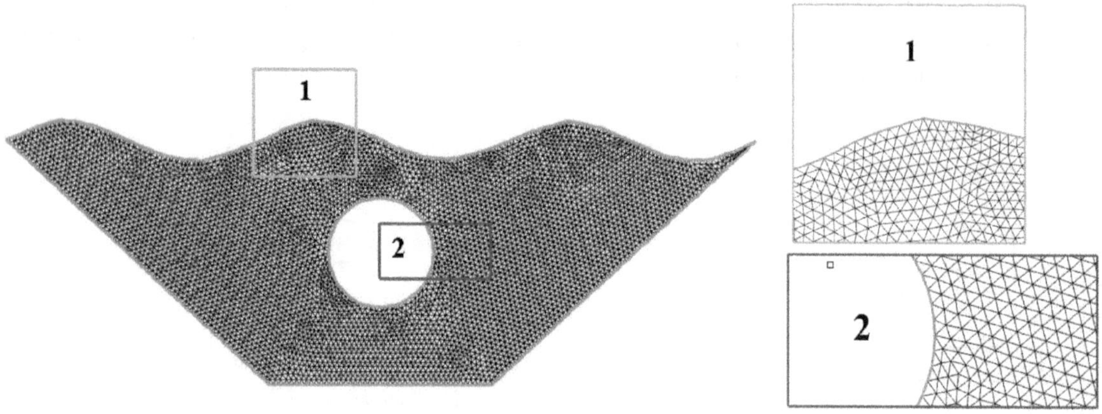

Figure 63. Comparison of the local Nu number for different grid sizes when

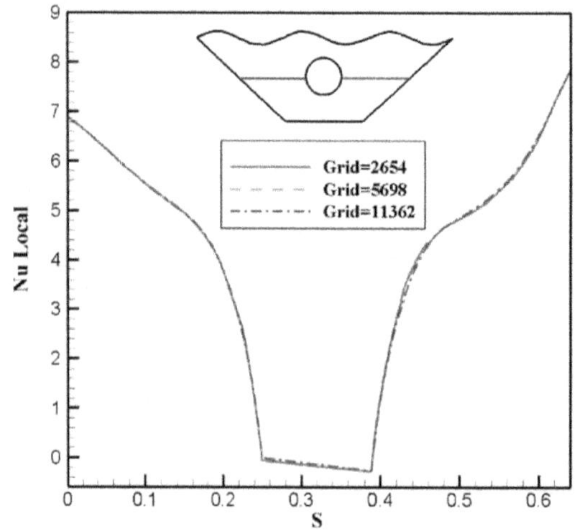

The difference in Nu values can generally be detected. Because it strikes a better balance between price and accuracy and offers a quicker solution, the 5698-element grid is chosen as the computational field for study. To evaluate the precision of the numerical solution method, it is essential to compare the results of the present study with the results of previous articles. Therefore, Figure 64(a) shows a comparison between this study and the results of research by Khanafer et al. (2003), It is clear that the results of both studies are in line with each other and are well-matched, so it can be concluded that the outputs obtained from the method under discussion are accurate and stable. In addition, the experimental findings of Wolff et al. (1988) has been compared for natural convection in a cubic cavity with different temperatures as the second validity of the present study. The comparison depicted that the temperature profiles in various locations are in excellent agreement, as shown in Figure 64(b).

Figure 64. (a) Validation of the present study with Khanfar et al. (2003). ϕ=0.05 and Pr=6.2 (Cu-Water). (b) Validation of the present study with the experimental work of Wolff et al. (1988) Z=0.5, Ra=1.689×10^5

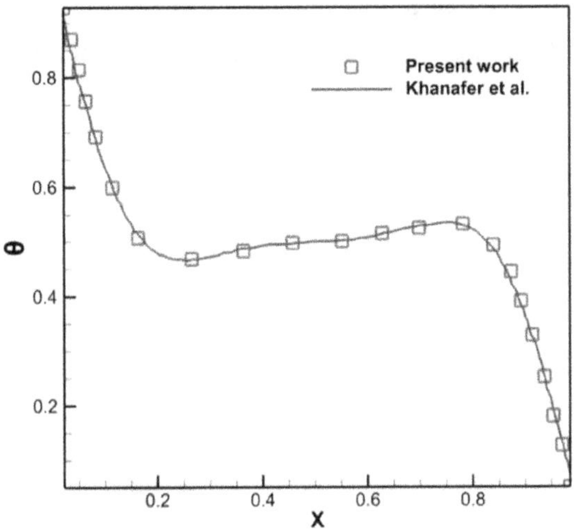

a. Validation of the present study with Khanfar et al. [19].
$\phi = 0.05$ *and* $\mathrm{Pr} = 6.2$ *(Cu-Water)*

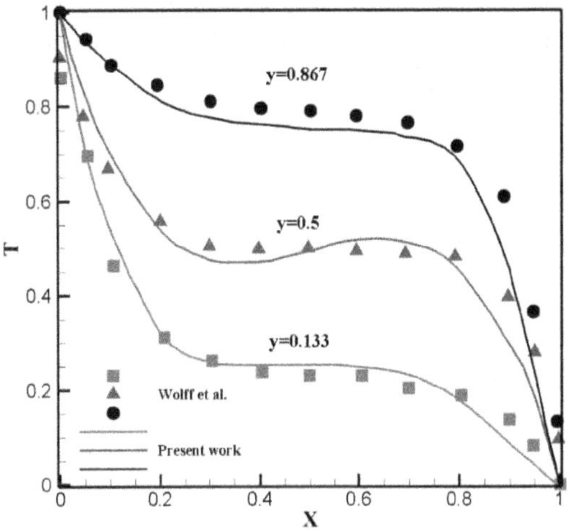

b. Validation of the present study with the experimental work of Wolff et al. [25]
Z=0.5, Ra = 1.689 × 10^5

5.3.2 Numerical Analysis

This study describes the effects of natural convection, porous medium porosity, Rayleigh dimensionless parameter, radiation strength, and effectiveness of magnetic fields on a nanofluid flow field under isothermal boundary conditions for a wavy trapezoidal porous cavity containing Mixture Hybrid nanofluid. The upper wall of the cavity was analyzed for different wave numbers (N = 0, 1, 2, 3, 4) using the highest local Nusselt and the best option, N = 3, was selected to perform the simulations. For each determinant, the resulting figures are presented with three different scales streamlined contours, isotherm profiles, and local Nu plots. The set points for determining porosity are 0.5, 0.7, and 0.9, while the Rayleigh number scale alternates between 100, 500, and 1500. The Ha number, which indicates the strength of the magnetic field, ranges from zero, which indicates that there is no external field, to 20 and 40, respectively, in this analysis. To better understand its impact, the radiation factor is changed from 0 to 0.5 and then to 1.

5.3.2.1 the Number of Waves Upper Wall Effect (N)

Stream functions and isotherms for a different number of upper wall waves (N = 0, 1, 2, 3, 4) are given in Figure 65, considering average values for porosity, Rayleigh number, radiation, Hartmann number and magnetic field angle. Inside the chamber, two vortices are created in opposite directions. The vortex on the right rotates clockwise, while the vortex on the left is positive and rotates counterclockwise. In the vicinity of the inclined and lower walls, a cold fluid flows. As it gets closer to the top of the chamber, fluid temperature and heat distribution increase around the hot circular cavity and the upper corrugated wall is observed. By examining the flow lines, it is clear that with the increase in the number of waves, the shape of the flow lines changes and spreads throughout the chamber, and the best heat transfer and temperature distribution occurs at N=3. Also, to choose the most appropriate geometry according to the number of waves, as seen in Figure 66, the local Nu number for the constant characteristic length on the desired surface is examined in terms of the number of waves. Since more significant Nusselt numbers indicate higher convective heat transfer, from the obtained results, the highest local Nu number corresponds to the geometry with N = 3. Therefore, the desired geometry is used as the basis of analysis to apply other parameters.

Figure 65. Stream functions and Isotherms for different number of waves (N) at Ɛ = 0.7, Ra = 500, Ha = 20, Rd = 0.5, and α = π/6

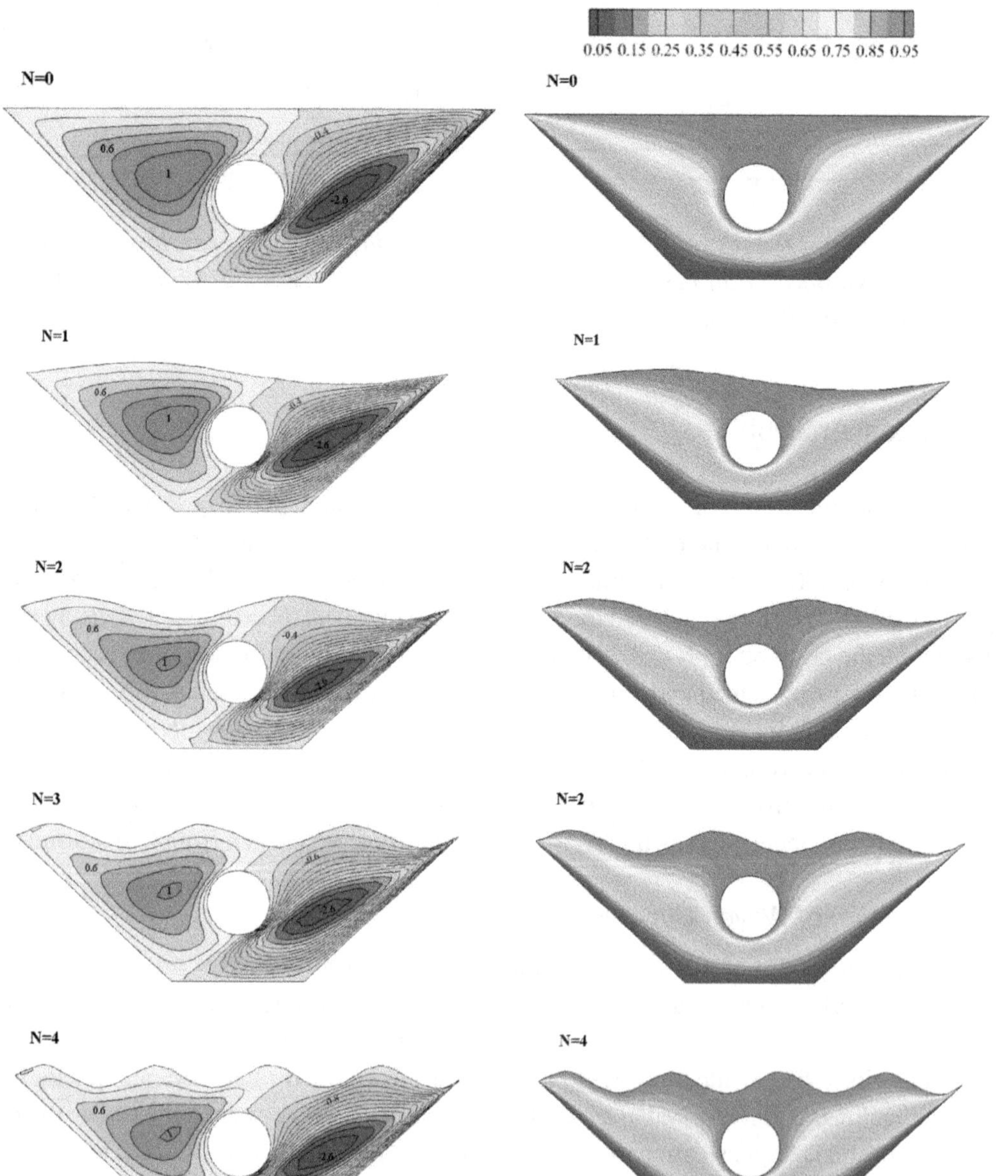

Figure 66. Local Nu number plot for different number of waves (N)

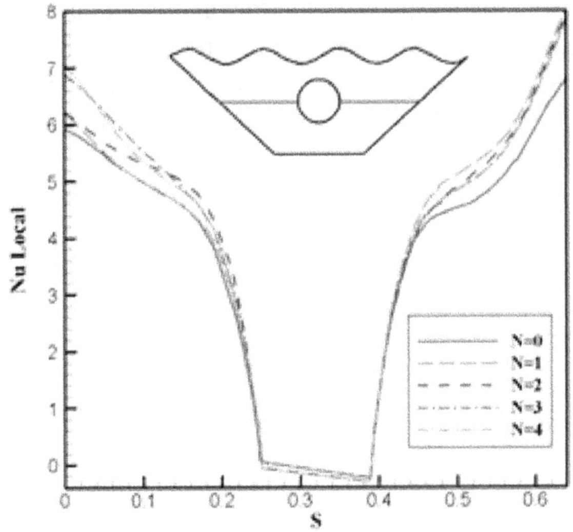

5.3.2.2 Qualification of Porosity Alteration Effectivity (ε)

The effect of changes in the fluid porosity ratio inside the chamber is shown in Figure 67. Streamlines and isotherms are obtained for porosities of 0.5, 0.7, and 0.9, keeping other effective variables constant in the mean value. As can be seen in isotherms, at a porosity of 0.5, more efficient heat transfer takes place from the hot upper wall to the lower chamber and around the inner hot cavity. to better understand this issue, the local Nusselt is investigated for three different porosity values in the direction of the specified characteristic length.

Figure 68 shows that the maximum local Nusselt number obtained for porosity is 0.5, so the fluid with the lowest porosity has a higher convective than other values because the higher the Nusselt numbers, the higher the convective heat transfer rate.

In Figure 69 appear the effect of temperature profiles along with the specified characteristic for various porosity values. As can be seen, the 0.7 and 0.9 porosity fluid diagrams overlap more than the 0.5 porosity diagrams, and at higher temperatures, close to 1, all three diagrams overlap. With a general look at the results of the study of different porosities in the chamber, it can be seen that in a fluid with less porosity, the convective heat transfer rate is higher, and the fluid flow is stronger than other values of porosity.

5.3.2.3 Qualification of Rayleigh Number Alteration Effectivity (Ra)

Since the Rayleigh number has three different values in this study, it has been investigated as one of the essential parameters in free convection. Figure 70 shows how the isothermal and stream lines work for it. Isotherm plots show how isotherm lines become more regular and less curved at smaller Rayleigh numbers. This shows that there is more convective heat transfer than conductive heat transfer. The ratio of buoyancy to heat loss is represented by the Rayleigh number. So at low Rayleigh numbers, flow is low due to buoyancy and the fluid transfers heat due to higher conductivity. Fluid flow and molecular movement increase with increasing Ra number. The streamlines in Figure 70 show that with the increase of the Ra number, the value of the stream function in the center of the vortices increases dramatically.

At Ra = 1500, the stream function reaches its maximum value. Due to the increase in the buoyancy force, the strength of the vortex and the speed of the flow in the chamber increases in these conditions. Three different Rayleigh number values are shown in Figure 71 along with the local Nusselt value corresponding to the selected surface. As can be seen, the graph with Ra=1500 has the highest Nu number and also has more curvature than the other two graphs, which indicates the improvement of heat transfer and stronger fluid flow.

Figure 67. Stream functions and isotherms for different values of porosity (Ɛ) at Ra = 500, Ha = 20, Rd = 0.5, and α = π/6

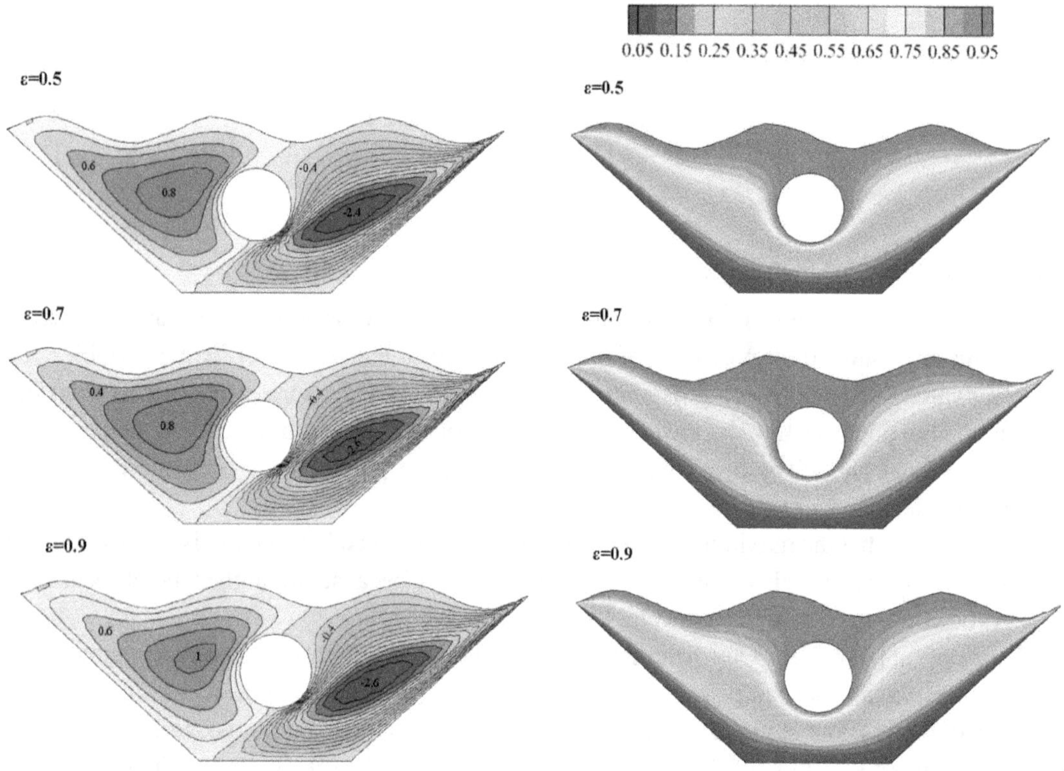

In Figure 72, the effect of temperature on the Rayleigh numbers for the three values considered for further investigation is investigated. However, the plot for Ra = 1500 is very different from the results for Ra = 100 and Ra = 500. As can be seen, as the temperature increases, the curvature of the graph increases and intersects the other graphs at two points, which indicates that with the increase of temperature and the increase of Rayleigh number, there is more disorder and thus more efficient heat transfer. comes. So, looking at these graphs, the claims mentioned above are supported.

Figure 68. Local Nu number plot for different porosity (Ɛ)

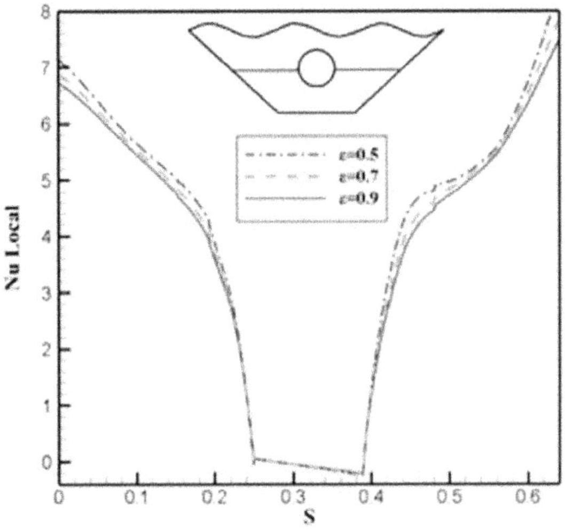

Figure 69. Temperature plot for different porosity (Ɛ)

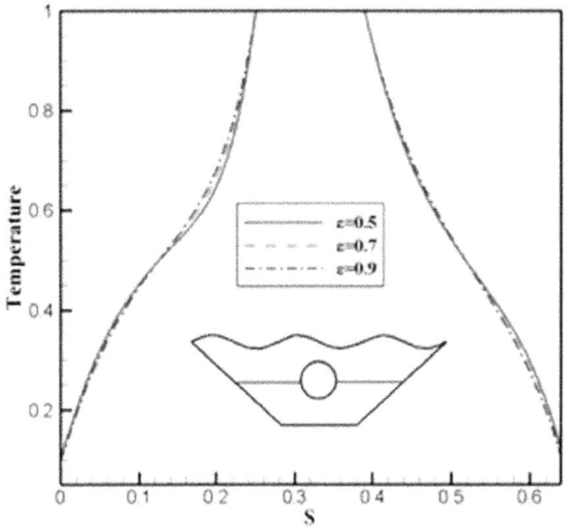

Figure 70. Stream functions and isotherms for different values of Rayleigh numbers (Ra) at Ɛ = 0.7, Ha = 20, Rd = 0.5, and α = π/6

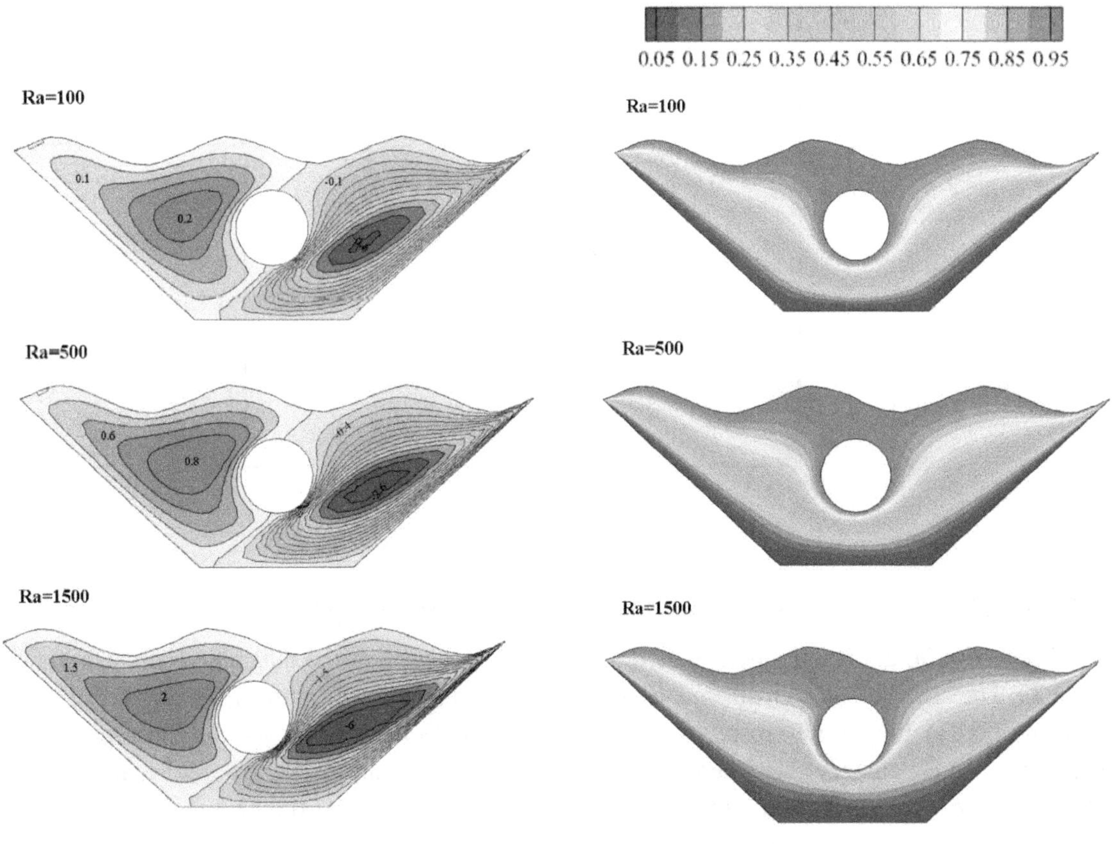

Figure 71. Local Nu number plot for different Ra numbers (Ra)

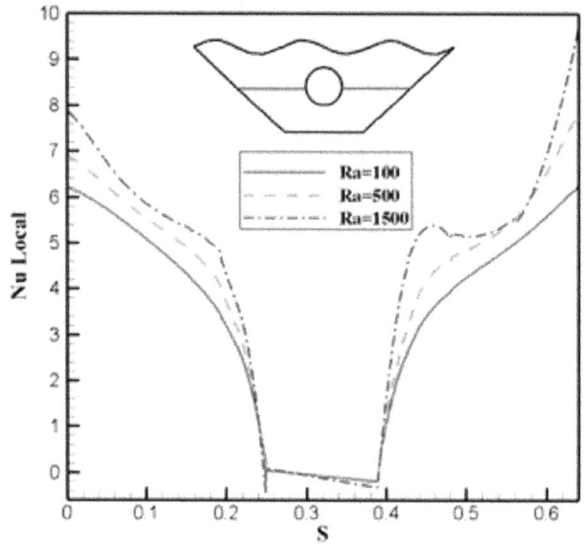

Figure 72. Temperature plot for different Ra numbers

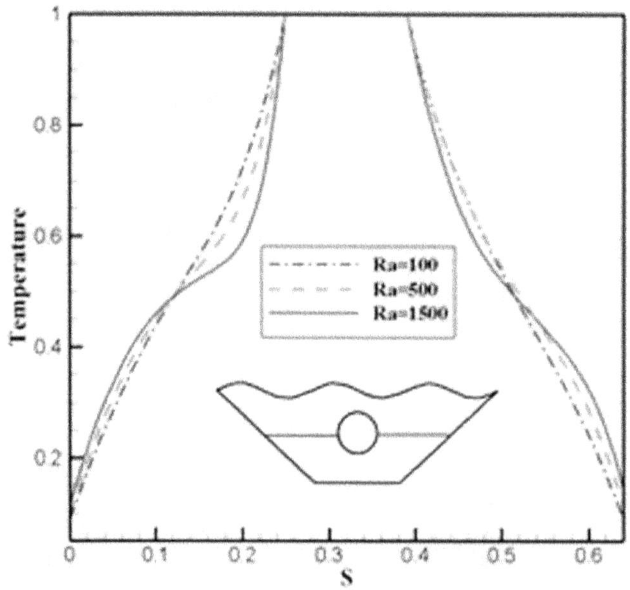

5.3.2.4 Qualification of Hartmann Number Alteration Effectivity (Ha)

The effects of different Ha numbers, which represent the strength of the magnetic field, are shown in Figure 73. The effect of Ha = 0, 20, and 40 on the nanofluid inside the cavity has been studied because increasing the Ha number will result in a stronger magnetic field. A closer look at the streamlines reveals that as the Ha number rises, the vortices become weaker and are drawn toward the chamber's side walls due to the Lorentz force created by applying a magnetic field. The Lorentz force that results as the magnetic field's strength rises slows the motion of the fluid masses. Additionally, as the fluid displacement declines, the buoyancy also does so, causing slower vortices and diminished power, as in the case of the Ha = 40. As a result, Ha = 0 is where the stream function's most extreme value is in Figure 73.

Figure 73. Stream functions and Isotherms for different values of Hartman numbers (Ha) at $\mathcal{E} = 0.7$*,* $Ra = 500$*,* $Rd = 0.5$*, and* $\alpha = \pi/6$

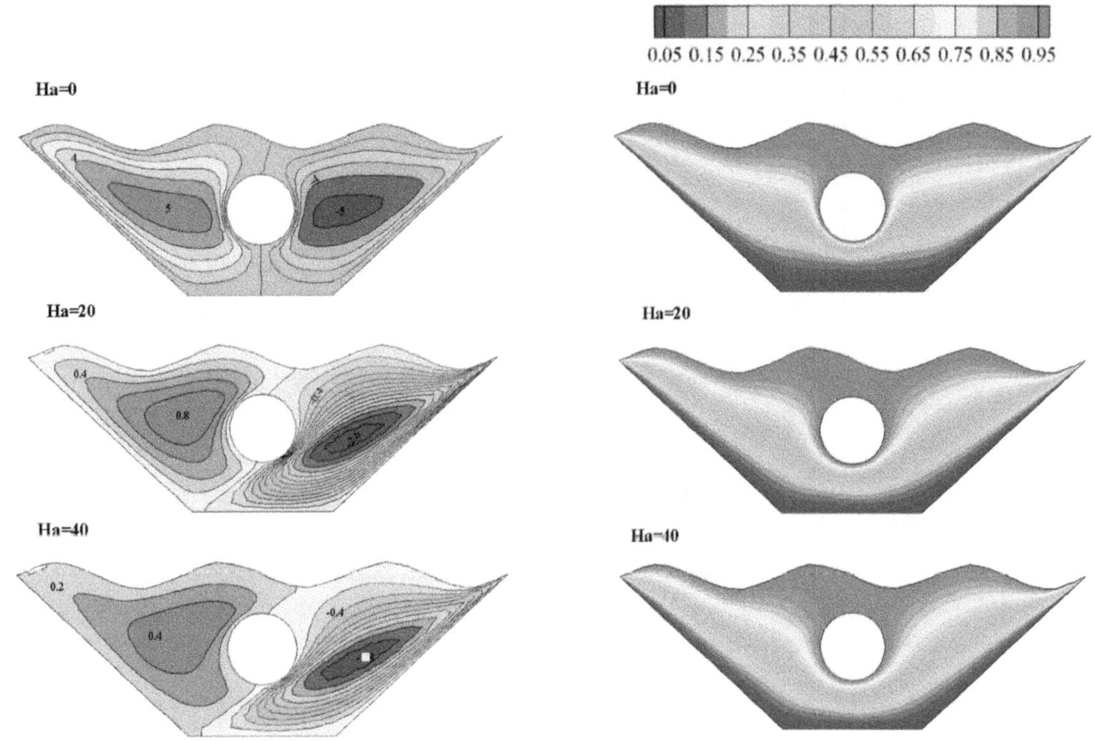

It is evident in isothermal profiles that in the absence of a magnetic field, heat transfer from the corrugated surface and the hot cavity inside the chamber to the fluid inside the chamber is more effective, indicating a more significant proportion of convective heat transfer than conductive heat transfer. The highest Nusselt number was obtained for Ha = 0, which is very different from other values applied to the magnetic field. Figure 74 illustrates the impact of graphs of various values of the Ha number on local Nusselt. This outcome may support these findings. Figure 75 depicts how the Hartman number affects the temperature at various values. The graphs of Ha = 20 and Ha = 40 are obtained with a slight curvature and produce similar results. Nevertheless, the graph's curvature is significantly increased without a magnetic field, intersecting the other graphs at two points. The absence of a magnetic field results in a temperature diagram with a large curvature, which denotes a more effective convective process.

Figure 74. Local Nu number plot for different Hartman numbers (Ha)

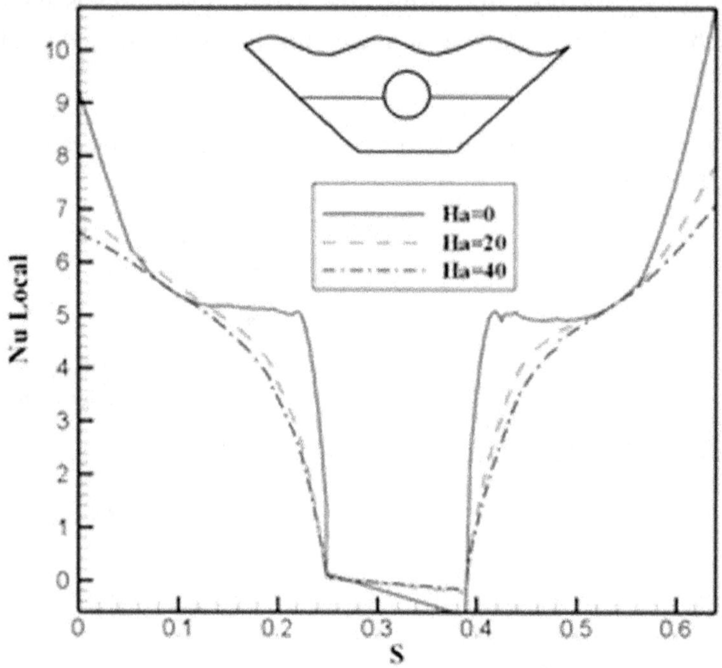

Figure 75. Temperature plot for various Hartman numbers (Ha)

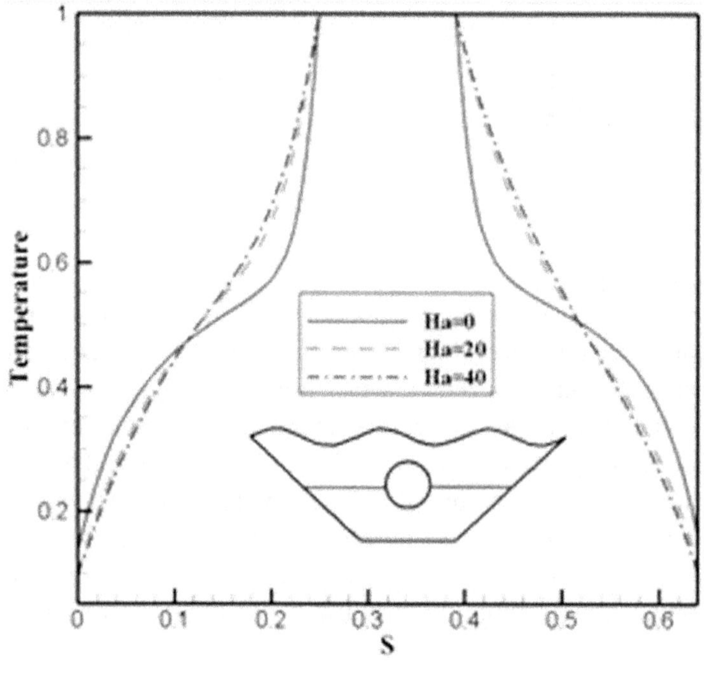

5.3.2.5 Qualification of Radiation Factor Alteration Effectivity (Rd)

A wavy porous enclosure makes it easier to assess Radiation when it is included in the governing equations. Rd is used as a standard for characterizing this mode of heat transfer phenomenon because it is a dimensionless form of radiation intensity. Due to the importance of the radiation parameter in heat transfer, its effect in three different values with constant values for other effective parameters has been investigated in this study. The resulting isothermal lines and streamlines are shown in Figure 76. The reason for these ponders to increase the heat transfer rate from the heat absorbed by the corrugated surface of the chamber and the inner circular hot cavity to the fluid inside the chamber. However, by examining the results, it is observed that the streamlines for all three parameter values of Radiation provide similar results.

Figure 76. Stream functions and Isotherms for different values of Radiation (Rd) at $\varepsilon = 0.7$*, Ra = 500, Ha = 20, and* $\alpha = \pi/6$

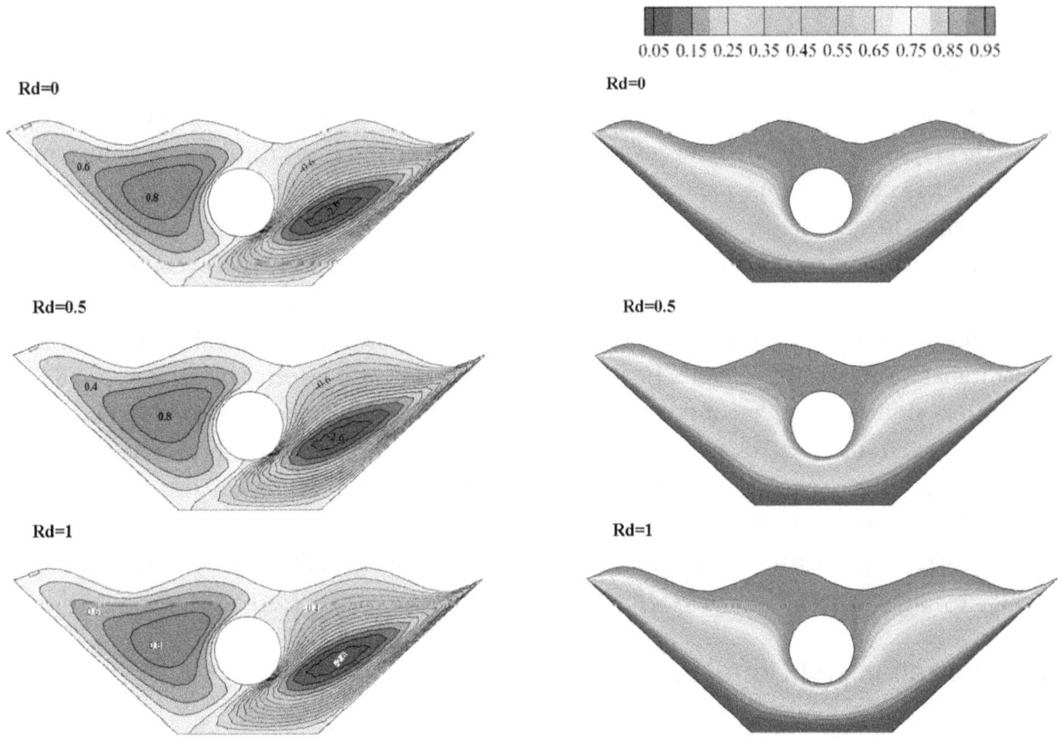

Similarly, in the case of isothermal lines, since the Rd is related to the heat transfer of Radiation, there is not much effect on the isotherms. For better investigation, different amounts of Radiation are shown on the local Nusselt number along the characteristic specified in Figure 77. The results show that no significant effect is observed with increasing the amount of Radiation on the obtained Nusselt number. Figure 78 also examines the effect of Radiation on temperature, and all three graphs show similar results and are almost identical. Therefore, in general, different amounts of Radiation do not significantly affect the rate of fluid flow and heat transfer within the chamber in this numerical study.

Figure 77. Local Nu number plot for different radiation (Ra)

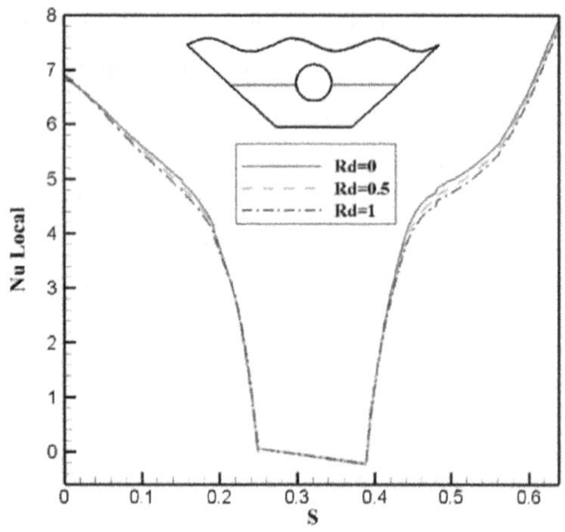

Figure 78. Temperature plot for different radiation (Rd)

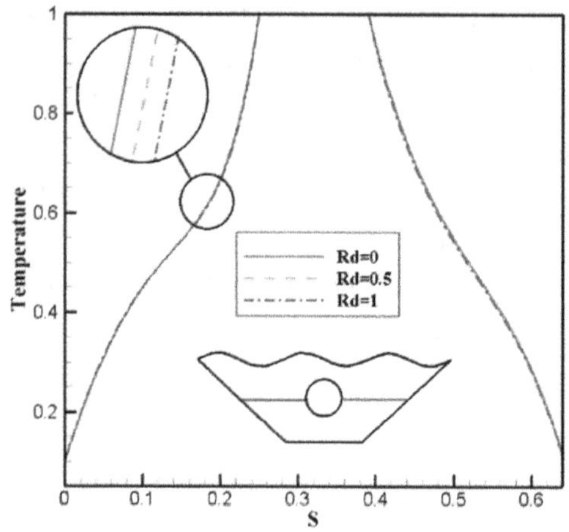

5.3.2.6 Angle of Magnetic Field Effectivity (α)

This section applies the effects of three different angle of the magnetic field to the desired geometry. Figure 79 displays the fluid inside the cavity, isotherms, and streamlined results. Applying a magnetic field with $\alpha = \pi/4$ causes stronger and larger vortices to form on both sides, indicating an improvement in the desired hybrid nanofluid's stream functions in terms of velocity and strength. Less convective heat transfer occurs compared to conductive heat transfer as the alpha angle decreases because of the vortices' size and velocity decrease. The impact of variations on the local Nu number is examined in Figure 80.

The results show that the alpha angle ($\pi / 4$) has the highest local Nusselt, indicating that the maximum heat transfer occurs at this angle. Figure 81 also depicts how it affects temperature. The graph of $\alpha = \pi / 4$ differs slightly from that of other values. The density of the isothermal lines decreases, as seen from the isothermal lines, and as temperature gradient decreases, so does the heat transfer rate within the chamber. Therefore, one of the essential variables in heat transfer within cavities is the angle of the magnetic field, which may be helpful in future engineering designs, Although in some sources (Shekaramiz et al., 2021), the distribution of nanofluid inside the chamber and the amount of heat transfer are not affected by changes in the magnetic field angle.

5.3.3 Optimization and Correlation Development

A better analysis that can succinctly illustrate the efficacy of each additional variable, both individually and collectively, is required due to a large number of significant parameters in this problem. This study investigated five parameters: the radiation factor, the porosity, the magnetic field, angle of magnetic field and the strength of natural convection using nondimensionalized numbers at three specifications. In order to cover a fully crossed design, 81 trials would typically be needed, which is a burdensome number. This is avoided by the Taguchi method, which also offers a more detailed view of the problem at hand while introducing an alternative solution to reduce the number of trials significantly. Even though its original purpose was to allow for higher quality goods and products, this is why it has gained significant applicability in many disciplines and contexts. The system, parameter, and tolerance designs are the three essential steps this method requires to fully achieve satisfying and economical procedures. Following these guidelines reduced the number of trials from 81 to just 16 runs, significantly reducing the computational cost and onerous postprocessing. Table 15 provides the specific data for trials that were ultimately designed.

Figure 79. Stream functions and Isotherms for different angle of magnetic field (α) at Ɛ = 0.7, Ra = 500, Ha = 20, and Rd = 0.5

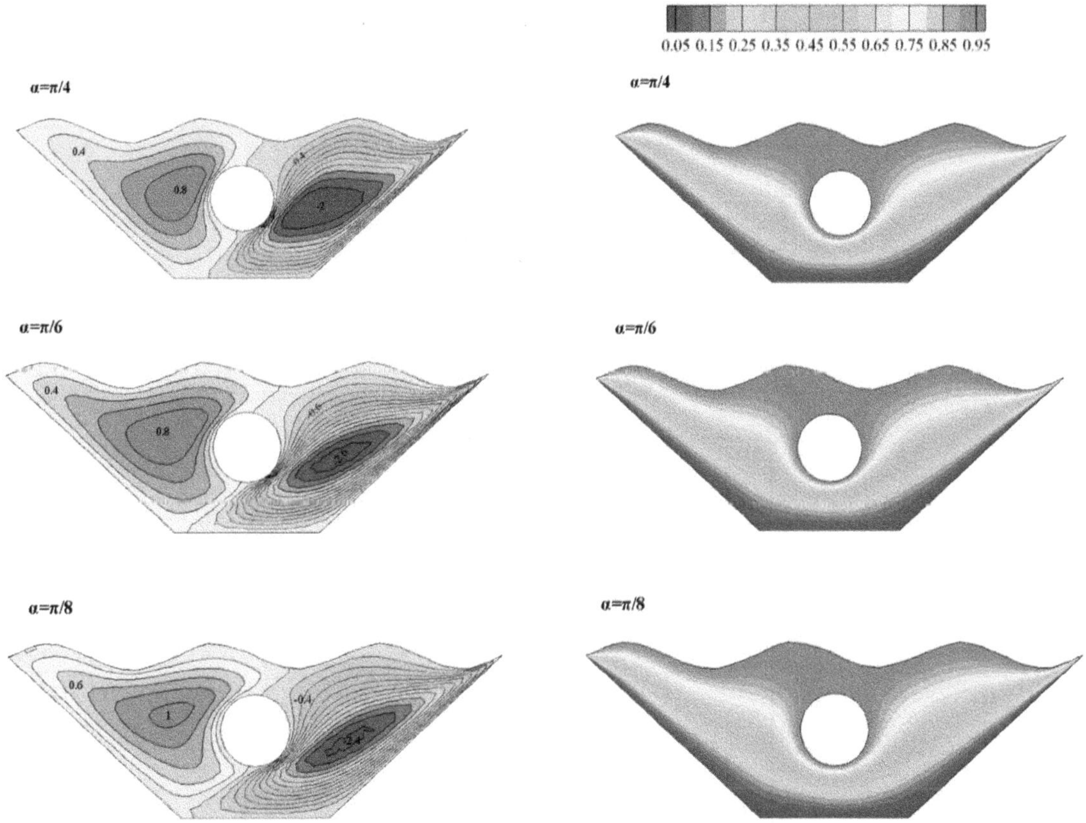

Response surface methodology, or RSM, has been extensively used in various optimization issues due to its simple mathematical foundations and high output accuracy. The final correlations for the response variables eliminate the need for time-consuming numerical trials and provide a more refined perspective on how each determinant interacts with the others. The use of RSM may be more advantageous because the problem's taken-into-account parameters span continuous zones. The average Nusselt number and five dimensionless control factors, Ha, Rd, Ra, Ɛ, α and, make up the response variable for the anticipated correlation from RSM integration with GFEM. Compared to the numerical output dealt with in the next section, the rounding, divergence, and iteration figures may produce some potential errors. A fundamental component of RSM development is obtaining a correlation that gradually becomes more in line with the data. Therefore, polynomial expressions take precedence due to their direct interaction and adaptability.

Figure 80. Local Nusselt number plot for different angle of magnetic field (α)

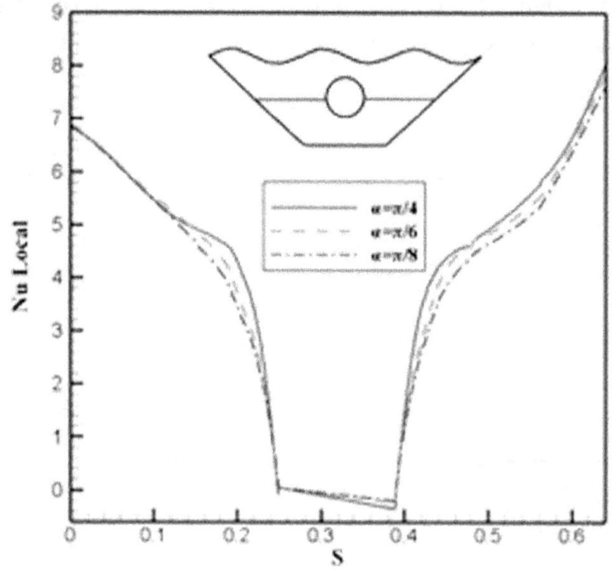

Figure 81. Temperature plot for different angle of magnetic field (α)

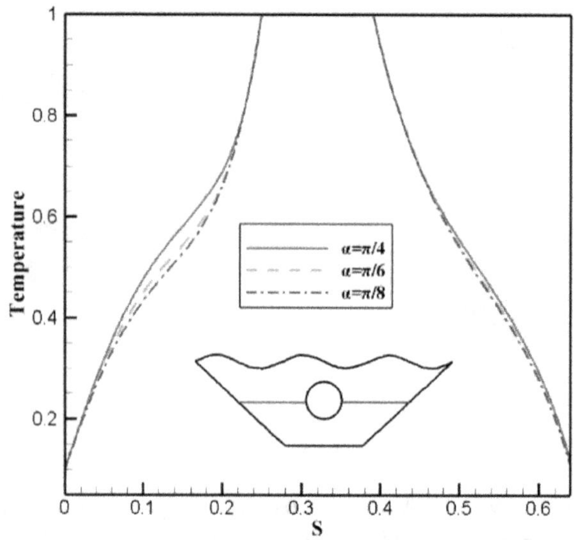

In cases of the preliminary agreement, the order of polynomials tends to increase to consider more complex issues. A typical second-order polynomial is defined as follows.

$$y = a_0 + \sum_{i=1}^{n} a_i x_i + \sum_{i=1}^{n} a_{ii} x_i^2 + \sum_{i=1}^{n} \sum_{j=1}^{n} a_{ij} x_i x_j \big|_{i<j} + \xi \tag{14}$$

Table 15. Planned tests by Taguchi strategy

Case Number	Rayleigh Number	Hartman Number	Porosity Parameter	Radiation	α
1	500	0	0.65	0.7	60
2	1500	0	0.9	0.35	45
3	500	30	0.9	0	30
4	100	30	0.8	0.7	45
5	1500	30	0.65	1	22.5
6	100	0	0.5	0	22.5
7	100	40	0.9	1	60
8	1000	30	0.5	0.35	60
9	500	40	0.8	0.35	22.5
10	1000	40	0.65	0	45
11	100	15	0.65	0.35	30
12	1500	40	0.5	0.7	30
13	500	15	0.5	1	45
14	1000	0	0.8	1	30
15	1000	15	0.9	0.7	22.5
16	1500	15	0.8	0	60

In this relation, Xj and Xi are the design variables, the coefficients are denoted by α, and the number of variables and errors is specific by ζ.

Figure 82 shows the three-dimensional graphs of the Nu_{avg} analysis with two pairs of parameters interactively from the RSM method. First, the interactive effect of Ha and Ra on the Nuavg is shown as the Ra number increases and the Nu_{avg} value increases. In contrast, at higher Hartmann numbers, the Nu_{avg} decreases. The following interactive effect is the porosity and radiation factor. According to the diagram, reducing the amount of porosity and the amount of radiation factor leads to a higher Nu_{avg}. In the last graph, the interactive effect of α and Rd can be seen, which increases the α and, on the other hand, decreases the radiation, resulting in a higher Nu number. With the help of the desirability function approach, an estimated response is converted into a scale-free value called desirability. Optimization goals can be maximizing, minimizing, or achieving the desired response value. Between 0 and 1 is the ideal range. An unacceptable configuration for the chosen response is represented by a 0 or elementary desirability, taking the value zero.

In contrast, the ideal case is illustrated by desirability taking the value 1. Figure 83 shows the desirability graphs influenced by two-parameter interactive factors for predicting the average Nusselt value. Based on this, the average Nusselt values according to the desirability of 1 in each pair of interactive parameters are displayed on the figure by flags.

To show the uncertainty in the point predictions, interval estimates can be applied to the optimization graphic. Interval estimates can also be used to understand how uncertainty affects achieving process goals. The bright yellow zone in Fig. 84 denotes that every range of intervals satisfies the required standards. On the other hand, the dark gold zone refers to the region where a piece of an interval estimate does not meet the criterion requirements but the point estimate does.

Figure 82. Impacts of different working parameters on average Nu number by RSM

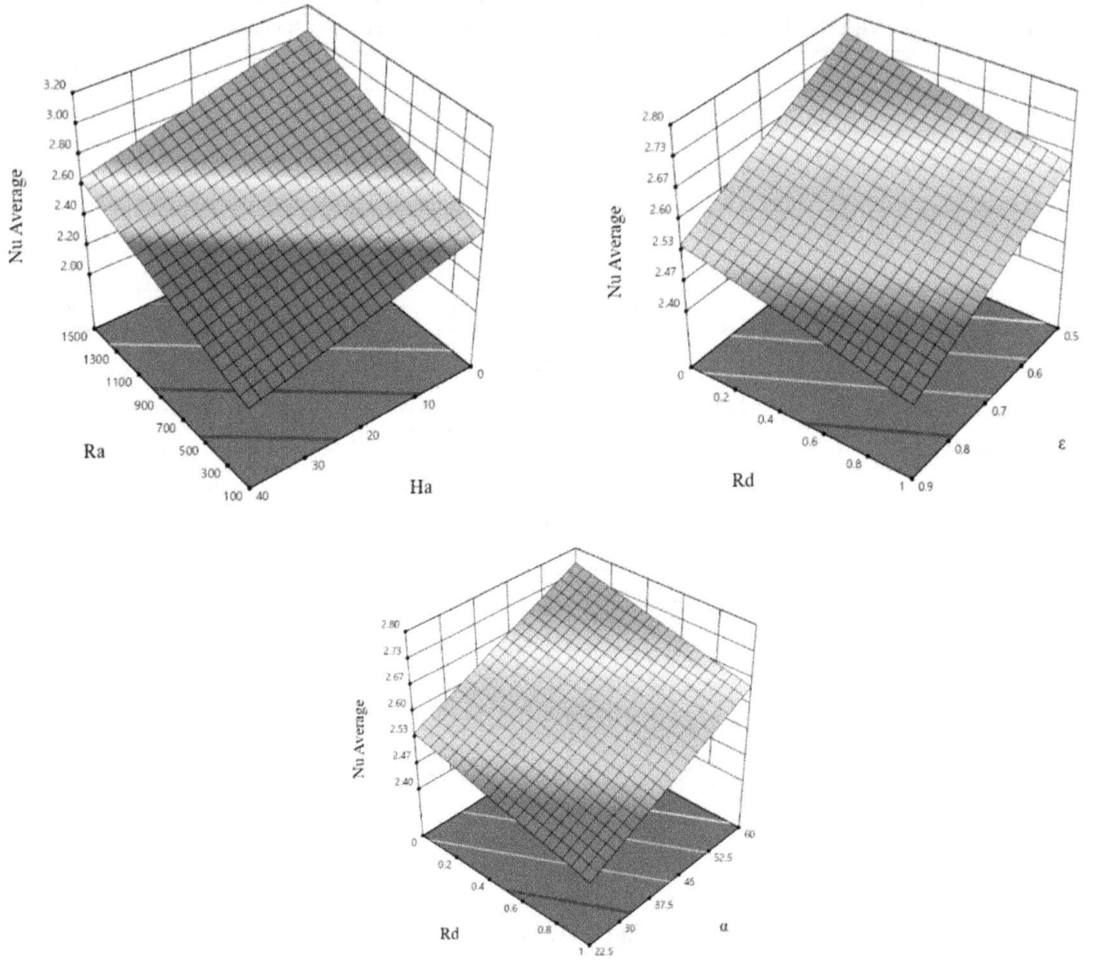

Equation 15 is presented to obtain the average Nu, a comprehensive relationship according to the dimensionless variables investigated in this study.

$$Nu_{average} = 2.69 - 0.5255\left(\varepsilon\right) - 0.0096\left(Ha\right) + 0.00035\left(Ra\right) - 0.114\left(Rd\right) + 0.00571(\alpha) \quad (15)$$

ANOVA analysis of variance is used to confirm and evaluate this relationship and analyze its performance, and its results are shown in Table 16. If analysis of variance is used, it is possible to understand which factor had the greatest impact. In this table, the coefficient of determination is used to compare RSM with real data. This coefficient is a reliable indicator. This study's value reaches 0.9122, which can justify the use of RSM and confirm its agreement with the real results.

Figure 85 shows the actual and predicted values for the Nu_{avg} relative to the hypothetical line. According to the obtained results, the RSM correlation estimate shows a good fit compared to the real data according to the existing complexities and the number of influencing factors.

Figure 83. Desirability diagram of different working parameters on average Nu number by RSM

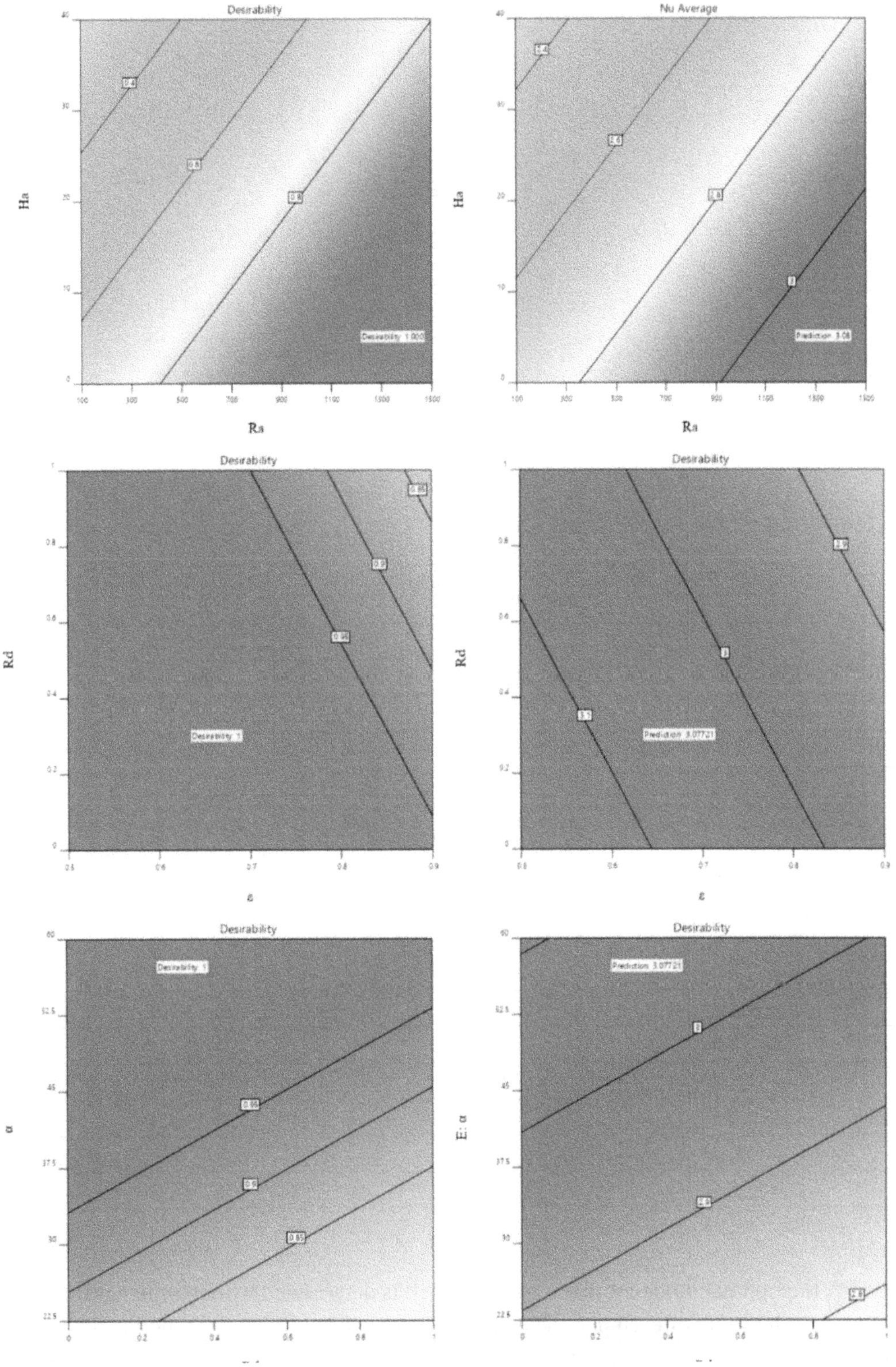

Figure 84. Overlay plots

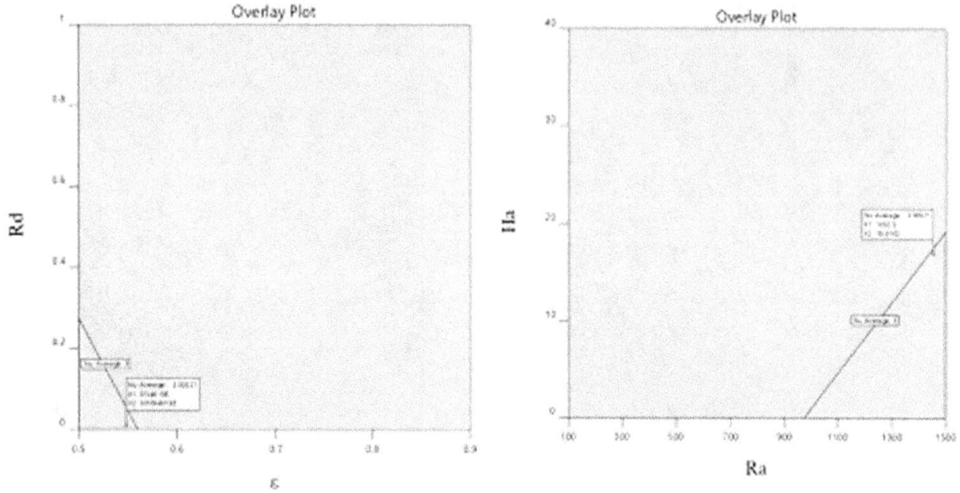

Table 16. The variance analysis (ANOVA)

Source	Sum of Squares	df	Mean Square	F-Value	P-Value	
Model	1.14	5	0.2272	20.79	< 0.0001	significant
A-Ra	0.5523	1	0.5523	50.55	< 0.0001	
B-Ha	0.3441	1	0.3441	31.49	0.0002	
C-ε	0.1015	1	0.1015	9.29	0.0123	
D-Rd	0.0295	1	0.0295	2.70	0.1312	
E-α	0.1083	1	0.1083	9.91	0.0104	
R^2=0.9122						

Table 17. Result of the optimization strategy by RSM

Parameter	Optimum Value
Ra	1214.46
Ha	2.86
ε	0.63
Rd	0.24
α	59.35

In Table 17, the optimal mode for the average Nusselt is determined by obtaining and validating the correlation so that the value of each parameter can be calculated at the optimal point. By putting these numbers in the obtained equation introduced earlier, the Nu_{avg} can be calculated, and the best optimal point is obtained.

Figure 85. Anticipated vs. real of the average Nusselt number

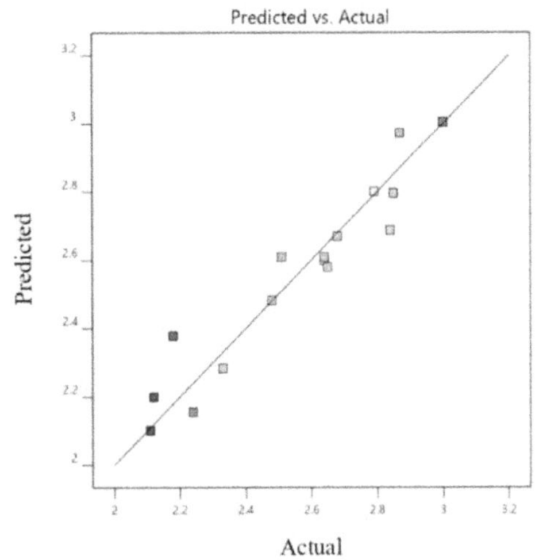

6. EXAMPLE 6: OPTIMIZATION APPROACH FOR MoS_2-WATER ETHYLENE GLYCOL MIXTURE NANOFLUID FLOW IN A WAVY ENCLOSURE

6.1 Problem Definition

6.1.1 Physical Configuration of the Problem

In the current investigation, the two-dimensional enclosure with a star-shaped cavity in its middle is considered. The enclosure is filled with MoS_2 nanoparticles, with different volume fractions ($0.01 \leq \varphi \leq 0.03$) and shape factors ($3 \leq Sf \leq 16.15$), and water/Ethylene Glycol mixture (with a mixing ratio of 50%-50% by volume) as a base fluid. Table 18 exhibits the thermophysical specifications of the involved constituents. The inner and outer walls of the medium are presumed to be fixed at the higher temperature (T_h) and lower temperature (T_c), respectively. Also, the mathematical formulation of the inner shape is $r=10\sin(6\theta)+45$. Due to the gradient in temperatures of the enclosure surfaces, the fluid experiences the buoyancy force.

Figure 86 illustrates the geometries of interest and coordinate system. As depicted, the cavity is subjected to an external magnetic field with uniform magnitude (B_0) in the horizontal direction, which induces the Lorentz force and hence forced convection. Accordingly, the Hartman number (Ha) is presumed to vary in the range of $0 \leq Ha \leq 60$. Moreover, the influences of thermal radiation ($0 \leq Rd \leq 0.8$) on natural convection are evaluated. The main assumptions considered in the current research are presented below.

- The laminar and steady flow of the nanofluid in the cavity is taken into account.
- The nanofluid is incompressible, homogeneous, and Newtonian.
- The medium is isotropic and thoroughly saturated with the fluid.
- The base fluid and nanoparticles are under thermal equilibrium.
- The thermophysical features of the constituents are presumed fixed (see Table 17), except for density, for which the Boussinesq model is employed.
- The induced electric field, Joule heating effect, and viscous dissipation are ignored.

Table 18. Characteristics of nanoparticles employed in this study (Barnoon, Toghraie, Dehkordi, & Afrand, 2019; Luo et al., 2016)

Physical Properties	Ethylene Glycol/Water	MoS_2
$\rho\ (\frac{kg}{m^3})$	1063.8	5060
$c_p\ (\frac{J}{kg.^\circ K})$	3630	397.21
$k\ (\frac{W}{m.^\circ K})$	0.387	904.4
$\beta \times 10^{-5}(k^{-1})$	58	2.84

Figure 86. The diagram of the physical configuration

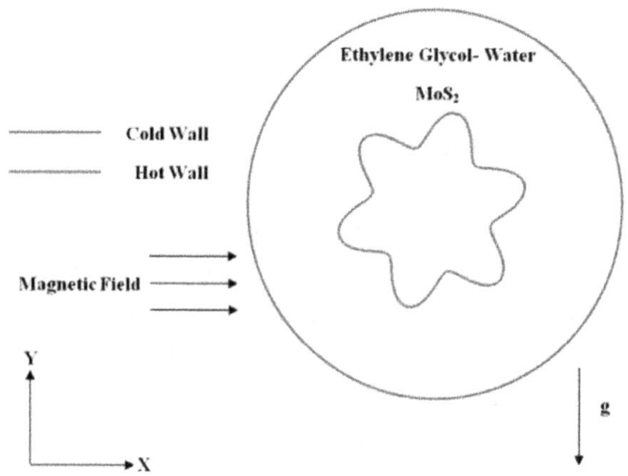

6.1.2 Governing Equations

According to the above assumptions, the conservation equations of mass, momentum and energy for the flow and temperature fields of the mixed nanofluid are stated in chapter 4. The boundary conditions of the problem are as follows:

At the inner surfaces: $T^{*}=T_h$ (16)

At the outer surfaces: $T^{*}=T_c$ (17)

At all surfaces: $u=v=0$ (18)

After transformations, the boundary conditions will be as follows:

At the inner walls: $T=1$ (19)

At outer walls: $T=0$ (20)

At all walls: $U=V=0$ (21)

6.2 Numerical Simulation

Solving flow equations is based on Galerkin standard finite element method (GFEM). In this study, the second-order interpolation (linear or cubical) between the triangular and quadrilateral elements is used to solve equations. The open-source FlexPDE software for numerical simulation is used and developed. After the first mesh has been generated, the solution residual is calculated by the software. Subsequently, if necessary, the meshing structure is improved to reach better accuracy. The flowcharts of the computational algorithm and numerical approach employed in the present work are illustrated in Figures .87a and 87b, respectively.

Figure 87. a) The flowchart of the computational algorithm, b) the flowchart of the numerical procedure

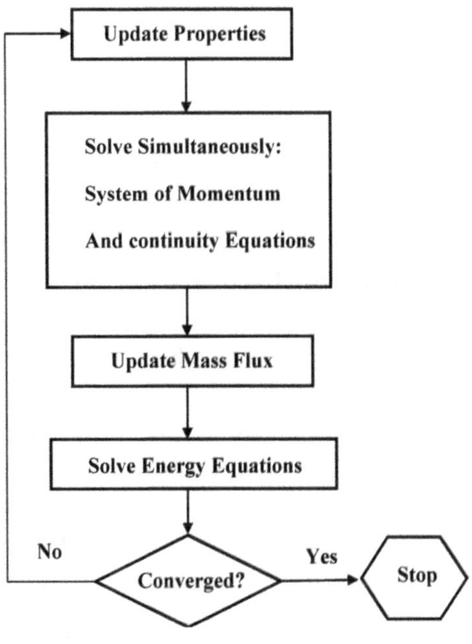

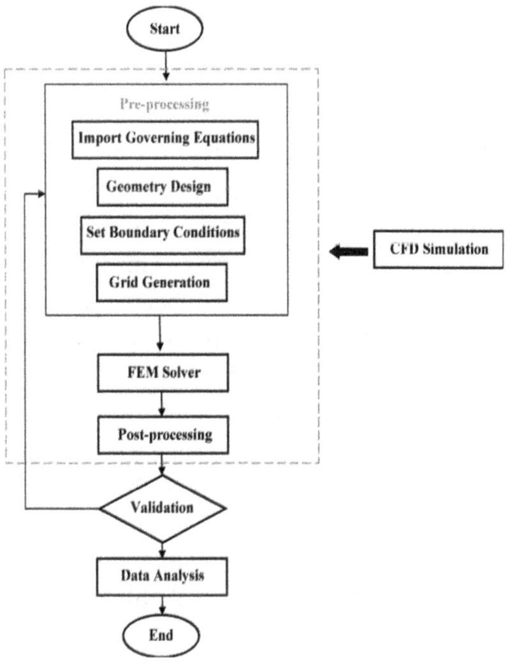

6.3 Impacts of Active Parameters

6.3.1 Heat Transfer

This subsection indicates numerical results for the hydrothermal behavior of water/EG base MoS_2 nanofluid within an enclosure subjected to an externally applied uniform magnetic field. The lines of the stream and temperature, and the behavior of heat transfer with several magnitudes of nanoparticles volume fraction (φ=0.01, 0.03, and 0.06), shape factor (spherical: 3, Platelets: 5.7, and lamina: 16.15), the Hartmann number (Ha=0, 30, and 60), and the radiation parameter (Rd=0, 0.4, and 0.8) are comprehensively assessed.

6.3.1.1 Performance Criterion by Variation of Nanoparticles' Volume Fraction

Figure 88 depicts the streamline and temperature field for different nanofluid solid volume fractions inside the enclosed space when Sf=5.7, Ha = 30, and Rd = 0.4. The flow pattern displays clockwise and counter-clockwise cells with oval nuclei on the right and left sides of the cavity, orderly. The formation of these cells results from the buoyancy force caused by temperature differences inside the cavity. As observable, the addition of nanoparticles to base fluid causes a diminution in the severity and velocity of cells inside the cavity. In other words, the strength of streamlines becomes gradually weak with an increment in nanoparticles' volume fraction. Moreover, the size of the cells degenerates with improving values of volume fraction. Temperature lines are highly intense in areas close to hot walls that such intensity and temperature gradient decrease by elevating solid volume fraction. The primary reason for temperature gradient reduction with increasing solid volume fraction is an increment of nanofluid thermal conductivity compared to the base fluid.

Figure 89(a)-(c) shows the change of Nu, temperature, and velocity with φ, respectively. It can be implied from Fig. 89(a) that due to the star-like geometry of the inner hot wall, the local Nusselt number profile experiences a periodic oscillation for all of the prescribed values of volume fraction. Also, Nu rises when solid concentration improves. In point of fact, enhancing φ increases the nanofluid's thermal conductivity, leading the conduction heat transfer to intensify. On the other hand, the enhancement of φ escalates the heat transfer convection mechanism. However, the impact of increasing the convection mode is more dominant than reducing the conduction mechanism. As a result, the Nusselt number augments by enhancing the volume fraction. Additionally, an increment in the nanofluid volumetric fraction leads the heat transfer rate to improve. This phenomenon can be elucidated by the fact that the coefficient of nanofluid's thermal expansion increases by an enhancement in volumetric fraction. By augmenting the heat transfer rate, there will be an increment in forced convective heat transfer which can be noticed in Fig. 89(a) in the format of Nu. As 6% of the nanoparticles are added to the fluid, the Nu boosts up to 26%.

Fig. 89(b) displays the temperature profile against the vertical average position. The temperature is zero at the cold outer surface while it approaches its maximum magnitude at the inner hot wall at a given volumetric fraction. According to this figure, increasing volume fraction reduces the temperature at a specific vertical mean position. As a matter of fact, the viscosity increases through growing the volume percentage of particles, leading the flow temperature to lessen. Fig. 89(c) indicates that vertical velocity declines when the nanoparticles' volume fraction grows on both sides of the heated wavy cavity. As φ

ascends, the nanofluid viscosity enhances. Owing to the higher viscosity, there would be an increment in the fluid resistance, causing the velocity of the working mixture nanofluid to lessen. The velocity remains zero at the inner and outer walls due to the no-slip condition but reaches the maximum value between them.

Figure 88. Flowlines (left) and temperature lines (right) for two different quantities of φ

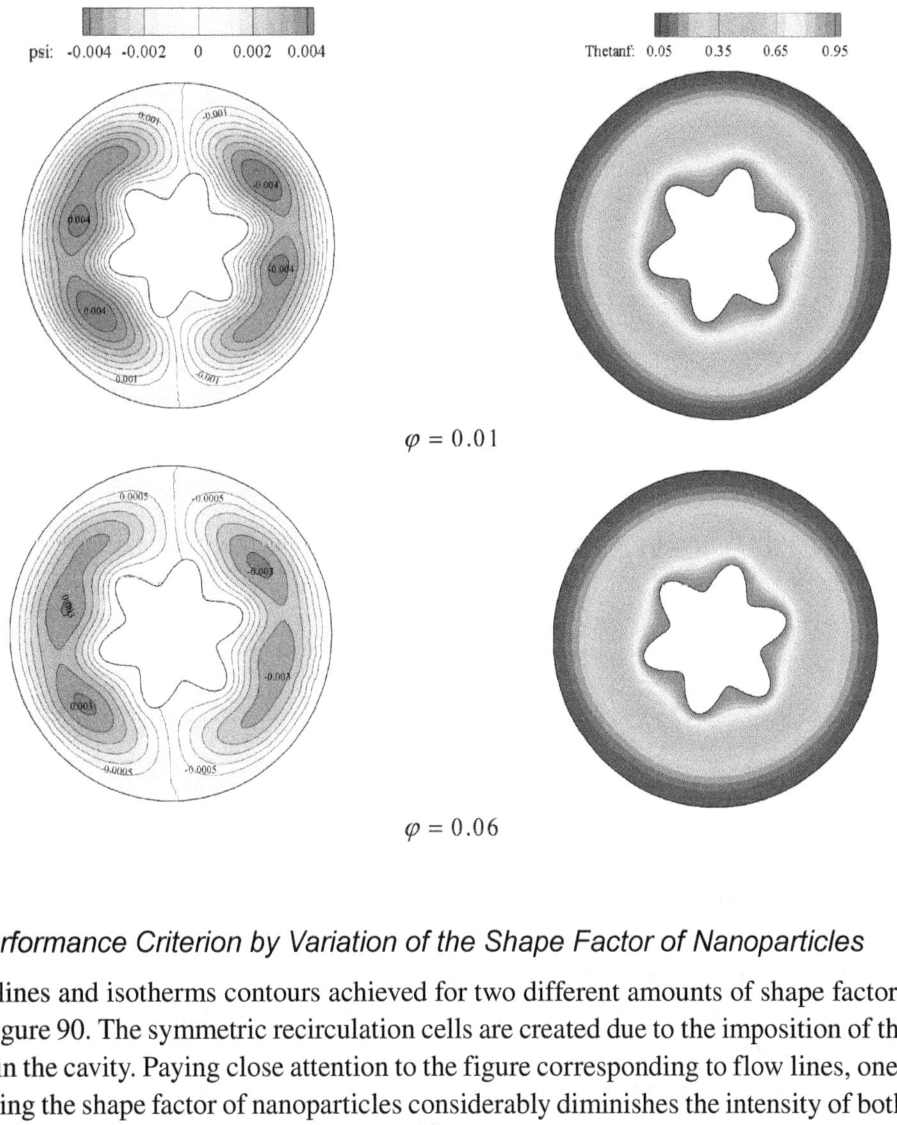

6.3.1.2 Performance Criterion by Variation of the Shape Factor of Nanoparticles

The streamlines and isotherms contours achieved for two different amounts of shape factors are represented in Figure 90. The symmetric recirculation cells are created due to the imposition of the boundary conditions in the cavity. Paying close attention to the figure corresponding to flow lines, one could infer that enhancing the shape factor of nanoparticles considerably diminishes the intensity of both clockwise and counter-clockwise recirculation cells inside the cavity. For the lower quantity of the shape factor, the buoyancy effect is much more robust, leading more intensive loops in the enclosure to appear in the plot and boluses size to augments. It is observed in the temperature lines that elevating the shape factor of nanoparticles causes a relative reduction of isotherm lines, and as a result, temperature difference diminution near the inner hot wall. Therefore, it could be expected that with regard to the temperature gradient decrement by enhancing the nanofluid shape factor, heat transfer inside the cavity also moderately declines, owing to the increment of the conduction mechanism relative to convection.

Figure 89. The change of (a) Nu at the inner heated surface of enclosure, (b) temperature, and (c) velocity with vertically average location with respect to nanoparticles volume fraction when Sf=5.7, Ha = 30, and Rd = 0.4

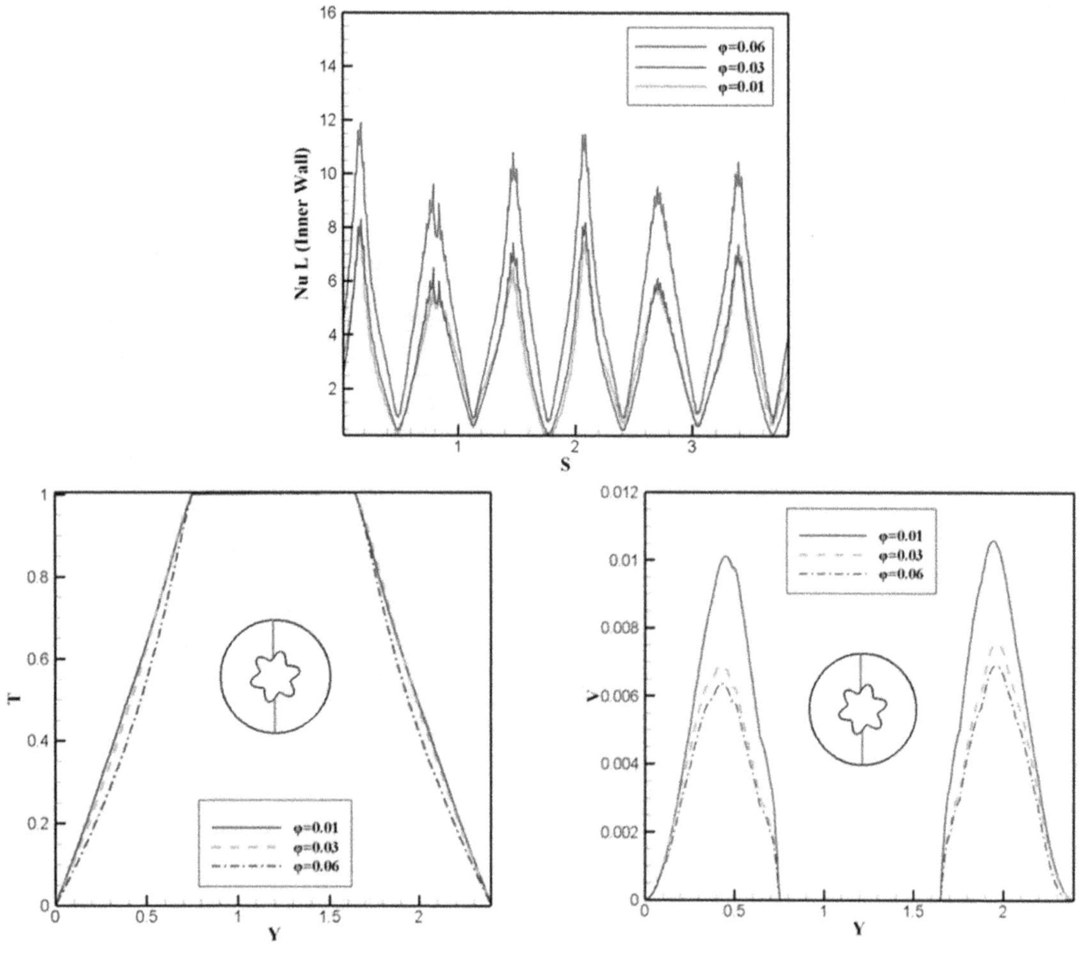

Figure 91(a)-(c) illustrates the Nu, temperature, and velocity along vertically average distance at several magnitudes of nanoparticles' shape factor, respectively. As shown from the figure, the local Nusselt number profile undergoes a periodic oscillation for all of the selected magnitudes of shape factor, owing to the geometry of the inner hot wall. Also, the Sf has a considerable effect on the heat transfer characteristics such that the Nu rises with enhancing the shape factor. As a matter of fact, the thermal conductivity enhances with the growth of the shape factor, leading to heat transfer performance improvement. Furthermore, it can be concluded that improving the shape factor attenuates both temperature and velocity. However, the velocity increment is much more significant than that of temperature. As can be observed, by increasing the shape factor of the nanoparticles from 3 to 16.5, the temperature and velocity diminish by 4% and 53%, respectively, when the vertical mean position is 0.5.

Figure 90. Flow lines (left) and temperature lines (right) for two different quantities of Sf

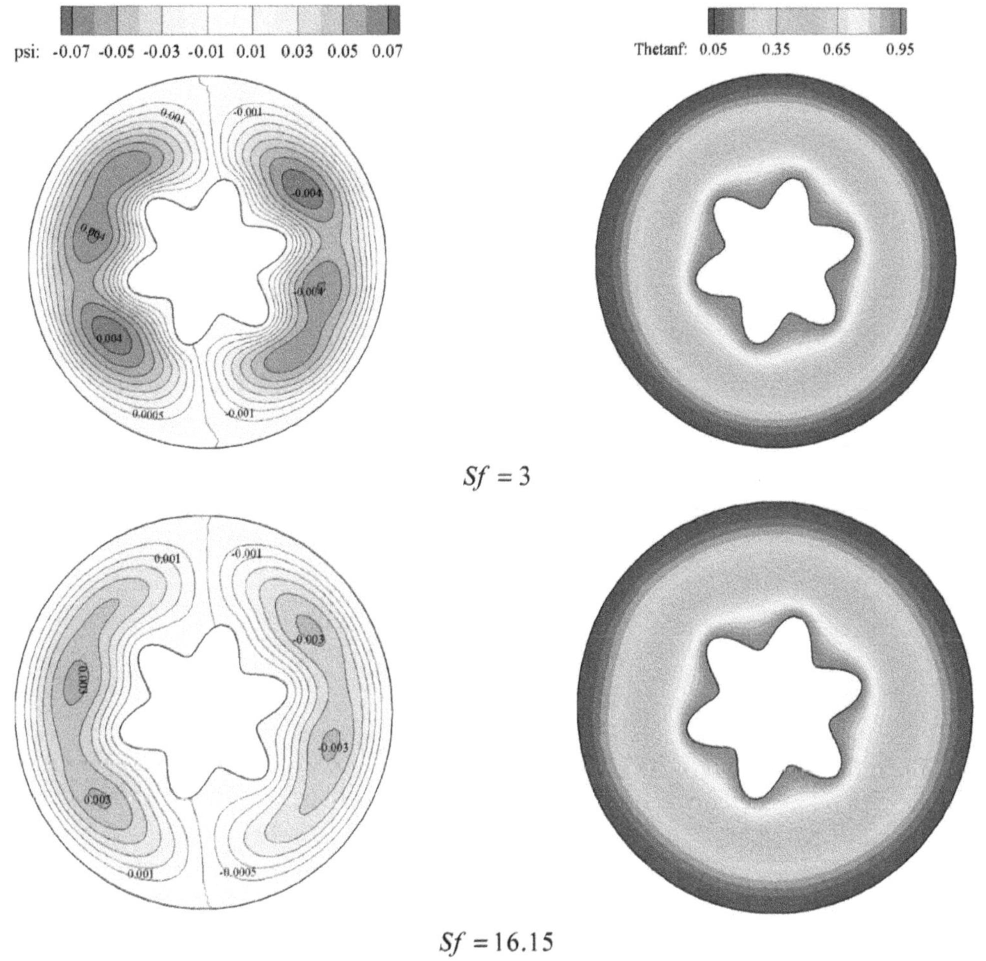

6.3.1.3 Performance Criterion by Variation of Ha

Figure 92 exhibits the lines of the flow and temperature in the presence/absence of the magnetic field. As observable, the presence/absence of a local magnetic field results in a different pattern of flow, especially in the center of the loops. It can also be deduced from the flow lines that by increasing the magnetic field effects with the presence of nanoparticles in the cavity, the severity and strength of both clockwise and counter-clockwise cells significantly degenerate, but their size becomes extensive. On the other hand, temperature contours are not affected considerably for the investigated range of Hartmann number. The highest temperatures occurred adjacent to the hot inner wavy wall in all cases. Lines of isotherm imply that the predominant heat transfer mechanism is force convection caused due to the heated mode of the inner wall. In contrast, free convection has a minor role in the enclosure.

The distributions of Nu, temperature, and velocity along the vertical mean position, taking into account the effect of the magnetic field parameter, are demonstrated in Fig. 93(a)-(c), respectively. Regarding the figure, the local Nusselt number varies in a periodic manner, which depends on the star-like geometry of the inner wall. Furthermore, an increment in the Ha results in weakening the heat and mass transfer.

Figure 91. The change of (a) Nu at the inner heated surface of enclosure, (b) temperature, and (c) velocity with vertically average location with respect to nanoparticles shape factor when φ=0.03, Ha=30, and Rd=0.4

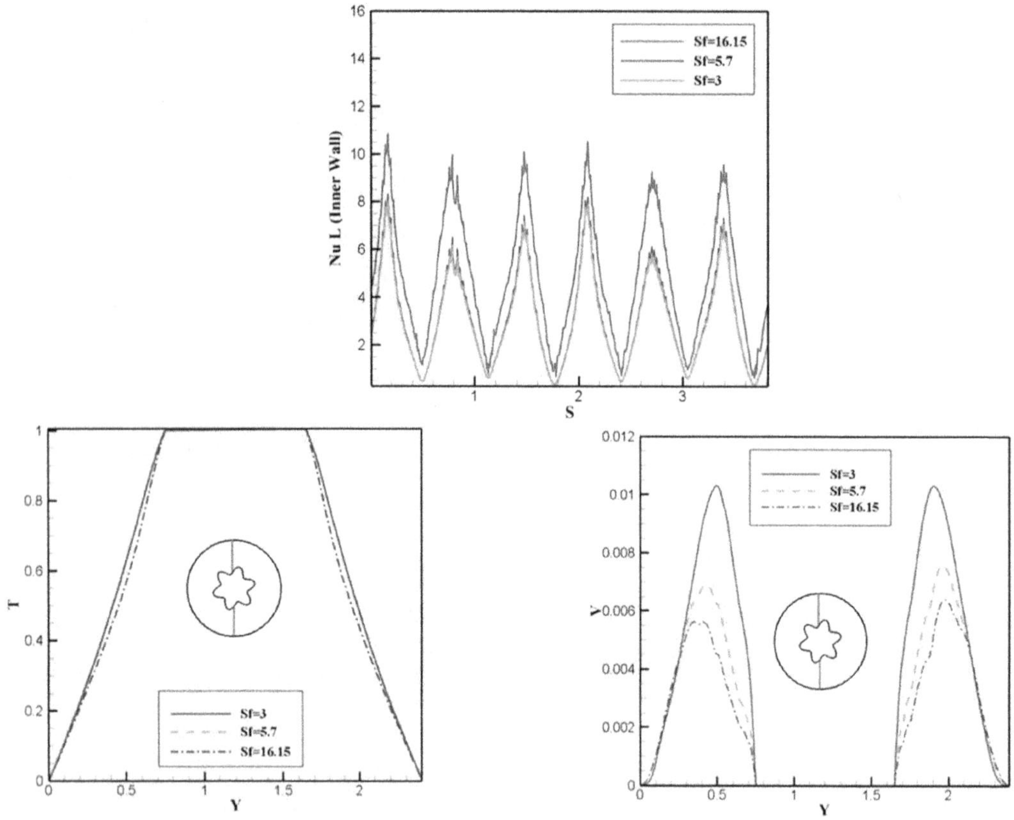

Accordingly, the convective flow and local Nusselt number considerably diminish by increasing the Ha since the Lorentz force resists the fluid flow. Increasing the Hartmann number also debilitates temperature and velocity, which is substantial for the velocity and subtle for temperature. It can be interpreted that although the Ha has no significant impact on heat flow, it plays an influential part in controlling fluid flow. With respect to Figure 93(a)-(c), when the vertical mean position equals 0.5, increasing Ha from 0 to 60 reduces the temperature, and velocity by 4% and, 89% respectively.

6.3.1.4 Performance Criterion by Variation of Rd

The influence of the Rd on the treatment of streamlines and isotherms inside the enclosure is portrayed in Figure 94. As shown, isotherms behavior becomes prevailing in the appearance of thermal radiation. This observation can be justified by the fact that the nanoparticles dispersed in the base fluid can better absorb and emit heat in the case that thermal radiation is taken into consideration. Therefore, the fluid inside the cavity causes higher heat transfer. It can also be observed that the flow field lines structure gets more vigorous and fills most parts of the enclosure by improving the thermal radiation parameter. As a matter of fact, considering the radiation mechanism improves the heat transfer, leading to an increment in the fluid displacement inside the cavity. Consequently, the streamlines in the presence of radiation heat transfer become much denser than that in the absence of radiation heat transfer.

Figure 92. Flow lines (left) and temperature lines (right) for two different quantities of Ha

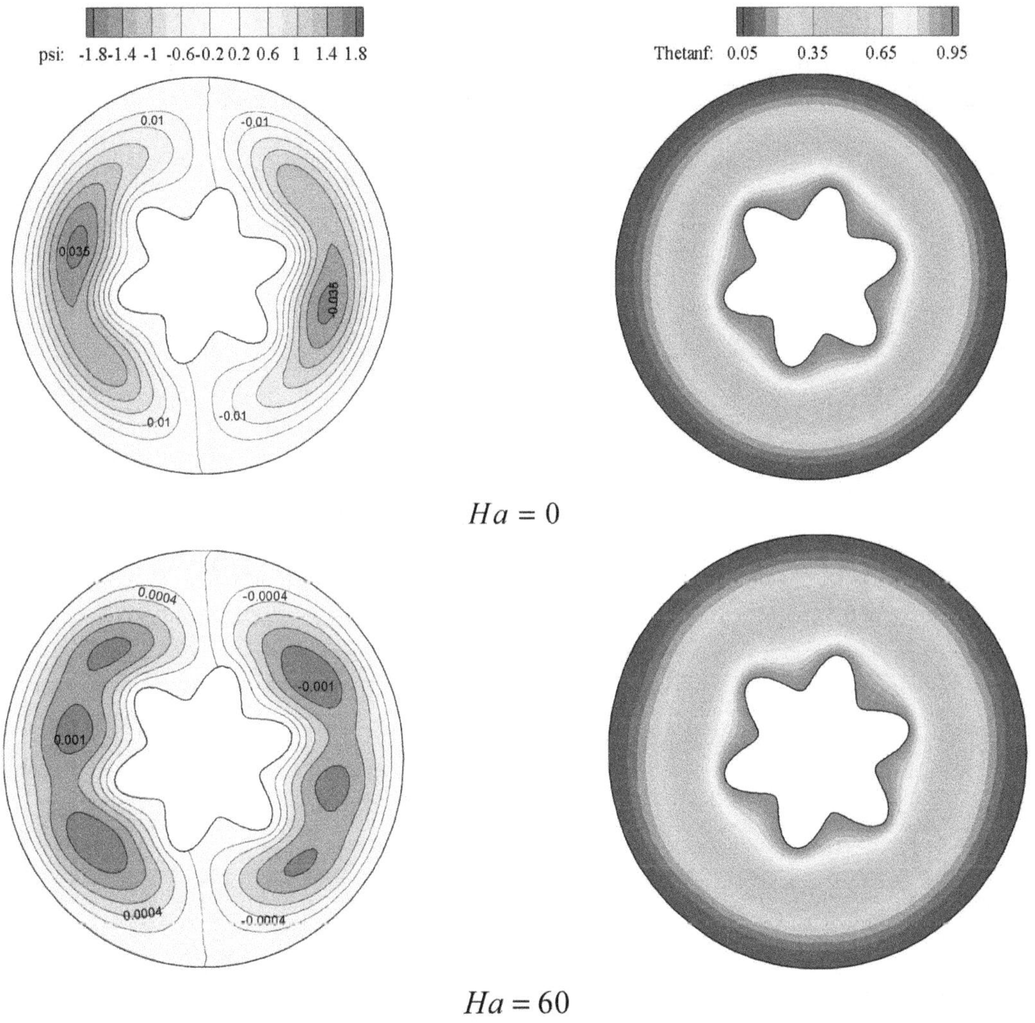

Figure 95(a)-(c) indicates the Nu, temperature, and velocity changes along the vertical mean position for φ=0.03, sf=5.7, and Ha=30 at various radiation parameters, respectively. The fluid within the enclosure causes higher heat transfer by enhancing the radiation parameter. As a matter of fact, the absorption coefficient and the heat excretion of the nanoparticles in the fluid enhance when the radiation parameter promotes. Hence, the nanoparticles can better attain the heat from the inner hot surface and transfer it to the outer cold surface, leading the thermal energy transition from both cold and hot surfaces to improve. In other words, the higher the radiation energy, the greater the rate of heat exchange. With respect to the figure, in the presence of radiation, the displacement of fluid increases, resulting in a proliferation in the fluid velocity inside the enclosure. In the area where the inner hole is placed, radiation and convection heat transfers do not exist. Thus, the temperature and velocity profiles in this region are not shown. It also can be seen that when the radiation parameter augments, the local Nusselt number boosts. Regarding the figure, when the s=1.5, as the Rd increases from 0 to 0.8, the local Nusselt number enhances to 57%. This observation further verifies the significance of the Rd for promoting heat transfer compared with other control factors.

Figure 93. The change of (a) Nu at the inner heated surface of enclosure, (b) temperature, and (c) velocity with vertically average location with respect to Ha when φ=0.03, sf=5.7, and Rd=0.4

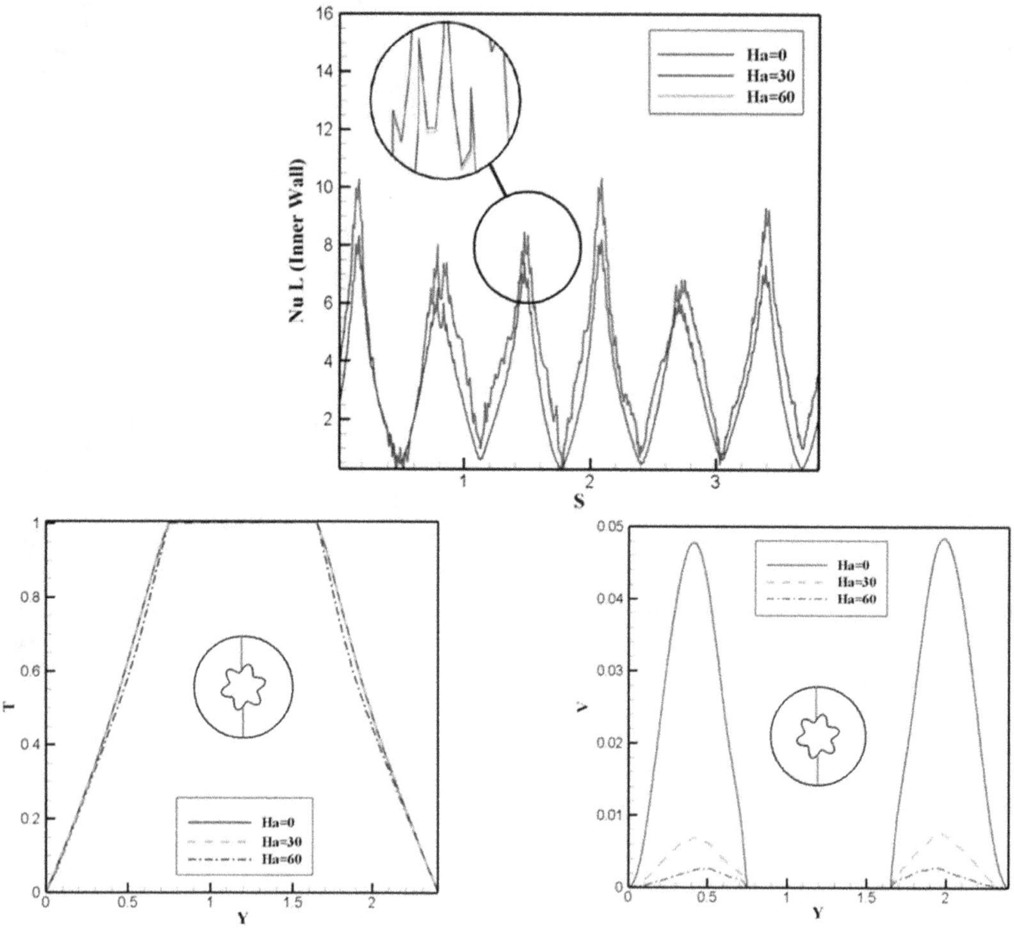

6.3.1.5 Effect of Control Factors on Nu_{ave}

Fig. 96 portrays the influences of crucial non-dimensional parameters including Ha, Rd, Sf, and φ on the $Nua_{ve.}$ Regarding Fig. 96(a), one can observe that the Nua_{ve} steadily lessens with an increase of Ha while remarkably augments with an increment of Rd. As a result, the radiation parameter is more effective than the magnetic field parameter in controlling the heat transfer behavior of the nanofluid inside the enclosure. As the magnetic field strength increases, the convection flow component of the fluid debilitates due to the function of the Lorentz force. Also, increasing Ha amplifies the conduction heat transfer inside the enclosure, which has a reducing effect on the $Nua_{ve.}$ Hence, magnetic field intensity gives a drop to the average Nusselt number. Furthermore, as the Rd grows, the nanoparticles' heat gain and excretion coefficients rise, resulting in enhanced heat transfer rate and $Nua_{ve.}$ According to Fig. 96(a), the Nua_{ve} reaches its highest value when the magnetic field is not considered, and the radiation parameter equals 0.8. In Fig. 96(b), a change of Nua_{ve} can be seen when simultaneous variation is provided to both nanoparticles' shape factor and volume fraction. As shown, the Nua_{ve} improves with a steep slope as the shape factor and volume percentage of nanoparticles increase. The reason for this can be elucidated by the fact that there would be an improvement in the buoyancy effect and the flow circulation in the enclosure by increasing the φ and Sf. The maximum $Nuav_{e\,o}$ccurs at the shape factor of 16.15 and the volume fraction of 0.06, which are their corresponding highest values.

Figure 94. Flow lines (left) and temperature lines (right) for two different quantities of Rd

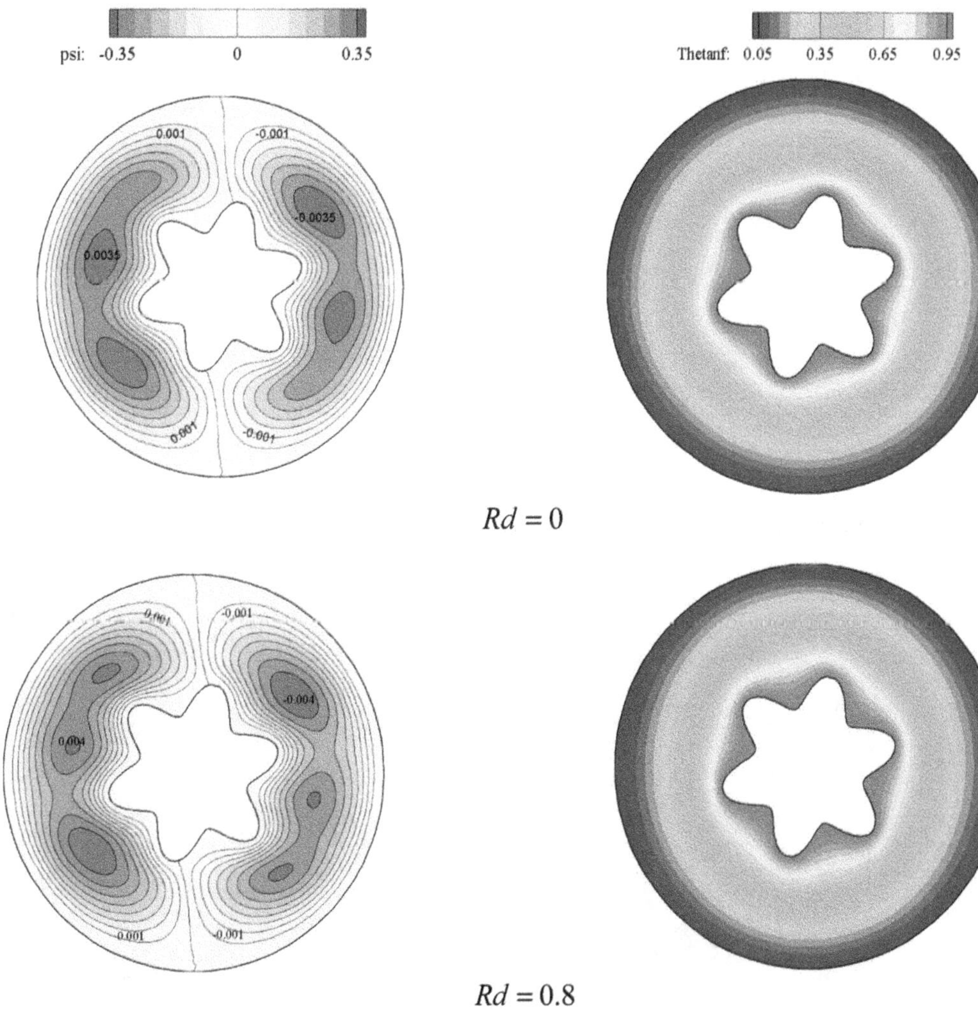

Figure 95. The change of (a) local Nusselt number at the inner heated surface of enclosure, (b) temperature, and (c) velocity with vertically average location with respect to radiation parameter when φ=0.03, sf=5.7, and Ha=30

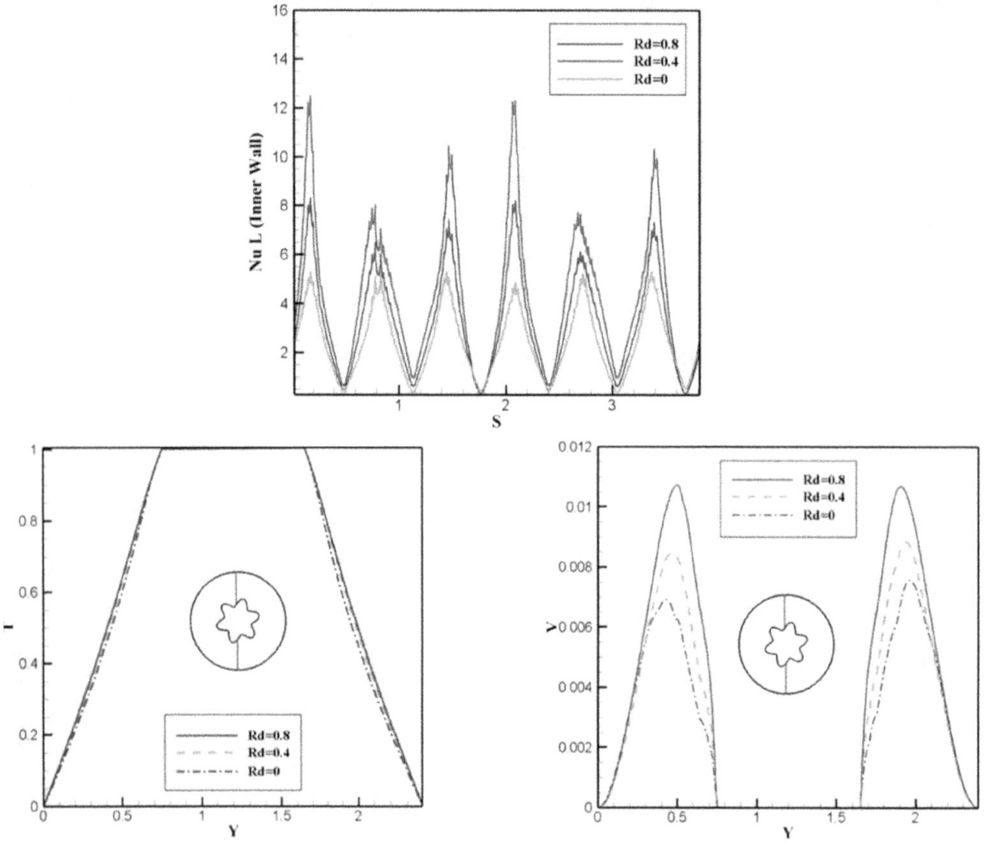

Figure 96. 3-D diagrams of Nu_{ave} changes as a function of (a) Ha and Rd, and (b) Sf and φ

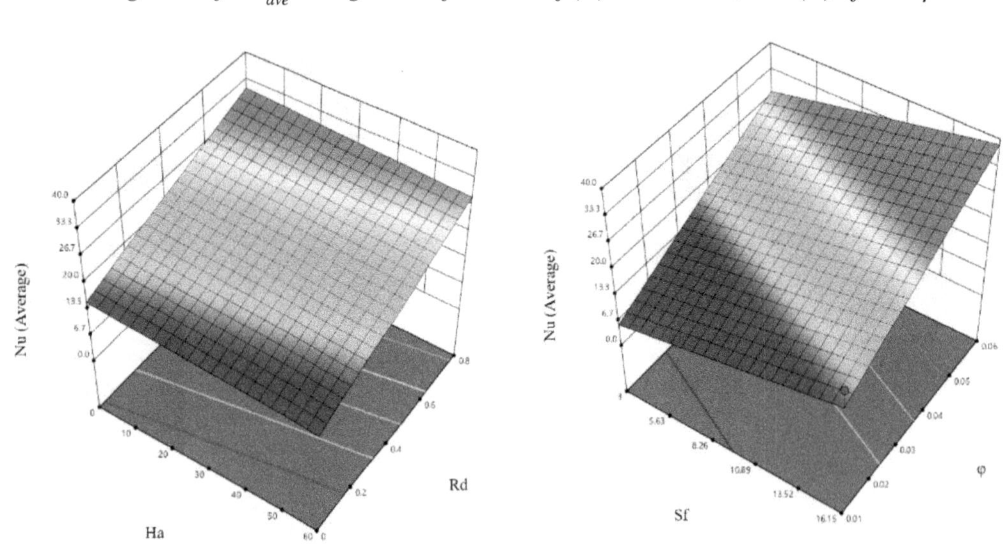

6.3.2 Optimization and Derivation an Nu_{ave} Formula

In general, the objective of the optimization is to attain the best design corresponding to any given criteria or constraints (Can et al., 2022; Pimenov et al., 2022; Said, Le, Sharma et al, 2022; Said, Nguyen, Sharma et al, 2022). In order to optimize the engineering tests of the design procedure, the effective Taguchi method is chosen. This technique not only can reduce the cost and time consumption of simulations by providing an experimental plan for the examination but also can determine the best level of each design variable. The Taguchi technique incorporates statistical and mathematical methods regarding orthogonal arrays and signal-to-noise ratio (SNR) to assess the principal and interaction effects of the control parameters (Taguchi, 1987). The orthogonal arrays are employed by the Taguchi approach to expedite the trial design procedure by virtue of decreasing the number of tests rather than all the feasible cases. Furthermore, the signal-to-noise ratio (SNR) quantifies the performance variability of the control parameters in the Taguchi technique. In order to perform the Taguchi approach, firstly, the optimization functions, control parameters, and their levels are determined. Then, an appropriate orthogonal array is chosen and the tests are carried out. Ultimately, the factors corresponding to optimal conditions are obtained. Fulfilling the maximum heat transfer rate by an optimum condition is the feasible advantage of the optimization approach implemented in the present study. The Taguchi optimization strategy consists of system design, parameter, and tolerance designs. The studied variables comprise Ha, Rd, φ, and Sf affecting Nu_{ave}, which is considered as the performance parameter. Each aforementioned design parameter is studied at three different levels. It should be noted that for the present variables and their corresponding levels, a complete factorial analysis requires 81 trials. Meanwhile, the application of Taguchi analysis diminishes the number of tests to nine. The information about the influential variables in the nine proposed samples is tabulated in Table 19.

Table 19. The trials designed by Taguchi technique

Case Number	Hartman Number	Radiation	Shape Factor	Volume Fraction
1	30	0.8	3	0.03
2	60	0.8	5.7	0.01
3	60	0	16.15	0.03
4	30	0.4	16.15	0.01
5	0	0.4	5.7	0.03
6	0	0	3	0.01
7	60	0.4	3	0.06
8	0	0.8	16.15	0.06
9	30	0	5.7	0.06

The employed Taguchi design method aims to augment heat transfer in this research. Hence, response surface methodology (RSM) is applied to determine the maximum magnitudes of the Nu_{ave}. Acquiring a formula to assess the influence of design factors on the objective function is substantial in RSM. The following correlation derived from the simulation data under optimum conditions demonstrates the relationship for an accurate estimation of the Nu_{ave} as the function of the considered operating factors.

$$Nu_{ave} = -7.835 - 0.0426(Ha) + 19.804(Rd) + 0.8079(Sf) + 447.21(\varphi) \tag{22}$$

Table 20 lists the optimum values of the control parameters, which can be applied to Equation 103 to obtain the maximum Nu_{ave}.

Table 20. The results achieved from optimization process by RSM

Parameter	Optimum Value
Ha	0
Rd	0.8
Sf	16.15
φ	0.06

6.3.2.1 Analysis of Variance (ANOVA)

In order to reveal the relative significance of the involved variables on the final response, ANOVA is employed to the results of the Taguchi technique. All the means are calculated and a hypothesis is formulated to implement the ANOVA method. Subsequently, the sum of squares (SS), Degrees of Freedom (DF), the Mean Squares, and the F Statistic are calculated. Finally, a p-value can be derived using the F-Statistic. The main outputs obtained through ANOVA for Nu_{ave} are given in Table 21, consisting of the sum of squares (SS), degree of freedom (DF), mean square, F ratio, and P ratio. The weightage of each control factor for the performance parameter is given under the F-value column, whereas P-values denote the probability used in this analysis. The larger F-value shows that the change of the control parameter leads to a considerable variation in the output response. In addition, it can be witnessed that the R^2 value is 0.8579, which confirms that there is suitable compatibility between the estimations obtained by the correlated equation and the numerical data under the defined optimal combination for the average Nusselt number.

Table 21. The outcomes of ANOVA

Source	Sum of Squares	df	Mean Square	F-Value	P-Value	
Model	1335.22	4	333.80	6.04	0.0549	significant
A-Ha	9.80	1	9.80	0.1773	0.6954	
B-Rd	376.52	1	376.52	6.81	0.0595	
C-Sf	188.91	1	188.91	3.42	0.1383	
D-φ	759.99	1	759.99	13.74	0.0207	
R^2=0.8579						
Adequate precision= 8.8085						

Figure 97 also compares predicted optimal results and actual data obtained by numerical simulation. Scrutiny of this figure leads one to conclude that the precision between the estimated magnitude and numerical simulation of the Nu_{ave} for the optimal combination is satisfactory, with a maximum error of 11%. Fig. 98 shows the variation of desirability as a function of considered controlling factors. As can be seen, increasing both Radiation parameter and shape factor enhances the desirability. Furthermore, desirability augments by increasing the nanoparticles' volume fraction and decreasing the Hartman number. Interval estimates can be attached to the optimization plot to demonstrate uncertainty in the point predictions. Interval estimates can also aid comprehend the effect of uncertainty on attaining process objectives.

Figure 97. Predicted optimal results versus actual data obtained for the Nu_{ave}

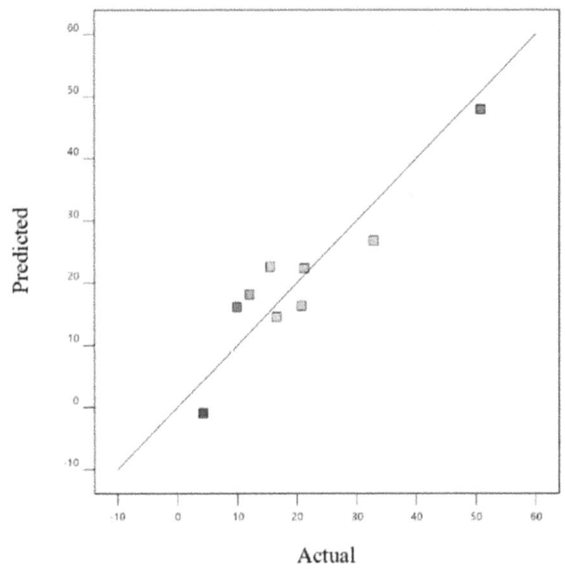

Figure 98. The effect of controlling parameters on the desirability

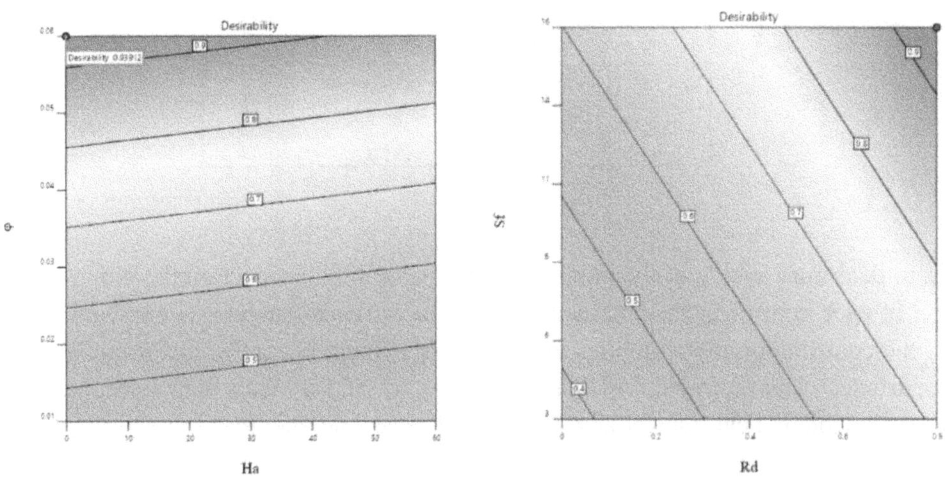

In Figure 99, the bright yellow zone indicates that the entire range of all intervals meets the prescribed criteria. On the other hand, the dark gold zone corresponds to where the point estimate satisfies the criteria requirements, but the portion of an interval estimate does not.

Figure 99. Overlay plots

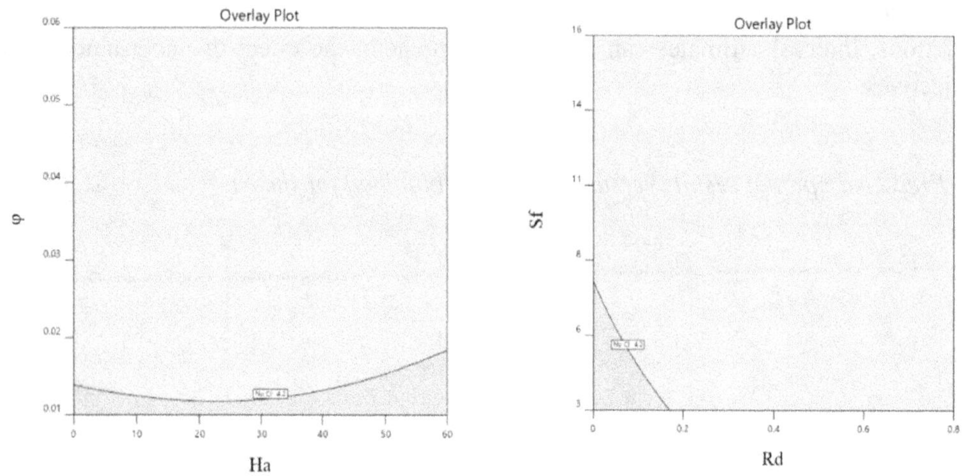

Figure 100. Contribution percentage of each influencing factor on the Nu_{ave}

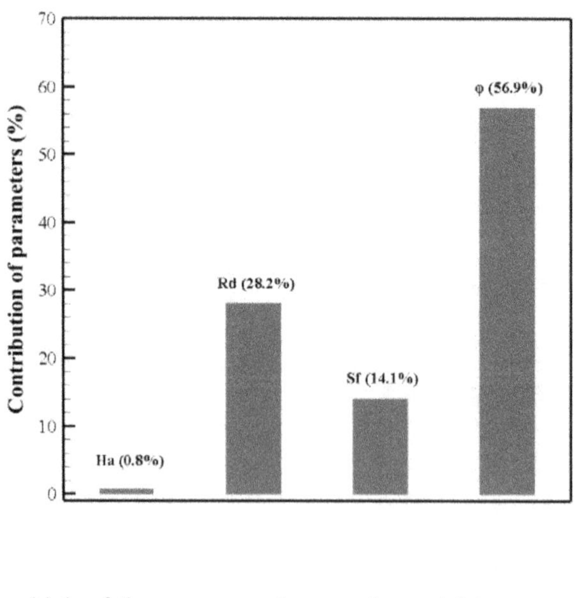

In order to determine which of the parameter has a substantial impact on the Nu_{ave}, the percentage contribution of each control factor obtained through the variance analysis is depicted in Figure 100. Scrutiny of the contribution ratios implies that the most influential control factor on the Nu_{ave} belongs to the nanoparticles' volume fraction with a contribution of 56.9%. Thus, nanoparticles' volume fraction plays a major role in heat transfer augmentation. The second and third most influential factors are the radiation parameter and the shape factor of nanoparticles, with a contribution ratio of 28.2% and 14.1%, orderly. Ultimately, the Hartmann number has the weakest percentage contribution (i.e., 0.8%). The results found from *ANOVA* can give insight into the design of thermal energy systems.

REFERENCES

Abderrahmane, A., Manoongam, A., Alizadeh, A. A., Younis, O., Zekri, H., Isa, S. S. P. M., Baghaei, S., Jamshed, W., & Guedri, K. (2023). Investigation of the free convection of nanofluid flow in a wavy porous enclosure subjected to a magnetic field using the Galerkin finite element method. *Journal of Magnetism and Magnetic Materials*, *569*, 170446. doi:10.1016/j.jmmm.2023.170446

Alkanhal, T. A., Sheikholeslami, M., Usman, M., Haq, R. U., Shafee, A., Al-Ahmadi, A. S., & Tlili, I. (2019). Thermal management of MHD nanofluid within the porous medium enclosed in a wavy shaped cavity with square obstacle in the presence of radiation heat source. *International Journal of Heat and Mass Transfer*, *139*, 87–94. doi:10.1016/j.ijheatmasstransfer.2019.05.006

Barman, P., & Rao, P. (2021). Effect of aspect ratio on natural convection in a wavy porous cavity submitted to a partial heat source. *International Communications in Heat and Mass Transfer*, *126*, 105453. doi:10.1016/j.icheatmasstransfer.2021.105453

Barnoon, P., Toghraie, D., Dehkordi, R. B., & Afrand, M. (2019). Two phase natural convection and thermal radiation of Non-Newtonian nanofluid in a porous cavity considering inclined cavity and size of inside cylinders. *International Communications in Heat and Mass Transfer*, *108*, 104285. doi:10.1016/j.icheatmasstransfer.2019.104285

Biswas, N., Mondal, M. K., Mandal, D. K., Manna, N. K., Gorla, R. S. R., & Chamkha, A. J. (2022). A narrative loom of hybrid nanofluid-filled wavy walled tilted porous enclosure imposing a partially active magnetic field. *International Journal of Mechanical Sciences*, *217*, 107028. doi:10.1016/j.ijmecsci.2021.107028

Buongiorno, J. (2006). *Convective transport in nanofluids*. Academic Press.

Can, A., Selimefendigil, F., & Öztop, H. F. (2022). A review on soft computing and nanofluid applications for battery thermal management. *Journal of Energy Storage*, *53*, 105214. doi:10.1016/j.est.2022.105214

Ghasemi, B., & Aminossadati, S. M. (2009). Natural convection heat transfer in an inclined enclosure filled with a water-CuO nanofluid. *Numerical Heat Transfer Part A*, *55*(8), 807–823. doi:10.1080/10407780902864623

Haq, R. U., Soomro, F. A., Hammouch, Z., & Rehman, S. U. (2018). Heat exchange within the partially heated C-shape cavity filled with the water based SWCNTs. *International Journal of Heat and Mass Transfer*, *127*, 506–514. doi:10.1016/j.ijheatmasstransfer.2018.07.101

Hosseinzadeh, K., Moghaddam, M. E., Asadi, A., Mogharrebi, A. R., & Ganji, D. D. (2020). Effect of internal fins along with hybrid nano-particles on solid process in star shape triplex latent heat thermal energy storage system by numerical simulation. *Renewable Energy*, *154*, 497–507. doi:10.1016/j.renene.2020.03.054

Hosseinzadeh, K., Mogharrebi, A. R., Asadi, A., Paikar, M., & Ganji, D. D. (2020). Effect of fin and hybrid nano-particles on solid process in hexagonal triplex latent heat thermal energy storage system. *Journal of Molecular Liquids, 300*, 112347. doi:10.1016/j.molliq.2019.112347

Khanafer, K., Vafai, K., & Lightstone, M. (2003). Buoyancy-driven heat transfer enhancement in a two-dimensional enclosure utilizing nanofluids. *International Journal of Heat and Mass Transfer, 46*(19), 3639–3653. doi:10.1016/S0017-9310(03)00156-X

Kuznetsov, A. V., & Nield, D. (2010). The onset of double-diffusive nanofluid convection in a layer of a saturated porous medium. *Transport in Porous Media, 85*(3), 941–951. doi:10.100711242-010-9600-1

Luo, X. H., Li, B. W., & Hu, Z. M. (2016). Effects of thermal radiation on MHD flow and heat transfer in a cubic cavity. *International Journal of Heat and Mass Transfer, 92*, 449–466. doi:10.1016/j.ijheatmasstransfer.2015.08.104

Mondal, P., & Mahapatra, T. R. (2021). MHD double-diffusive mixed convection and entropy generation of nanofluid in a trapezoidal cavity. *International Journal of Mechanical Sciences, 208*, 106665. doi:10.1016/j.ijmecsci.2021.106665

Nield, D. A., & Kuznetsov, A. V. (2011). The Cheng–Minkowycz problem for the double-diffusive natural convective boundary layer flow in a porous medium saturated by a nanofluid. *International Journal of Heat and Mass Transfer, 54*(1-3), 374–378. doi:10.1016/j.ijheatmasstransfer.2010.09.034

Pimenov, D. Y., Mia, M., Gupta, M. K., Machado, Á. R., Pintaude, G., Unune, D. R., Khanna, N., Khan, A. M., Tomaz, Í., Wojciechowski, S., & Kuntoğlu, M. (2022). Resource saving by optimization and machining environments for sustainable manufacturing: A review and future prospects. *Renewable & Sustainable Energy Reviews, 166*, 112660. doi:10.1016/j.rser.2022.112660

Qi, C., Liang, L., & Rao, Z. (2016). Study on the flow and heat transfer of liquid metal based nanofluid with different nanoparticle radiuses using two-phase lattice Boltzmann method. *International Journal of Heat and Mass Transfer, 94*, 316–326. doi:10.1016/j.ijheatmasstransfer.2015.11.068

Qi, C., Yang, L., & Wang, G. (2017). Numerical study on convective heat transfer enhancement in horizontal rectangle enclosures filled with Ag-Ga nanofluid. *Nanoscale Research Letters, 12*(1), 1–10. doi:10.118611671-017-2095-8 PMID:28476081

Said, Z., Le, D. T. N., Sharma, P., Dang, V. H., Le, H. S., Nguyen, D. T., Bui, T. A. E., & Nguyen, V. G. (2022). Optimization of combustion, performance, and emission characteristics of a dual-fuel diesel engine powered with microalgae-based biodiesel/diesel blends and oxyhydrogen. *Fuel, 326*, 124987. doi:10.1016/j.fuel.2022.124987

Said, Z., Nguyen, T. H., Sharma, P., Li, C., Ali, H. M., Ahmed, S. F., ... Truong, T. H. (2022). Multi-attribute optimization of sustainable aviation fuel production-process from microalgae source. *Fuel, 324*, 124759. doi:10.1016/j.fuel.2022.124759

Saleh, H., Roslan, R., & Hashim, I. (2011). Natural convection heat transfer in a nanofluid-filled trapezoidal enclosure. *International Journal of Heat and Mass Transfer, 54*(1-3), 194–201. doi:10.1016/j.ijheatmasstransfer.2010.09.053

Sheikholeslami, M. (2019). Numerical approach for MHD Al2O3-water nanofluid transportation inside a permeable medium using innovative computer method. *Computer Methods in Applied Mechanics and Engineering*, *344*, 306–318. doi:10.1016/j.cma.2018.09.042

Shekaramiz, M., Fathi, S., Ataabadi, H. A., Kazemi-Varnamkhasti, H., & Toghraie, D. (2021). MHD nanofluid free convection inside the wavy triangular cavity considering periodic temperature boundary condition and velocity slip mechanisms. *International Journal of Thermal Sciences*, *170*, 107179. doi:10.1016/j.ijthermalsci.2021.107179

Sheremet, M. A., Cimpean, D. S., & Pop, I. (2017). Free convection in a partially heated wavy porous cavity filled with a nanofluid under the effects of Brownian diffusion and thermophoresis. *Applied Thermal Engineering*, *113*, 413–418. doi:10.1016/j.applthermaleng.2016.11.033

Sheremet, M. A., & Pop, I. (2014). Conjugate natural convection in a square porous cavity filled by a nanofluid using Buongiorno's mathematical model. *International Journal of Heat and Mass Transfer*, *79*, 137–145. doi:10.1016/j.ijheatmasstransfer.2014.07.092

Shulepova, E. V., Sheremet, M. A., Oztop, H. F., & Abu-Hamdeh, N. (2020). Mixed convection of Al2O3–H2O nanoliquid in a square chamber with complicated fin. *International Journal of Mechanical Sciences*, *165*, 105192. doi:10.1016/j.ijmecsci.2019.105192

Taguchi, G. (1986). *Introduction to quality engineering: designing quality into products and processes*. Academic Press.

Taguchi, G. (1987). *Quality Resources*. Taguchi Techniques for Quality Engineering.

Taguchi, G. (1990). *Introduction to Quality Engineering*. Asian Productivity Organization.

Venter, G., Haftka, R., & Starnes, J. Jr. (1996, August). Construction of response surfaces for design optimization applications. In *6th Symposium on Multidisciplinary Analysis and Optimization* (p. 4040).10.2514/6.1996-4040

Wolff, F., Beckermann, C., & Viskanta, R. (1988). Natural convection of liquid metals in vertical cavities. *Experimental Thermal and Fluid Science*, *1*(1), 83–91. doi:10.1016/0894-1777(88)90051-9

Chapter 6
Introduction to Phase Change Materials

ABSTRACT

The topic of energy storage is crucial because of the rising demand for energy, the finite nature of fossil fuels, the rise in environmental contaminants, and other factors. Phase change material (PCM) latent heat storage systems are an efficient energy storage method that has attracted a lot of attention lately. The introduction of phase change materials and the relations between them for thermal energy storage have been covered in this chapter. The economic rationale behind the usage of these materials is equally as significant as the environmental, and technological concerns will come after these materials undergo phase shift, storing thermal energy and releasing it as needed.

INTRODUCTION

The need to embrace renewable energy sources is growing stronger every day as greenhouse gas emissions and gasoline prices rise. Direct sun radiation has long been seen as a readily available energy source in most places of the world. Globally, scientists are searching for new, sustainable energy sources. One of the things that is nearly as vital as the development of new energy sources is the creation of energy storage devices. The challenge facing technology today is storing energy in a way that makes sense and can be regularly transformed into the needed form. In addition to narrowing the gap between supply and demand for energy, energy storage also improves the efficiency and dependability of systems that generate energy and is crucial to energy storage. These devices save energy and money, enabling the storage of precious fuel. Phase change material (PCM) storage is one method of thermal energy storage. These materials store thermal energy when they transition from a solid to a liquid state or the other way around. These materials either absorb heat from the environment or return it to it as they phase shift from solid to liquid or from liquid to solid. Even after thousands of phase change cycles, the phase change material is still able to hold onto this latent heat energy. Phase change materials (PCM) have drawn a lot of attention recently as one of the energy storage technologies because of their high energy storage capacity and ability to store energy at nearly constant temperature (energy storage as heat) during a phase

DOI: 10.4018/979-8-3693-0932-2.ch006

change process (Maccarini et al., 2018). This chapter has covered a broad overview of phase change material-related subjects and presented formulas that are relevant to it.

1. THERMAL ENERGY STORAGE METHODS

Thermal energy storage, often known as heat and cold storage, is the storage of heat or cold for later use. The storage method must be reversible in order to recover heat or cold after some time. There are two types of thermal energy storage systems: 1- Sensible thermal energy storage 2- Latent thermal energy storage (Sarbu & Sebarchievici, 2018).

1.1 Sensible Thermal Energy Storage

The most prevalent method of storing heat energy is in the form of sensible heat, heat is transported to the storage environment, resulting in an increase in temperature in the storage environment. Because a sensor can detect this temperature rise, the stored heat is referred to as sensible heat. In fact, energy is stored in the storage of sensible thermal energy by the temperature fluctuations of an intermediate storage, such as water, air, oil, rock beds, bricks, sand, or soil. The quantity of energy stored by sensible heat devices is related to the difference in storage temperature between the initial and final temperatures, the mass of the intermediate storage employed, and its heat capacity (Koçak et al., 2020).

1.2 Latent Thermal Energy Storage

Latent heat is the heat transfer that occurs when a substance changes state from one phase to another. The change in latent heat is typically significantly greater than the change in perceptible heat at a given temperature, which is determined by its specific heat (Jouhara et al., 2020). For instance, when water transitions to the vapor phase. The solid-liquid phase change, or melting and freezing, can store a substantial amount of energy if the correct material is used. If the phase chamber has a larger capacity (typically liquid), the pressure does not fluctuate considerably, and the melting and freezing of the stored substance occurs at a constant temperature. When heat is delivered to the phase change material during the melting process, the material maintains its temperature at the melting temperature, which is known as the phase change temperature.

2. PHASE CHANGE MATERIAL (PCM)

Organic or inorganic substances with the potential to absorb and store huge amounts of heat energy are known as phase change materials. Thermal energy storage happens during the phase shift process (the transition from solid to liquid or vice versa) in these materials. When these materials change phases from solid to liquid or from liquid to solid, they absorb or return heat to the environment. Even after hundreds of phase change cycles, the phase change material can keep this latent heat energy without changing. When compared to sensible heat storage materials, phase transition materials have a higher density. They can absorb or release a huge amount of energy at a steady temperature and have greater thermal energy storage. Materials used to store thermal energy generated by phase transition must have a high latent heat and, depending on the application, good or low thermal conductivity. The physical and chemical properties of the phase transition materials utilized influence the design and application of stored latent heat. Based on the state change during energy intake and phase transition, PCM may be classified into three types: liquid-solid, liquid-gas, and solid-solid. The most common kind is solid-liquid phase change materials, which are classified into three types: organic compounds (paraffin and fatty acids), mineral substances (hydrate salts and metals), and eutectic substances (Fallahi et al., 2017). Eutectics are materials created by combining two or more different PCMs in such a way that their melting and freezing temperatures are harmonious, symmetrical, and non-segregated.

3. TYPES OF PHASE CHANGE MATERIALS

1. Organic materials: These types of phase change materials are made of organic and natural materials and these materials are divided into three main groups. Materials such as paraffins, fatty acids, and organic mixes possess high latent heat of fusion, are recyclable, and are stable both chemically and thermally. Additionally, they are not corrosive to metals. However, they also have drawbacks. These substances have low heat conductivity, low phase change enthalpy, and poor compatibility with polymer. They are also combustible (Kwok et al., 2010).
2. Inorganic PCM: These materials can store or release heat energy in the solid phase by using one or more energy storage techniques, such as magnetic transformations, crystallographic structure transformations, disorder transformations, and transformations between amorphous and crystalline structures (Ferrer et al., 2015). Among the advantages of this type of PCMs, we can mention high latent heat of melting in proportion to volume, being available and economical in terms of economy, being non-flammable, high thermal conductivity and rapid phase change in the melting state. The significant volume change during phase change, excessive cooling during solid-liquid transition, the necessity to add material due to corrosion after numerous cycles of melting and freezing, and the instability of their thermal potential are all disadvantages of these types of PCMs. These materials are appropriate for industrial operations that need to recover heat from heat loss (at high temperatures).
3. Eutectics: These materials include a combination of two or more types of organic-organic, inorganic-inorganic, or inorganic-organic phase change materials (Li et al., 2020). All components of eutectic materials become liquid at the same time as the melting process, and they form a crystal mixture during the freezing step (Mehling & Cabeza, 2008). These materials have the advantages of quick phase change, larger volumetric heat capacity than organic PCMs, and the ability to turn into a liquid gel in a molten state by adding a little water. The lack of information on thermal and chemical properties due to the newness of this technology is one of the downsides of these materials. In general, salt water solution with eutectic composition is used to store energy at temperatures below zero degrees Celsius.

4. PROPERTIES OF PHASE CHANGE MATERIALS (PCM)

The following properties must be present in phase change materials used in the design of thermal storage systems (Mehling & Cabeza, 2007).

- Thermal properties: To evaluate the thermal properties of PCMs, three features can be mentioned: acceptable temperature for phase change, high latent heat, and good heat transport. When selecting the proper PCM, keep its melting temperature in mind so that it falls within the desired range. For example, in many conditions of use, the heat of the day and the cold of the night are used to balance these materials; as a result, the range of their change of state should be between 10 and 30 degrees Celsius, especially in cases of use in buildings (Raj & Velraj, 2010), PCMs that melt at temperatures below 15 degrees are typically used for cooling, and materials that melt at temperatures above 90 degrees Celsius are typically used for cooling. All materials that melt between these two temperatures can be employed in general mechanisms and melt with solar heat. In summary, the physical properties include suitable phase change temperature, high latent heat of phase change and good heat transfer.
- Physical properties: To evaluate the physical properties of PCMs, four factors can be mentioned: phase stability, high density, low volume change, and low vapor pressure. The phase stability parameter aids in maintaining a controlled thermal reserve during phase changes (melting and freezing). High density (high density) allows for more storage in a smaller volume. Finally, the minor volume change and low vapor pressure eliminate any difficulties caused by volume change or evaporation of constituent materials during phase shift. In other words, the physical properties of phase change materials include favorable phase balance, high density, low volume change, low vapor pressure, renewable phase change.
- Kinetic properties: Among the most important kinetic properties, we can mention not having the problem of supercooling, having a high speed of nucleation and high speed of crystal formation. The problem of supercooling is observed especially in hydrated salts. The existence of this problem is one of the effective parameters to prevent the spread of phase change materials. This problem can be ignored in small amounts, but if this problem is between 5 and 10 degrees or more, it effectively renders PCM ineffective.
- chemical properties: To ensure that PCMs used, they should have crucial chemical features such as chemical stability, compatibility with construction materials, non-toxicity, non-corrosiveness, and non-flammability in addition to physical properties (Barreneche et al., 2014). Phase change materials should have a long life. There are two ways to describe this factor. First, they should not self-destruct after a certain period of time. Second, they must retain their state and stability in order to be economically viable due to multiple times of state transition.
- Thermodynamic properties: PCMs should have a high latent heat capacity, which allows them to store more thermal energy in a smaller volume. This property enables them to manufacture light and small covers while maintaining adequate efficiency. Furthermore, they must have a high specific heat capacity (Jelle & Kalnæs, 2017).
- Economic properties: The pricing and availability of PCMs, as well as all materials and processes in the globe, is the determining factor. These materials should be chosen in such a way that they are economically viable for use.

5. HOW PHASE CHANGE MATERIALS WORK

Phase change materials have the ability to change their state in a specific temperature range, in the sense that they keep their temperature for the duration of the state transition. In fact, these materials function to store thermal energy by being heated parallel to the environment during the heating process until they reach their melting point (phase change). Despite the fact that the temperature of the surrounding environment continues to rise after reaching this temperature, the temperature of these materials and, of course, the surrounding environment remains constant and resists the rise due to the fact that it is changing phase. During this time, which normally lasts many hours, the phase change material absorbs a huge quantity of heat from the surroundings but does not use it to raise its temperature, but instead uses the received heat to change its phase. Its phase transitions from solid to liquid while maintaining its temperature and surrounding environment constant.

6. PCM DISADVANTAGES

6.1 Thermal Conductivity

One of the reasons restricting the usage of PCMs is their low thermal conductivity, which might limit and make their use as cooling devices impracticable. Three major approaches have been proposed to address this issue: (1) constructing PCM composites with different metals or unique graphites to improve heat transmission, (2) employing metal materials in the shape of spheres, plates, propellers, and fins to improve heat transfer inside PCMs, and (3) using nano materials in PCM structure. Various types of nanocarbons (carbon nanotubes) can be employed in the structure of PCMs, as well as stable metals (gold and copper), Al2O3, CuO, metal oxides (silica, alumina, zirconium), and metal nitrides (aluminum, nitride) (Zhai et al., 2013). Although various studies have been undertaken to enhance the conductive heat transfer coefficient of PCMs, certain studies demonstrate that an excessive increase in this coefficient lowers PCM energy storage capacity. This is especially significant in applications where low heat transmission rates are required.

6.2 Phase Separation of PCMs

To clarify this problem, hydrated salts can be used as an example. These salts suffer from a decrease in their energy storage density. In this way, the repetition of the heating and cooling cycle leads to a continuous decrease in their thermal energy storage efficiency. This problem is one of the major problems of hydrated salts, which is expressed as inhomogeneity in melting or phase separation. This problem causes that the water released from the solid material during the phase change is not enough to change the phase (melting) of all the solid materials, and due to the difference in density, salts with less water or no water settle at the bottom of the container. This process is irreversible and the repetition of this process leads to the formation of hydrated salt lower than the original salt at the time of melting, so that they lose their ultimate thermal ability. This problem can also be seen in eutectic materials; Where these materials are divided into their components, it has been determined in various researches that the presence of additional water to remove anhydrous salts increases the stability of salts. But in practice, creating strong coatings to prevent contact with the outside is one of the most promising methods to prevent the reduction of the energy storage density of hydrated salts (Lane, 1992).

6.3 Super Cooling in PCM

Another issue with PCMs, particularly with hydrated salts, is supercooling, which lowers the materials' efficiency. Because of this issue, the material must be cooled to a temperature below its phase change temperature in order to revert to the solid phase. A great deal of research has been done to address this issue. Generally speaking, it can be said that agitating these substances while they melt will lessen both this issue and the decomposition issue. It has been demonstrated in numerous tests that the issue of supercooling in hydrated salts may be avoided by adding various materials with the capacity to enhance the crystalline structure and the stability of the hydrated structure of PCMs. In extreme circumstances, the supercooling issue essentially stops the material from recovering to its solid state and halts the working process. This problem can be handled to a great degree with ingredients that aid in the nucleation and preservation of the crystal of these salts (Farid & Khalaf, 1994).

6.4 Stability in PCMs

The useful life of a PCM is by far the most significant limiting factor. Put another way, the primary feature of these materials is the number of cycles they can withstand without deteriorating or changing. The two primary causes of these materials' instability may be their inherent fragility and incapacity to preserve their chemical composition, or the corrosion of PCM into the materials around it. Therefore, it is crucial that PCMs be designed with resistant materials or encapsulations to keep them isolated from their surroundings (Zalba et al., 2003).

7. APPLICATIONS OF PCMs

Thermal energy storage using PCMs has been applied in several industries and is not just limited to solar energy heating and cooling or specialized devices. The main classification of PCM applications is as follows:

- Construction applications: According to studies, using PCM in a building can lower daytime temperatures by roughly 4 degrees Celsius, albeit this difference in temperature is dependent on the surrounding air temperature. Research has concentrated on the usage of PCMs in various building components. According to a study, adding PCMs to building walls lowers the amount of outside heat that enters the structure. It can be acquired within the home (Thantong & Chantawong, 2017). In order to store heat transferred to these surfaces during the day and release it during the night, PCMs are used in the construction of the building's facade and in the solar chimneys. These systems not only lower the interior temperature during the day (the PCMs in the building's facade prevent heat from entering the house's interior components), but they also release the heat stored in these materials as the building cools down at night. It makes the building warmer. This action essentially lowers the amount of energy required to maintain the ideal building temperature (Li & Liu, 2014). In general, it can be claimed that PCMs can be employed as a thermal buffer in order to raise the minimum nighttime temperature and decrease the thermal peak during hot periods. PCMs can be chosen in the melting range of 30 to 50 degrees Celsius when energy conservation is the primary objective or unique environmental conditions apply (Faraj et al., 2020). The ideal melting temperature range for PCMs is roughly 20 to 28 degrees Celsius.

- Batteries: One type of equipment that is extremely sensitive to temperature changes is batteries, whose performance is impacted by variations in temperature. According to research by Hemri et al., PCMs placed adjacent to batteries may realistically keep the temperature of the batteries below 60 degrees Celsius even in the worst-case scenarios by preventing the temperature of the higher batteries from being transferred to the side batteries. This occurred in spite of the fact that the proximity of batteries would quickly heat up and set off a series of chain reactions if these materials were not used. Somasundaram et al. were able to ascertain the impact of PCMs on regulating the temperature of cylindrical lithium-ion batteries during discharge to galvanostatic mode (i.e., when the battery is allegedly being discharged to its maximum capacity) by a simulated numerical analysis. mention Paraffin-based PCM and a graphite matrix were utilized in this experiment. The findings demonstrated that batteries' temperatures can drop by up to 5 degrees Celsius when PCM is present (Hémery et al., 2014; Shabani & Biju, 2015).
- Domestic water heating systems: Heating residential water is one of the most energy-intensive industries in today's globe, especially during the winter months. During the winter, up to 70% of the energy consumed by houses might be used for heating and water consumption. On the other hand, one of the most effective applications of solar energy may be found in domestic water heating systems (DWHS). One disadvantage of employing this method is its inability to maintain and store thermal energy in itself. It is also unpredictable, as are other renewable systems, and there is a supply-demand imbalance. As a result, it is critical to identify techniques to improve this capacity. DWHS systems work on the principle that by placing screens against the sun during the day, they absorb heat energy into the liquid inside them and release it when needed. The system's efficiency is determined by the type of liquid and its ability to store thermal energy. Because PCMs can store thermal energy, they can be a considerable improvement for heating water throughout the day and especially at night. A great deal of study has been conducted to demonstrate these qualities (Velraj et al., 1999).
- Heating of greenhouses: Energy-saving methods are critical in all sectors, from transportation to buildings; this issue is especially critical in the greenhouse business. Because greenhouses now consume a substantial amount of energy. On the other hand, because greenhouses ensure a sustained crop, the growing need for energy in the greenhouse sector cannot be lowered. Greenhouses typically lose 20 to 40% of their energy owing to insufficient insulation. Greenhouse optimization might comprise the combination of numerous parameters. These factors can be paired with other operators in greenhouses, such as solar power generation systems, to improve efficiency, or they can be used directly for thermal management (Vieira et al., 2016).

8. RELATIONS GOVERNING PCM

Considering the abovementioned assumptions and the geometrical configuration of the system, the continuity equation is given as follows (Hosseinzadeh et al., 2020):

$$\frac{\partial u_i}{\partial x_i} = 0 \tag{1}$$

The momentum equation can be written as follows:

$$\frac{\partial u_i}{\partial t} + u_j \frac{\partial u_i}{\partial x_j} = \frac{1}{\rho}\left[\mu_{eff} \nabla^2 u_i - \frac{\partial p}{\partial x_i} + \rho \beta_e \left(T - T_{ref}\right) g_i\right] \tag{2}$$

The energy equation can be formulated as follows:

$$\frac{\partial T}{\partial t} + u_i \frac{\partial T}{\partial x_i} = \frac{\partial}{\partial x_i}\left(\frac{k}{\rho C_p}\frac{\partial T}{\partial x_i}\right) + \frac{L_f}{\rho C_p}\frac{\partial S}{\partial t} \tag{3}$$

The solid fraction equation is as follows:

$$\begin{cases} S = (T_m + \frac{T_0}{2} - T) / T_0 & (T_m - T_0) < T < (T_m + T_0) \\ S = 1 & T < (T_m - T_0) \\ S = 0 & T > (T_m + T_0) \end{cases} \tag{4}$$

Temperature is shown by T, and S, T_0, and T_m indicate the solid fraction, melting interval, and melting temperature, orderly. Due to omitting convection, the energy equation in the annuls and fins can be rewritten as follows:

Conservation of energy in PCM zone:

$$\begin{gathered} (\rho C_p)_{thnf} \frac{dT}{dt} = \nabla(k_{thnf} \nabla T) + L_{thnf} \frac{dS}{dt} - \frac{\partial q_r}{\partial y} \\ q_r = -\frac{4\sigma_e}{3\beta_R}\frac{\partial T^4}{\partial y}, \quad T_4 \cong 4T_\infty^3 T - 3T_\infty^4, \quad Rd = \frac{4\sigma_e T_\infty^3}{3\beta_R k_f} \end{gathered} \tag{5}$$

Conservation of energy in the fin zone:

$$\rho C_p \frac{dT}{dt} = \nabla(k \nabla T) \tag{6}$$

The two forms of energy considered in the current research, latent and sensible, can be calculated by the following relations:

$$E_{sensible} = mC_pT = \rho VC_pT \tag{7}$$

$$E_{latent} = m(1-S)L_f = \rho V(1-S)L_f \tag{8}$$

Also, the whole amount of energy in the computational region can be easily computed by:

$$E_{total} = E_{sensible} + E_{latent} = \rho\int\left(C_pT + (1-S)L_f\right)dV \tag{9}$$

Additionally, the intermediate PCM temperature is measured during the process by utilizing the following formula:

$$T_{ave} = \frac{\int TdA}{\int dA} \tag{10}$$

REFERENCES

Barreneche, C., Navarro, H., Serrano, S., Cabeza, L. F., & Fernández, A. I. (2014). New database on phase change materials for thermal energy storage in buildings to help PCM selection. *Energy Procedia*, *57*, 2408–2415. doi:10.1016/j.egypro.2014.10.249

Fallahi, A., Guldentops, G., Tao, M., Granados-Focil, S., & Van Dessel, S. (2017). Review on solid-solid phase change materials for thermal energy storage: Molecular structure and thermal properties. *Applied Thermal Engineering*, *127*, 1427–1441. doi:10.1016/j.applthermaleng.2017.08.161

Faraj, K., Khaled, M., Faraj, J., Hachem, F., & Castelain, C. (2020). Phase change material thermal energy storage systems for cooling applications in buildings: A review. *Renewable & Sustainable Energy Reviews*, *119*, 109579. doi:10.1016/j.rser.2019.109579

Farid, M. M., & Khalaf, A. N. (1994). Performance of direct contact latent heat storage units with two hydrated salts. *Solar Energy*, *52*(2), 179–189. doi:10.1016/0038-092X(94)90067-1

Ferrer, G., Solé, A., Barreneche, C., Martorell, I., & Cabeza, L. F. (2015). Review on the methodology used in thermal stability characterization of phase change materials. *Renewable & Sustainable Energy Reviews*, *50*, 665–685. doi:10.1016/j.rser.2015.04.187

Hémery, C.-V., Pra, F., Robin, J.-F., & Marty, P. (2014). Experimental performances of a battery thermal management system using a phase change material. *Journal of Power Sources*, *270*, 349–358. doi:10.1016/j.jpowsour.2014.07.147

Hosseinzadeh, K., Moghaddam, M. E., Asadi, A., Mogharrebi, A. R., & Ganji, D. D. (2020). Effect of internal fins along with hybrid nano-particles on solid process in star shape triplex latent heat thermal energy storage system by numerical simulation. *Renewable Energy*, *154*, 497–507. doi:10.1016/j.renene.2020.03.054

Jelle, B., & Kalnæs, S. (2017). Phase change materials for application in energy-efficient buildings. *Cost-effective energy efficient building retrofitting*, 57-118.

Jouhara, H., Żabnieńska-Góra, A., Khordehgah, N., Ahmad, D., & Lipinski, T. (2020). Latent thermal energy storage technologies and applications: A review. *International Journal of Thermofluids*, *5*, 100039. doi:10.1016/j.ijft.2020.100039

Koçak, B., Fernandez, A. I., & Paksoy, H. (2020). Review on sensible thermal energy storage for industrial solar applications and sustainability aspects. *Solar Energy*, *209*, 135–169. doi:10.1016/j.solener.2020.08.081

Kwok, K. W., Leung, K. M., Flahaut, E., Cheng, J., & Cheng, S. H. (2010). Chronic toxicity of double-walled carbon nanotubes to three marine organisms: Influence of different dispersion methods. *Nanomedicine (London)*, *5*(6), 951–961. doi:10.2217/nnm.10.59 PMID:20735229

Lane, G. A. (1992). Phase change materials for energy storage nucleation to prevent supercooling. *Solar Energy Materials and Solar Cells*, *27*(2), 135–160. doi:10.1016/0927-0248(92)90116-7

Li, D., Wu, Y., Wang, B., Liu, C., & Arıcı, M. (2020). Optical and thermal performance of glazing units containing PCM in buildings: A review. *Construction & Building Materials*, *233*, 117327. doi:10.1016/j.conbuildmat.2019.117327

Li, Y., & Liu, S. (2014). Experimental study on thermal performance of a solar chimney combined with PCM. *Applied Energy*, *114*, 172–178. doi:10.1016/j.apenergy.2013.09.022

Maccarini, A., Hultmark, G., Bergsøe, N. C., & Afshari, A. (2018). Free cooling potential of a PCM-based heat exchanger coupled with a novel HVAC system for simultaneous heating and cooling of buildings. *Sustainable Cities and Society*, *42*, 384–395. doi:10.1016/j.scs.2018.06.016

Mehling, H., & Cabeza, L. F. (2007). Phase change materials and their basic properties. In *Thermal energy storage for sustainable energy consumption: fundamentals, case studies and design* (pp. 257–277). Springer. doi:10.1007/978-1-4020-5290-3_17

Mehling, H., & Cabeza, L. F. (2008). Heat and cold storage with PCM. *Heat and Mass Transfer*, 11–55.

Raj, V. A. A., & Velraj, R. (2010). Review on free cooling of buildings using phase change materials. *Renewable & Sustainable Energy Reviews*, *14*(9), 2819–2829. doi:10.1016/j.rser.2010.07.004

Sarbu, I., & Sebarchievici, C. (2018). A comprehensive review of thermal energy storage. *Sustainability (Basel)*, *10*(1), 191. doi:10.3390u10010191

Shabani, B., & Biju, M. (2015). Theoretical modelling methods for thermal management of batteries. *Energies*, *8*(9), 10153–10177. doi:10.3390/en80910153

Thantong, P., & Chantawong, P. (2017). Experimental study of a solar wall collector with PCM towards the natural ventilation of model house. *Energy Procedia, 138*, 32–37. doi:10.1016/j.egypro.2017.10.041

Velraj, R., Seeniraj, R., Hafner, B., Faber, C., & Schwarzer, K. (1999). Heat transfer enhancement in a latent heat storage system. *Solar Energy, 65*(3), 171–180. doi:10.1016/S0038-092X(98)00128-5

Vieira, D. R., Calmon, J. L., & Coelho, F. Z. (2016). Life cycle assessment (LCA) applied to the manufacturing of common and ecological concrete: A review. *Construction & Building Materials, 124*, 656–666. doi:10.1016/j.conbuildmat.2016.07.125

Zalba, B., Marín, J. M., Cabeza, L. F., & Mehling, H. (2003). Review on thermal energy storage with phase change: Materials, heat transfer analysis and applications. *Applied Thermal Engineering, 23*(3), 251–283. doi:10.1016/S1359-4311(02)00192-8

Zhai, X., Wang, X., Wang, T., & Wang, R. (2013). A review on phase change cold storage in air-conditioning system: Materials and applications. *Renewable & Sustainable Energy Reviews, 22*, 108–120. doi:10.1016/j.rser.2013.02.013

Chapter 7
The Application of FlexPDE in Latent Heat Thermal Energy Storage Problems

ABSTRACT

One of the fundamental requirements of human society is the implementation of a strategy to manage energy usage. One of these techniques is the use of renewable resources. Due to their storage capacity, energy storage systems (TESS) can lessen the instability of renewable resources. There are three types of thermal energy storage: energy stored as sensible heat, energy stored as latent heat (LHTESS), and energy stored through chemical reactions. Because of the high capacity for energy storage and latent heat storage, phase change materials (PCM) have attracted a lot of attention recently. PCM are broken down into three groups: gas-liquid, solid-liquid, and solid-solid. A drawback of PCM is its poor thermal conductivity. Fins in the PCM structure and the usage of nanofluids are two solutions to this issue. Different configuration of fins is T, Y, V, and tree-like. This chapter cover the aforementioned subjects.

INTRODUCTION

The ever-increasing growth of the world's population and economy has significantly increased the demand for thermal energy. This increase has serious and harmful effects on the environment. The emission of greenhouse gasses, especially carbon dioxide, is the main cause of these consequences. This phenomenon, together with the destructive effects on the ozone layer, can cause global warming. These factors alone will more than increase energy demand by 2050. As mentioned earlier, with the increasing consumption of fossil fuels and the deterioration of weather conditions, it seems urgent to adopt a new method to control the current conditions. The 18% increase in energy consumption in the last decade and the projected 30% increase in energy consumption in the coming years is also a reason to use an alternative method. Renewable energy is a reliable way to solve the environmental problems we face. Today, efforts are being made to generate energy from renewable sources. For example, in the field of global electricity generation, renewable energy sources have generated a total of 24.5% of the energy.

DOI: 10.4018/979-8-3693-0932-2.ch007

Buildings are one of the most important factors in increasing energy demand. Heating and air conditioning (HVAC) systems account for nearly 60% of energy consumption in buildings. For this reason, cooling and heating equipment is one of the most important and largest sources of energy consumption in the world today. It should be noted that energy management is important not only in buildings, but also in many other industries. The erratic behavior of renewable resources is one of the problems that affects their reliability. Thermal energy storage systems (TESS) can reduce the unsteady nature of renewable resources due to their storage capability. In recent years, TESS has reduced the impact and thermal load of various devices, which has also increased the efficiency of thermal devices. The features of thermal energy storage systems (TESS) include high energy storage capacity, constant working temperature and high heat transfer speed. Thermal energy storage is divided into three categories:

1. Energy storage in the form of sensible heat
2. Energy storage as latent heat
3. Thermal energy storage by chemical reaction

In energy storage in the form of sensible heat, there is no phase change and energy storage can be seen as an increase in temperature. In the method of storing thermal energy in the form of latent heat, thermal energy is stored latently and relying on the phase change of materials. Hidden energy storage is about 5 to 14 times more than visible energy storage. The energy storage system through chemical reaction is such that during a chemical reaction, a certain amount of energy is required to create a bond or separate compounds from each other. It is worth mentioning that in this method the chemical reaction must be completely reversible. The Latent Heat Thermal Energy Storage System (LHTESS) is an effective mechanism that stimulates the storage material to change its phase from solid to liquid (melting) or liquid to solid (freezing). The applications of LHTESS include solar applications, cooling of electronic devices, steam generation and peak load reduction. It should also be noted that LHTESS can operate in isothermal conditions, minimizing the cost and complexity of maintenance. In recent years, the use of phase change materials (PCM) as one of the energy storage technologies due to their high energy storage capability and the ability to store energy during a phase change process at an almost constant temperature, i.e. energy storage in the form of heat Nahan has been very much considered. The ability and application of PCM has been investigated in many industries such as solar power plants for regulating the temperature of plates, solar dryers in agricultural industries, batteries, construction and domestic water heaters. Surveys and forecasts show that the gap between energy supply and demand can be covered by using PCM. In a survey, it has been determined that walls made with 30% PCM materials retain heat equivalent to walls with 5 times their thickness. PCM can be divided into three categories, solid- liquid, liquid-gas, and solid-solid, based on the change of state when receiving energy and phase change. Among these three categories, solid-liquid phase change is the most widely used. In addition to the above, PCM can also be used as passive heat storage. Among the positive properties of PCM, we can mention the ability to be used as a source with constant temperature, low temperature drop, suitable melting point, high energy melting and chemical stability. Thermal conductivity, phase separation of PCM, supercooling in PCM and stability in PCM are among the problems of PCM. The biggest disadvantage of PCM is its low thermal conductivity, which can make the use of PCM as a cooling device limited and impractical under certain conditions. There are three main methods to solve the low thermal conductivity of PCM:

1. Making PCM composite with different metals or special graphites to increase heat transfer
2. Using metal materials in the form of spheres, plates, propellers, fins to increase heat transfer in PCM
3. Use of nanomaterials in PCM structure

Sahan and Paksoy (2018) used new malleable PCM composites in TESS. Smaller volume and high storage capacity are mentioned as the advantages of this system. The efficiency of thermal systems increases in two active and passive ways. In the passive method, no external energy is needed and fins are used. These fins are placed in different arrangements and made of different materials. Extended surfaces are a good example of a passive method. As mentioned, one of the techniques used by researchers to increase the performance of PCM-based TESS systems is metal fins in different types of length, circular and pin. Tiari et al. (2016) with high melting point PCM in a finned heat pipe. Their investigation showed that longer fins lead to higher base wall temperature. Also, with the increase in the number of fins, the freezing speed and the base temperature of the container increase. Ahmed et al. (2017) investigated the effect of changing fin thickness in a heat exchanger. Hosseinzadeh et al. (2019) applied tree-shaped fins to LHTESS to accelerate freezing in the system. Their investigation showed that the high angle of fin branching increases the freezing rate. Alizadeh et al. (2019) investigated the use of Y-shaped fins in PCM solidification in LHTESS. The acceleration of the process by increasing the length of the fin and the angle of the two branches was the result of this study. In their study, Hosseinzadeh et al. (2019) showed that changing the thickness of fins reduces the freezing time by 9.93% compared to longitudinal fins. Optimizing the arrangement of blades can lead to achieving the best performance. Bhagat et al. (2018) investigated the performance of LHTESS by optimizing fin geometrical parameters such as fin thickness, fin number and fin height. The results of this study show that fin thickness and number of fins have a significant effect on efficiency. Special configurations of fins include Y, V, T fins and tree-like arrays. Ismail et al. (2001) investigated numerically and experimentally the increase of LHTESS with fins in the freezing process. Their investigation showed that with the increase in fin length, the freezing speed also increases. In most of the research done by adding fins to PCM based storage systems, rectangular fins were used. However, among the new reviews, Y-shaped fins and tree-like fins are more interesting due to their efficiency and better performance. In their study, Liu et al. (2019) used a triangular fin to increase freezing in the LHTESS shell and tube exchanger. Their research showed that the use of triangular fins reduces the freezing time by 38.3% compared to regular rectangular fins. Another method used to increase the thermal conductivity and accelerate the heat transfer of PCM is the use of nanoparticles. The rationale for adding nanoparticles to the PCM is that adding fins to the tubes reduces the added weight of the space available for the PCM, thereby limiting the design and energy capacity of the LHTESS. Additionally, as a result of fins entering the entire PCM space, natural convection, which as a favorable phenomenon contributes to the fluidity and circulation of materials inside the tubes, is damaged. Therefore, the use of nanoparticles alone or in combination with other methods is fully justified. Nanoparticles are added to PCM due to the ability to increase the thermal conductivity of the system. One approach is the dispersion of nanoparticles in pure PCMs, known as NEPCM. NEPCMs have higher thermal conductivity than conventional PCM, thus increasing the melting/freezing rate. Elbahjaoui an El Qarnia (2017) in his investigation based on limited volume on the NEPCM solidification process in a rectangle, showed that by making NEPCM, the efficiency of the system increases in terms of reducing the discharge time. Kant et al. (2017) showed in their study that the use of graphene nanoparticles in the system improves the speed of melting. Lohrasbi et al. (2016) added copper nanoparticles as PCM to water to accelerate the freezing process in a finned LHTESS. Their results show that using copper nanoparticles with volume fractions of 2.5 and 5% leads to a reduction of 8.8 and 16% in freezing time. In the study conducted by Gurzin et al. (2019) it was shown that increasing the volume fraction of nanoparticles to 4% causes a 15% reduction in freezing time. In this study, they investigated the reduction of freezing time by using NEPCM. Hybrid nanofluid is a new type of nanofluid that consists of composite nanoparticles. Hybrid nanofluids have received much attention due to their ability to increase the heat transfer rate. The process of forming hybrid nanofluid is in two ways: 1- Hybrid nanofluid is made by distributing two or more different types of nanoparticles in the base fluid separately. 2- Nanoparticles spread in the base fluid and stick together. PCM reinforced with hybrid nanoparticles is known as HNEPCM. In an experimental study, Madhesh et al. (2014) investigated the effect of hybrid nanoparticles on heat transfer. The results of their investigation show that the addition of hybrid nanoparticles increases the heat transfer coefficient by 50%. Hosseinzadeh et al. (2019) investigated the effect of hybrid Cu–TiO_2 nanoparticles along with thermal radiation on the freezing process in LHTESS reinforced with tree-like branched fins. Its investigation showed that increasing the parameter related to thermal radiation and the angle of the fins branch decreases the total energy and average temperature. Alizadeh et al. (2019) investigated the freezing process of HNEPCM in a LHTESS with Koch snowflake cross section. Their results showed that by increasing the shape factor of nanoparticles, the freezing process is done faster. Chamkha et al. (2017) investigated the melting process of PCM in a cavity containing hybrid nanoparticles. Their investigation shows that the volume fraction of nanoparticles significantly affects the fluid fraction because the heat conduction in the system is enhanced.

1. EXAMPLE 1: INVESTIGATION OF LHTESS FILLED BY HYBRID NANO-ENHANCED PCM WITH KOCH SNOWFLAKE FRACTAL CROSS SECTION IN THE PRESENCE OF THERMAL RADIATION

1.1 Problem Definition

Figure 1 demonstrates the geometry and boundary conditions of present study. The working fluid flows in the inner tube with Koch snowflake cross section and the Hybrid nano-enhanced PCM fills the space between the inner tube and outer circular storage tank. The Koch Snowflake structure is built by starting with an equilateral triangle (zeroth fractal iteration), removing the inner third of each side, building another equilateral triangle at the location where the side was removed, and then repeating the process indefinitely. Each repetition is denoted as a fractal iteration. For example, iteration zero is represented as n = 0, iteration one as n = 1, and so on. The zeroth and first three fractal iterations of the Koch Snowflake fractal are depicted in Fig. 1.

The number of sides for the zeroth iteration is that of an equilateral triangle. Eq. (1) is utilized to determine the number of sides. The length of the side of the Koch Snowflake, denoted as l, can be calculated by using Eq. (2), which *a* represents the length of the initial equilateral triangle from iteration. Moreover, the radius of outer storage tank is equal to 20 mm. The inside wall is kept at constant temperature of 240 K and the initial temperature of Hybrid nano-enhanced PCM is equal to 275 K. The physical properties of use PCM and nanoparticles are summarized in Table. 1.

$$N = 3\times 4^n \tag{1}$$

$$L = a*3^{-n} \quad (2)$$

The governing equations for heat and mass transfer, taking into account the two-dimensional unsteady freezing process inside the system as well as the effect of radiation, are stated in LHTESS in Chapter 6.

Figure 1. Two-dimensional geometry and boundary conditions of present study

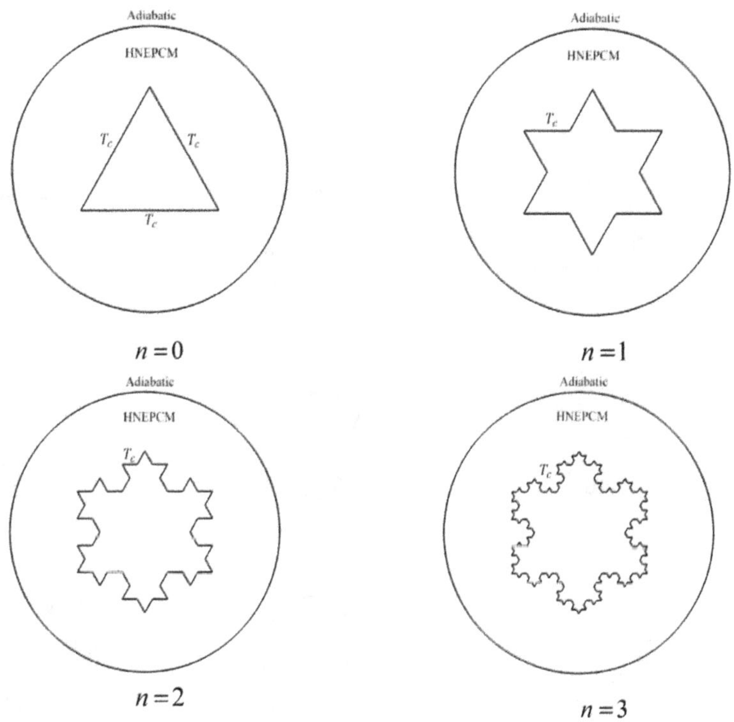

Table 1. The physical properties of PCM and nanoparticles

Property	PCM	Cu	TiO_2
$\rho\,(\frac{kg}{m^3})$	997(Liquid phase)	8960	4250
	916 (Solid phase)		
$C_p\,(\frac{J}{kgK})$	4148 (Liquid phase)	385	686.2
	2030 (Solid phase)		
$k\,(\frac{W}{mK})$	0.6 (Liquid phase)	400	8.9538
	2.22 (Solid phase)		
$L_f\,(\frac{J}{kg})$	335,000	-	-

1.2 Numerical Simulation

The Galerkin Finite Element Method (FEM) is employed to simulate the unsteady heat and mass transfer during solidification process. The parameters are calculated on the nodes which are located on the corners of cell. The final calculations are solved by the Newton-Raphson method. This method solves the system of equations by an implicit backward difference approach. The solidification process is simulated by utilizing the adaptive mesh as shown in Fig. 2.

Figure 2. Adaptive mesh refinement procedure when Rd=0.5 and s=5.7

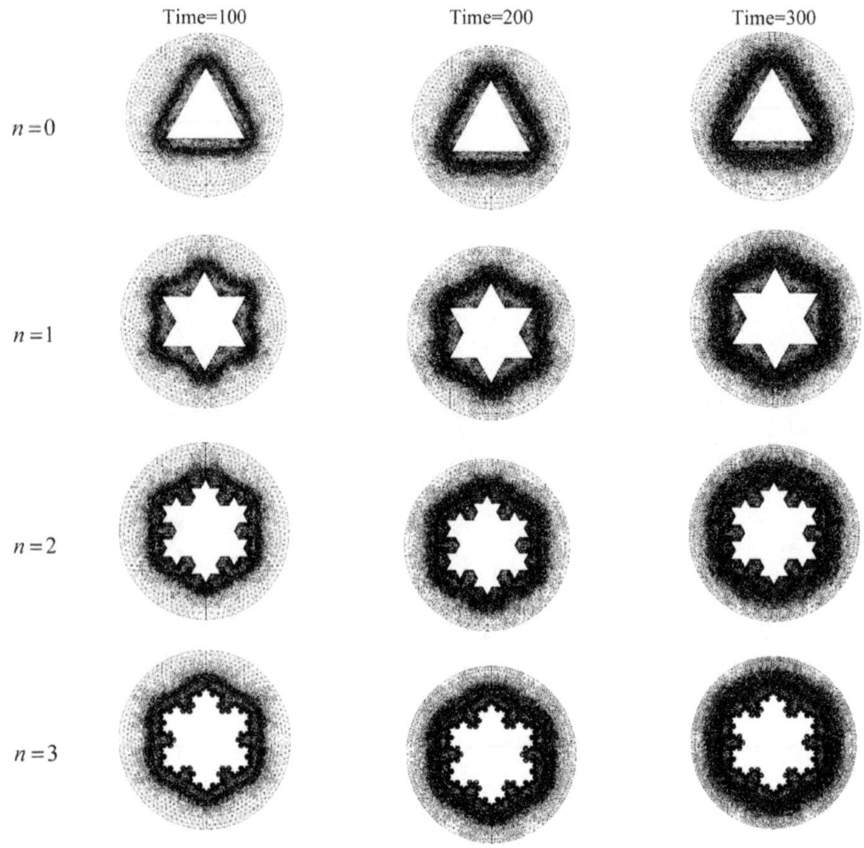

The numerical method is validated by comparing the results with previous study (Ismail et al., 2001). It can be observed from Figure 3 that an excellent agreement is obtained.

1.3 Impacts of Active Parameters

In order to accelerate the solidification process in a LHTESS with a Koch snowflake fractal cross section, HNEPCM and thermal radiation effect are employed in present work. The unsteady governing equations of heat and mass transfer during the solidification process are solved by applying the FEM. In this study, the influences of nanoparticles shape factor (s= 3.7, 5.7, 16.1), the thermal radiation parameter (Rd= 0, 0.5, 1) and the fractal pattern iteration (n= 0, 1, 2, 3) on the solidification process are studied.

Figure 3. Comparison between present study and experimental work
by Ismail et al. (2001)

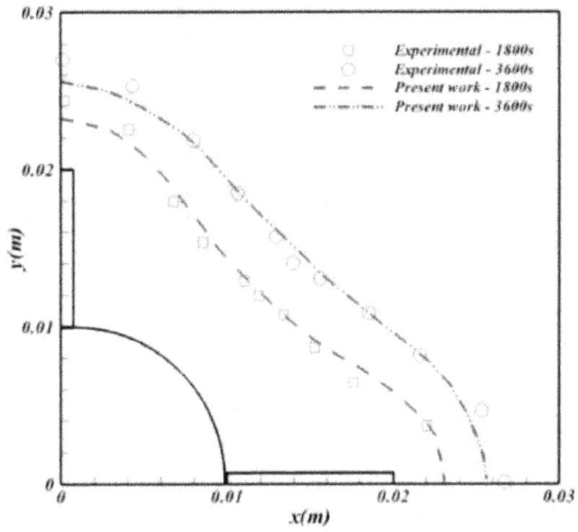

Figure 4 represents the isotherm and solid fraction contours during the solidification process for different values of thermal radiation parameter (Rd= 0, 0.5, 1) and fractal pattern iteration (n= 0, 1, 2, 3). This figure depicts that by augmenting the thermal radiation parameter, the temperature of HNEPCM becomes closer to the inner tube temperature due to enhanced heat transfer rate.

Also, it can be seen that higher values of solid fraction are obtained by raising the radiation parameter and fractal pattern iteration, in same time. In this case, the enhanced thermal penetration depth causes the solidification process to be completed faster. Therefore, by enhancing the thermal radiation parameter, the full solidification time reduces considerably. Among these cases, applying n=3 and Rd=1 results in fastest solidification process which takes 626 s to be completed, when the Platelets type of nanoparticles is considered in the process (s=16.1).

Figure 5(a, b, c) demonstrates the influence of nanoparticles shape factor on average temperature, solid fraction and total energy, respectively. Three types of nanoparticles including the Hexahedron, the Platelets and the Lamina with different shape factors are investigated in present study. As shown in Figure 5a, raising the shape factor results in enhancing the thermal conductivity of HNEPCM and as a result, the average temperature of the system decreases during the process. It can be observed from Figure 5b, due to reduced average temperature of the system, the rate of solidification increases and the solid fraction experiences closer values to 1. Moreover, total energy of the system is a decreasing function of nanoparticles shape factor as depicted in Figure 5c.

Figure 4a. Temperature (Left) and solid fraction (Right) contour plots in n=0, s=5.7

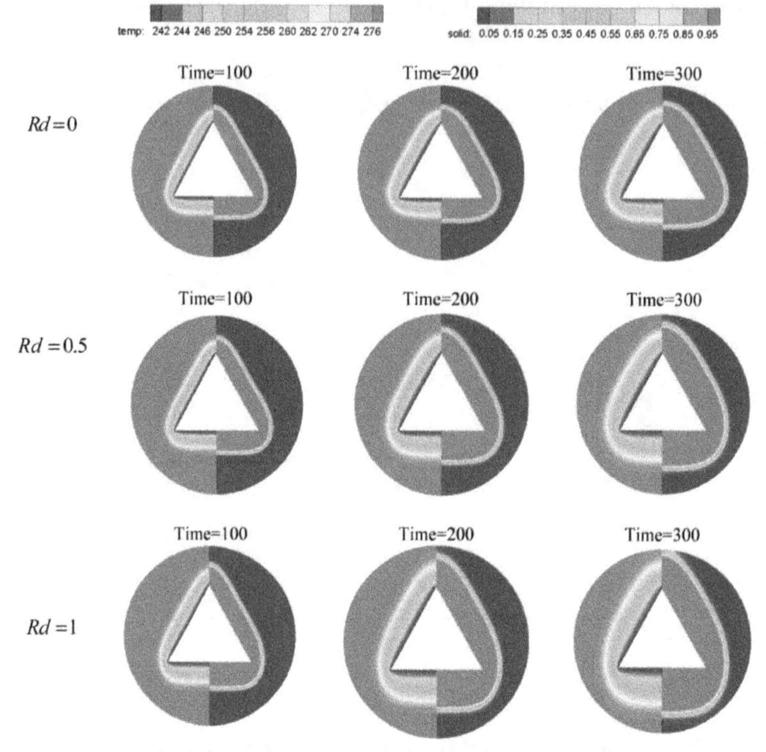

Figure 4b. Temperature (Left) and solid fraction (Right) contour plots in n=1, s=5.7

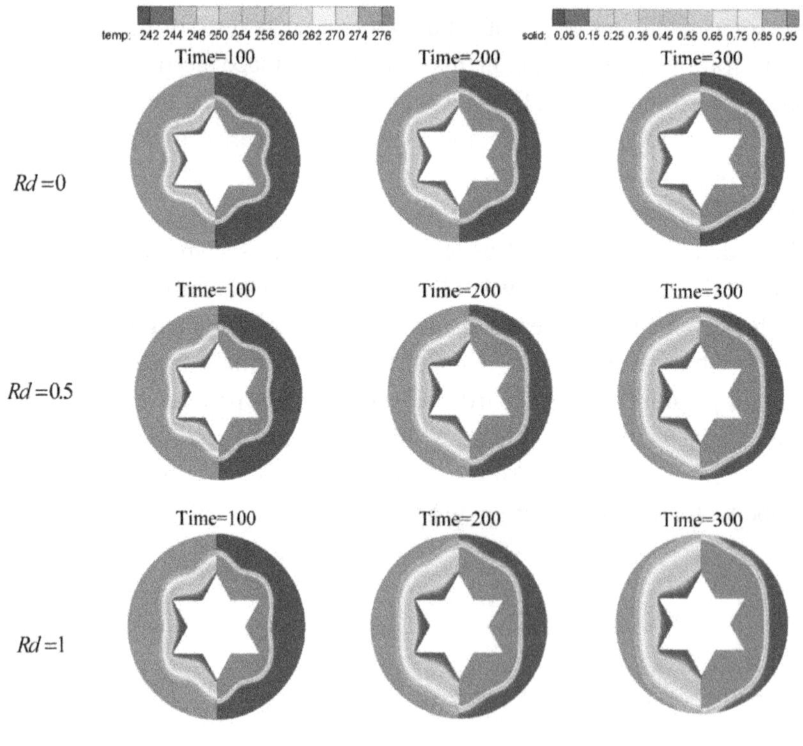

Figure 4c. Temperature (Left) and solid fraction (Right) contour plots in n=2, s=5.7

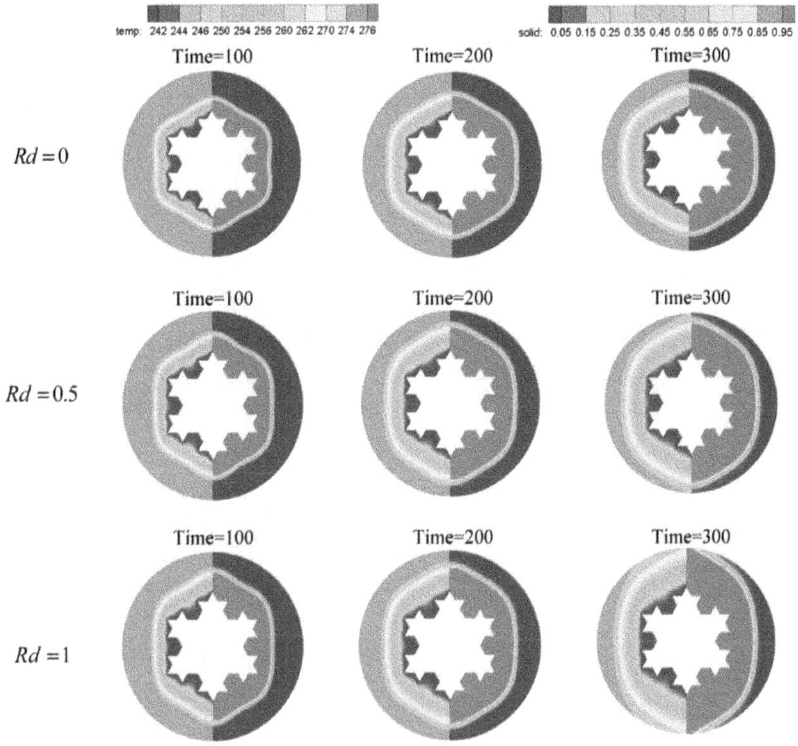

Figure 4d. Temperature (Left) and solid fraction (Right) contour plots in n=3, s=5.7

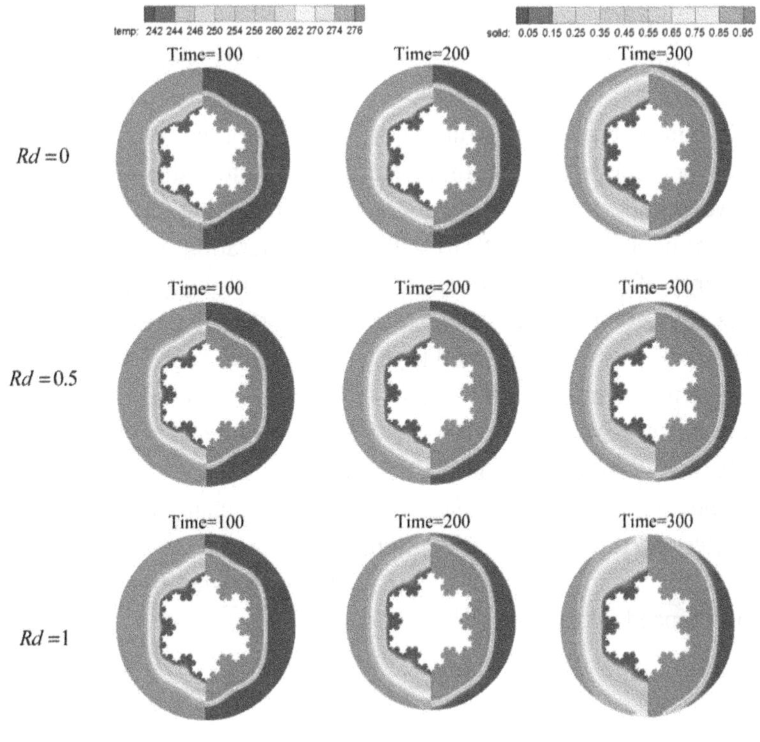

Figure 5. Effects of nanoparticles shape factor on (a) solid fraction, (b) average temperature, and (c) total energy in LHTESS with Koch snowflake cross section when Rd=0.5 and n=3

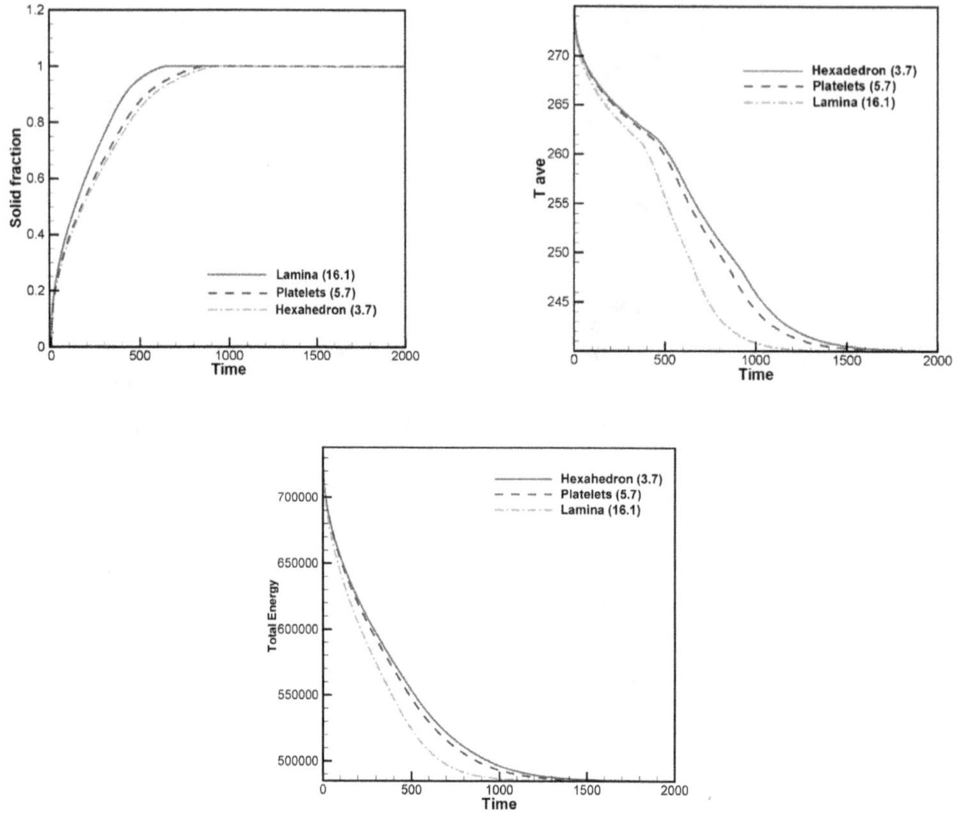

The variations of average temperature, solid fraction and total energy by changing the thermal radiation parameter are illustrated in Fig. 6(a, b and c). This figure demonstrates that increasing the thermal radiation causes a drop in average temperature of the system. So, the solid fraction increases during the process and the solidification process is completed in lower times due to enhancing the heat transfer rate. In addition, lower values of total energy are obtained by raising the thermal radiation parameter.

Figure 7(a, b and c) depicts the effect of fractal pattern iteration on the average temperature, solid fraction and total energy of the system. In all cases, increasing the iteration (n) leads to enhance the occupied area by the HNEPCM and also the penetration depth which results in decreasing the average temperature of the system. Also, it is expected to have higher values of solid fraction during the process due to reduced average temperature of the system. Moreover, it can be seen that the variation of solid fraction, average temperature and total energy between the zeroth and first iteration of Koch snowflake is more considerable than when the iteration varies from 1 to 3.

Figure 8 shows the variation of full solidification time by changing the nanoparticles shape factor, radiation parameter and fractal pattern iteration. It can be seen from this figure that enhancing the nanoparticles shape factor, radiation parameter and fractal pattern iteration will accelerate the solidification process due to enhanced thermal penetration depth.

Figure 6. Effects of radiation parameter on a) solid fraction, b) average temperature, and c) total energy over computational domain when s=5.7and n=3

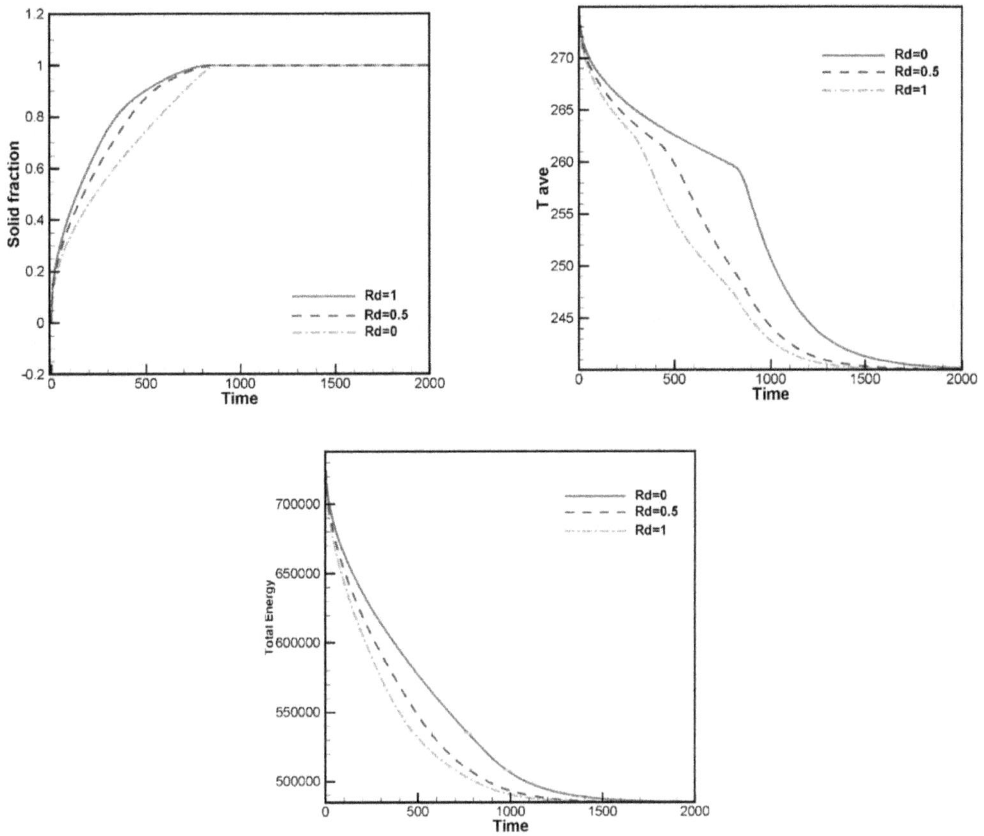

Also, a correlation for full solidification time is extracted as a function of s, Rd and n which is presented as below:

$$Time = 1550.64 - 443.8(n) - 5.16(s) - 244.22(Rd) + 2.68(n)(s) + 75.06(n)(Rd) + 1.23(s)(Rd) + 76.75(n^2) - 1.14(s^2) - 14.54(Rd^2) \quad (3)$$

It can be concluded that by applying the Hybrid nanoparticles to the base PCM along with Koch snowflake cross section, the penetration depth enhances and as a result, the solidification process is completed in lower times.

Figure 7. Effects of fractal pattern iteration on a) solid fraction, b) average temperature, and c) total energy over computational domain when s=5.7 and Rd=0.5

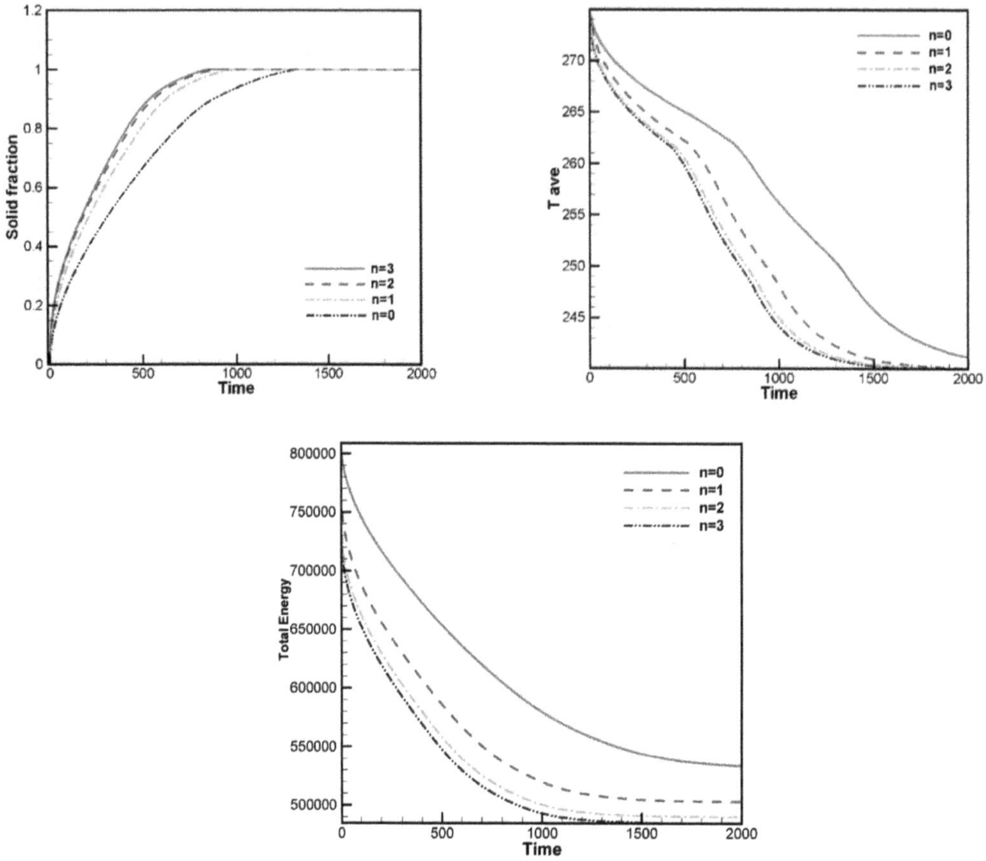

2. EXAMPLE 2: SOLIDIFICATION PROCESS OF HNEPCM IN A LHTESS WITH TREE-LIKE BRANCHING FIN IN THE PRESENCE OF THERMAL RADIATION

2.1 Problem Definition

Figure 9 demonstrates the geometry and boundary conditions of present study. The working fluid flows in the inner tube and the HNEPCM fills the space between the inner tube and outer circular storage tank. In order to enhance the thermal penetration depth, the tree-like branching fins are attached to the outer side of inner tube.

Table 2 illustrates the geometry parameters of applied fins. The inside wall is kept at constant temperature of 240 K and the initial temperature of HNEPCM is equal to 285 K. It should be note that the solidification is a time-dependent process, so it should be simulated unsteadily. In present work, an unsteady simulation with the time-step equal to 1×10^{-5}(s) is carried out.

The governing equations for heat and mass transfer in LHTESS, taking into account the two-dimensional unstable freezing process inside the system, as well as the effect of radiation, are comprehensively stated in Chapter 6.

Figure 8. Full solidification time for different values of Rd, n and s

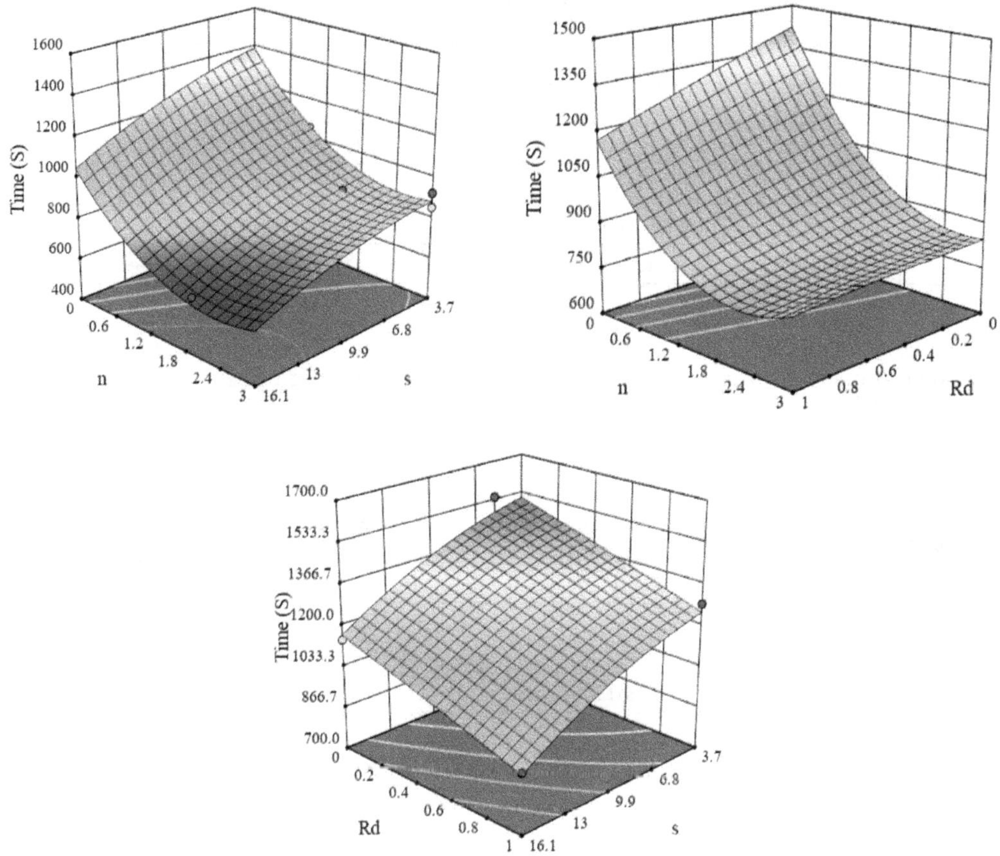

Figure 9. (a) Three-dimensional view of LHTESS with tree-like branching fin; (b) two-dimensional geometry and boundary conditions of present study

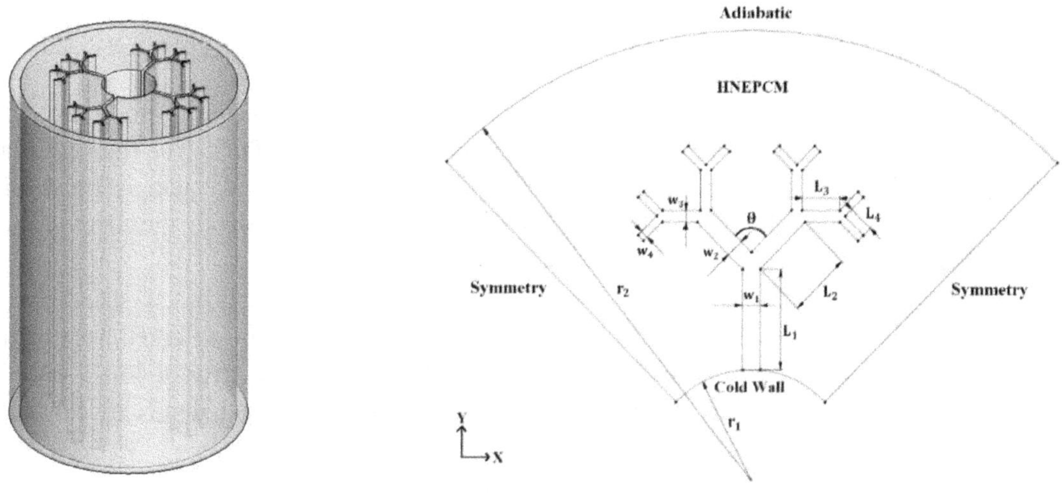

Table 2. The geometry parameters of tree-like branching fin

Geometry Parameters	Values	Geometry Parameters	Values
θ	30°, 60°, 90°	w_2	0.3535 (mm)
L_1	6.25 (mm)	w_3	0.25 (mm)
L_2	4.42 (mm)	w_4	0.177 (mm)
L_3	3.125 (mm)	r_1	30 (mm)
L_4	2.21 (mm)	r_2	7.5 (mm)
w_1	0.5 (mm)		

The physical properties of use PCM and nanoparticles are summarized in Table 3.

Table 3. The physical properties of PCM, nanoparticles and fin

Property	PCM	Cu	TiO_2	Fin
$\rho\,(\frac{kg}{m^3})$	997 (Liquid phase) 916 (Solid phase)	8960	4250	8933
$C_p\,(\frac{J}{kgK})$	4148 (Liquid phase) 2030 (Solid phase)	385	686.2	385
$k\,(\frac{W}{mK})$	0.6 (Liquid phase) 2.22 (Solid phase)	400	8.9538	400
$L_f\,(\frac{J}{kg})$	335,000	-	-	-

2.2 Numerical Simulation

The FEM is applied to simulate the unsteady heat and mass transfer during solidification process. The parameters are calculated on the nodes which are located on the corners of cell. The final calculations are solved by the Newton-Raphson method. This method solves the system of equations by an implicit backward difference approach. The solidification process is simulated by utilizing the adaptive mesh as shown in Figure 10.

In present work, the numerical method is validated by comparing the results based on Galerkin FEM with previous experimental study published by Ismaeil et al. (2001). It can be observed from Figure 11 that this comparison represents an excellent agreement with the maximum error of 4.9%. Also, it can be concluded that the assumption of neglecting natural convection in the simulation is logical and close to reality.

Figure 10. Adaptive mesh refinement procedure when Rd=0.5 and sf=5.7

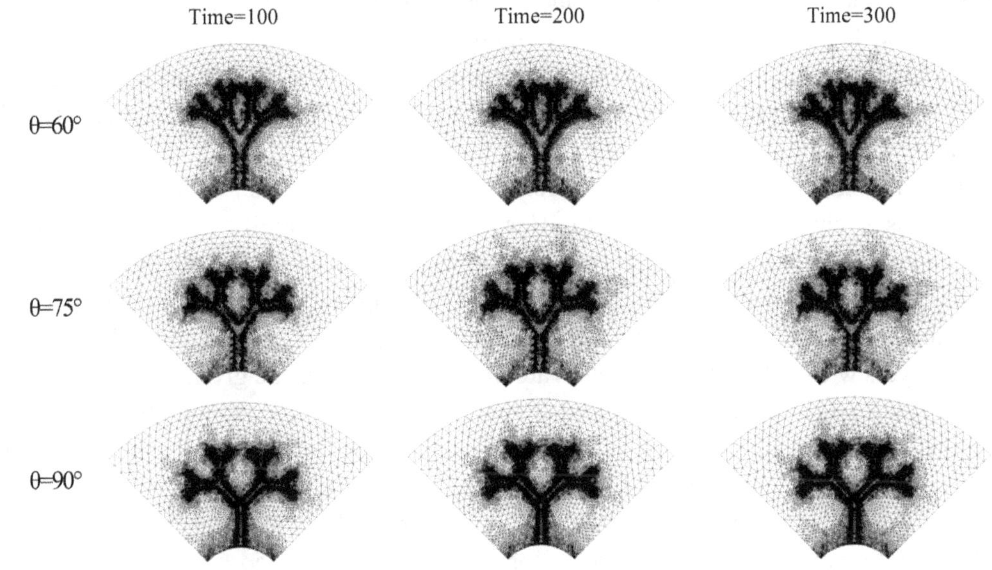

Figure 11. Comparison between present study and experimental work by Ismaeil et al. (2001)

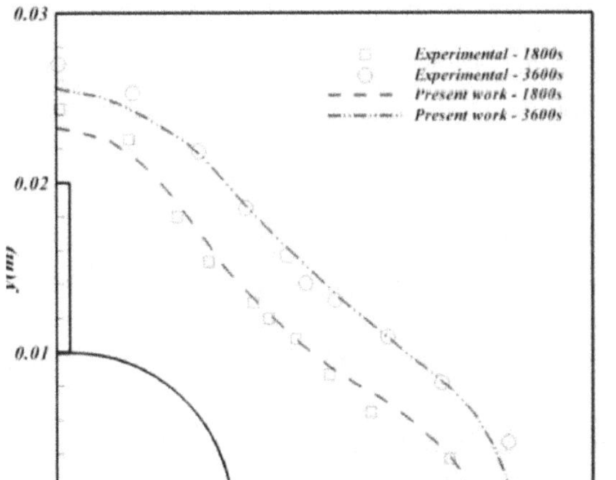

2.3 Impacts of Active Parameters

This study investigates the numerical simulation of unsteady solidification process in a LHTESS with tree-like branching fin considering the effects of HNEPCM and thermal radiation. The unsteady governing equations of heat and mass transfer during the solidification process are solved by applying the FEM. The shape factor of nanoparticles (sf= 3.7, 5.7, 16.1), the thermal radiation parameter (Rd=0, 0.5, 1) and the bifurcation angle of fins (θ= 60°, 75°, 90°) are the parameters which affect the solidification process in this research.

The isotherm and solid fraction contours during the solidification process for different values of thermal radiation parameter (Rd= 0, 0.5, 1) and fin bifurcation angle (θ= 60°, 75°, 90°) are shown in Figures 12, 13 and 14. It can be seen from these figures that increasing the radiation parameter and fin bifurcation angle results in higher values of solid fraction in same time. In this situation, the solidification process is completed faster due to an increase in thermal penetration depth. The effect of thermal radiation parameter on the solidification rate is more considerable than fin bifurcation angle. Moreover, as the value of thermal radiation parameter increases, the full solidification time reduces considerably. Among these cases, choosing θ=90° and Rd=1leads to the fastest solidification process which takes 601 s to be completed, when the Platelets type of nanoparticles are applied in the process (sf=16.1).

Figure 15(a, b, c) illustrates the effect of nanoparticles shape factor on average temperature, solid fraction and total energy, respectively. In this research, three types of nanoparticles including hexagons, platelets, and layers with different shape factors are investigated, which are given in chapter 3. As shown in Figure 15a, increasing the shape factor leads to enhance the thermal conductivity of HNEPCM and as a result, the average temperature of the system decreases during the process. It can be observed from Figure 15b, as the average temperature of the system decreases by augmenting the shape factor, the rate of solidification increases and the solid fraction moves toward 1. Also, the total energy of the system is a decreasing function of nanoparticles shape factor as shown in Figure 15c.

The effect of thermal radiation parameter on the average temperature, solid fraction and total energy variations during the solidification process is illustrated in Fig. 16(a, b and c). It can be seen that augmenting the thermal radiation parameter causes the average temperature of the system to be reduced. As a result, the solid fraction increases during the process and the solidification process is completed in lower times due to enhancing the heat transfer rate. In addition, it can be concluded that lower values of total energy are obtained by augmenting the thermal radiation parameter.

Figure 17(a, b and c) depicts the influence of fin bifurcation angle on the average temperature, solid fraction and total energy of the system. In all cases, increasing the bifurcation angle leads to enhance the occupied surface by the fin and also the penetration depth which causes a reduction in average temperature of the system. Also, it is expected to have higher values of solid fraction during the process due to decreasing the average temperature of the system by raising the bifurcation angle. Also, it can be found that choosing the fin bifurcation angle equal to 90° is the most effective case for reducing the full solidification time.

Figure 18 shows the response surfaces of full solidification time for nanoparticles shape factor, radiation parameter and fin bifurcation angle. It can be seen from this figure that enhancing the nanoparticles shape factor, radiation parameter and fin bifurcation angle will accelerate the solidification process due to enhanced thermal penetration depth. Applying the RSM, a correlation for full solidification time is extracted as a function of sf, Rd and θ which is presented as below:

$$\begin{aligned} Time = &+844.86 - 1.93(\theta) - 18.6015(sf) - 178.074(Rd) + 0.01468(\theta)(sf) \\ &-0.18715(\theta)(Rd) - 1.24191(sf)(Rd) - 0.0260(\theta^2) + 0.75(sf)^2 + 89.168(Rd^2) \end{aligned} \tag{4}$$

It can be concluded that by adding the hybrid nanoparticles along with tree-like branching fin to the base PCM, the penetration depth enhances and as a result, the solidification process is completed in lower times.

Figure 12. Temperature (Top) and solid fraction (Bottom) contour plots in θ=60°, sf=5.7

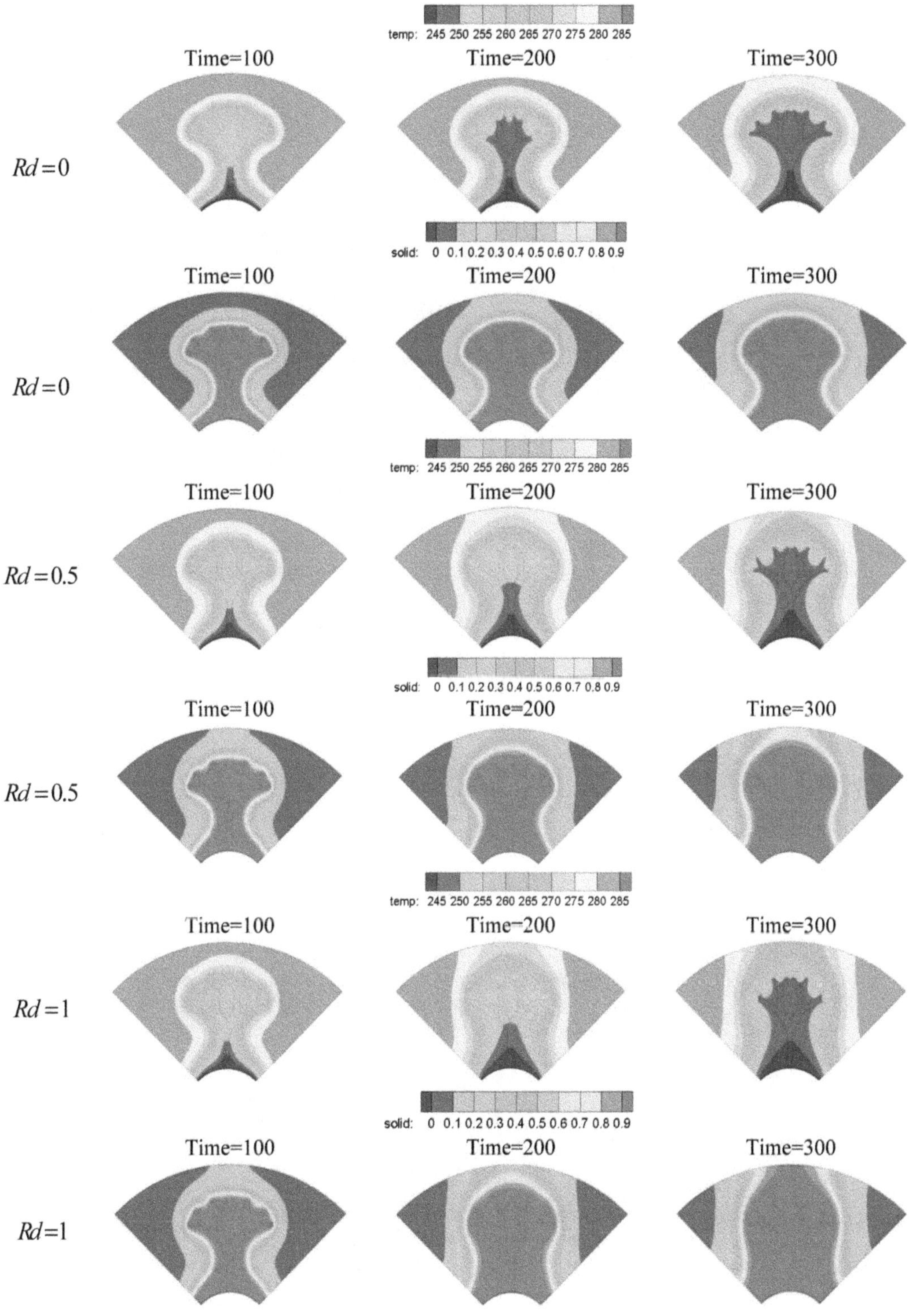

Figure 13. Temperature (Top) and solid fraction (Bottom) contour plots in θ=75°, sf=5.7

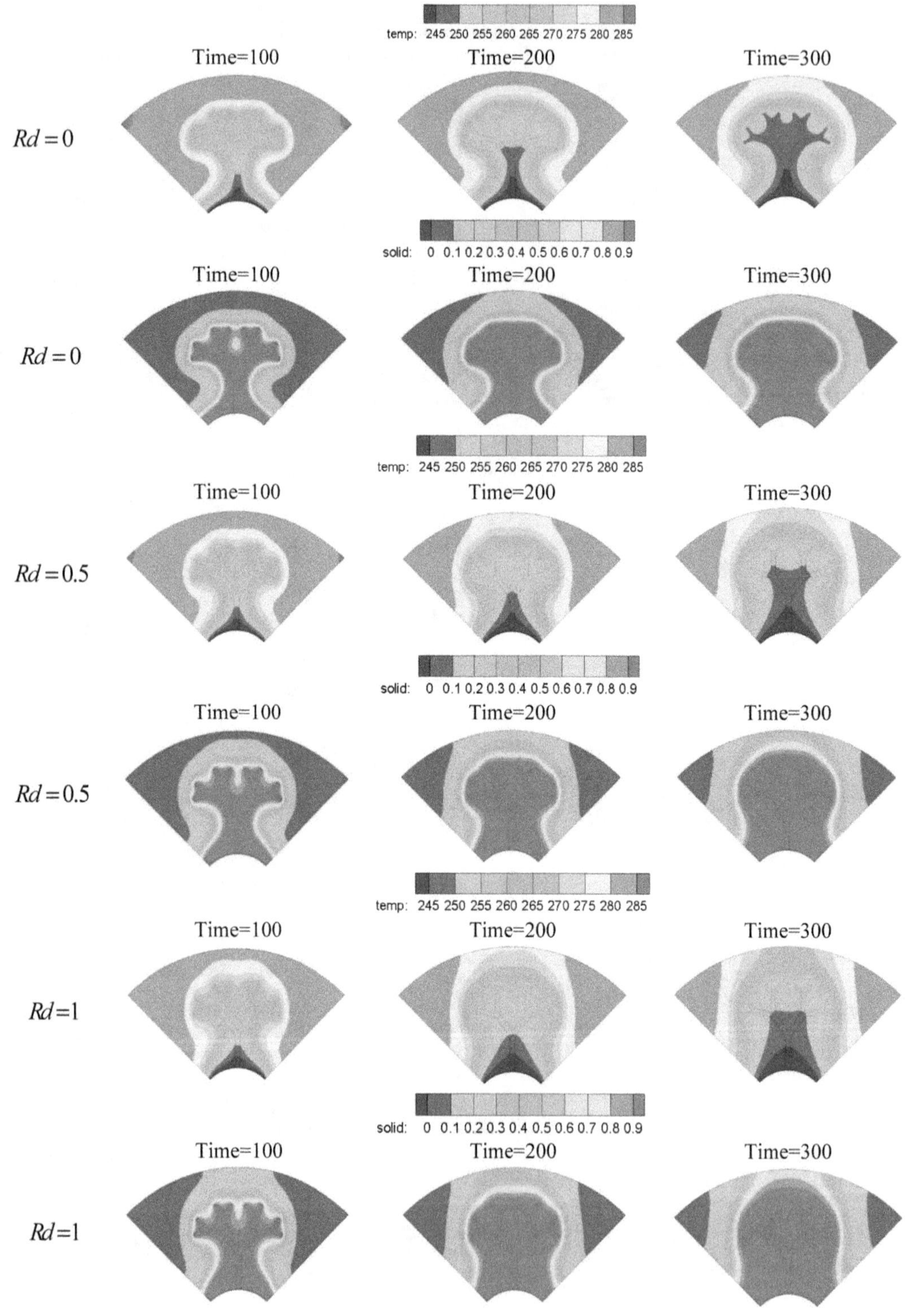

Figure 14. Temperature (Top) and solid fraction (Bottom) contour plots in θ=90°, sf=5.7

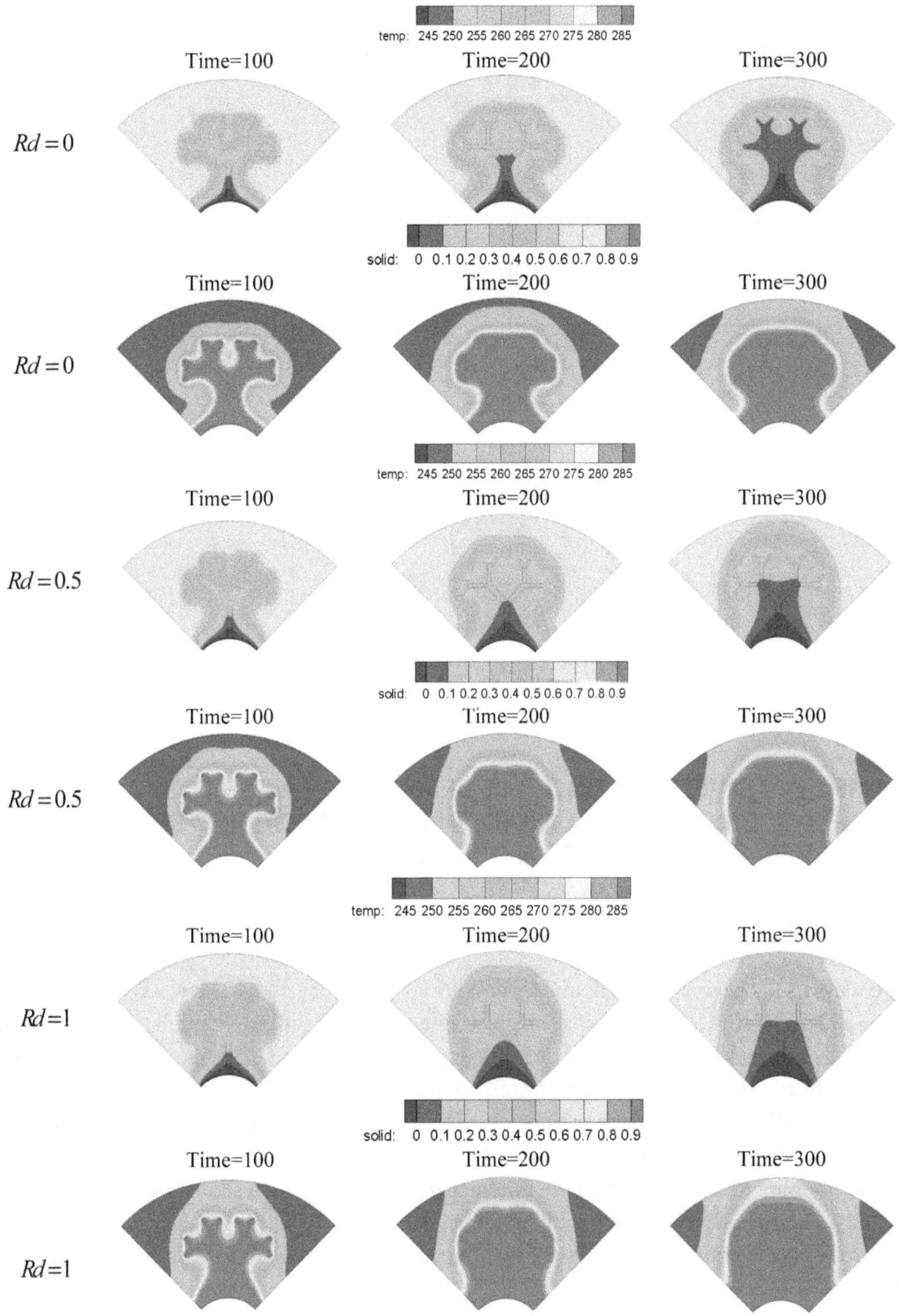

Figure 15. Effects of nanoparticles shape factor on (a) solid fraction, (b) average temperature, and (c) total energy in LHTESS with tree-like branching fin when Rd=0.5 and θ=75°

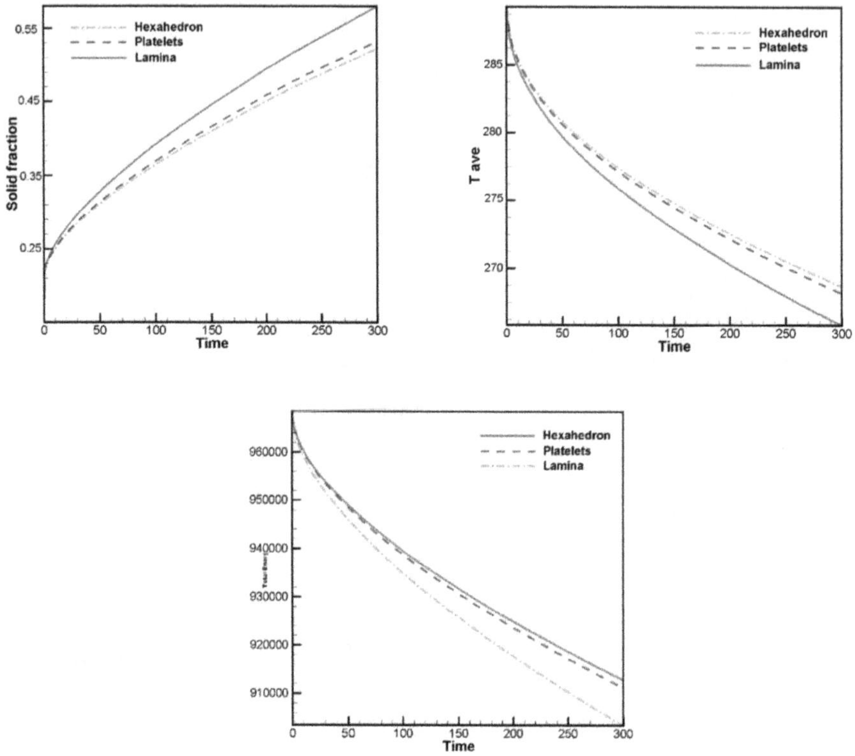

3. EXAMPLE 3: INVESTIGATING THE EFFECTS OF HYBRID NANOPARTICLES ON SOLID-LIQUID PHASE CHANGE PROCESS IN A Y-SHAPED FIN-ASSISTED LHTESS BY MEANS OF FEM

3.1 Problem Definition

Figure 19 depicts the geometry and boundary conditions of current study. The working fluid flows in the inner tube and the HNEPCM fills the space between the inner tube and outer circular storage tank. In order to enhance the thermal penetration depth, the Y-shaped fins are attached to the outer side of inner tube.

Table. 4 illustrates the geometric parameters of Y-shaped fins.

The inside wall is kept at constant temperature of 240 K and the initial temperature of HNEPCM is equal to 275 K. The physical properties of PCM and hybrid nanoparticles are summarized in Table 5. It is worth mentioning that solidification is a time-dependent process, so it should be simulated unsteadily. In this study, an unsteady simulation with a time-step equal to $2\times10^{-4}(s)$ is performed.

The governing equations considering the two-dimensional unsteady freezing process inside the system are stated in Chapter 6.

Figure 16. Effects of radiation parameter on a) solid fraction, b) average temperature, and c) total energy over computational domain when sf=5.7 and θ=75°

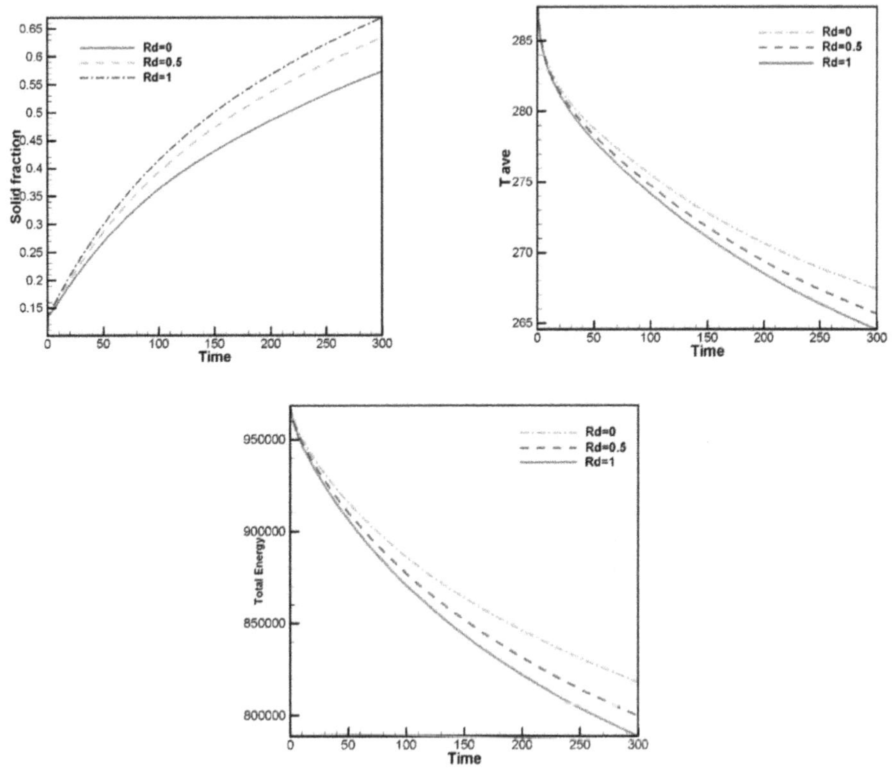

3.2 Numerical Simulation

The Standard Galerkin Finite Element Method is applied to simulate the unsteady heat and mass transfer during the solidification process. Implementing cubic interpolation over triangles, this method solves the problem. The parameters are calculated on the nodes which are located on the corners of cell. By substituting the definitions, dividing the dependencies on variables, integrating over cells and finally making differential equations based on variables, the Galerkin couple matrix is formed. The final calculations are solved by the Newton-Raphson method. This method solves the system of equations by an implicit backward difference approach. The solidification process is simulated by utilizing the adaptive mesh, as shown in Figure 20.

When the adaptive mesh is employed, the grids will be refined where the variation of variables is rapid so that the desired accuracy is obtained. In this study, the numerical method is validated by comparing the results with previous study (Ismail et al., 2001). It can be observed from Figure 21 that an excellent agreement with the maximum error of 4.9% is obtained.

Figure 17. Effects of fin bifurcation angle on a) solid fraction, b) average temperature, and c) total energy over computational domain when sf=5.7 and Rd=0.5

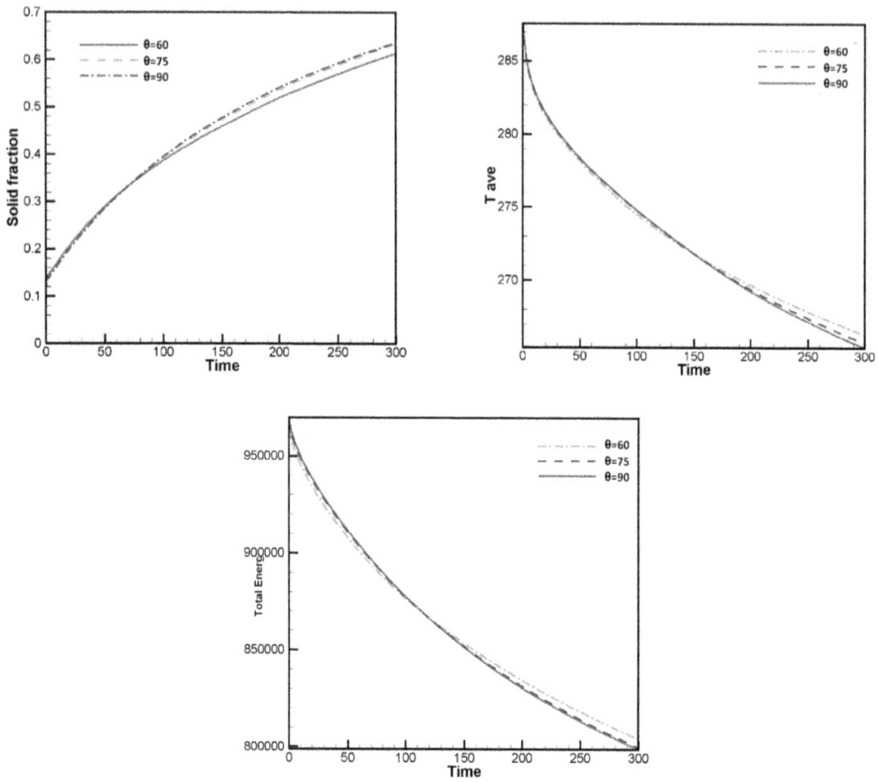

3.3 Impacts of Active Parameters

This study investigates the numerical simulation of unsteady solidification process in a Y-shaped fin assisted LHTESS considering the effect of HNEPCM. The unsteady governing equations of heat and mass transfer during the solidification process are solved by applying the FEM. The effects of different parameters such as: the shape factor of nanoparticles (*Sf*= 3.7, 5.7, 16.1), the bifurcation angle of Y-shaped fins ($\theta = \pi/6, \pi/4, \pi/3$) and the fin length (L= 13.5mm, 15.75mm, 18mm) are investigated.

A comparison between the solidification front of LHTESS enhanced with Y-shaped fin and LHTESS without fin is carried out in Figure 22. It can be observed that in the absence of Y-shaped fins, the solidification process begins only near the inner cold wall.

Figure 18. Response surfaces of full solidification time for different values of Rd, θ and sf

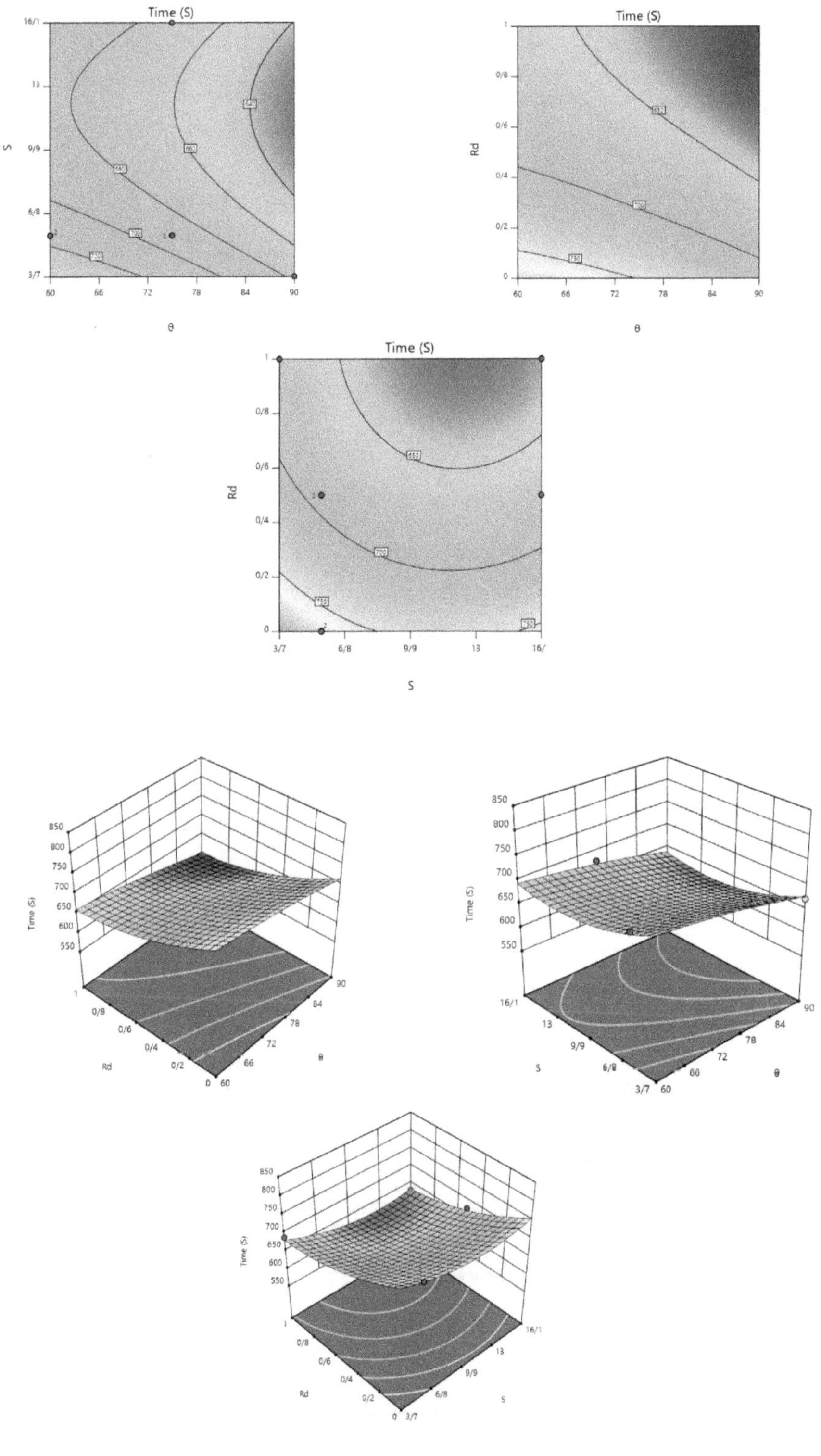

Figure 19. (a) Three-dimensional view of Y-shaped assisted LHTESS; (b) two-dimensional geometry and boundary conditions of present study

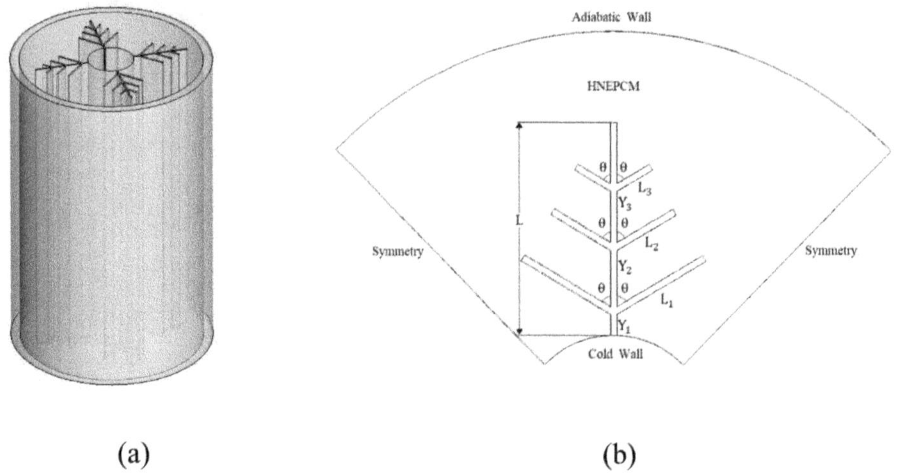

Table 4. The geometric parameters of Y-shaped fin

Geometry Parameters	Values	Geometry Parameters	Values
θ	$\pi/6$, $\pi/4$, $\pi/3$	L_3	2.7, 3.15, 3.6 (mm)
L	13.5, 15.75, 18 (mm)	Y_1	1.35, 1.575, 1.8 (mm)
L_1	6.75, 7.875, 9 (mm)	Y_2	5.4, 6.3, 7.2 (mm)
L_2	4.5, 5.25, 6 (mm)	Y_3	3.375, 3.9375, 4.5 (mm)

Table 5. The physical properties of PCM, nanoparticles, and fin

Property	PCM	Cu	TiO_2	Fin
$\rho\,(\frac{kg}{m^3})$	997 (Liquid phase) 916 (Solid phase)	8960	4250	8933
$C_p\,(\frac{J}{kgK})$	4184 (Liquid phase) 2030 (Solid phase)	385	686.2	385
$k\,(\frac{W}{mK})$	0.6 (Liquid phase) 2.22 (Solid phase)	400	8.9538	400
$L_f\,(\frac{J}{kg})$	335,000	-	-	—

Figure 20. Adaptive mesh refinement procedure when L=18mm, Sf=5.7, θ=π/4

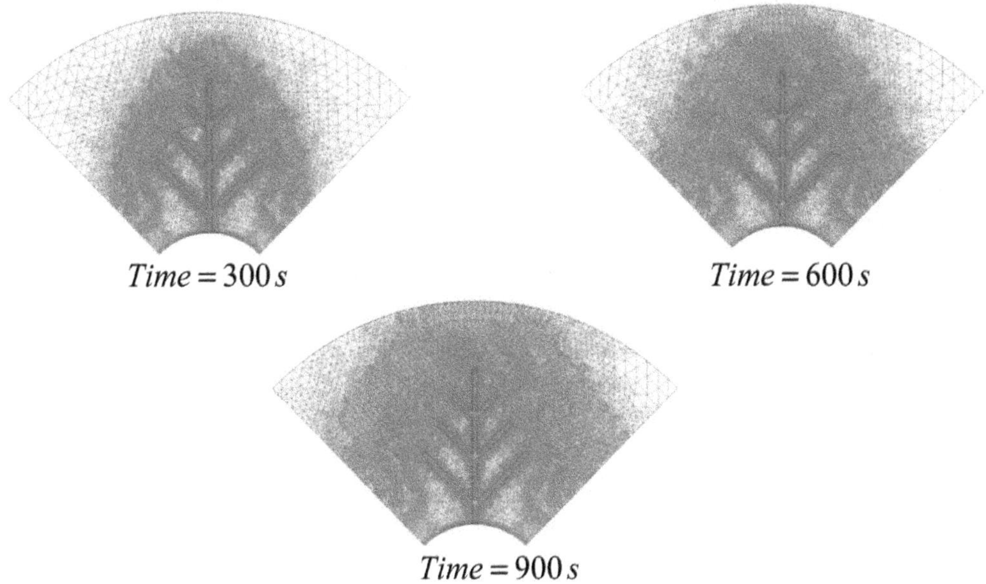

Figure 21. Comparison between present study and experimental work
by Ismaeil et al. (2000)

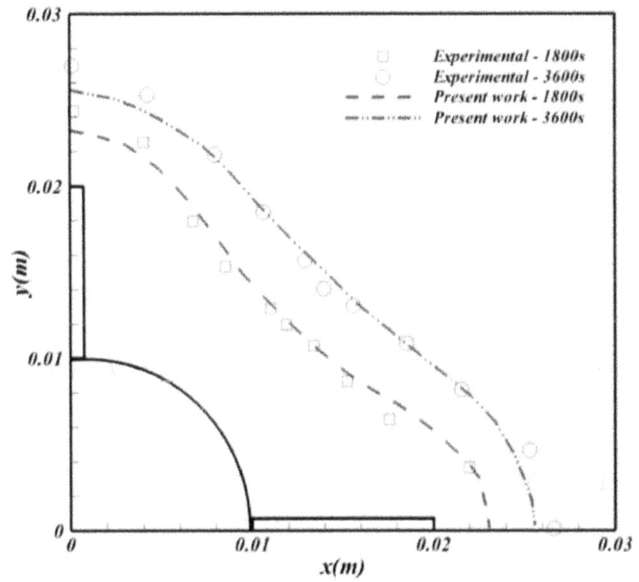

However, attaching Y-shaped fins to the inner wall causes the thermal penetration depth to be enhanced to the HNEPCM region. Consequently, the solidification process begins in both regions including near the cold walls and at the adjacency of the Y-shaped fin surface. Therefore, applying Y-shaped fins to the LHTESS causes a significant enhancement in heat transfer rate and a considerable acceleration in solidification process. The isotherm and solid fraction contours during the solidification process for different values of bifurcation angle ($\theta = \pi/6, \pi/4, \pi/3$) are shown in Figure 23, 24 and 25. It can be seen from these figures that increasing the fin length (L) and fin bifurcation angle (θ) results in higher values of solid fraction in same time. In this situation, the solidification process is completed faster due to an increase in thermal penetration depth. The effect of fin length on the solidification rate is more considerable in higher values of θ. Moreover, as the value of θ increases, the full solidification time reduces considerably. Among these cases, choosing $\theta=\pi/3$ and L=18mm leads to the fastest solidification process which takes 1675 s to be completed, when the Platelets type of nanoparticles are applied in the process (Sf=5.7).

Figure 22. Comparison between solidification front in LHTESS without fin and LHTESS with Yshaped fin at three time steps

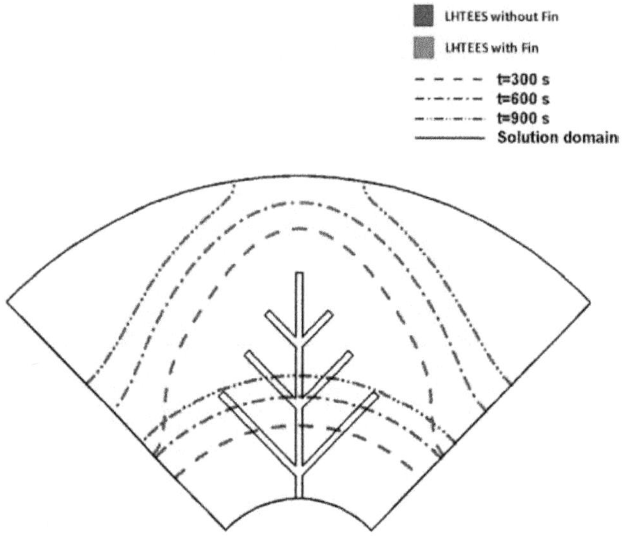

Figure 26(a, b, c) illustrates the effect of nanoparticles shape factor on average temperature, solid fraction and total energy, respectively. In this study, three types of nanoparticles including the Hexahedron, the Platelets and the Lamina with different shape factors are studied. As shown in Fig. 26a, increasing the shape factor leads to enhance the thermal conductivity of HNEPCM and as a result, the average temperature of the system decreases during the process.

It can be observed from Fig. 26b, as the average temperature of the system decreases by augmenting the shape factor, the rate of solidification increases and the solid fraction moves toward 1. Also, the total energy of the system is a decreasing function of nanoparticles shape factor as shown in Figure 26c.

The average temperature, solid fraction and total energy variations during the solidification process are illustrated in Figures 27-29. In all cases, increasing the bifurcation angle leads to enhance the occupied surface by the fin and also the penetration depth which causes a reduction in average temperature of the system. Also, it is expected to have higher values of solid fraction during the process due to decreasing the average temperature of the system by raising the bifurcation angle. In addition, it can be concluded that lower values of total energy are obtained by augmenting the bifurcation angle of the fins.

Figure 23. Temperature (Top) and solid fraction (Bottom) contour plots in $\theta=\pi/6$, $Sf=5.7$

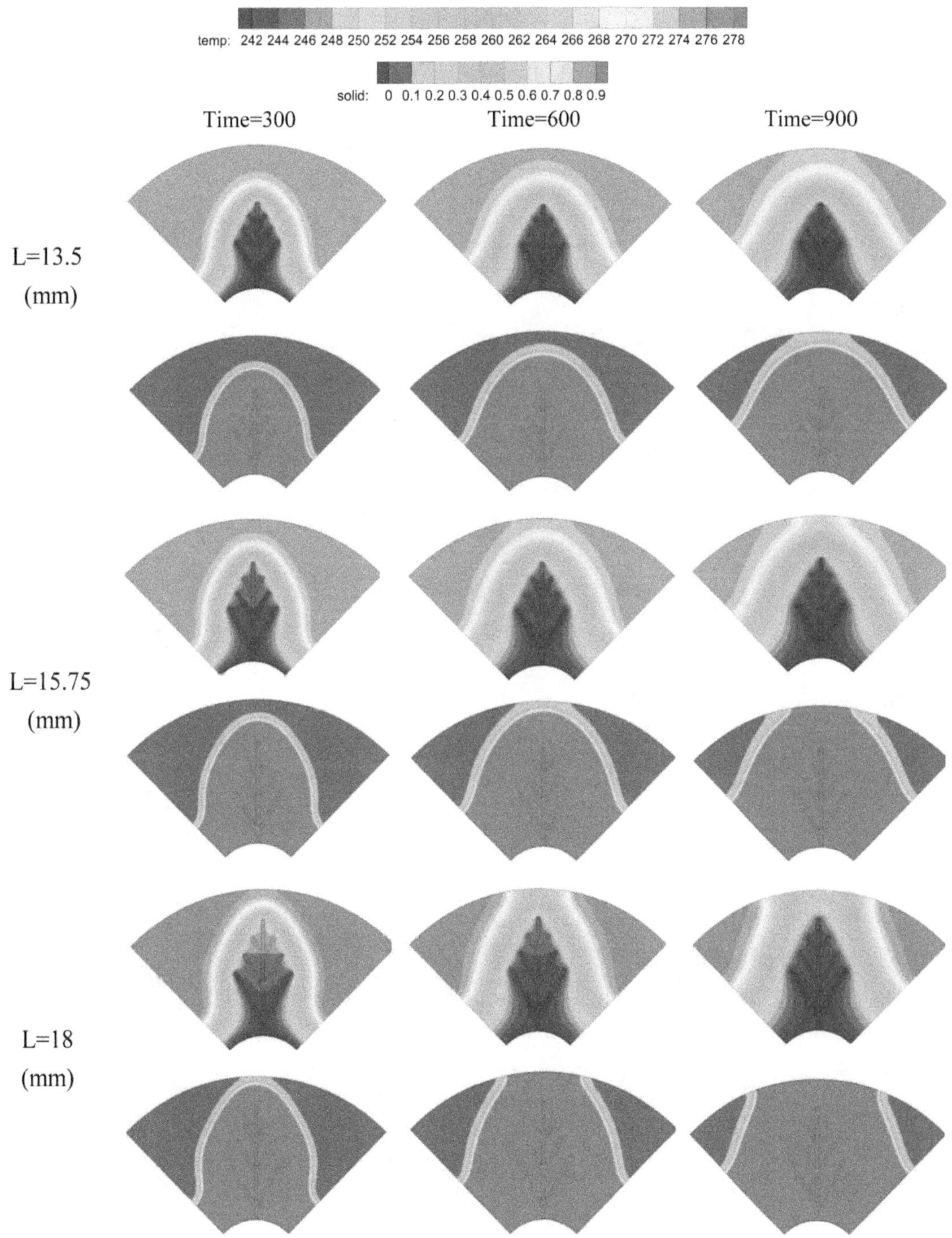

Figure 24. Temperature (Top) and solid fraction (Bottom) contour plots in $\theta=\pi/4$, $Sf=5.7$

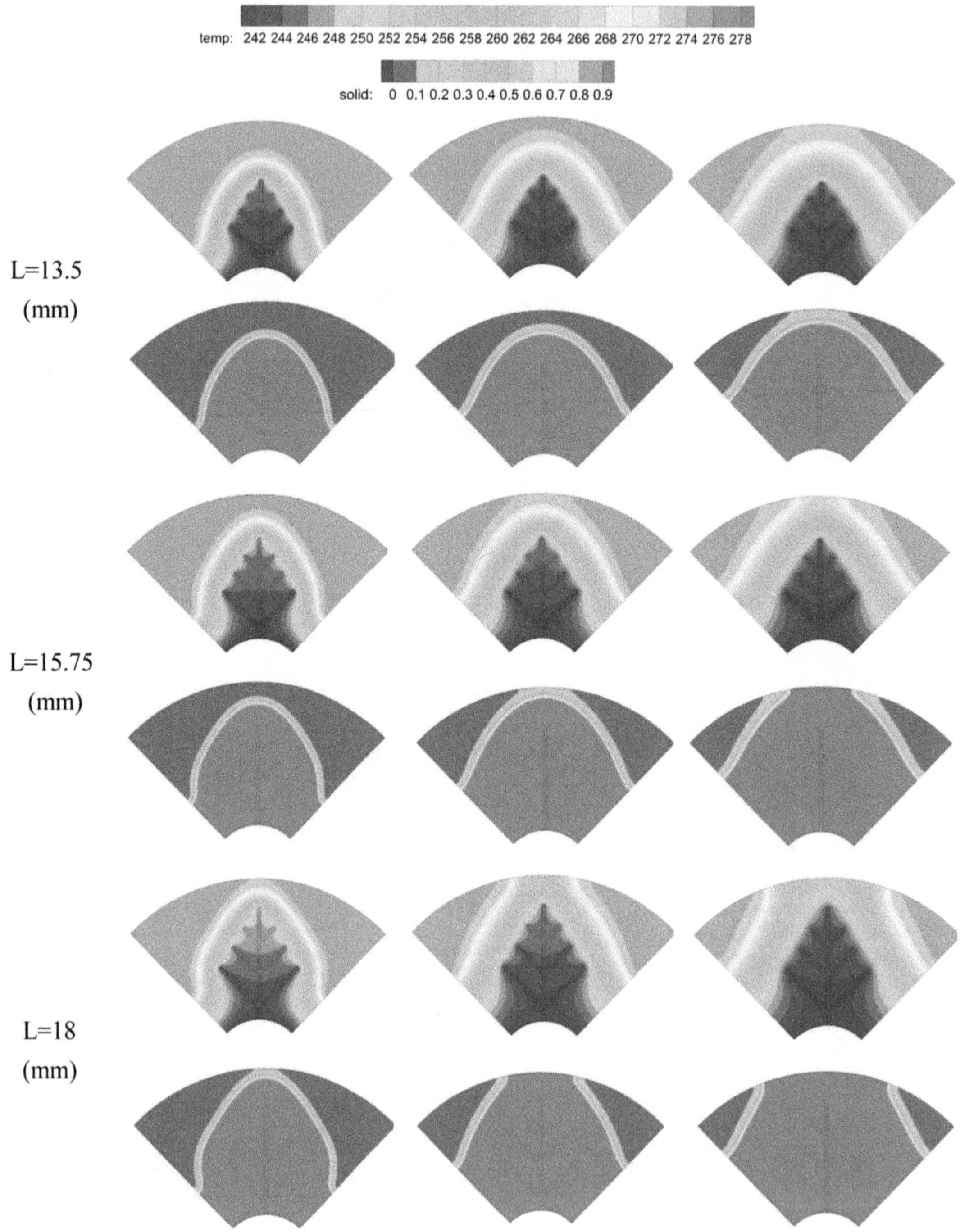

Figure 25. Temperature (Top) and solid fraction (Bottom) contour plots in $\theta=\pi/3$, $Sf=5.7$

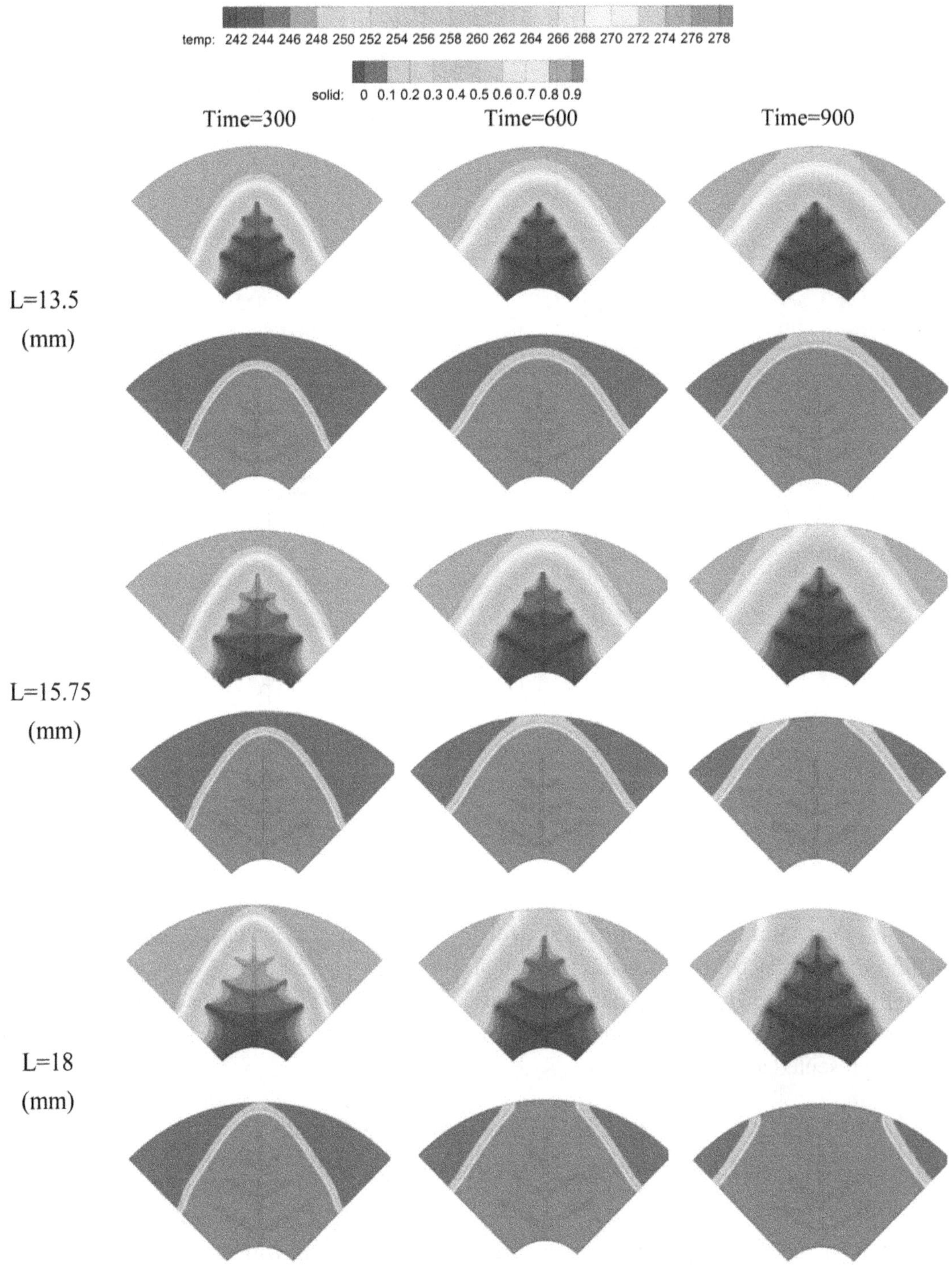

Figure 26. Effects of nanoparticles shape factor on (a) average temperature, (b) solid fraction, and (c) total energy in Y-shaped fin assisted LHTESS

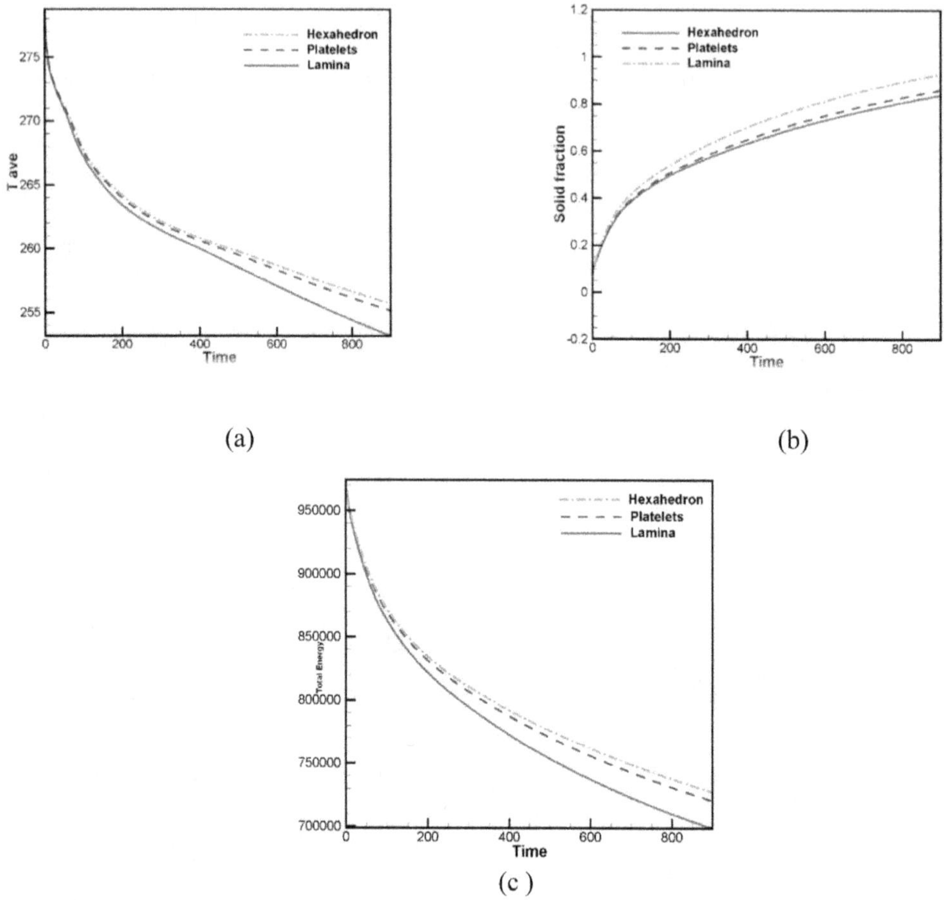

Figure 30 shows full solidification time as a function of nanoparticles shape factor, fin length and fin bifurcation angle. It can be seen from this figure that enhancing the nanoparticles shape factor, fin length and fin bifurcation angle will accelerate the solidification process due to enhanced thermal penetration depth. Applying the RSM, a correlation for full solidification time is extracted as a function of L, θ and Sf which is presented as below:

$$\begin{aligned} Time &= 7077.05 - 149.18(Sf) - 360.07(L) - 18.88(\theta) + 3.65(Sf)(L) \\ &+0.23(Sf)(\theta) - 0.368(L)(\theta) + 1.47284(Sf)^2 + 7.16(L^2) + 0.17(\theta^2) \end{aligned} \tag{5}$$

It can be concluded that by adding the hybrid nanoparticles along with Y-shaped fin configuration to the base PCM, the penetration depth enhances and as a result, the solidification process is completed in lower times.

Figure 27. Effects of L and θ on average temperature variations over computational domain when Sf=5.7

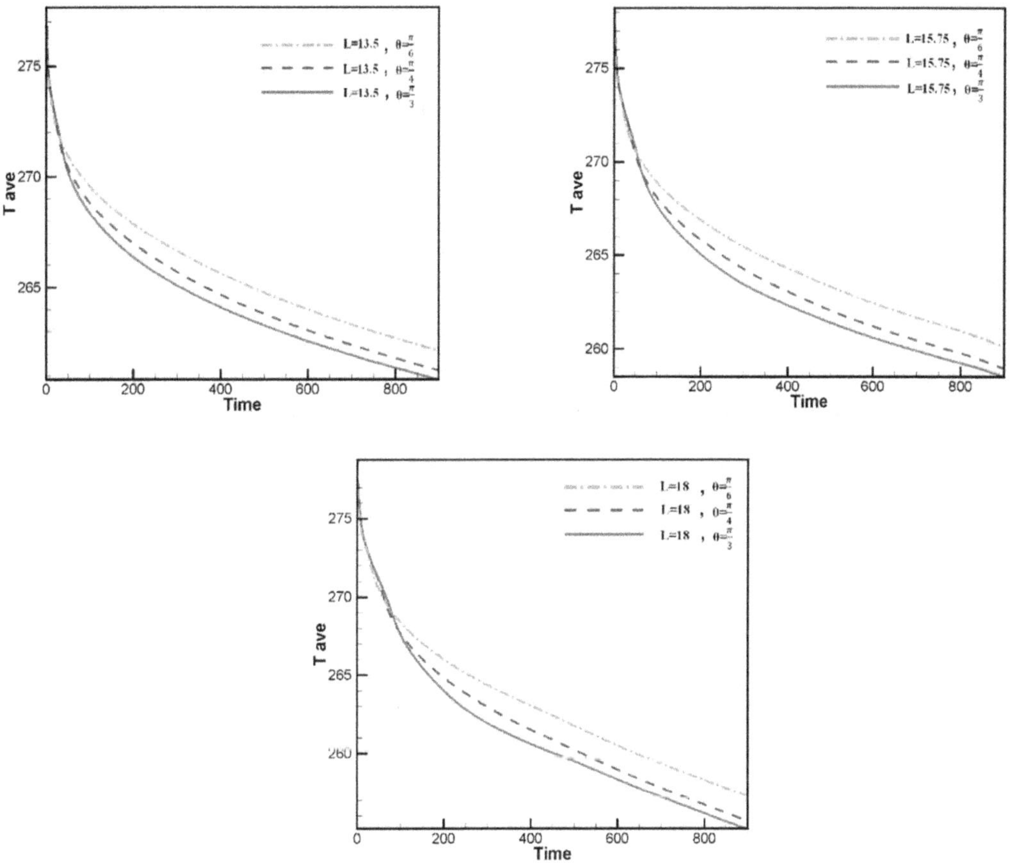

4. EXAMPLE 4: EFFECT OF NANOPARTICLE SHAPE FACTOR AND SNOWFLAKE CRYSTAL STRUCTURE ON DISCHARGING ACCELERATION LHTESS CONTAINING (Al_2O_3-GO) HNEPCM

4.1 Problem Definition

Figure 31 represents the schematic and boundary conditions of this problem. The cold fluid flows in the internal tube with snowflake crystal structure and the HNEPCM solidifies in the external shell. The internal wall is fixed at 240 K and the initial temperature of HNEPCM is 275 K.

Moreover, as the HNEPCM zone is liquid at the inception of procedure, its primary solid fraction is regarded to be 0. The thermophysical properties of PCM and nanoparticles are tabulated in Table 6.

As shown, the thermophysical properties of PCM are presented for both phases, because phase change occurs during the solidification procedure and neglecting the properties variation is not allowed. As the solidification procedure depends on time, an unsteady simulation with a time-step of $5\times10^{-5}(s)$ is regarded in this work. By considering the two-dimensional unsteady solidification procedure inside the system and also the radiation effect, the governing equations are given in chapter 6.

Figure 28. Effects of L and θ on solid fraction over computational domain when Sf=5.7

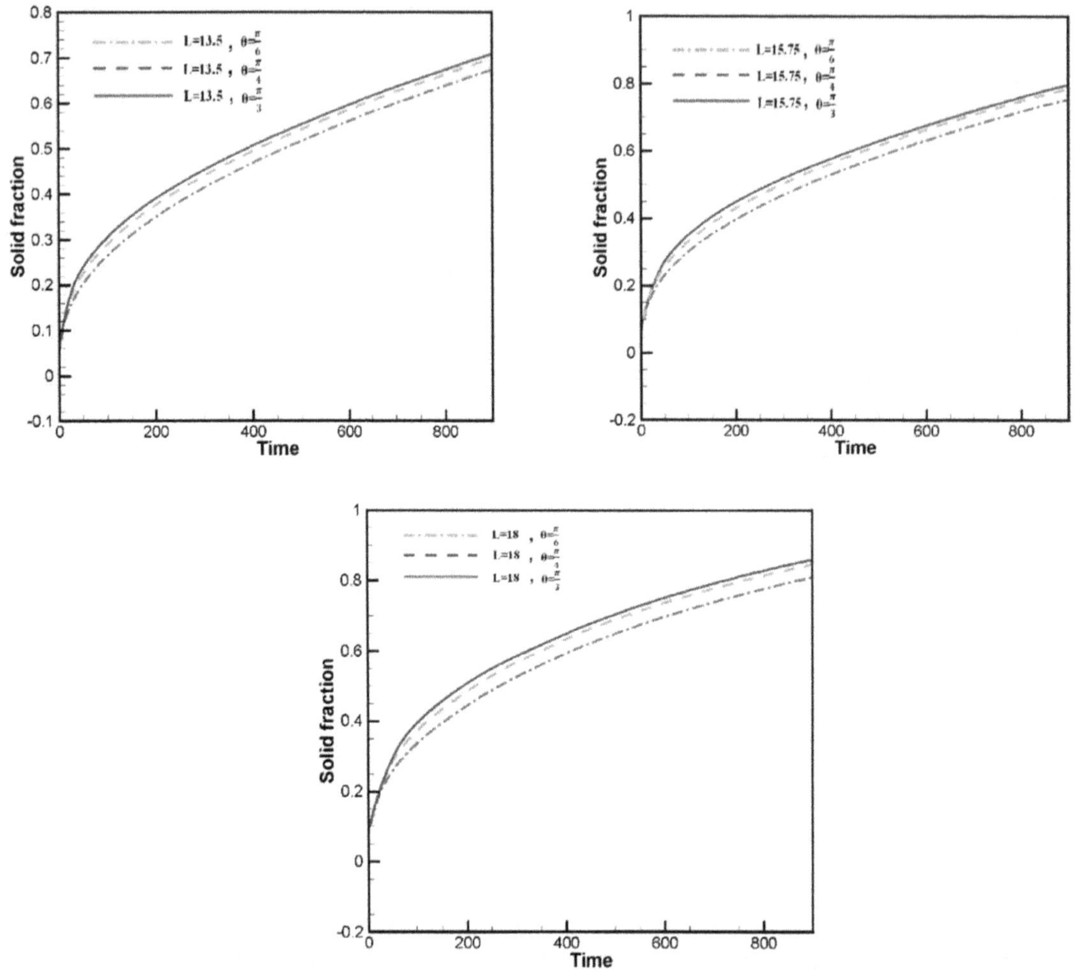

4.2 Numerical Simulation

In present study, the simulation of solidification process is carried out using SGFEM. In order to solve the phase change procedure, cubic interpolation is applied over triangles. In the problems with triangle grid, the calculations are performed on the corner of cells which is called as node. In this method, the differential equations are made by replacing the descriptions, applying the dependencies to the variables, integration over cells and differentiating the final system regarding to variables. Then, these differential equations complete the Galerkin matrix. As the solidification is a time-dependent process, an implicit backward method is employed for time integration and solving the system. Moreover, in this simulation an adaptive grid refinement strategy is used to refine the grid when the variable alterations are considerable. The refinement procedure continues during the process until the required precision which is determined by the user is obtained (Figure 32).

Figure 29. Effects of L and θ on total energy variations over computational domain when Sf=5.7

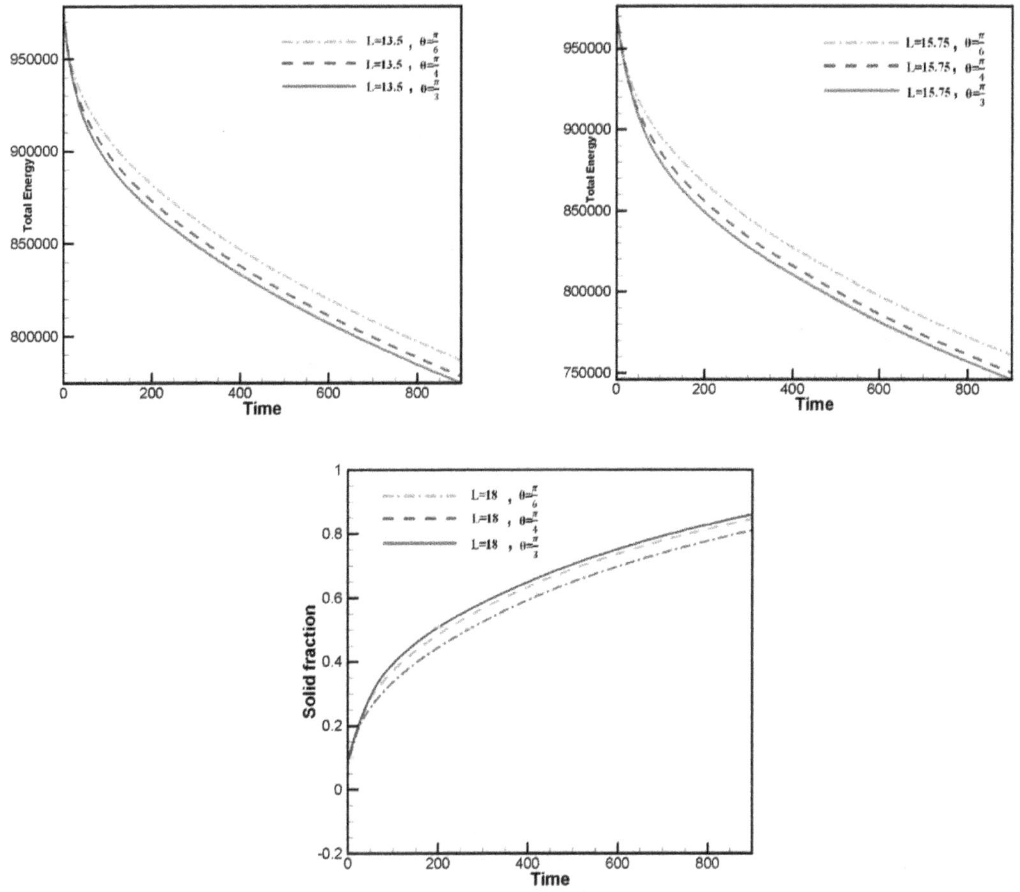

Moreover, a comparison between present numerical results and previous investigation (Ghadikolaei et al., 2018) is done to approve the accuracy of this simulation, as depicted in Fig. 33. This figure compares the solidification fronts in the system at three time steps and indicates an excellent agreement between the results, so that the maximum reported error is equal to 4.9%.

4.3 Impacts of Active Parameters

In present study, HNEPCM along with thermal radiation effect are employed to expedite the solidification procedure in a LHTESS with snowflake crystal structure inspired from the nature. Hybrid nanoparticles are dispersed to PCM with two different volume fractions. A numerical procedure based on SGFEM is used to simulate the solidification procedure. Role of nanoparticle volume fraction (φ= 0, 0.02, 0.04), nanoparticle shape factor (sf= 3.7, 5.7, 16.1) and the thermal radiation parameter (Rd= 0, 0.5, 1) on solidification procedure is investigated, in this research.

Figure 30. Full solidification time for different values of L, θ, and Sf

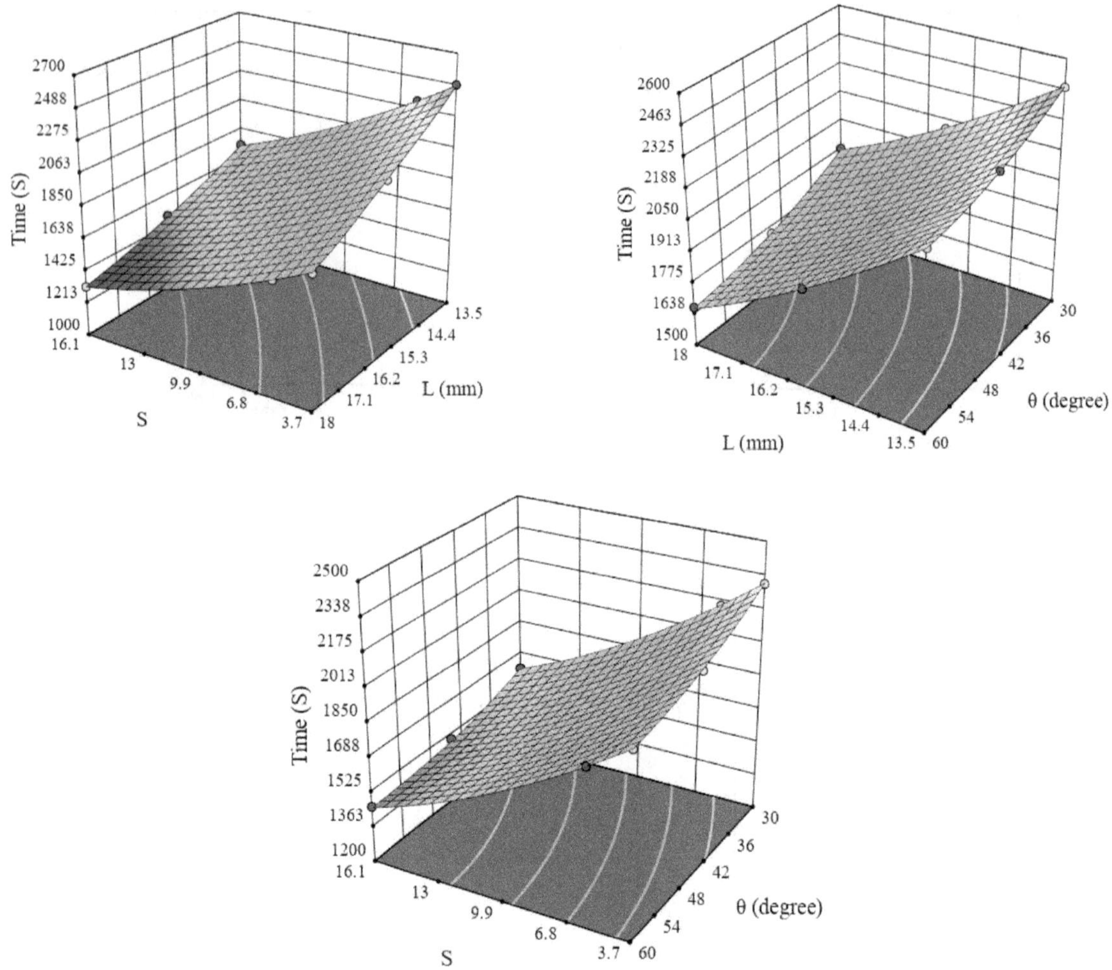

Figure 31. (a) Three-dimensional view of LHTESS with snowflake crystal structure, (b) two-dimensional geometry and boundary conditions of present study

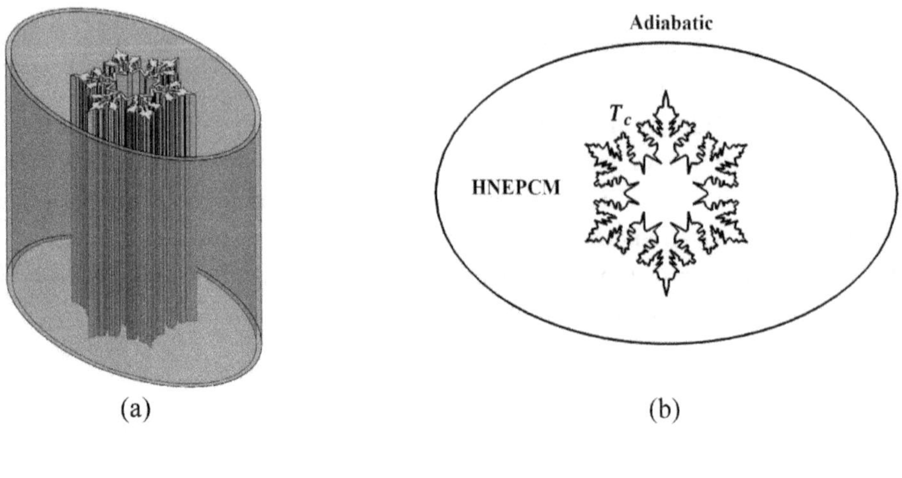

Table 6. The physical properties of water as PCM, hybrid nanoparticles

Property	PCM	Al_2O_3	GO
$\rho(\frac{kg}{m^3})$	997 (Liquid phase)	3970	1800
	916 (Solid phase)		
$C_p(\frac{J}{kgK})$	4184 (Liquid phase)	765	717
	2030 (Solid phase)		
$k(\frac{W}{mK})$	0.6 (Liquid phase)	40	5000
	2.22 (Solid phase)		
$L_f(\frac{J}{kg})$	335,000	-	-

Figure 32. Adaptive mesh refinement procedure when Rd=0.5 and sf=5.7

Figure 33. Comparison of present study with experimental study published by Ismaeil et al. (2001)

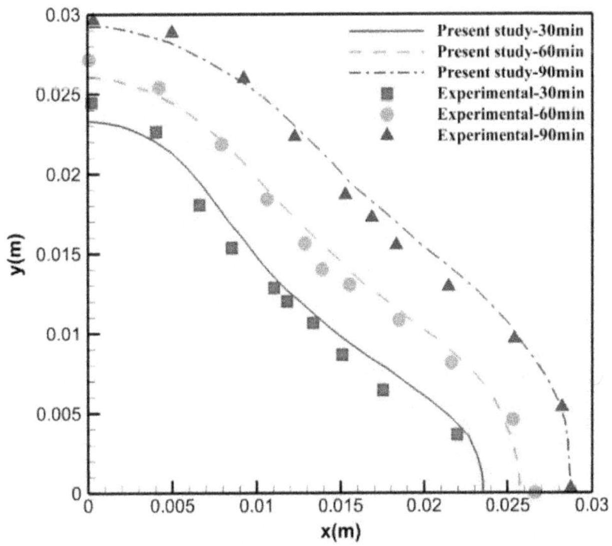

The influence of radiation parameter on the solidification procedure is illustrated in Figure 34 as isotherm and solid fraction contours, for three various time steps. It should be noticed that thermal radiation effect is taken into account by considering a heat source in the energy equation based on Rosseland approximation. As shown in Figure 34, raising the thermal radiation parameter causes the temperature of HNEPCM to be closer to the temperature of inner tube with snowflake structure, due to enhanced heat transfer rate.

This figure also depicts that higher solid fraction is obtained by enhancing the thermal radiation parameter. Because, increasing the thermal radiation results in augmenting the thermal penetration depth, which means that solidification procedure is completed in lower time. As a result, it can be said that the FST reduces considerably.

Figure 34. Temperature (Left) and solid fraction (Right) contour plots in ϕ=0.04, sf=5.7

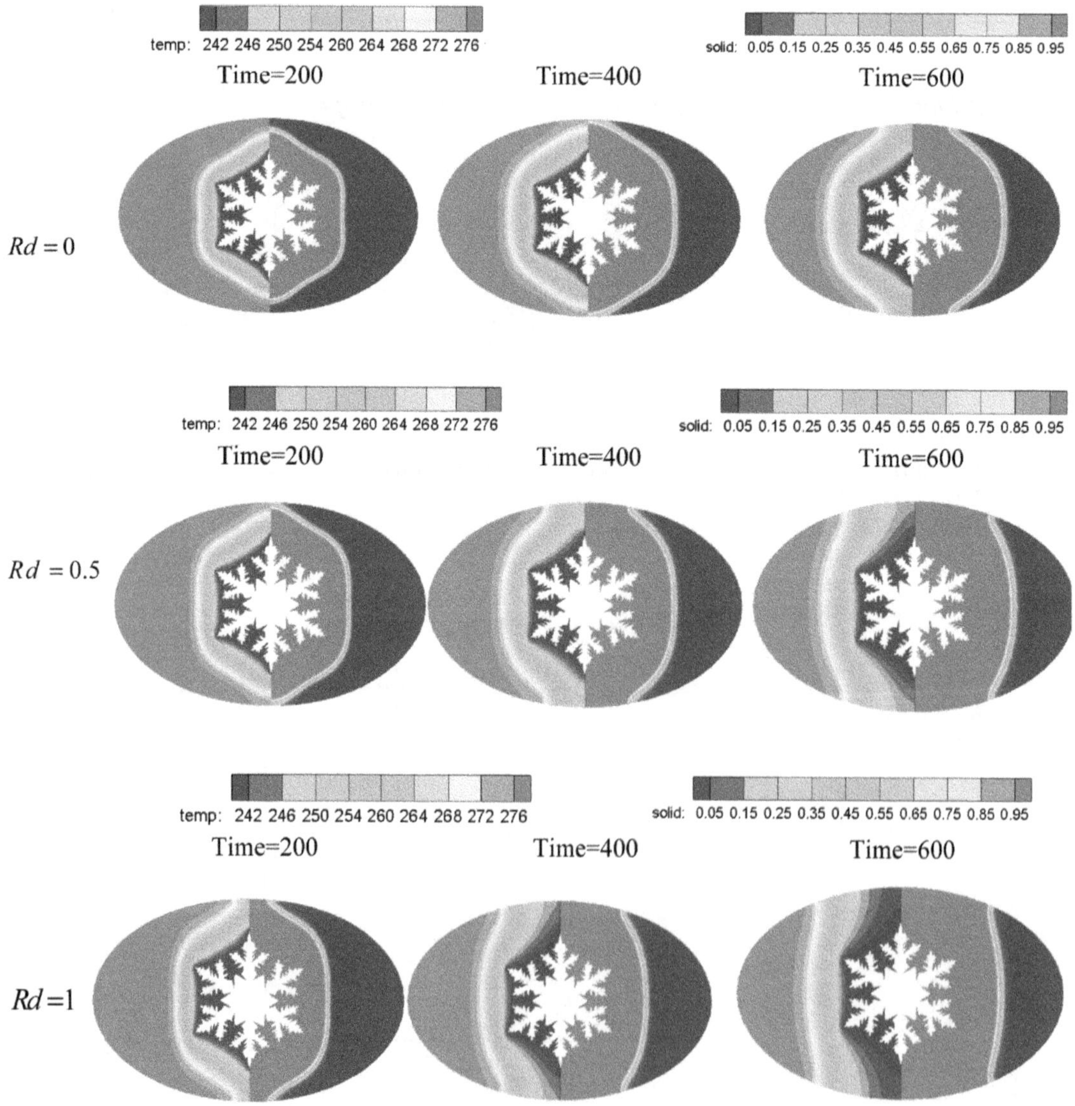

Figure 35 and Table. 7 display the influence of hybrid nanoparticles on the performance of LHTESS during the solidification procedure. It can be observed that the solid fraction is profoundly affected by adding hybrid nanoparticles to LHTESS. Because, the influence of adding nanoparticles on procedures with conduction heat is more considerable than when the natural convection is the predominant procedure. Therefore, as the conduction heat is predominant in solidification procedure, raising the nanoparticle volume fraction leads to augment the solid fraction. Figure 35 also indicates that the solidification rate will not change considerably by adding nanoparticles at the inception of the procedure, but by progress in time, the influence of hybrid nanoparticle dispersion into PCM on solidification rate is more remarkable. In addition, according to Table. 7, applying hybrid nanoparticles with volume fraction of 0.04 leads to the highest solidification rate so that an increment of 24.1% is obtained in this rate.

Table 7. Effect of nanoparticle volume fraction on FST and improvement

Nanoparticle Volume Fraction (φ)	FST (s)	Increment in Solidification Rate (%)
0	3218	-
0.02	2789	13.3
0.04	2440	24.1

Figure 35. Effect of nanoparticle volume fraction on solidification rate at three various time steps

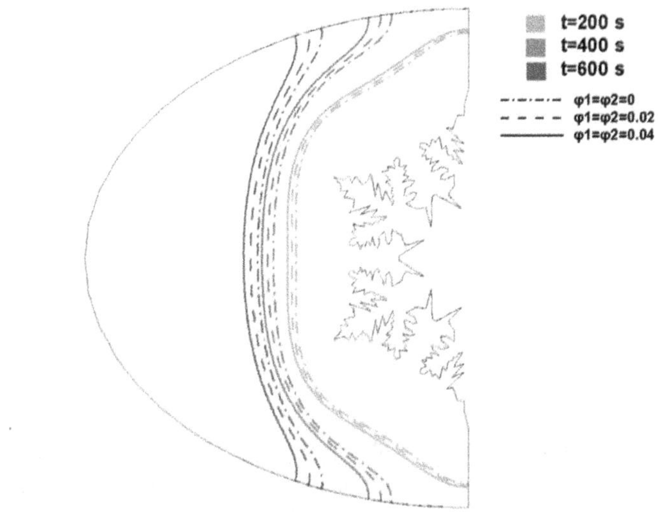

The variations of average temperature, total energy and solid fraction by changing the nanoparticles volume fraction over the solution domain are illustrated in Figuer 36 (a, b, c). It can be seen from Figure 36(a) that the average temperature of the system experiences lower values by augmenting the nanoparticles volume fractions, which means that the temperature of solution domain becomes closer to the cold wall. According to equation in chapter 3, raising the nanoparticles volume fraction leads to decrease the latent heat of fusion and heat capacity. Therefore, the total energy of the system which is defined as sum of these two heats (Eq. 6) drops considerably. Obtained information from Figure 36(c) and Figuer 35 is the same. Both figures indicate that increasing the nanoparticles volume fraction leads to increase the solidification rate. In this situation, full solidification occurs in lower time.

Figure 36. Effects of nanoparticles volume fraction on a) average temperature, b) total energy, and c) solid fraction over computational domain when sf=5.7 and Rd=0.5

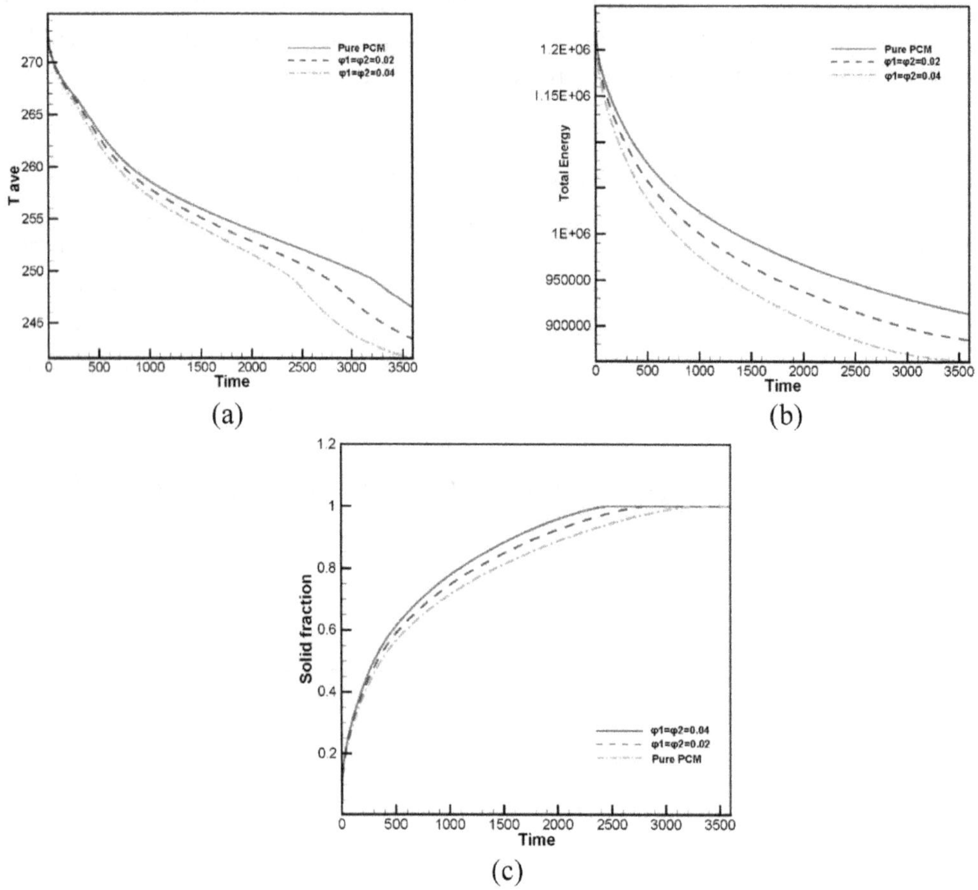

Nanoparticles shape factor is the second aspect of hybrid nanoparticle characteristics which is reviewed in this paper. Figure 37(a, b, c) demonstrates the influence of nanoparticles shape factor on average temperature, total energy and solid fraction, respectively. Three shapes of nanoparticles such as: the Hexahedron, the Platelets and the Lamina with different shape factors are investigated.

Thermal conductivity of HNEPCM and in sequence, thermal penetration depth of the system increase by augmenting the nanoparticles shape factor. Therefore, the average temperature of the system decreases during the process, as displayed in Figure 37(a). Figure 37(b) shows that energy storage capacity experiences its maximum value at the inception of solidification procedure. Also, among applied shapes, using Hexahedron nanoparticles (sf=3.7) leads to higher values of total energy. Finally, as the average temperature and total energy of the system decrease by raising the nanoparticles shape factor, the solid fraction increases and the solidification procedure is accelerated.

Figure 37. Effects of nanoparticles shape factor on (a) average temperature, (b) total energy, and (c) solid fraction in LHTESS with snowflake structure when Rd=0.5 and ϕ=0.04

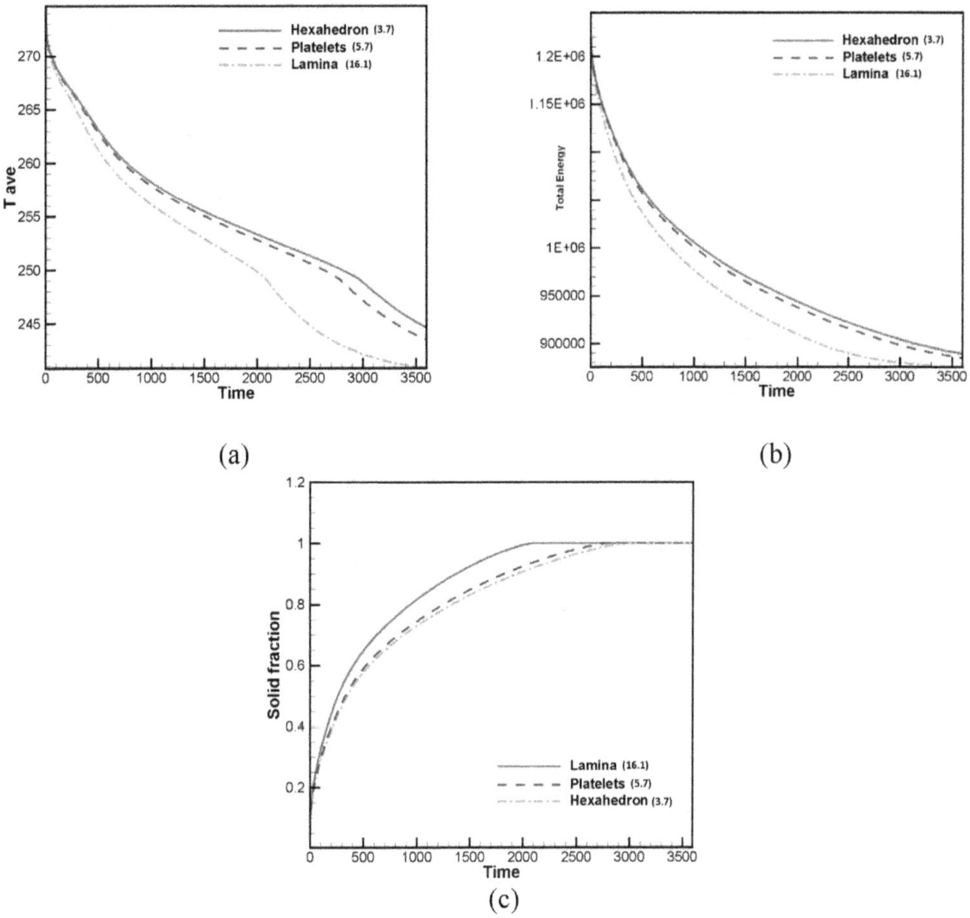

Figure 38 represents the variation of FST and total energy by changing the nanoparticles volume fraction, nanoparticles shape factor and radiation parameter. It can be observed that enhancing all these parameters will accelerate the solidification procedure due to enhanced thermal conductivity and thermal penetration depth. Also, two correlations for FST and total energy of the system are extracted as a function of *sf*, *Rd* and *ϕ* which are presented as below:

$$Time = 3702.98 - 17065.88(\phi) - 23.82(sf) \\ -852.99(Rd) + 1618.11(\phi)(sf) + 17121.46(\phi)(Rd) + 38.72(sf)(Rd) \tag{6}$$

$$Total\ Energy = 923733 - 908339(\phi) - 8.57(sf) - 962.56(Rd) + 17.07(\phi)(sf) \\ +10833.66(\phi)(Rd) + 1.93(sf)(Rd) + 185770(\phi^2) + 0.41(sf^2) + 173.55(Rd^2) \tag{7}$$

Figure 38. FST (Left) and Total energy (Right) as a function of Rd, ϕ and sf

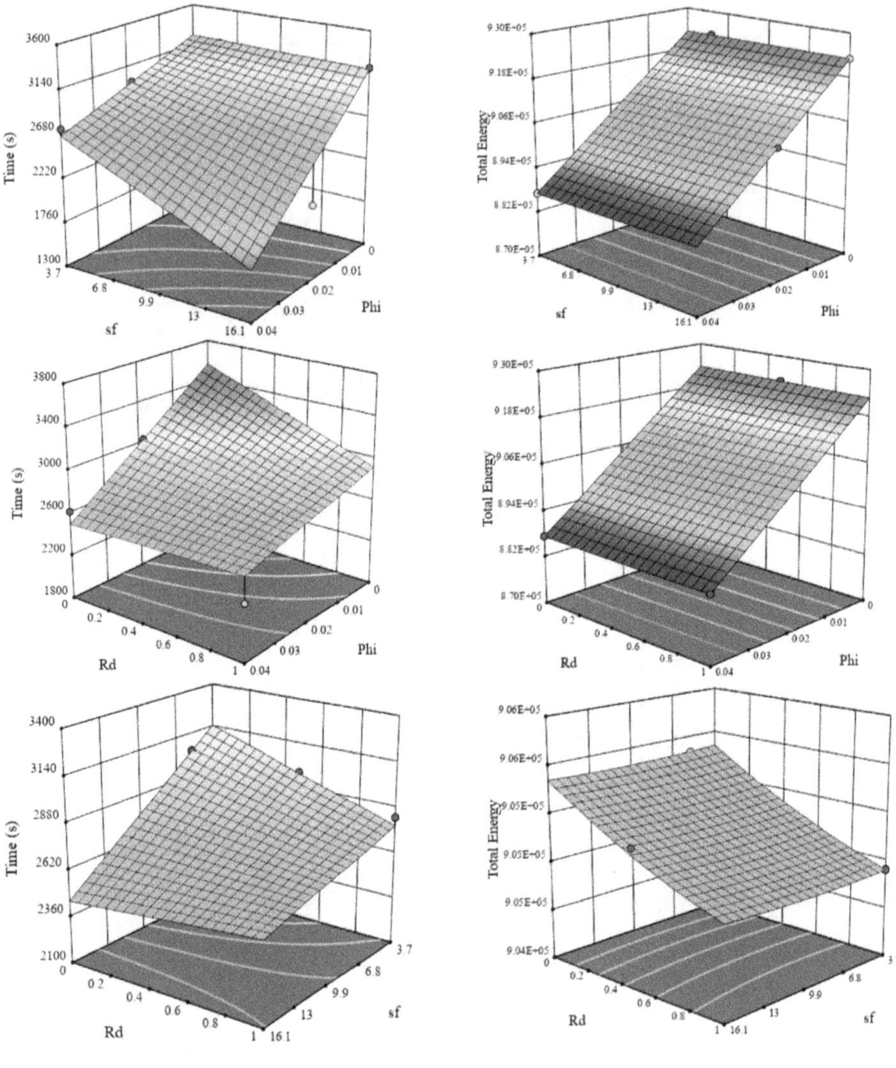

5. EXAMPLE 5: SOLIDIFICATION PROCESS OF PHASE CHANGE MATERIAL IN A LHTESS IN THE PRESENCE OF FINS WITH VARIABLE THICKNESS AND HYBRID NANOPARTICLES

5.1 Problem Definition

Figure 39 demonstrates the geometry and boundary conditions of present study. The Heat Transfer Fluid (HTF) flows in the inner tube and the PCM solidifies in the storage tank shell. The inside wall is kept at constant temperature of 240 K and the initial temperature of PCM is equal to 278 K. Also, as the PCM region is liquid at the beginning of process, its initial solid fraction is considered equal to 0. The inner and outer diameter of the LHTESS are equal to 20 and 80 mm, respectively. In order to enhance the thermal penetration depth in the PCM region, fins with variable thickness are connected to the inner tube. The half fin angle (β), the fin height relative to the annular gap (h) and the number of fins (m) are the geometric parameters which are considered to obtain the best fin configuration (Table. 8).

Figure 39. (a) Three-dimensional view of fin with variable thickness assisted LHTESS, (b) two-dimensional geometry and boundary conditions of present study

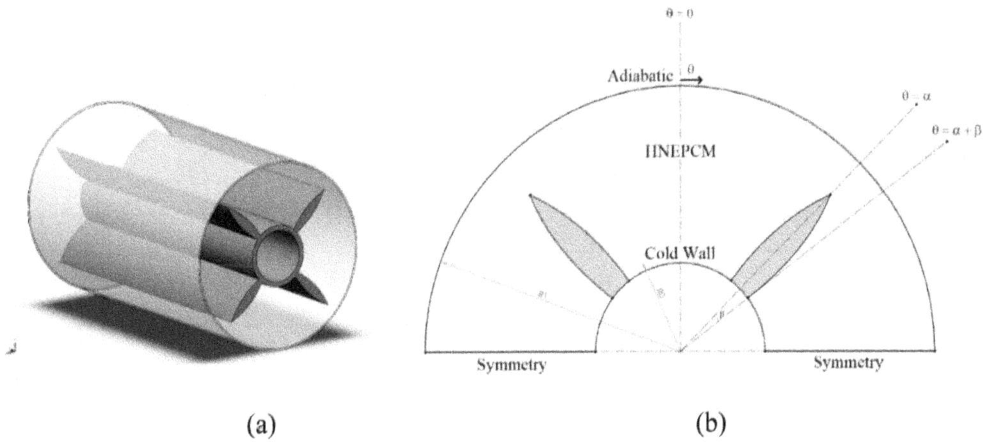

Table 8. Geometric parameters of fin with variable thickness

Geometry Parameter	Range
β	[2°, 4°, 6°, 8°]
h	[0.25, 0.5, 0.75]
m	[2, 3, 4]

In this study, $Al_2O_3 - Go$ /water is used as HNEPCM and the physical properties of used PCM, hybrid nanoparticles and copper fin are summarized in Table. 9. As it can be seen in this table, the physical properties of used PCM are listed for both liquid and solid phases, because the solidification is a phase changing process and the variation of properties cannot be neglected.

It should be note that the solidification is a time-dependent process, so it should be simulated unsteadily. In present work, an unsteady simulation with the time-step equal to $2\times10^{-5}(s)$ is carried out. According to previous studies and abovementioned points, the governing equations of solidification process in present work are based on following assumptions:

- The solidification process in the LHTESS is considered to be two-dimensional in Cartesian coordinates (x, y).
- The simulation of solidification process in the LHTESS is considered to be unsteady.
- Conduction heat transfer mechanism is the predominant process in the simulation and natural convection is neglected.

- The physical properties of PCM change during the solidification due to phase change in the process.

Based on above assumptions, the governing equations of heat and mass transfer during the solidification process in the LHTESS in two-dimensional Cartesian coordinates (x, y) are fully given in chapter6.

Table 9. The physical properties of water as PCM, hybrid nanoparticles, and fin

Property	PCM	Al_2O_3	GO	Fin
$\rho\,(\frac{kg}{m^3})$	997 (Liquid phase)			8960
	916 (Solid phase)			
$C_p\,(\frac{J}{kgK})$	4184 (Liquid phase)			385
	2030 (Solid phase)			
$k\,(\frac{W}{mK})$	0.6 (Liquid phase)			400
	2.22 (Solid phase)			
$L_f\,(\frac{J}{kg})$	335,000	-	-	-

5.2 Numerical Simulation

The Standard Galerkin Finite Element Method is used to simulate the unsteady heat and mass transfer during solidification process. Implementing cubic interpolation over triangles, this method solves the problem. The parameters are calculated on the nodes which are located on the corners of cell. By substituting the definitions, dividing the dependencies on variables, integrating over cells and finally making differential equations based on variables, the Galerkin couple matrix is formed. The final calculations are solved by the Newton-Raphson method. This method solves the system of equations by an implicit backward difference approach. The solidification process is simulated by utilizing the adaptive mesh as shown in Fig. 40. When the adaptive mesh is employed, the grids will be refined where the variation of variables is rapid, such that the desired accuracy is obtained.

In present work, the numerical method is validated by comparing the results based on Galerkin Finite Element Method with previous experimental study published by Ismaeil et al. (2001). It can be observed from Figure 41 that this comparison represents an excellent agreement with the maximum error of 4.9%. Also, it can be concluded that the assumption of neglecting natural convection in the simulation is logical and close to reality.

Figure 40. Adaptive mesh refinement in three time steps including 100, 200, and 400 (s)

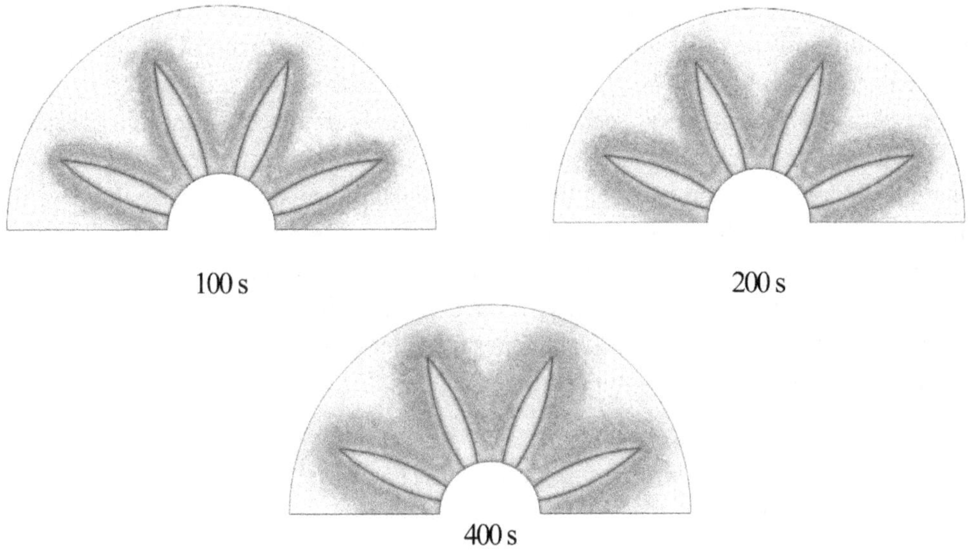

Figure 41. Comparison of present study with experimental study published by Ismaeil et al. (2001)

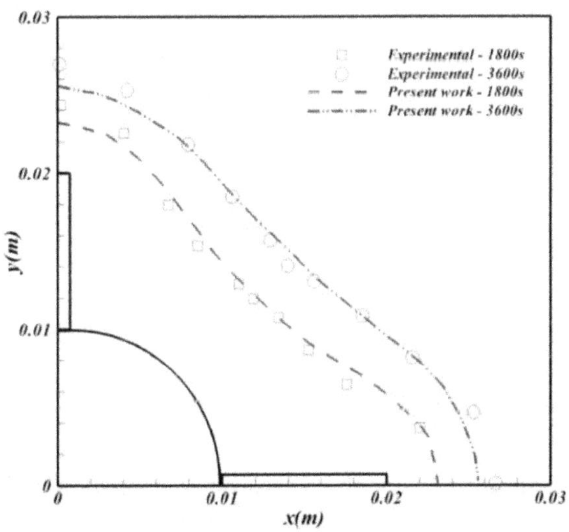

5.3 Impacts of Active Parameters

5.3.1 The Reason of Using Fin in LHTESS

LHTESS is used to store the released thermal energy during the solid-liquid phase change of PCM. The low thermal conductivity of pure PCM makes it difficult to achieve higher heat transfer rates during the solidification process. Therefore, it is necessary to employ a proper technic for overcoming this defect. In this study, fins with variable thickness are used in LHTESS to compensate the low thermal conductivity of pure PCM. As shown in Fig. 42, when no fins are employed in the LHTESS, the solidification process starts only near the inner cold wall. But, applying fins to the LHTESS leads to enhance the thermal penetration depth to PCM region. As a result, in addition to regions near the cold wall, the solidification process begins in the adjacency of fin surface. Therefore, the heat transfer rate enhances which causes a considerable expedition in solidification process. This acceleration is more remarkable in systems with the long distance between the heat source and heat sink, because the pure PCM with low thermal conductivity and high thermal resistance fills this space which decelerates the solidification process. It is worth mentioning that for comparing the solidification front in cases with fins and without fin, the PCM mass should be equal in two cases. Since applying fins to the system causes less space in LHTESS for PCM which reduces its mass, the inner diameter of LHTESS without fin is considered to be larger than the other one, as shown in Figure 42.

Figure 42. Comparison between solidification front in LHTESS without fin and LHTESS with variable thickness fin at three time steps

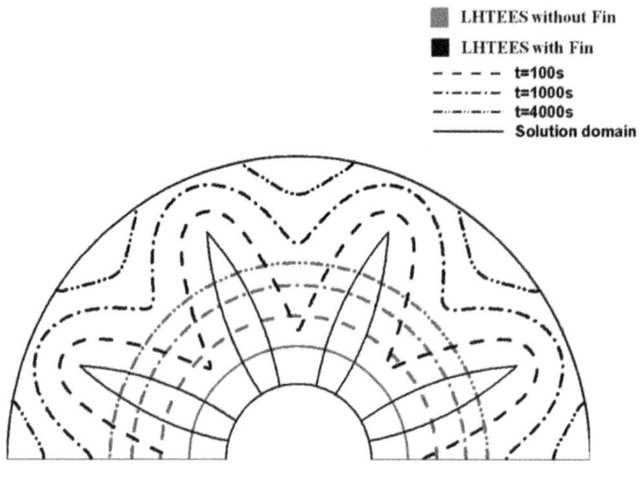

5.3.2 Optimization of Fins With Variable Thickness

In this section, RSM is used to optimize the geometric parameters of applied fins with variable thickness including the half fin angle (β), the fin height relative to the annular gap (h) and the number of fins (m) during the solidification process, as shown in Figure 39 (b). In this optimization, the objective function is to minimize the full solidification time, in order to expedite the phase change process. As shown in Figure 43, the fin height and the number of fins profoundly affect the solidification rate, but varying the half fin angle will not change the solidification rate, considerably. It can be concluded from Fig. 5 that increasing the half fin angle results in enhancing the heat transfer rate only due to an increase in heat transfer area, but the thermal penetration depth will not be enhanced to PCM region by increasing the half fin angle. Therefore, increasing the half fin angle leads to a slight increase in the solidification rate. It can be concluded that selecting the maximum half fin angle of studied range (β=8°) leads to the optimal case from the viewpoint of minimum full solidification time.

Figure 43. Response surfaces of full solidification time for geometric parameters of fin with variable thickness (β,h,m)

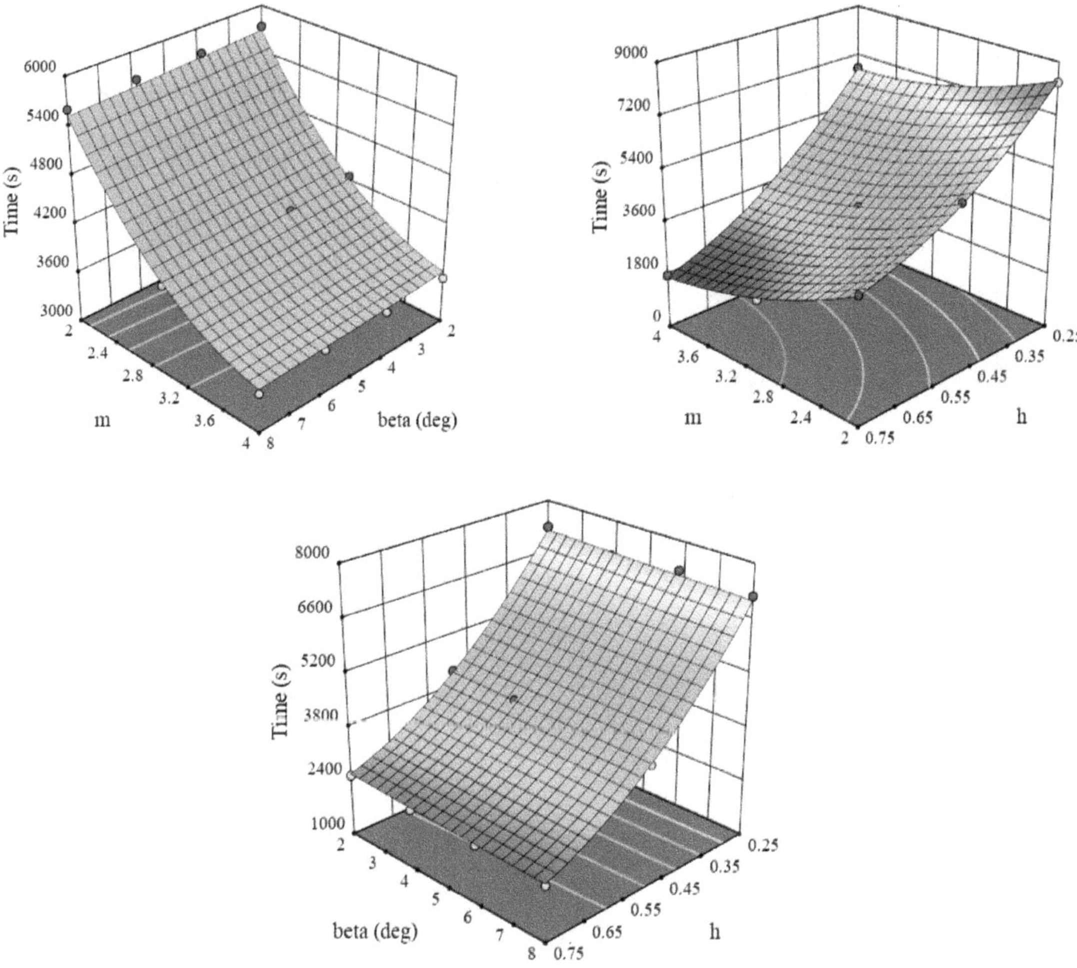

Figure 44 represents the effect of fin height relative to the annular gap on solidification rate. It can be observed from this figure that the solidification rate experiences higher values by increasing the fin height due to enhanced thermal penetration depth to PCM. Therefore, the maximum fin height relative to the annular gap in the studied range (h=0.75) is its optimized value for accelerating the solidification process.

The variation of full solidification time by varying the number of applied fins is represented in Figure 45. As mentioned before, applying fins enhance the heat transfer rate from two viewpoints. Firstly, adding conductive fins will compensate the low thermal conductivity of PCM during the solidification process. Secondly, the number of areas in which the solidification process is started will be increased due to enhanced thermal penetration depth. It can be concluded that increasing the number of fins intensifies these effects and improves the heat transfer rate. Therefore, the optimized value of the number of fins for achieving the minimum full solidification time is equal to the maximum in the studied range (m=4).

Figure 44. Effect of fin height relative to the annular gap on full solidification time when a) m=2, b) m=3, and c) m=4

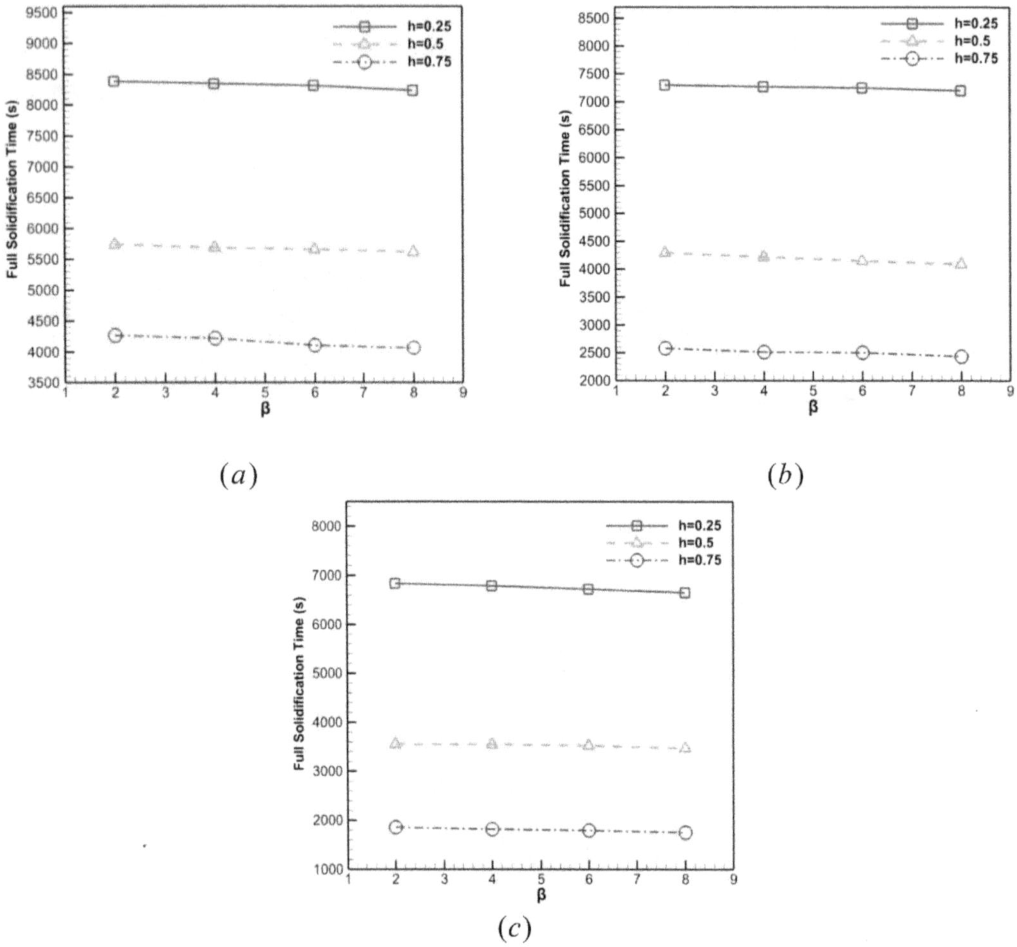

According to abovementioned discussions, the best fin configuration which leads to expedited solidification process include the highest half fin angle (β=8°), the highest fin height relative to the annular gap (h=0.75) and the maximum number of fins (m=4) in the investigated range. It should be mentioned that by applying these geometric parameters to the desired fin, the solidification process takes 1750.98s to be completed. The temperature and solid fraction contour plots for optimized fin configuration (m=4, β=8°, h=0.75) during the solidification process are presented in Figure 46, for three different time steps including 100s, 200s and 400s. It is worth mentioning that accelerating the liquid-solid phase change process and obtaining uniform temperature distribution is desired in this study. Therefore, it can be concluded that the thermal penetration depth can be enhanced into the PCM uniformly by employing optimized fin configuration. The optimized fin system also accelerates the solidification process in such a way that no unsolidified PCM remains near the outer wall. Since remaining unsolidified PCM with high thermal resistance and low thermal conductivity in the LHTESS weakens the solidification process, optimized fin with variable thickness are employed to reduce this PCM mass and overcome the problem as shown in Figure 46.

Figure 45. Effect of the number of fins on full solidification time when a) h=0.25, b) h=0.5, and c) h=0.75

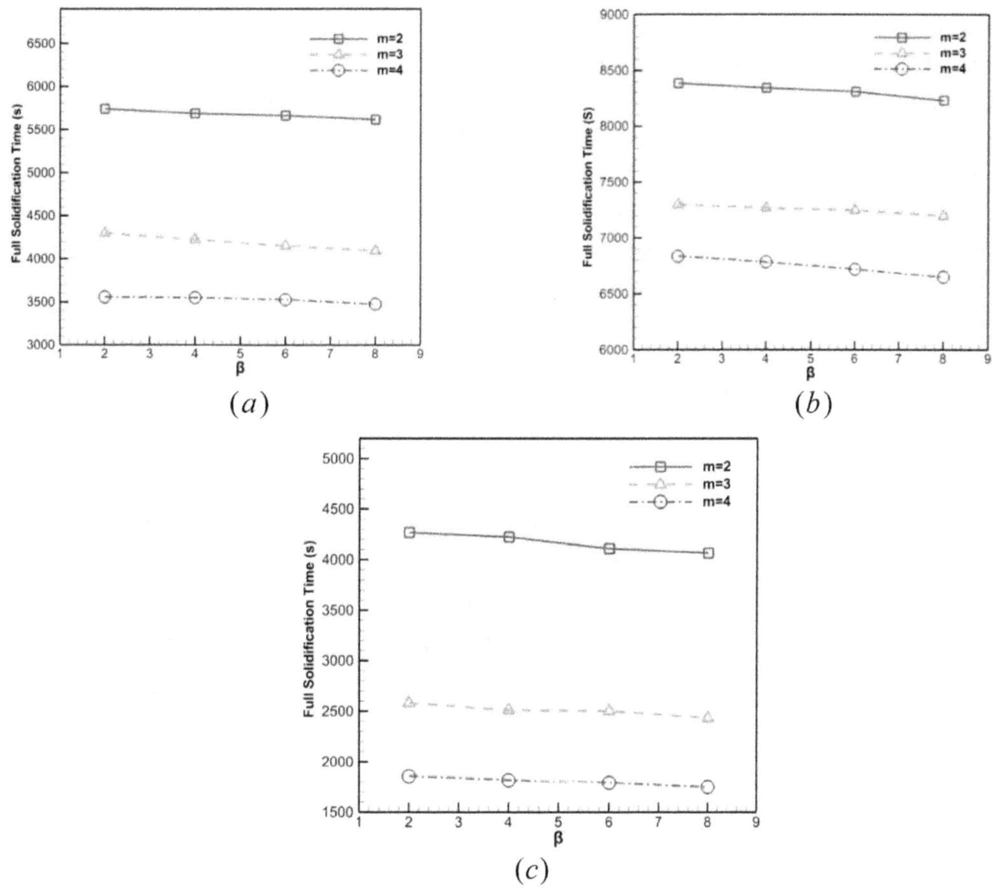

Figure 46. Solid fraction (top) and temperature (bottom) contour plots in three time steps during the solidification process for optimized fin configuration (β=8°, h=0.75, m=4)

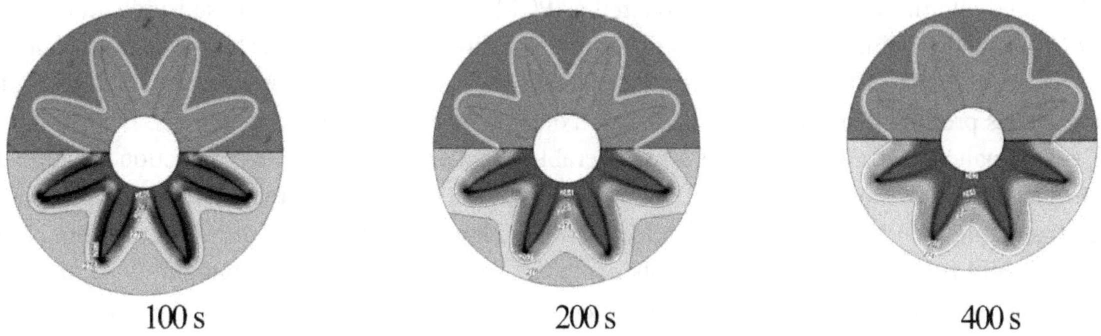

From other hand, as mentioned before, the shape of employed fin in this study is considered as the novelty of current investigation. In order to prove the advantage of using this fin shape rather than simple longitudinal fin, a numerical simulation is carried out with simple longitudinal fins which their area and length are equal to the optimized ones. Figure 47 shows the temperature and solid fraction contour plots of LHTESS enhanced with simple longitudinal fins. Moreover, a comparison between the LHTESS enhanced with optimized variable thickness fins and LHTESS enhanced with simple longitudinal fins is carried out from the viewpoint of full solidification time, as shown in Figure 48. It can be observed from this figure that using optimized fins with variable thickness causes the solidification process to be completed 9.93% faster than the case of simple longitudinal fins. This Increment in solidification rate is why this fin shape is considered as the novelty of current study.

Figure 47. Solid fraction (top) and temperature (bottom) contour plots in three time steps during the solidification process for simple longitudinal fin

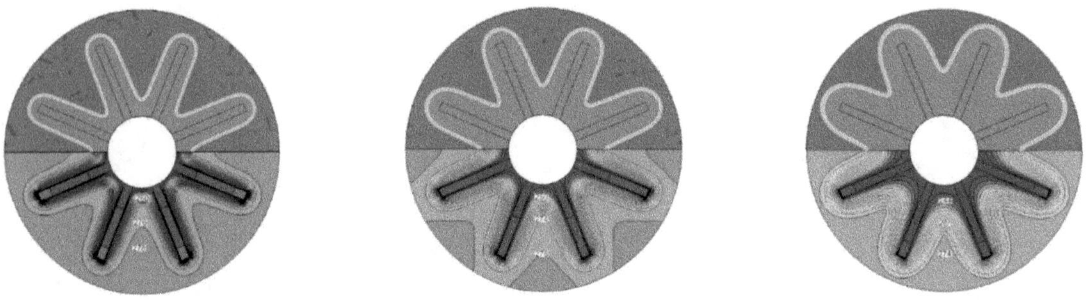

100 s　　200 s　　400 s

5.3.3 Hybrid Nanoparticle Dispersion Into the Pure PCM

In this section, the effect of adding hybrid nanoparticles to pure PCM and making Hybrid Nano-Enhanced PCM is studied. In this study, Al_2O_3-GO hybrid nanoparticles with three volume fractions of 0, 2.5, 5% are dispersed into the water which is considered as PCM. No slip between the base fluid and nanoparticles is assumed. Figure 49 and Table 10 illustrate the influence of Al_2O_3-GO hybrid nanoparticles on the performance of LHTESS during the solidification process. It can be observed that adding hybrid nanoparticles profoundly affects the solid fraction during the solidification process.

Because nanoparticle dispersion has a considerable effect on processes with conduction heat transfer mechanism, but it has no considerable effect when the natural convection is the predominant procedure. Since the conduction heat transfer mechanism is the predominant process in the solidification, higher values of solid fraction are obtained by enhancing the nanoparticles volume fraction. Also, it can be found from this figure that enhancing the nanoparticles volume fraction will not change the solidification rate, considerably, at the beginning of the process.

But, by time progressing, the effect of dispersing hybrid nanoparticles into the pure PCM on the solidification rate becomes more significant. This is because of this fact that at the beginning of the process, the solidification happens only near the cold wall which has a high temperature difference with unsolidified PCM.

Figure 48. Comparison between solid fraction of LHTESS enhanced with optimized fin with variable thickness and LHTESS enhanced with simple longitudinal fin during the process

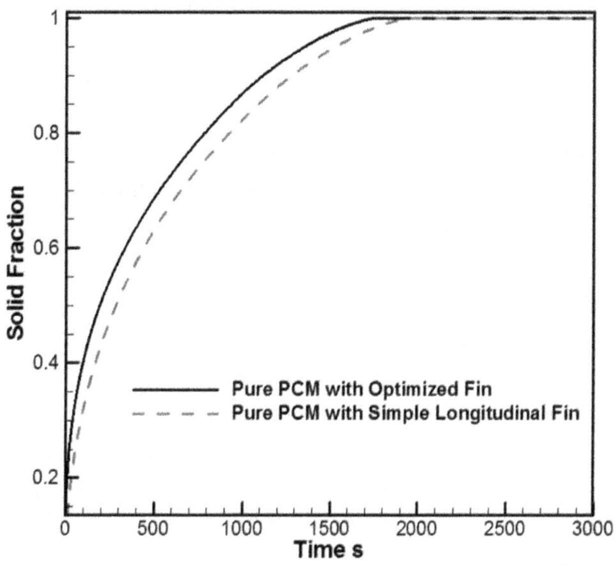

Table 10. Effect of hybrid nanoparticles volume fraction on full solidification time and acceleration rate

Nanoparticles Volume Fraction (ϕ)	Full Solidification Time (s)	Increment in Solidification Rate (%)
0.0%	11200	-
2.5%	6421.91	42.63
5.0%	4150	62.94

Figure 49. Effect of hybrid nanoparticle volume fractions on solidification front in three time steps

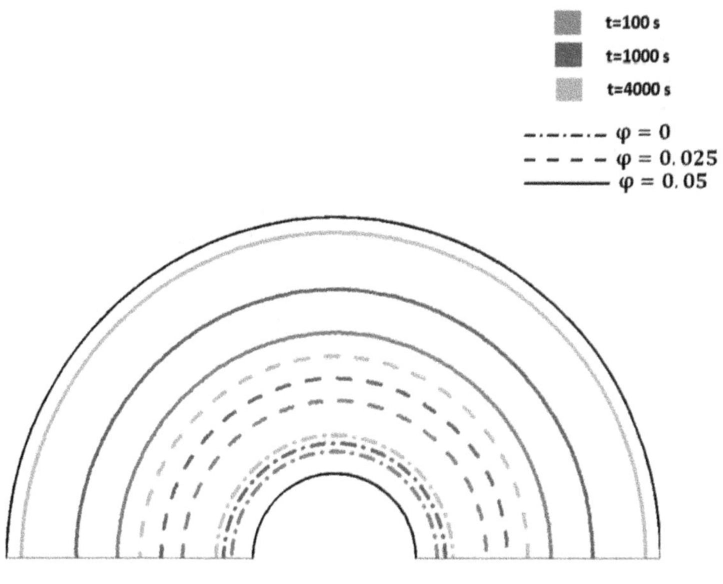

This temperature difference leads to accelerate the solidification process and no enhancement techniques are needed. But, by time progressing, ice with low thermal conductivity fills the space between the cold wall and unsolidified PCM which causes a high thermal resistance in the system and weakens the solidification process and needs a suitable technique to be overcome. This is why the effect of nanoparticle dispersion on the solidification rate is more considerable with time progressing.

5.3.4 Comparison Between Applying Fins With Variable Thickness and Hybrid Nanoparticle Dispersion as the Enhancement Techniques

In this section, a comparison between enhancement techniques including fin addition and nanoparticle dispersion is investigated form the viewpoint of both solidification rate and maximum energy storage capacity. Figures 50 and 51 are presented to compare the outcomes of using fin with variable thickness and hybrid nanoparticle as the enhancement techniques. Figure 50 shows the solid fraction during the solidification process and the slop of this curve is defined as the solidification rate. It can be observed that variable thickness fin-assisted LHTESS experiences higher solidification rate than the case of hybrid nano-enhanced LHTESS, so it needs lower time for full solidification. By applying the optimized fin configuration, full solidification time becomes 6.396 times faster than the case of LHTESS without fin and also 2.37 times faster than the case of hybrid nanoparticle dispersion with 0.05 volume fraction. This figure also represents that employing both enhancement techniques including optimized fin and hybrid nanoparticle dispersion simultaneously leads to the highest solidification rate.

Figure 50. Comparison between solid fraction of LHTESS with pure PCM without fin, pure PCM with optimized fin, HNEPCM with 2.5 and 5% volume fractions during the solidification process

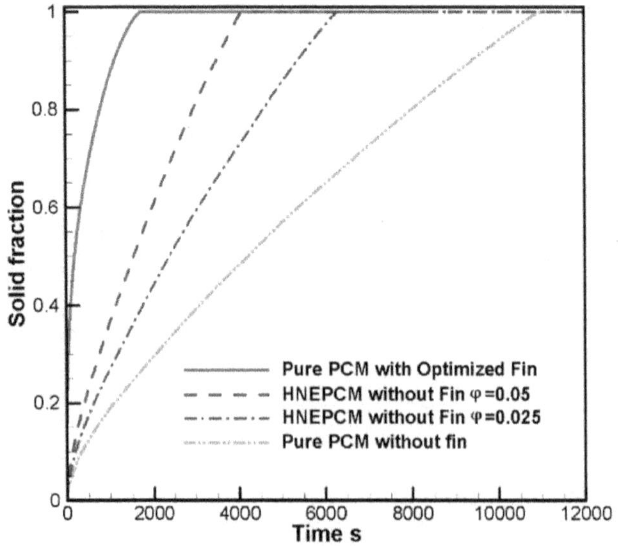

Figure 51 illustrates the total energy of the system during the process. According to this figure, the maximum energy storage capacity of the system is obtained at the start of the solidification process which is equal to the highest value of total energy. Also, it can be observed that in all cases including LHTESS with optimized fin, LHTESS with hybrid nanoparticle dispersion and LHTESS without fin and hybrid nanoparticle, the maximum energy storage capacity is the same. But, the solidification rate experiences it maximum value in the case of LHTESS with optimized fin. Therefore, regarding to solidification rate and maximum energy storage capacity, employing optimized fin enhances the performance of LHTESS more than dispersing hybrid nanoparticles.

Figure 51. Comparison between total energy of LHTESS with pure PCM without fin, pure PCM with optimized fin, HNEPCM with 2.5 and 5% volume fractions during the solidification process

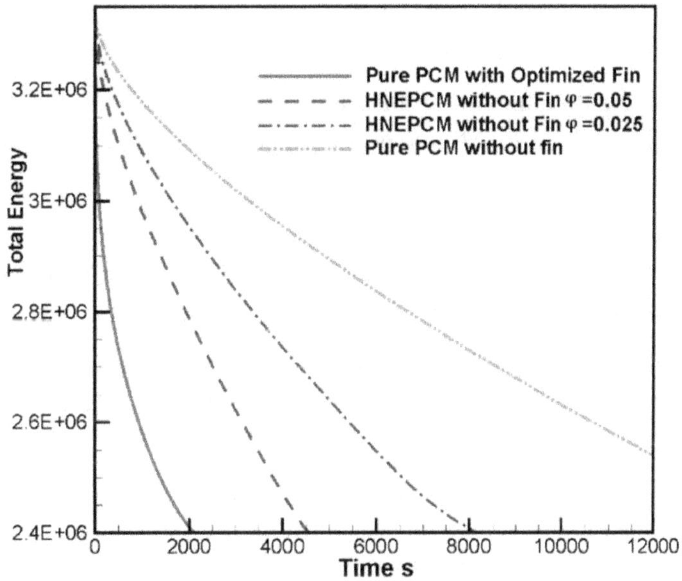

REFERENCES

Ahmad, W., Syed, K. S., Ishaq, M., Hassan, A., & Iqbal, Z. (2017). Numerical study of conjugate heat transfer in a double-pipe with exponential fins using DGFEM. *Applied Thermal Engineering*, *111*, 1184–1201. doi:10.1016/j.applthermaleng.2016.09.171

Alizadeh, M., Hosseinzadeh, K., & Ganji, D. D. (2019). Investigating the effects of hybrid nanoparticles on solid-liquid phase change process in a Y-shaped fin-assisted LHTESS by means of FEM. *Journal of Molecular Liquids*, *287*, 110931. doi:10.1016/j.molliq.2019.110931

Alizadeh, M., Hosseinzadeh, K., Mehrzadi, H., & Ganji, D. D. (2019). Investigation of LHTESS filled by Hybrid nano-enhanced PCM with Koch snowflake fractal cross section in the presence of thermal radiation. *Journal of Molecular Liquids*, *273*, 414–424. doi:10.1016/j.molliq.2018.10.049

Bhagat, K., Prabhakar, M., & Saha, S. K. (2018). Estimation of thermal performance and design optimization of finned multitube latent heat thermal energy storage. *Journal of Energy Storage*, *19*, 135–144. doi:10.1016/j.est.2018.06.014

Chamkha, A. J., Doostanidezfuli, A., Izadpanahi, E., & Ghalambaz, M. J. A. P. T. (2017). Phase-change heat transfer of single/hybrid nanoparticles-enhanced phase-change materials over a heated horizontal cylinder confined in a square cavity. *Advanced Powder Technology*, *28*(2), 385–397. doi:10.1016/j.apt.2016.10.009

Elbahjaoui, R., & El Qarnia, H. (2017). Thermal analysis of nanoparticle-enhanced phase change material solidification in a rectangular latent heat storage unit including natural convection. *Energy and Building*, *153*, 1–17. doi:10.1016/j.enbuild.2017.08.003

Ghadikolaei, S. S., Hosseinzadeh, K., & Ganji, D. D. (2018). Investigation on ethylene glycol-water mixture fluid suspend by hybrid nanoparticles (TiO2-CuO) over rotating cone with considering nanoparticles shape factor. *Journal of Molecular Liquids*, *272*, 226–236. doi:10.1016/j.molliq.2018.09.084

Gorzin, M., Hosseini, M. J., Rahimi, M., & Bahrampoury, R. (2019). Nano-enhancement of phase change material in a shell and multi-PCM-tube heat exchanger. *Journal of Energy Storage*, *22*, 88–97. doi:10.1016/j.est.2018.12.023

Hosseinzadeh, K., Alizadeh, M., & Ganji, D. D. (2019). Solidification process of hybrid nano-enhanced phase change material in a LHTESS with tree-like branching fin in the presence of thermal radiation. *Journal of Molecular Liquids*, *275*, 909–925. doi:10.1016/j.molliq.2018.11.109

Hosseinzadeh, K., Alizadeh, M., Tavakoli, M. H., & Ganji, D. D. (2019). Investigation of phase change material solidification process in a LHTESS in the presence of fins with variable thickness and hybrid nanoparticles. *Applied Thermal Engineering*, *152*, 706–717. doi:10.1016/j.applthermaleng.2019.02.111

Ismail, K. A. R., Alves, C. L. F., & Modesto, M. S. (2001). Numerical and experimental study on the solidification of PCM around a vertical axially finned isothermal cylinder. *Applied Thermal Engineering*, *21*(1), 53–77. doi:10.1016/S1359-4311(00)00002-8

Kant, K., Shukla, A., Sharma, A., & Biwole, P. H. (2017). Heat transfer study of phase change materials with graphene nano particle for thermal energy storage. *Solar Energy*, *146*, 453–463. doi:10.1016/j.solener.2017.03.013

Liu, S., Peng, H., Hu, Z., Ling, X., & Huang, J. (2019). Solidification performance of a latent heat storage unit with innovative longitudinal triangular fins. *International Journal of Heat and Mass Transfer*, *138*, 667–676. doi:10.1016/j.ijheatmasstransfer.2019.04.121

Lohrasbi, S., Sheikholeslami, M., & Ganji, D. D. (2016). Discharging process expedition of NEPCM in fin-assisted latent heat thermal energy storage system. *Journal of Molecular Liquids*, *221*, 833–841. doi:10.1016/j.molliq.2016.06.044

Madhesh, D., Parameshwaran, R., & Kalaiselvam, S. (2014). Experimental investigation on convective heat transfer and rheological characteristics of Cu–TiO2 hybrid nanofluids. *Experimental Thermal and Fluid Science*, *52*, 104–115. doi:10.1016/j.expthermflusci.2013.08.026

Şahan, N., & Paksoy, H. (2018). Novel shapeable phase change material (PCM) composites for thermal energy storage (TES) applications. *Solar Energy Materials and Solar Cells*, *174*, 380–387. doi:10.1016/j.solmat.2017.09.022

Tiari, S., Qiu, S., & Mahdavi, M. (2016). Discharging process of a finned heat pipe–assisted thermal energy storage system with high temperature phase change material. *Energy Conversion and Management*, *118*, 426–437. doi:10.1016/j.enconman.2016.04.025

Chapter 8
The Application of FlexPDE in Triplex Latent Heat Thermal Energy Storage Problems

ABSTRACT

Apart from shell-and-tube heat exchangers, which are the most frequent kind for a wide range of applications, triplex-tube heat exchangers, also known as triple concentric tube heat exchangers (TCTHX), have recently gained popularity. This sort of heat exchanger is made up of three tubes, the middle of which houses PCM while the other two house heat transfer fluid (HTF). This position improves heat transfer between HTF and PCM via an additional interface of inner and outer tubes through which HTF flows.

INTRODUCTION

A heat exchanger with three concentric tubes nestled inside one another is called a triple concentric tube heat exchanger (TCTHX). in contrast to conventional single- or double-tube heat exchangers, this design permits more intricate heat exchange combinations. Applications for triple concentric tube heat exchangers (TCTHX) can be found in many different industries, such as chemical processing, food and beverage processing, HVAC systems, and renewable energy systems. Phase change materials (PCMs) can be integrated into triplex concentric tube heat exchangers to improve their thermal performance. Materials known as phase change materials have the ability to absorb and release significant amounts of heat when a phase shift occurs, such as from a solid to a liquid and vice versa. Adding PCMs to heat exchangers is one way to increase thermal energy storage capacity and efficiency. The melting process of the phase change material in Triplex heat exchanger is simulated in this chapter. This converter has three tubes: inner, outer, and center. The fluid is inserted in the inner and outer tubes of these converters, while the phase altering substance is placed in the intermediate tube. Fins are inserted in both the inner and middle surfaces of this chapter, and many configuration examples are investigated. It is critical to pay attention to the geometric features of TCTHX. It is critical to know the radius of the inner, middle, and outer tubes, as well as their thickness, as well as the length and thickness of the fin.

DOI: 10.4018/979-8-3693-0932-2.ch008

1. EXAMPLE 1: EFFECT OF TWO DIFFERENT FINS (LONGITUDINAL -TREE LIKE) AND HYBRID NANO-PARTICLES (MoS_2–TiO_2) ON SOLIDIFICATION PROCESS IN TRIPLEX LATENT HEAT THERMAL ENERGY STORAGE SYSTEM

1.1 Problem Definition

Figure 1 depicts the configuration of three-dimensional LHTES system for tree-like and conventional rectangular fins.

Figure 1. (a) Three-dimensional geometry of LHTESS with fin, (b) two-dimensional view and boundary conditions of present study

(a) (b)

Also, the computational domain of this system is displayed in Figure 2. In this study, a PCM-based energy storage unit is under scrutiny consisting of three concentric tubes the middle of which is filled with PCM and other two tubes serve as the container of HTF flowing across the whole system. This heat exchanger is facilitated with two types of fins namely, rectangular and tree-like fins, and Hybrid nanoparticles benefiting the solidification rate. Homogenously dispersed nanoparticles are composed of $MoS_2 - TiO_2$ while copper for its high conductivity is chosen for the material of fins in this study. Both the two kinds of fins are mounted such that they are attached to the inner tube only while six of them regardless of type are located vertically or tilted at an angle of 45 degrees.

Figure 2. Adaptive mesh refinement in two time steps including 60 (s) and 250 (s)

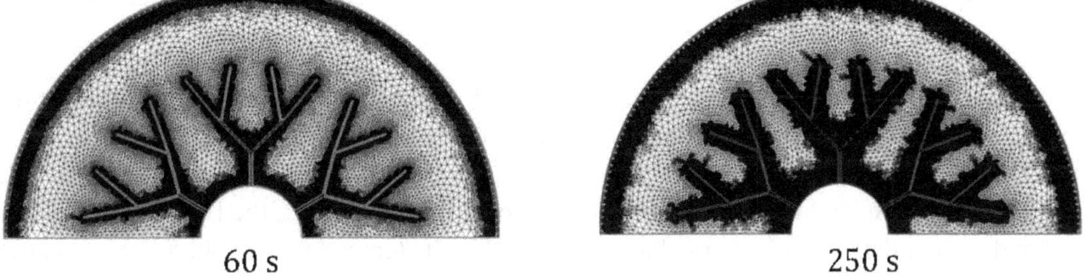

60 s 250 s

Furthermore, the dimension of each case is considered to be changeable so that the amount of PCM is exactly the same as the bare tube. The original dimensions of the system and fins' proportions are listed in Table 1. Regarding the PCM used in this system, water appeared to be of several desirable properties meriting its selection as the phase- changing material such as its reliability during numerous cycles of phase transition and isothermal condition it can provide when it reaches to solidification temperature. Table 2 exhibits all the properties of various materials employed in this study.

Table 1. Geometry parameters of LHTESS

Parameters	Values
α	$\pi/4$
β	$5\pi/18$
γ	$2\pi/9$
a	180[mm]
b	45[mm]
c	28.567[mm]
d	37.505[mm]
e	48.766[mm]
f	1.28[mm]

Table 2. The physical properties of water as PCM, hybrid nanoparticles and fin (Alizadeh et al., 2019)

Property	PCM	MoS_2	TiO_2	Fin
$\rho\,(\frac{kg}{m^3})$	997 (Liquid phase) 916 (Solid phase)	5060	4250	8960
$C_p\,(\frac{J}{kgK})$	4184 (Liquid phase) 2030 (Solid phase)	397.21	686.2	385
$k\,(\frac{W}{mK})$	0.6 (Liquid phase) 2.22 (Solid phase)	904.4	18	400
$L_f\,(\frac{J}{kg})$	335,000	-	-	-

The middle tube accommodating the PCM has been designated as the computational domain since using some simplifications, boundary conditions reflect the whole process with negligible error. At the beginning of the procedures, the temperature of the PCM is set to be 278 K equally, securing a liquid state of PCM at t=0. At later times, when the tracking of the process starts, the HTF is being invariably pumped to draw out the energy preserved in the PCM while its temperature is 240 K. This temperature is selected due to being well below water freezing point to ensure that the solidification will occur throughout the process. To consider the HTF influence in the solving domain, the walls of the annulus are assigned as the mediators for the HTF by setting constant temperature of 240 K as the boundary conditions. Another influential variable in numerical modeling when the transient phenomenon is under investigation, is time step which its appropriate selection ensures reliable results. To this end, 2×10^{-5} s has been adopted as the time interval for better accuracy. Regarding the simplifications and assumption considered in this study, there are several impactful simplifications by which the numerical procedure duration has been lowered by a sizeable amount while the accuracy of the analysis remained within the acceptable thresholds. All the principal assumptions read as follows:

1- The solving procedure applied within a confined two-dimensional modeling of the LHTES unit.
2- Although buoyancy force can be observed at the initial stages of the solidification, the dominancy of conduction outweighs convection influence, thus the convection mode of heat transfer is opted out of the governing equations.
3- The phase changing process is taking place under laminar, unsteady and incompressible conditions.
4- The HTF temperature variance is omitted resulting in the walls' constant temperature.
5- the properties of the materials are time-independent.
6- the properties of HNEPCM remain similar through the whole annulus.

The governing equations of heat transfer in the system are comprehensively reviewed in Chapter 6.

1.2 Numerical Simulation

FlexPDE which is an opensource software has been picked to perform the numerical procedures of the solidification phenomenon. This software has been utilized based on GFEM proving to be a reliable method in solving this type of problems. This methodology exploits residual-driven applications in which each variable is assigned with a weight function to balance the importance of individual parameters. Then, non-linear equations undergo a transformation to organize a system of linear equations which are far easier to be solved. Due to the fact that this GFEM is a numerical modeling of the original system, the error is inevitable, so the domain needs to be divided into sufficient parts to ensure the minimization of the errors. Hence, a dynamic mesh arrangement has been put into practice as demonstrated in Fig. 2. This adaptability paves the way for a considerably more accurate results while all the regions surpassing the error threshold will experience a downsizing and vice versa (Hosseinzadeh, Alizadeh, Tavakoli et al, 2019). To substantiate the dependability of this study, an experimental study by Ismail et al. (2001) has been chosen to be replicated by GFEM. Fig. 3 reveals the final comparison between the current methodology and the experimental study. From the result, it can be inferred that the results of both investigations are in excellent agreement, with having no more than 5% of error at any stage of the process. Thus, this renders current research authentic and reliable.

1.3 Impacts of Active Parameter

Implementing all the necessary steps to perform related computations, the results are divided into three subsections to better demonstrate the importance of individual heat promoters as follows:

Figure 3. Validation of present simulation with experimental study by Ismaeil et al. (2001)

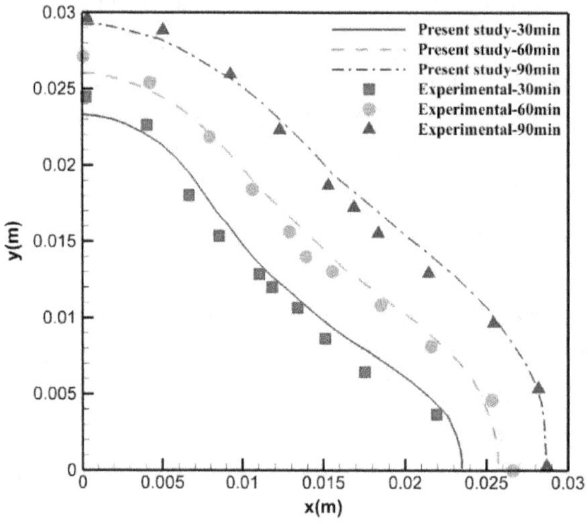

At first, finned tube with two layouts, namely tree-like and rectangular, along with plain tube are considered and analyzed. Then nanoparticles with different concentrations (0.025, 0.05) which dispersed into the annulus homogenously are come to investigation and finally concurrently applied fins and HNEPCM are put into practice.

1.3.1 Comparison of Bare Tube, Tree-Like Fins, and Rectangular Fins

To clarify the significance of fins and their design, Fig.4 is presented depicting three scenarios at three temporal stages. The PCM inside the annulus initially is melted and isothermal all over the space of the middle tube while HTF flowing in the inner and outer tubes is considered to be pumped at the constant temperature confidently below PCM freezing point to get the process started. At 60s, two thin layers are taking form for the bare tube, with solid-liquid interface being adjacent to the cooling walls, whereas for finned cases, a somewhat similar pattern can be observed, with solid-liquid interface being extended to the proximity of fins. Although representing comparable behaviors, the tree-like fins provide faster solidifying rate than rectangular ones thank to their bifurcations invading the operating field by larger margins. Another factor making this comparison more convincing is the amount of PCM being equally the same for all three schemes. Subsequently, the elapse of time for the plain tube is quite seamless due to the relatively small progress of solid-liquid interface forming just two concentric circles in the vicinity of tubes inside of which HTF is running. Other two finned cases continue to penetrate rapidly, however the difference between them is growing in favor of tree-like configuration. At 1000s, the difference is self-evident as tree-like finned TCTHX stands as the top choice regarding solidification rate, with having almost the whole annulus solidified but a small annular portion edged by the branches of tree-shaped fins. In contrast, the rectangular fins penetrate through working field more diagonally compared to tree-like scenario, but they lack the power to extend their influence on the V-shaped areas emerging in between. Last but not least, the bare tube shows no sign of departure from the uniformity it exhibits early on and still two circles occupying a negligible amount of annulus represent this case's ability to solidify PCM without any assistance. In order to further elucidate the dissimilarities among these approaches, three figures depicting average temperature, total energy and solid fraction against time are illustrated. Figure 5 indicates the rates at which the average temperatures (starting at 275 K) of the annulus are coming down over the solidification phase until they reach a plateau.

Figure 4. Solid fraction contours in three temporal stages during the solidification operation of LHTESS for pure PCM (left: no-fin storage, center: rectangular finned storage, right: tree-shaped finned storage)

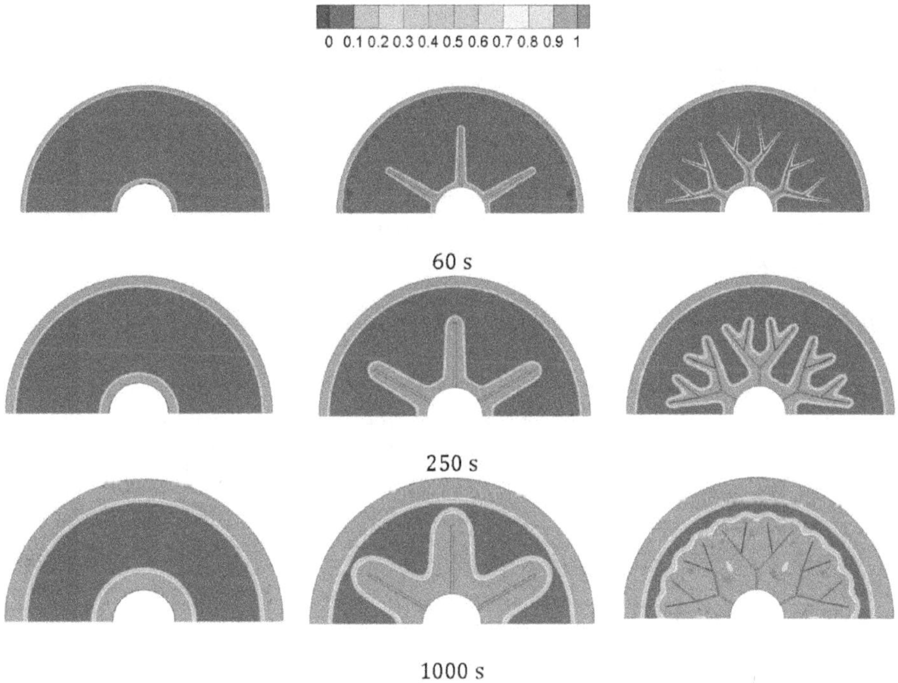

What can be inferred from this figure concerns the growing difference among three cases under investigation, with following an almost identical trend at the beginning and then diverging as the time goes by, especially a sharp contrast between finned cases and plain tube. Figure 6 showing the total energy of PCM over the course of freezing stage exactly confirms the observations of the previous figure, although there is a bold distinction between tree-like fins and rectangular ones as tree shape fins have a steep slope till 1500 s and then proceed with a much gentler slope afterwards.

This occurrence means if the energy release rate is a key factor in determining the best case for an application, the tree-like fins is of great priority. One of the most informative plots in this comparison can be solid fraction against time shown in Figure 7. As mentioned previously, the difference early on is somewhat insignificance since the effect of fins intrusion into the PCM space takes some time to be well-established and when its increasing influence comes of age, the solid fraction follows a more significant trend. While for the plain tube it takes approximately 7000 s to reach complete solidified state, this process ends at somewhere around 3500 s for rectangular fins and tree-like fins marks the best performance in terms of transition time when only a quarter of original configuration's time is needed for PCM to be solidified.

Figure 5. Temperature plot for different configuration of fins

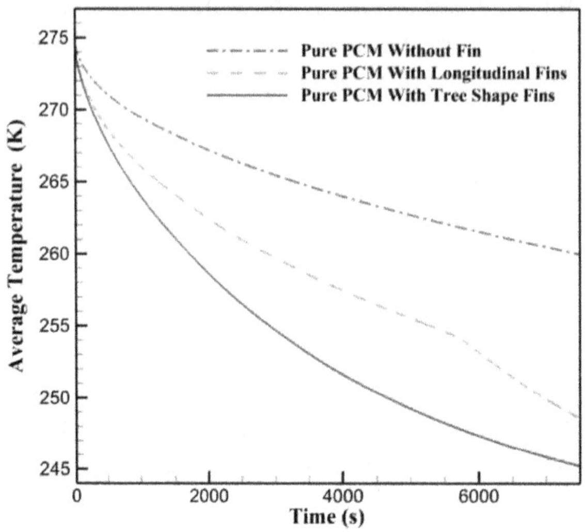

Figure 6. Total energy plot for different configuration of fins

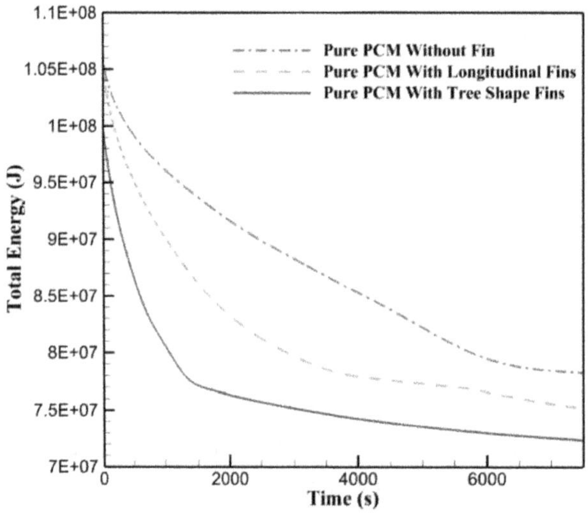

1.3.2 HNEPCM Inclusion Analysis at Different Concentrations

This part of the study represents the influence of HNEPCM and its concentration on the overall heat transfer of solidifying PCM as all the analysis done in subsections applied alike in terms of boundary and initial conditions. Three cases with having 0, 0.025 and 0.05 concentrations are put forward as depicted in Figure 8. Initially, the middle tube housing the PCM being isothermal liquid is enclosed by two tubes inside of which cold HTF is flowing to discharge the LHTES. As the time progresses to 60 s, there is scarcely any difference between three cases as two circles at the proximity of the annulus walls are formed for all scenarios alike.

Figure 7. Solid fraction plot for different configuration of fins

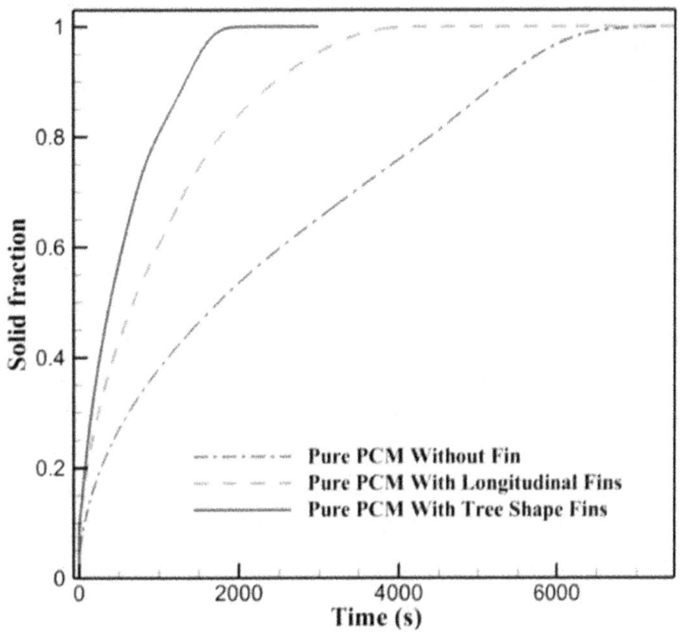

Figure 8. Solid fraction contours in three temporal stages during the solidification operation of LHTESS for various concentrations

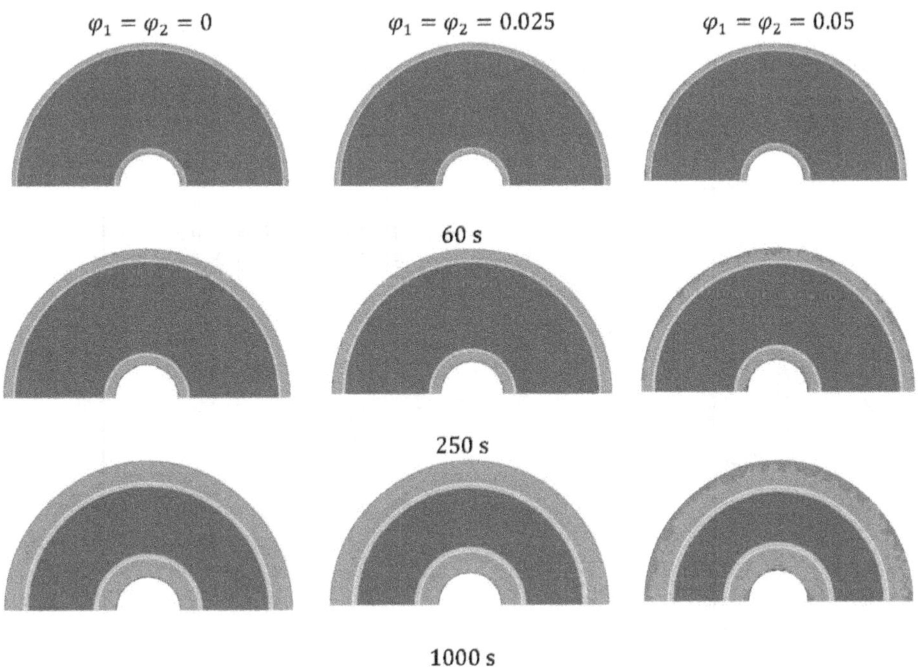

This trend seems to hold true until 250 s while only a negligible progress can be observed in the third case with hybrid nanoparticles of 0.05 concentration. Yet, nanoparticles addition appears to be of no use when it comes to solidification rate, however, this trend starts changing by looking at the third row of solid fraction contours, with the solid-liquid interface forging ahead in a more pronounced speed for cases with more fractions of nanoparticles.

The time required for the PCM to reach stabilized state are presented in Figures 9-11 while specific variables, namely average temperature, total energy and solid fraction differentiate them. Figures 9 and 10 reveals a similar pattern that both average temperature and total energy take.

This elucidates the fact that the more concentration of nanoparticles, the better result can be realized within the same conditions. Eventually, solid fraction plot corroborates all the previous findings since the lowest duration for complete solidification being almost 3500 s is claimed by HNEPCM with 0.05 volume fraction whereas for HNEPCM with 0.025 volume fraction and pure PCM it takes 5000 s and 7000 s, respectively. The main inference of this section is that HNEPCM and volume fraction of nanoparticles undoubtedly play a positive role in expediting the process studied here, although looking at the previous section of fins analysis, two major distinction is conspicuous. One is related to the initial impact of nanoparticles and fins where thank to penetration of fins into the heart of the annulus, the early monitoring of process illustrates better performance of fins by a vast difference compared to nanoparticles addition resulting in faster improvement of solid-liquid interface. The second one concerns the total amount of time for energy releasing process which places fins far ahead and make the fins a better individual promoter in solidification process. Despite aforesaid reasons, the fact that this work is fixated on solidification process in which the heat transfer is governed by conduction is a contributing factor for these conclusions and the melting behavior of the same systems is assumed to inconsistent with these inferences due to the convection role.

Figure 9. Temperature plot for different concentrations of HNEPCM

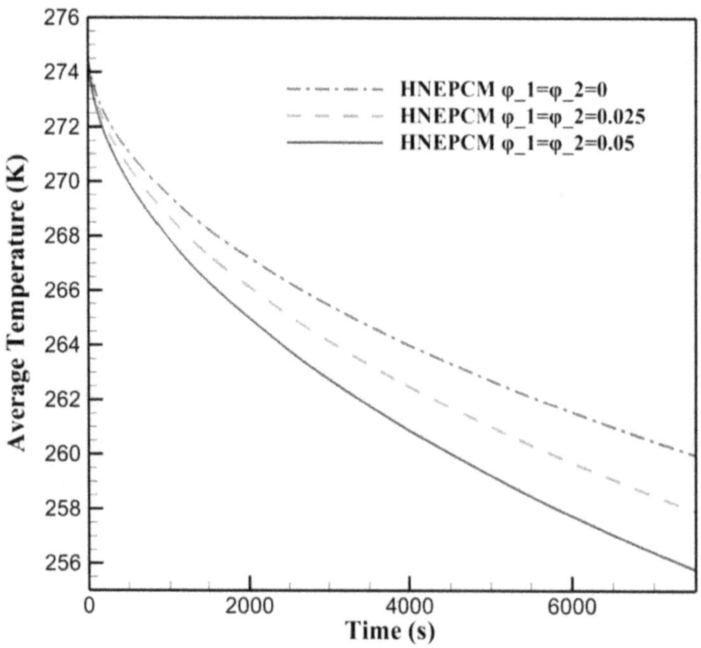

Figure 10. Total energy plot for different concentrations of HNEPCM

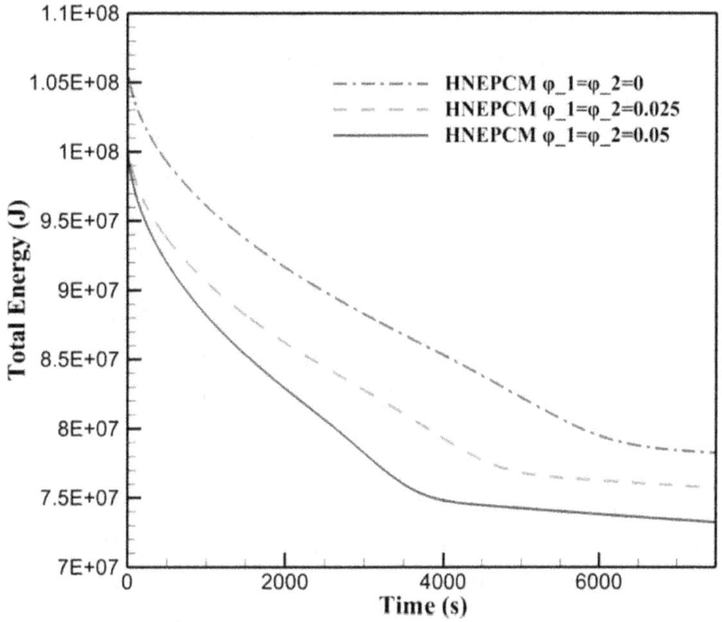

Figure 11. Solid fraction plot for different concentrations of HNEPCM

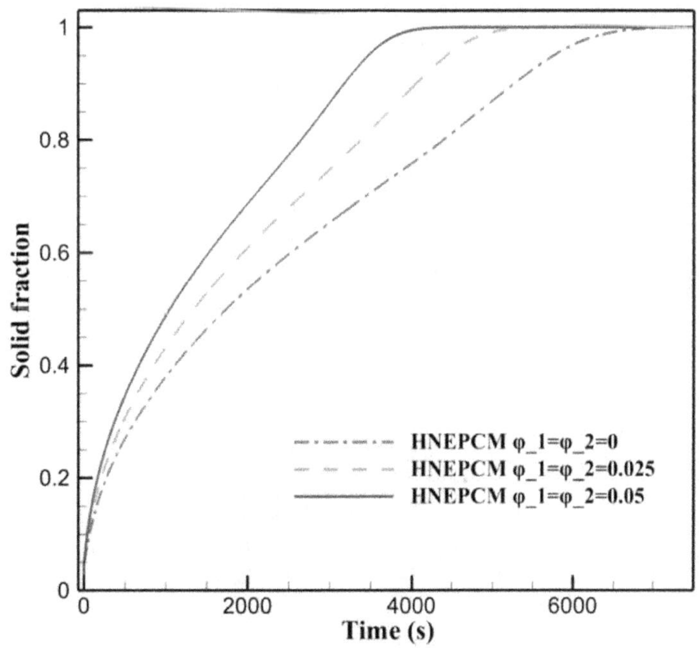

1.3.3 Side-by-Side Implementation of Tree-Like Fins and HNEPCM

Adding the best case of fins and hybrid nanoparticles into the LHTES brings some insight into the ultimate achievable enhancement among all the investigated parameters in this paper. Both solid fraction and temperature contours are presented by Figure 12 in which three stages of process have been tracked and put into perspective. From the temperature point of view, the first illustrated contour showing the storage system at 60 s reveals that there is just an insignificant deviation from the original temperature of the annulus except for the minor area close to the cooler walls and tree-like fins, however, the solid-liquid interface at the same time progresses by more appreciable margins due to the fact that, at the transition period, it is the temperature trailing behind solid fraction, considering the isothermality of the latent heat transfer phenomenon. The next contours corresponding to the state of the system at 250 s and 1000 s also pursue this trajectory, with the effect of concurrent incorporation of fins and nanoparticles starting to grow as time elapses. Reaching 250 s, the annulus yet lacks the entirety of solid state, but it offers a minor improvement to only-finned tube and a major surge compared to only-HNEPCM case, thus introducing tree-like fins as the better heat booster if the options are limited or the energy release rate at early stages is of more weight. For better understanding the combined implementation of both enhancement techniques from utilitarian point of view, Figure 13 make the judgment far easier. In this figure, the best-case scenario can be attributed to combined usage as a reduction of roughly 30% obtained compared to tree-like scenario occupied the second priority in this analysis. Looking at the final solid fraction time, this improvement signifies using tree-like fins along with HNEPCM, although during the initial phases of this process, this improvement is not representing similar conclusion.

Figure 12. The simultaneous influence of HNEPCM and fins on temperature contours in three temporal stages during the solidification process of LHTESS with fin

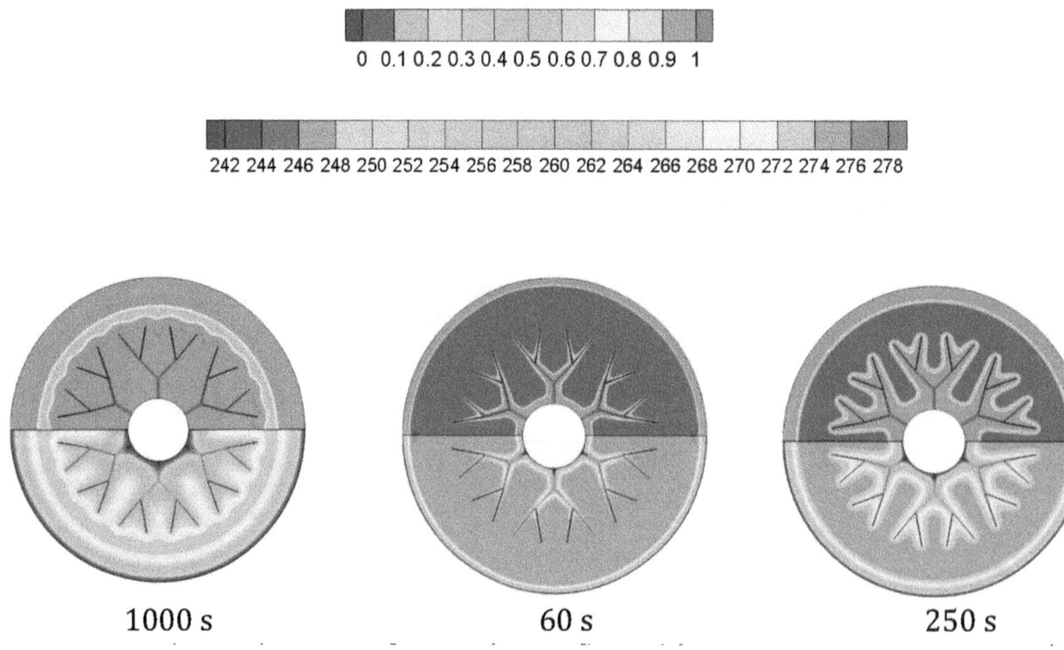

Figure 13. Solid fraction plot for different modes

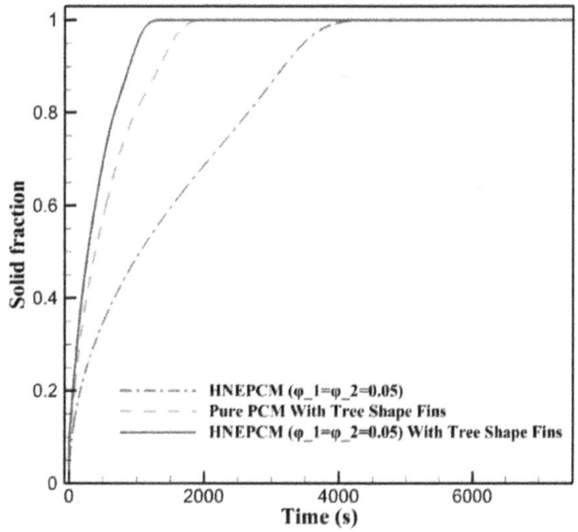

2. EXAMPLE 2: SOLIDIFICATION ENHANCEMENT IN TRIPLEX THERMAL ENERGY STORAGE SYSTEM VIA TRIPLETS FINS CONFIGURATION AND HYBRID NANOPARTICLES

2.1 Problem Definition

Figure 14 schematically displays the complete geometry and the cross-section of the longitudinal finned triplex-tube heat exchanger. The internal fins are produced from aluminum, stainless steel, and copper, which improves the heat transfer procedure. This heat exchanger involves internal, central, and external cylinders. Hybrid nano-enhanced phase change material (HNEPCM) occupies the annular gap between the middle and inner cylinders whereas the HTF moves through both the inner and outer cylinders which gives an appropriate condition for faster solidification. Nine fins with the thickness (e) and length (c,d) in three sets of triplets are placed inside the space between the middle and inner cylinders, while they are fixed to the inner cylinder. The vertical set was made from aluminum whereas by going in a clockwise direction around the inner tube, the second and third sets were made from stainless steel and copper, respectively.

The specific data regarding the sizes of the geometry are reported in Table. 3. Also, Table 4 indicates the specifications of hybrid nanoparticles and PCM, which are utilized in the current study. In the current study, MoS_2 and Fe_3O_4 are selected as hybrid nanoparticles (HNPs) as well as the water is selected as a PCM. For having a reliable presumption about the fluid phase of HNEPCM at the beginning stage, the temperature of HNEPCM expected to be 278 K consistently, that is reassuringly far from the water's melting point. The temperatures of inner and middle cylinders are selected to be 240 K which is being adequately lower than the water's melting point as the coolant liquid can be considered to be under constant temperature situations.

Figure 14. (a) 3D configuration of LHTESS with fins, (b) cross-section and boundary conditions of progressing investigation

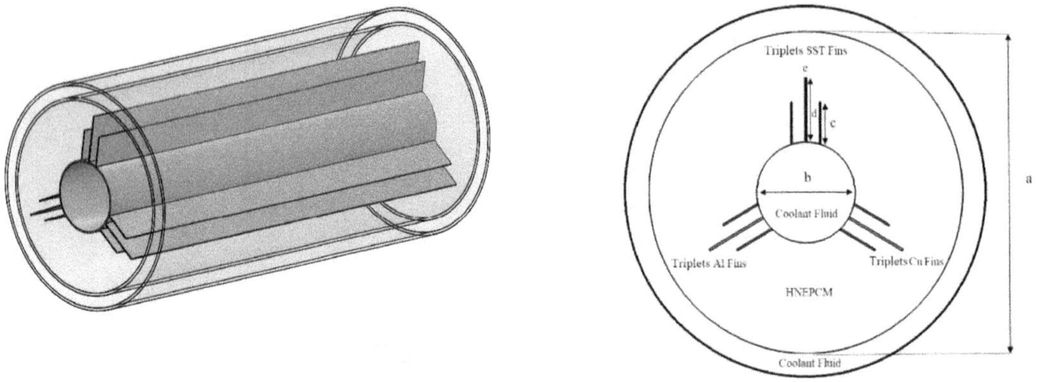

(a) **(b)**

During solidification, the heat transfer happens by just conduction in solid PCM and by both convection and conduction in liquid PCM because of the temperature gradient across the annulus. This, reciprocally, makes a buoyancy-driven flow in the fluid portion of the PCM. Although convection appears in the liquid part of the HNEPCM, it can be neglected to simplify the numerical procedure and reduce the computational costs when the solidification phenomenon is under scrutiny.

Table 3. Geometry parameters of LHTESS

Parameters	Dimensions [mm]
a	150
b	40
c	30
d	20
e	1

Besides, the subsequent assumptions were made: (1) The temperature variety in the coolant liquid is unimportant, (2) Viscous dissipations are insignificant, (3) No-slip conditions for speeds at the boundaries, (4) All thermo physical specifications of the HNEPCM are temperature-independent except density in the momentum equations, (5) Unsteady problem and the time step of $2\times10^{-5}(s)$ is selected, and (6) No heat gain and loss from the environments.

The governing equations of the HNEPCM solidification along with the required assumptions are described in below.

In the computational fluid dynamics (CFD) programming utilized for 2D simulating the PCM solidification process, the continuity, momentum, and energy governing relations are characterized by the general vector type accordingly. These relations are given in Chapter 6.

Table 4. The physical properties of water as PCM, hybrid nanoparticles, and fins

Property	PCM	MoS_2	Fe_3O_4	Fins
$\rho\,(\frac{kg}{m^3})$	997 (Liquid phase) 916 (Solid phase)	5060	5180	Cu: 8960 Al: 2702 SST: 7900
$C_p\,(\frac{J}{kgK})$	4184 (Liquid phase) 2030 (Solid phase)	397.21	670	Cu: 385 Al: 1050 SST: 560
$k\,(\frac{W}{mK})$	0.6 (Liquid phase) 2.22 (Solid phase)	904.4	9.7	Cu: 400 Al: 229.05 SST: 20.22
$L_f\,(\frac{J}{kg})$	335,000	-	-	-

2.2 Numerical Simulation

Among numerous computational methods, the Galerkin's Standard Finite Element Method (GFEM) has been used for solving the governing equations of the present investigation with concerning the boundary conditions. GFEM has a more relevant technique, which depends on the weighted residuals to solve nonlinear relations. In this analysis, the open-source coding software called FlexPDE is applied for the simulation. The aforementioned method makes ordinary differential equations of partial differential equations by utilizing a trial function. At that point, a coupled matrix by separating the dependents and variables into two particular categories takes a form, since the fruition of this action is supplemented by the Newton-Raphson technique. The strategy settles the basic form of formulas containing time-based integration by using the backward difference method.

2.3 Impacts of Active Parameters

2.3.1 Grid Study and Validation Results

The adaptive mesh has been used in the present examination due to the great importance of the grid arrangement in a numerical simulation. Figure 15 exhibits this structure, with mesh quality being exposed to improvement by utilizing density increment over solving time.

The present model has been performed in FlexPDE programming. In this programming, once the initial mesh is built, FlexPDE will keep on evaluating the solution error as well as improving the mesh as required to achieve the selected precision. In time-dependent cases, an adaptive refinement procedure will correspondingly be used for the initial estimations of the parameters, to improve the mesh where the parameters go through fast variation. Additionally, the cells which built during the adaptive refinement procedure are able to remerge. Also, the cells which built by the initial explicit density controls are fixed. Also, for ensuring, a grid independency analysis has been performed as Figure 16. So as to survey this investigation, the results are validated with Ismail et al. (2001) which are presented in Figure 17. The same physical properties, boundary conditions, and initial conditions in the practical investigation were utilized to approve the current outcomes. It clearly shows that the outputs determined by the current technique are in close agreement with the outputs of Ismail et al. (2001). This demonstrates the current numerical model is reliable enough to be utilized for studying the development of PCM solidification by the mixture of fins and hybrid nanoparticles in the triplex-tube TES framework.

Figure 15. Adaptive mesh refinement in two time steps containing 10 and 50 (s)

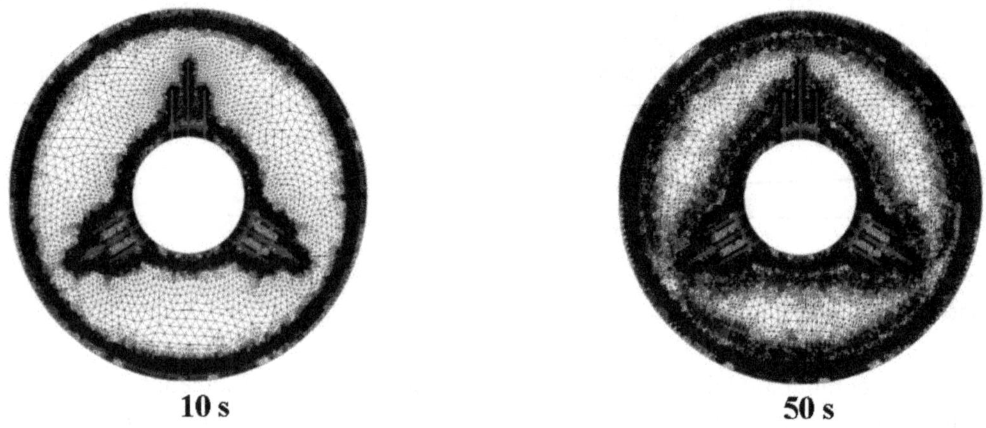

Figure 16. Grid independency analysis

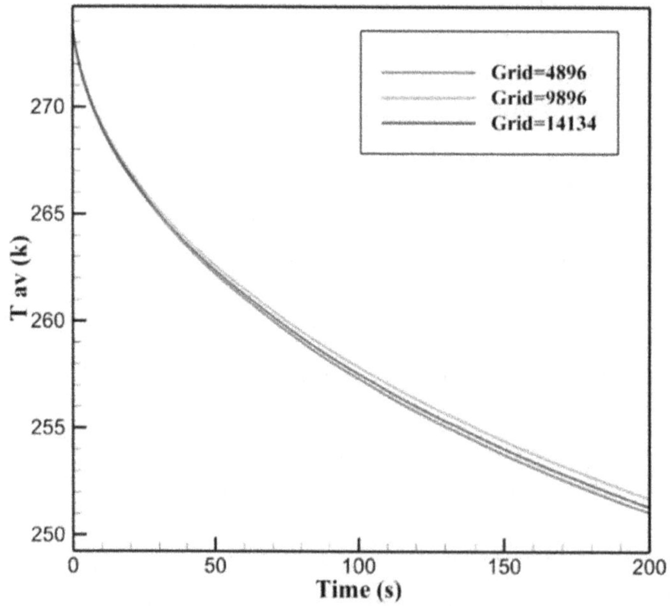

Figure 17. Validation of present study with previous experimental examination carried out by Ismaeil et al. (2001)

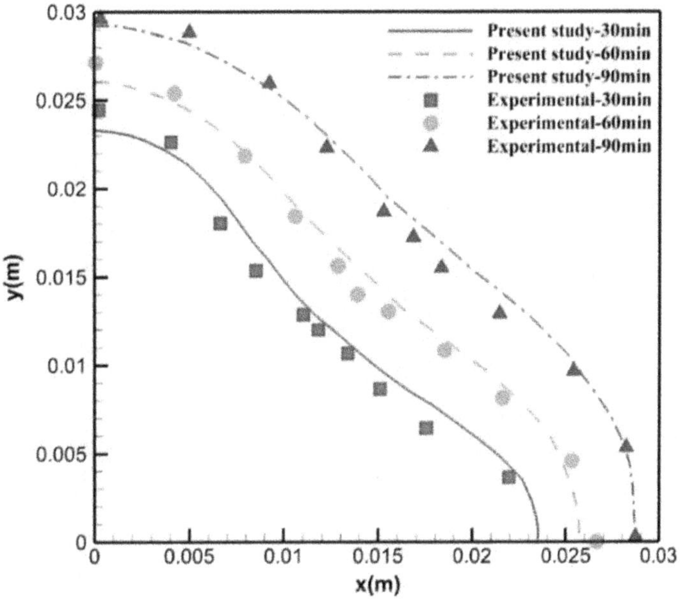

2.3.2 Performance Benchmark by Effective Parameters

A dynasty of numerical simulations was done to evaluate the improved solidification of water by hybrid nanoparticles and the three sets of triplets' fins made from aluminum, stainless steel, and copper in a triplex-tube thermal energy storage unit. The Finite Element Method (FEM) has been employed to solve the unsteady governing equations of mass and heat transfer during the solidification progression. The hybrid nanoparticles concentration (φ= 0, 0.025, 0.05), thermal radiation parameter (*Rd*= 0, 0.5, 1), and shape factor of nanoparticles (*sf* = 3, 8.6, 16.15) are the factors which have influences on the solidification procedure in this examination.

2.3.2.1 Influence of the Hybrid Nanoparticles Concentration

To exhibit the impact of dispersing nanoparticles on PCM solidification inside the annulus between the finned internal and central cylinders of the triplex-tube, the solid-liquid interface contours (Figure 18) and completed solidification process time (Figure 19) have been recorded over different time periods solidification. The outcomes concerning the impacts of nanoparticles present that as the nanoparticle volume concentration rises, the size of the solid layer grows slightly. This is credited to the fact that the presence of the combination of hybrid nanoparticles and fin, promotes a little bit faster propagation of solidification and higher conductive heat transfer over the liquid PCM. Results from Figure 18 suggest that under the same PCM volume reduction the combination of nanoparticles and fins accomplishes more desirable heat transfer to the PCM than the fins alone.

Also, Figure 19 displays the reduction in solidification process time by increasing the hybrid nanoparticles concentration.

Figure 18. Solid fraction contours in two temporal stages during the solidification process of LHTESS for various nanoparticle volume fraction of HNEPCM (left: after 10s, right: after 50s)

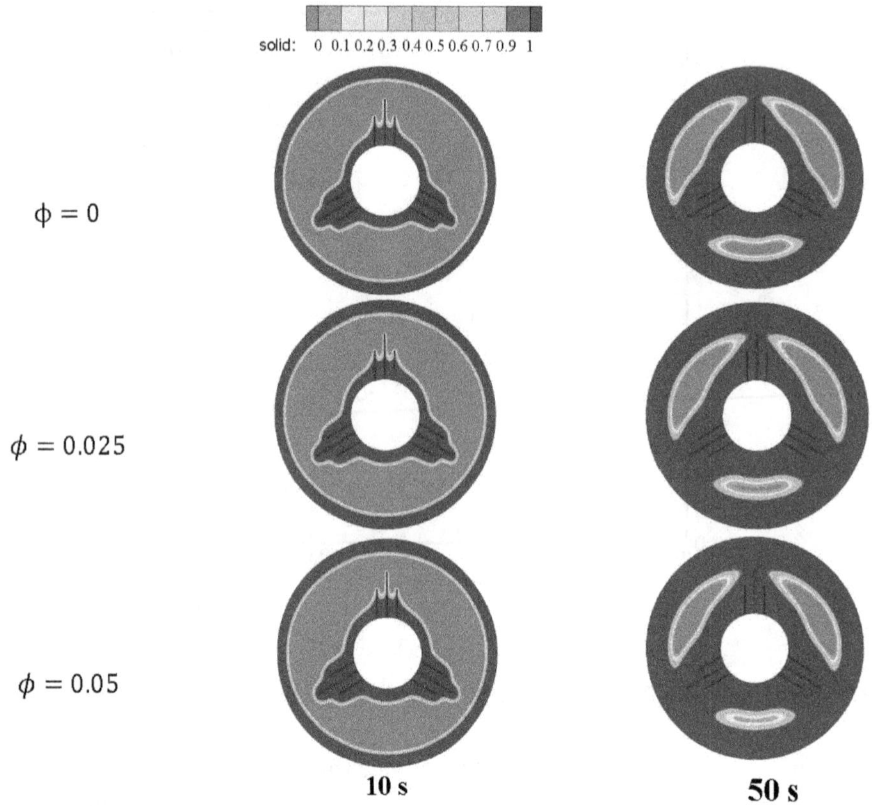

Figure 19. Solidification process for HNEPCM with various nanoparticle volume fraction: (a) solid fraction, (b) full solidification time (FST)

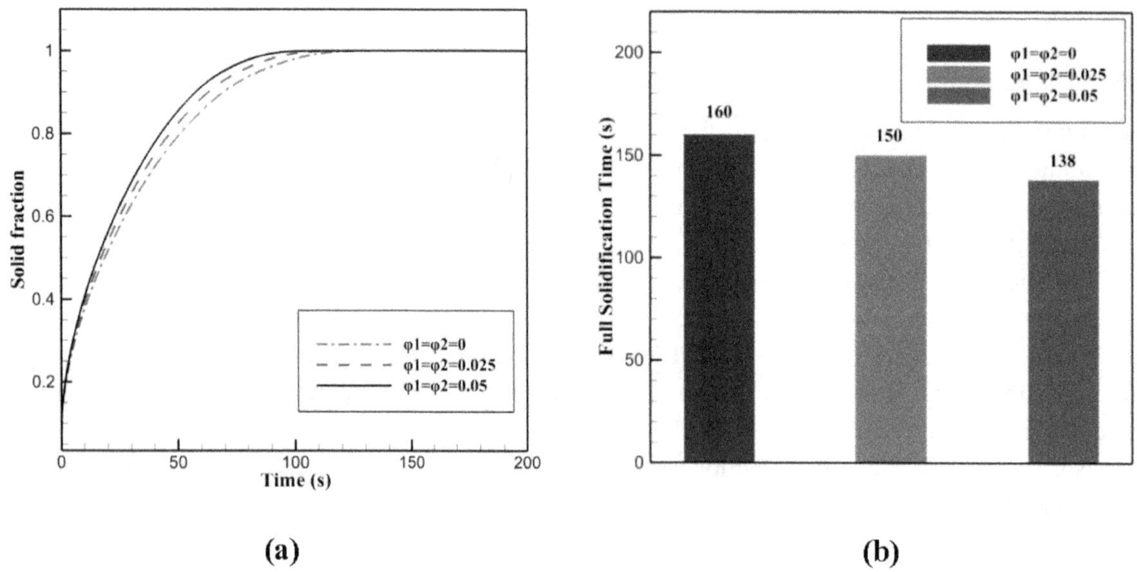

2.3.2.2 Influence of Thermal Radiation Parameter

The solid contour through the solidification procedure for various estimation of thermal radiation parameter (Rd = 0, 0.5, 1) are shown in Figure 20 for fixed amounts of φ=0.025 and Sf=8.6. It can be clearly observed that an increase in the radiation parameter causes higher estimations of solid fraction. In this circumstance, the solidification procedure is done quicker because of an addition to thermal penetration depth. Accordingly, from Figure 21, the full solidification process time decreases noticeably as the estimation of the thermal radiation parameter enhances. The solid layer growths in size to fill the major portion of the annulus by increasing the radiation parameter. The solidification is completed first in the lower part because of the buoyancy effect which keeps on expanding in this area, while the solidification is somewhat deferred in the upper part of the annulus because of the moderately more density of solid PCM in comparison with the liquid PCM.

Figure 20. Solid fraction contours in two temporal stages during the solidification process of LHTESS for various nanoparticle radiation parameter of HNEPCM (left: after 10s, right: after 50s)

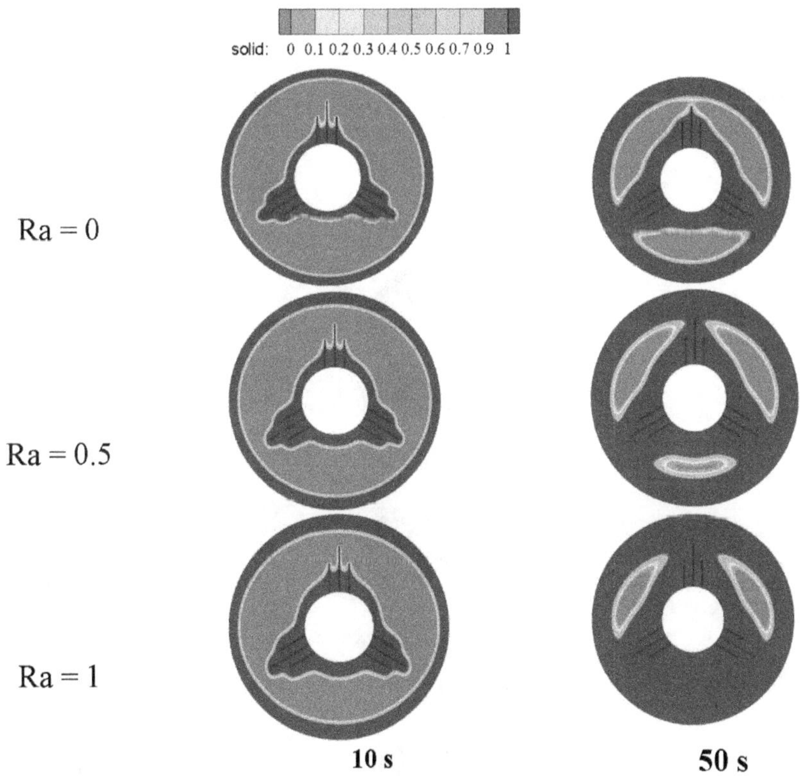

However, heat transfer seems to be notably developed by the presence of fins as the solid layers close to the fins are enhanced, especially longer fins. Heat transfer by conduction from the fin to the PCM causes the solid layer to become greater which finally fills the spaces between the fins.

Figure 21. Solidification process for HNEPCM with various Radiation parameter: (a) solid fraction, (b) full solidification time (FST)

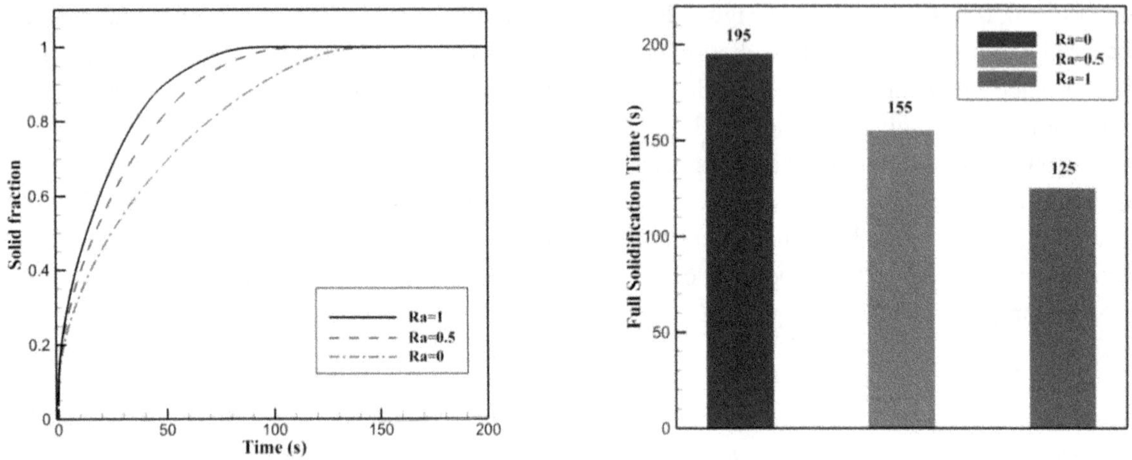

(a) **(b)**

Figure 22. Solid fraction contours in two temporal stages during the solidification process of LHTESS for various nanoparticle shape factor of HNEPCM (left: after 10s, right: after 50s)

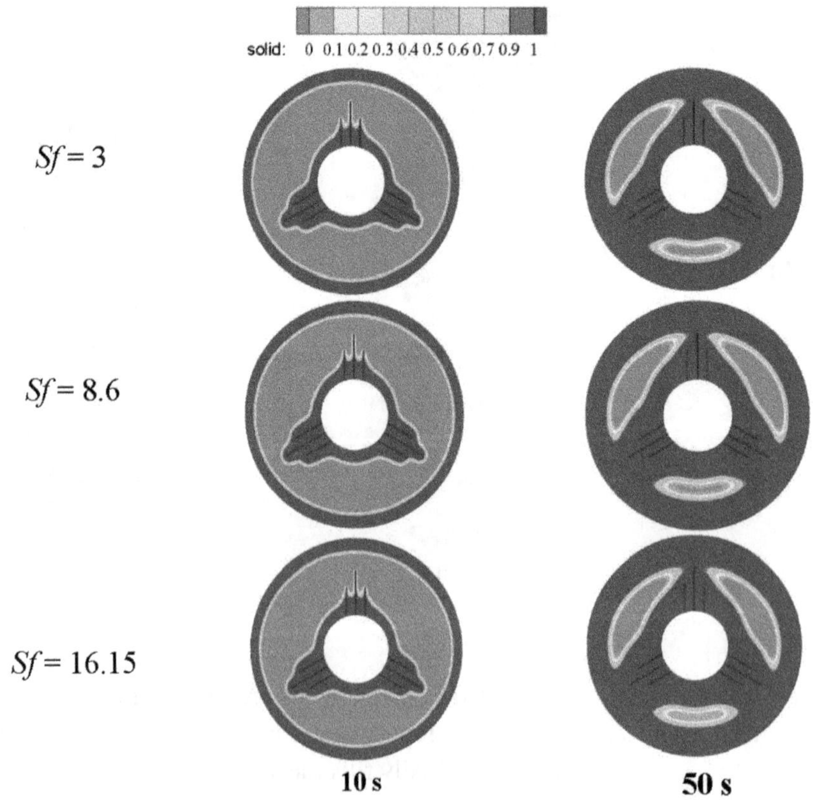

2.3.2.3 Influence of Shape Factor

The solid fraction evolution during solidification for the cases with various shape factor nanoparticles for fixed amounts of φ=0.025 and Rd=0.5 is illustrated in Figure 22. From this figure, the increase in the shape factor during this process causes an insignificant influence on either the location or the shape of the interfaces. Figure 23 clearly shows that the solidification time for all three cases are nearly the equivalent.

2.3.3 Average Temperature and Total Energy Analysis

The temperature of the HTF is a noteworthy factor which additionally should be considered in the design of TES units with PCMs as the storage media. The phase change rate remarkably has been improved as well as the duration of the PCM phase change has been shortened by enhancing the HTF's charging temperature. Figure 24 shows the instantaneous average temperature of HTF solid fraction over a full charging of the triplex-tube unit.

Figure 23. Solidification process for HNEPCM with various shape factor: (a) solid fraction, (b) full solidification time (FST)

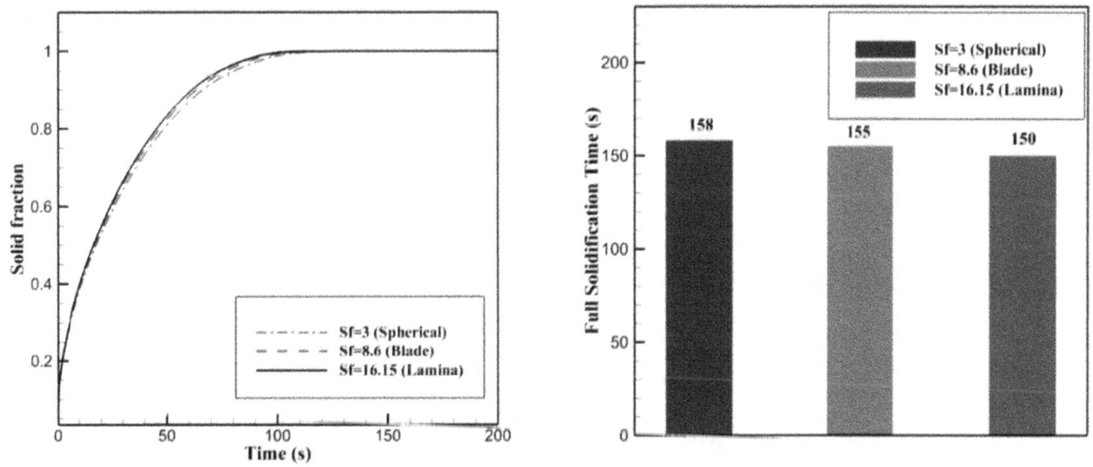

(a) **(b)**

As can be seen from Figure 24(a), by dispersing hybrid nanoparticles into the PCM, average temperature witnesses an enhancement in the transition rate; although in the beginning, no significant change occurs, but this gap is changing over time. Enhancing the volume concentration of hybrid nanoparticles from 0.025 to 0.05 is beneficial in terms of the temperature gradient.

Data from Figures 24 a and c indicate that the temperature gradient changes with increasing radiation parameter are very noticeable in contrast to the other two variable parameters. This speed up the phase change rate. Also, that the least amount of change is related to the change of shape factor parameter, which is almost negligible. Figure 25 is presented the total energy (J) versus time (s). It infers the rate at which high-energy liquid PCM transfers its energy to the cold liquid streaming in the outside as well as inside cylinders. Based on figure 25, the combined finned annulus with nanoparticles and radiation factor equivalents to 1 (Rd=1) occupies the first place concerning the energy damping rate. According to Figure 25(a), in the presence of HNEPCM with φ=0.025, Rd=1, and Sf=8.6 case facing a sharp drop from the beginning of the procedure to almost 70 s and then continuing a straight way to the end.

This trend discloses the considerable effect of rate promoters when utilizing at the same time. With respect to different cases displayed in Figure 25, the pattern follows a similar trend, but the noticeable rate of total energy drop is seen in the presence of HNEPCM and Rd=1 case. The closest cases to the best case are the case with φ=0.05, Rd=05, and Sf=8.6 as well as φ=0.025, Rd=0.5, and Sf=16.15 which still there are gaps between them as its first stage of energy transfer has been extended to about 100 s.

Figure 24. Average temperature of HNEPCM during solidification procedure: (a) various φ, (b) various Rd, and (c) various Sf

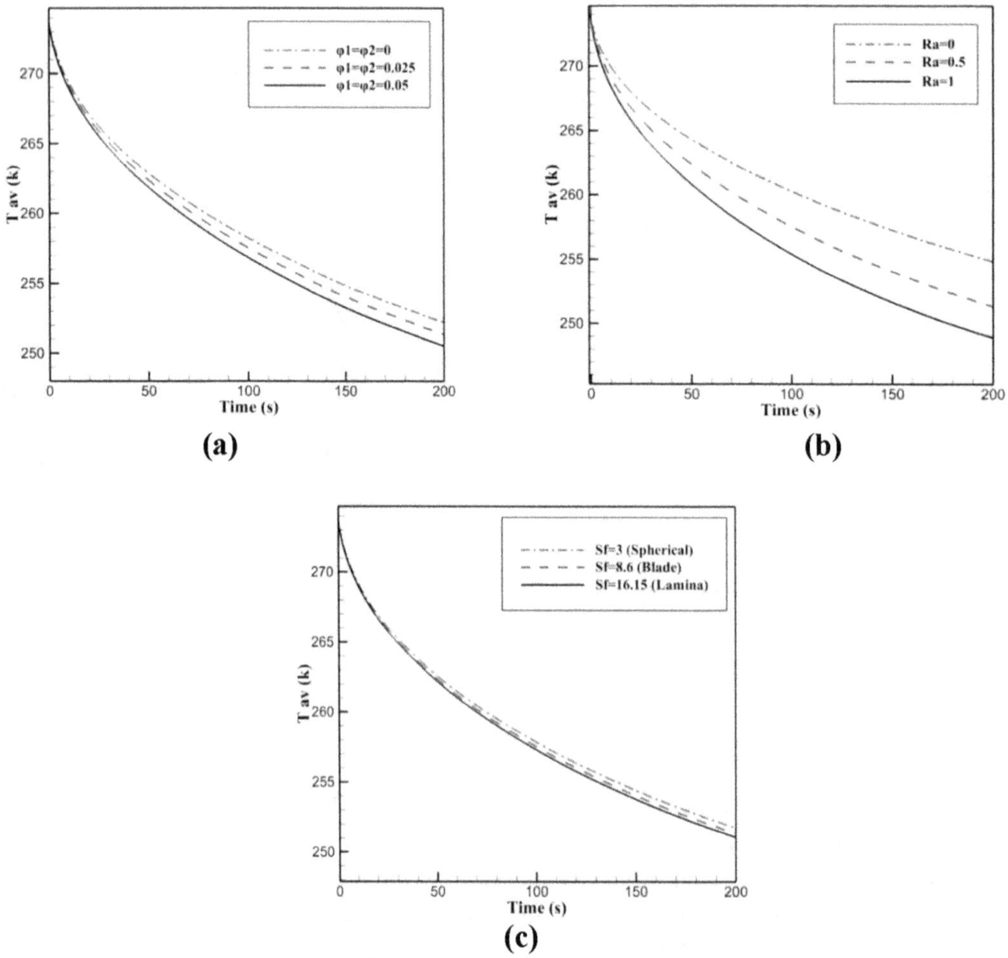

Figure 25. Total energy of HNEPCM during solidification process: (a) various φ, (b) various Rd, and (c) various Sf

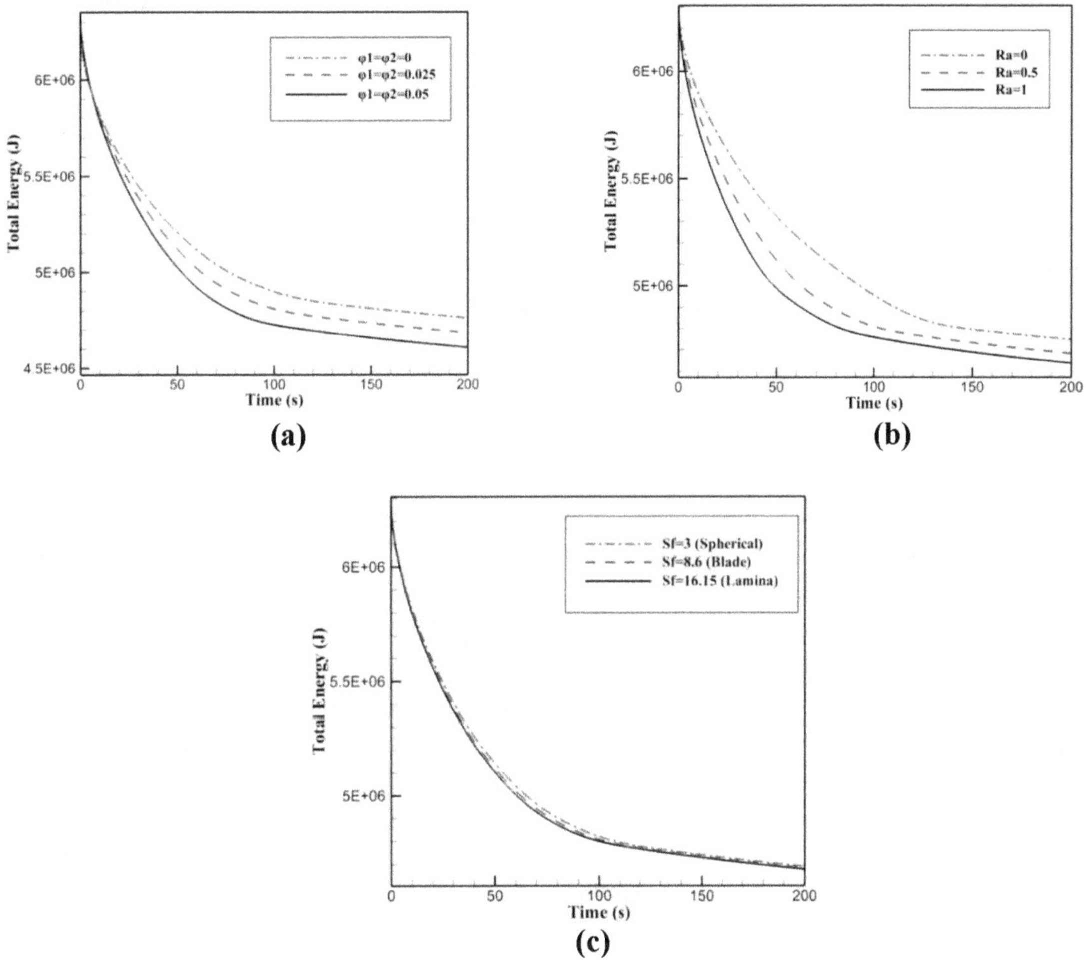

2.3.4 Optimization and Develop a Full Solidification Process Time Correlation

The Taguchi method has a broad scope of uses in engineering and industrial issues. Hence, it befitted a topic of interest in many investigations and is a regularly selected technique for optimizing design parameters. This methodology was initially prescribed to improve the quality of products using the application of engineering and statistical concepts. This method depends on two essential ideas; the first is that the quality misfortunes must be determined as deviations from the objectives, not conformance to arbitrary specifications, and the subsequent one is that to acquire high framework quality levels financially, the quality must be structured into the item. Taguchi strategy proposes a three levels procedure to acquire the desired product quality by designs, which are system design, parameter design, and tolerance design. The Nanoparticles volume fraction, Radiation parameter, and shape factor are the involve variables. Three factors each at three levels need $3^3 = 27$ runs in a full factorial trial, while Taguchi's factorial experiment technique decreases it to only 9 runs which is an incredible improvement. Based on the Taguchi design concept, the L_9 orthogonal array is selected for the simulations as reported in Table 5.

Table 5. Designed tests by Taguchi method

Case Number	Radiation Parameter (Ra)	Nanoparticle Volume Fraction (ϕ)	Shape Factor (sf)
1	0	0	3
2	0.5	0	8.6
3	1	0	16.15
4	1	0.025	3
5	0	0.05	16.15
6	0	0.025	8.6
7	0.5	0.05	3
8	0.5	0.025	16.15
9	1	0.05	8.6

The optimization approach that has been utilized is the Response Surface Methodology (RSM). This approach applies a series of mathematical techniques as well as statistical approaches for the analysis, and modeling of the problems. The usage of RSM to design optimization is planned for diminishing the cost of expensive analysis procedures (e.g. CFD analysis or FEM) and their related numerical noise. Venter et al. (1996) have examined the benefits of utilizing RSM for design optimization applications. So as to discover the impact of the variables on the response in a procedure, RSM is used to build a model between controllable input parameters and the response variables. Also, the errors that arise in the numerical applications of this method are related to the rounding of values, the non-convergence, and the number of iterations considered. The initial phase in RSM is to detect an appropriate approximation almost near to the real correlation. The most generally known models are first or second-order polynomials. A second-order structure can expressively develop the optimization procedure once a first-order structure is not practical because of the interaction between surface curvatures and variables. A second-order model is generally characterized as:

$$y = a_0 + \sum_{i=1}^{n} a_i x_i + \sum_{i=1}^{n} a_{ii} x_i^2 + \sum_{i=1}^{n}\sum_{j=1}^{n} a_{ij} x_i x_j \,|_{i<j} + \xi \tag{1}$$

where the design variables have been shown by x_i and x_j. Additionally, a is the tuning factor and n is the number of factors. Additionally, ξ illustrate the error. Furthermore, with aim of minimizing the full solidification process time (*FST*), the influences of nanoparticle volume fraction, radiation parameter, and shape factor have been employed in that correlation. Figure 26 deliberates the relations between the variable factors and their impact on the *FST*. It shows that *FST* has an inverse connection with ϕ, *Ra*, and *sf*. Equation (2) explains the relationship of the variable factors with the full solidification process time.

$$FST = 220.66 - 66.66(Ra) - 686.66(\phi) - 1.36(sf) \tag{2}$$

The final models were examined statistically by analysis of variance (ANOVA) for investigating whether the mathematical structures can acceptably define the proposed responses. Table 6 is provided for the variance analysis (ANOVA) which represents the scale of the discrepancy of predicted and actual outcomes. The coefficient of determination (R^2) value is 0.9563 which is considerable and illustrates a good agreement between predicted and real data.

Figure 26. Effects of various operating parameters on FST by RSM

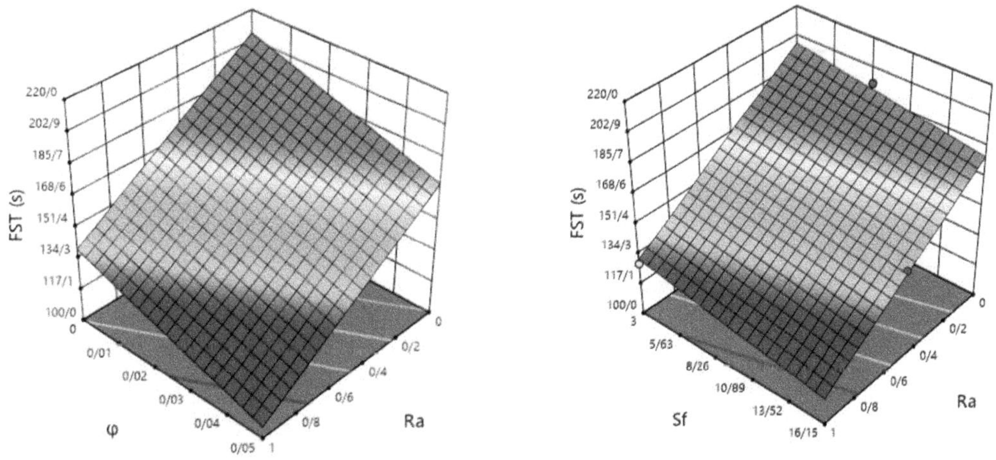

In Figure 27, the numerical data obtained earlier are plotted against the related estimations of the *FST* dependent on the correlated equation (Eq. 2). These outcomes determine how the current correlation enhances the earlier equations by considering the effect of nanoparticle volume fraction, Radiation parameter, and shape factor on the efficiency of solidification process. Note that the solid line represents the good agreement between the numerical outputs and the expectations for each case. In Figure 27, the numerical data obtained earlier are plotted against the related estimations of the *FST* dependent on the correlated equation (Eq. 2).

These outcomes determine how the current correlation enhances the earlier equations by considering the effect of nanoparticle volume fraction, Radiation parameter, and shape factor on the efficiency of solidification process. Note that the solid line represents the good agreement between the numerical outputs and the expectations for each case

The optimum values of the parameters that include in this correlation can be seen in Table 7. By applying these values to equation 2 the minimum *FST* can be calculated. In addition, Figure 28 displays the contribution of each parameter in influencing the solidification process time by the percentage.

Figure 27. Predicted vs. actual of the FST

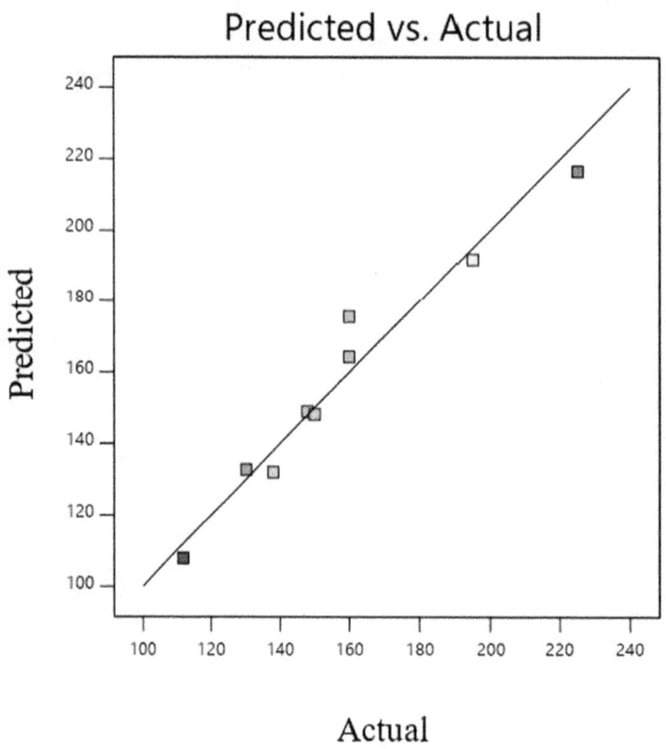

Table 6. The variance analysis (ANOVA)

Source	Sum of Squares	df	Mean Square	F-Value	P-Value	
Model	8920.15	3	2973.38	36.43	0.0008	significant
A-Ra	6666.67	1	6666.67	81.69	0.0003	
B-ö	1768.17	1	1768.17	21.67	0.0056	
C-Sf	485.32	1	485.32	5.95	0.0588	
R^2=0.9563						

Table 7. Outcome of the optimization procedure by RSM

Parameter	Optimum Value
Ra	1
Ø	0.05
sf	16.15

Figure 28. Contribution of each parameter on the FST by the percentage

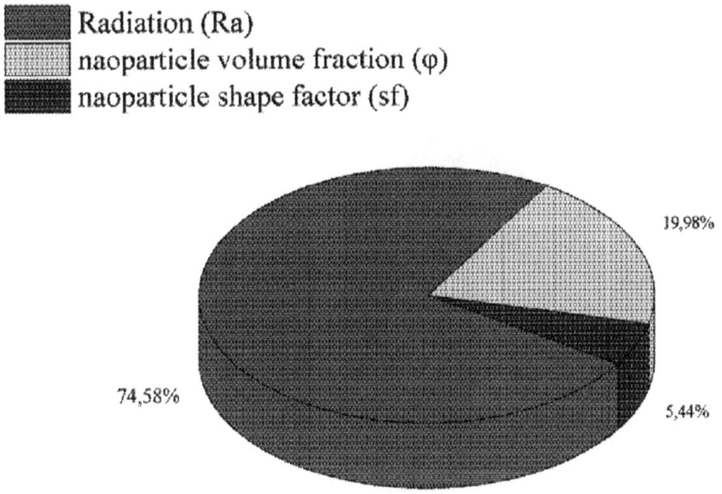

3. EXAMPLE 3: EFFECT OF INTERNAL FINS ALONG WITH HYBRID NANO-PARTICLES ON SOLID PROCESS IN STAR SHAPE TRIPLEX LATENT HEAT THERMAL ENERGY STORAGE SYSTEM BY NUMERICAL SIMULATION

3.1 Problem Definition

The configuration of the horizontal finned triplex-tube heat exchanger is depicted in Fig. 29 along with a two-dimensional illustration of this system. The fins inside the annulus are made of copper, which enhances the heat transfer process. There are three tubes inside this heat exchanger in which a star-shaped inner and outer tubes accommodate HTF as the coolant fluid and the middle space are filled with HNEPCM. This arrangement provides a suitable condition for faster solidification due to double-sided heat transfer. Twelve fins are housed inside the space between inner and middle tubes, while half of them are fixed to the star-shaped tube, and the remaining ones adhere to the middle tube.

The exact information concerning dimensions of the problem are displayed in Table. 8, and Table. 9 delineates the physical properties of hybrid nanoparticle and PCM. In this paper, water is chosen as a PCM and MoS_2-Fe_3O_4 as HNPs. For having a reliable assumption about the liquid phase of HNEPCM at the initial point, the temperature of HNEPCM assumed to be 278 K uniformly, which is reassuringly far from the melting point of water. The inside and middle tubes are set to be under a constant temperature of 240 K being sufficiently below the melting point of water since the coolant fluid can be deemed to be under fixed temperature conditions. The solving domain is considered to be the whole space loaded up with HNEPCM to have a more accurate result with a, b, c, and d symbolizing diameter of the middle tube, dimension of the star-shaped side,dimension of the fin length and fins' thickness, respectively. Furthermore, Due to the unsteady essence of this problem, the time step of $2\times10^{-5}(s)$ is picked, which seems adequate considering the complexity of the solidification process.

Figure 29. (a) Three-dimensional configuration of LHTESS with fin, (b) two-dimensional view and boundary conditions of ongoing study

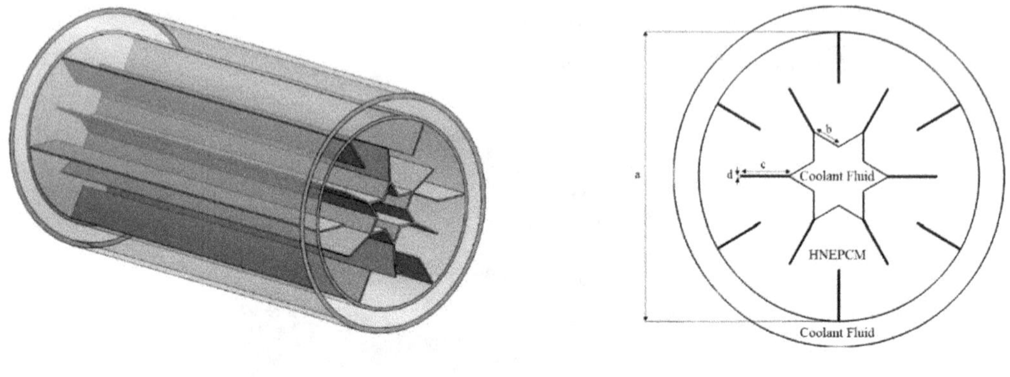

(a) (b)

Table 8. Geometry parameters of LHTESS

Parameters	Dimensions [mm]
a	120
b	24.53
c	42.5
d	1

The main assumption in this paper is related to the dominant form of heat transfer type which is supposed to have the form of conduction. Although convection appears in the liquid part of the HNEPCM, it can be neglected to simplify the numerical procedure and reduce the computational costs when solidification phenomenon is under scrutiny.

Other assumptions existed in this study are as follows:

1. A two-dimensional numerical simulation is selected and employed.
2. The nature of the phenomenon is assumed to be transient and incompressible.
3. All the physical properties related to HNEPCM is contingent upon time and temperature gradient.
4. No-slip boundary condition is set at the vicinity of the walls and fins.
5. Temperature fluctuation of coolant fluid are deemed negligible.

The governing equations of this problem are stated in Chapter 6.

Table 9. The physical properties of water as PCM, hybrid nanoparticles, and fin (Hosseinzadeh, Alizadeh, Alipour et al, 2019)

Property	PCM	MoS_2	Fe_3O_4	Fin
$\rho\,(\frac{kg}{m^3})$	997 (Liquid phase) 916 (Solid phase)	5060	5180	8960
$C_p\,(\frac{J}{kgK})$	4184 (Liquid phase) 2030 (Solid phase)	397.21	670	385
$k\,(\frac{W}{mK})$	0.6 (Liquid phase) 2.22 (Solid phase)	904.4	9.7	400
$L_f\,(\frac{J}{kg})$	335,000	-	-	-

3.2 Numerical Simulation

For solving the governing equations of the current study with corresponding boundary conditions, among many numerical approaches, GFEM has been utilized due to its' more applicable method, which is based on the weighted residuals to solve non-linear equations. The software used in this study is opensource FlexPDE to perform GFEM. GFEM transforms partial differential equations into ordinary differential equations using a trial function. Then, a coupled matrix by dividing the variables and dependents into two distinct classes takes a form, after which the fruition of this practice is complemented by the Newton-Raphson method. This method solves the simplest form of equations comprising time-based integration with the means of backward difference technique. As the grid arrangement is of great importance in a numerical simulation, the adaptive mesh has been engaged in the current study. Figure 30 demonstrates this arrangement, with mesh quality being subjected to enhancement using increasing density over solving period.

3.3 Impacts of Active Parameters

3.3.1 Influence of Fins Inclusion Into LHTESS

By comparing two different cases based on fins possession, a noticeable difference can be realized by adding fins into pure PCM to enhance the solidification rate, as shown in Figure 31 for various time steps (60, 160, and 260 s). As solidification initiates, adjacent areas near the walls start transforming into solid-phase, indicating growth in the solid-liquid interface. In the case without fins, two thin layers around walls emerge while finned cases show bigger layers around fins and walls. With time, the solidification march in annulus continues as expected due to cold fluid flowing from inside and outside constantly.

Figure 30. Adaptive mesh refinement in two time steps including 60 and 160 (s)

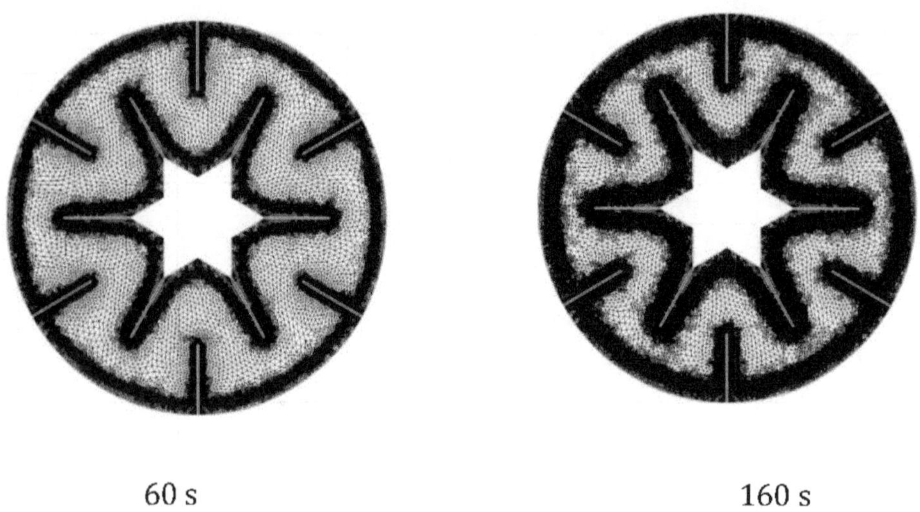

Figure 31. Solid fraction contours in three temporal stages during the solidification operation of LHT-ESS for pure PCM (left: no-fin storage, right: finned storage)

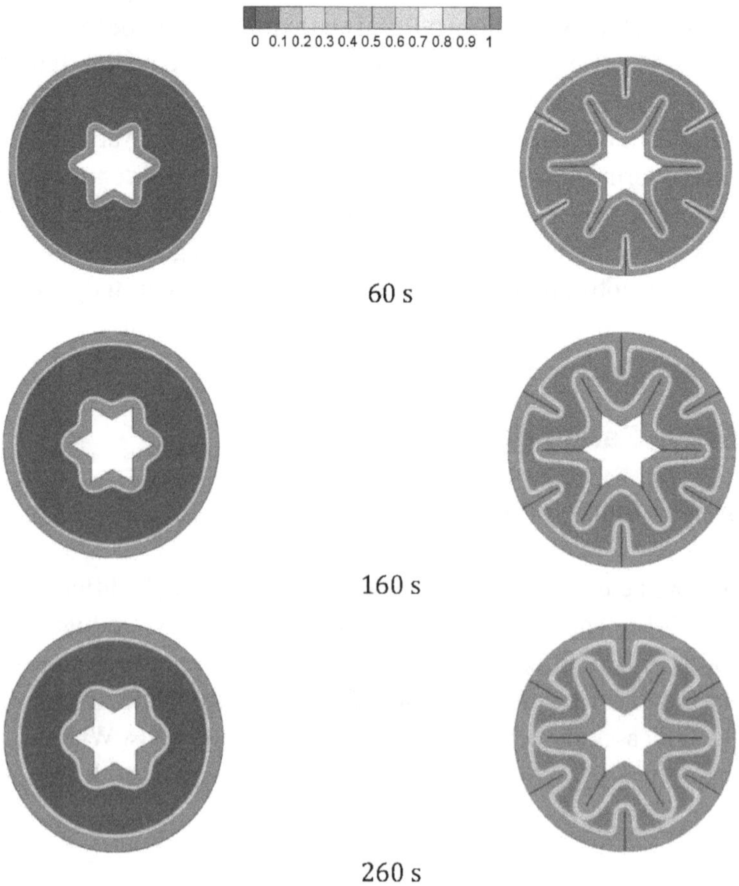

However, the sharp dichotomy between cases with fins and without fins grows more prominent as time goes by because with fins inside PCM, the available surface between PCM and tubes extends. Although this extra surface seems insignificant early on, it manifests itself during the process. Therefore, taking advantage of heat transfer basic principle provides a significant improvement in solidification rates. For a more accurate evaluation, the amount of PCM in both cases are the same. In Figure 32 the front of LHTESS at two different time steps for finned and no-fin cases is depicted to have a better understanding of the fins' effect. This figure corroborates the results driven from the previous figure, with having a roughly equal invasion of the solid-liquid interface at 80 s and witnessing a sharp difference at 360 s.

Figure 32. Comparison of solidification rate between LHTESS with and without fin

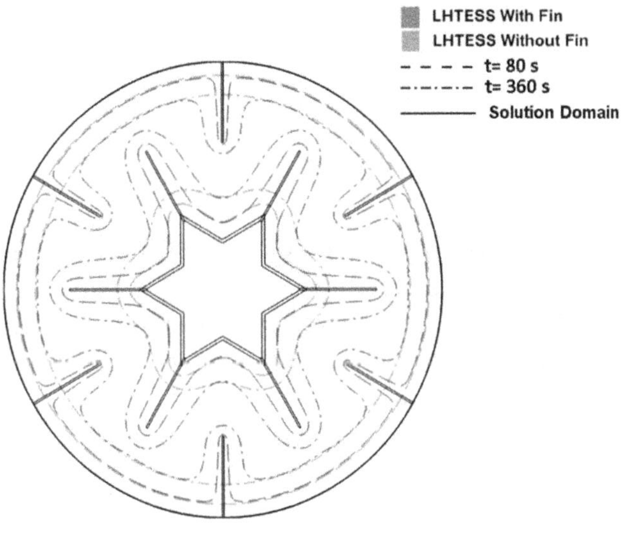

3.3.2 Assessment of HNEPCM Addition Without Fins

For evaluating the effect of hybrid nano addition on the melting rate in the annulus without fins consideration, Fig. 33 demonstrates the solid-liquid interface contours over three temporal stages (40, 140, and 400 s). Furthermore, different volume fractions, namely, 0, 0.025, and 0.05, have been included to fully showcase the promotion of the melting rate delivered by HNEPCM using various lines. The first time point sees somewhat similar solid-liquid interfaces during which all the three volume fractions are located adjacent to inner and middle wall tubes. This similarity between various volume fractions of HNEPCM hinders any insight into the influence of nano insertion, but as time proceeds, this similarity fades away, resulting in a more visible discrepancy between different volume fractions at last period. The increase in the solid-liquid interface can be attributed to the growth of the thermal conductivity of the HNEPCM which is more pronounced in the last stages of the transition process.

Figure 33. The effect of HNEPCM volume fractions on no-fin LHTESS

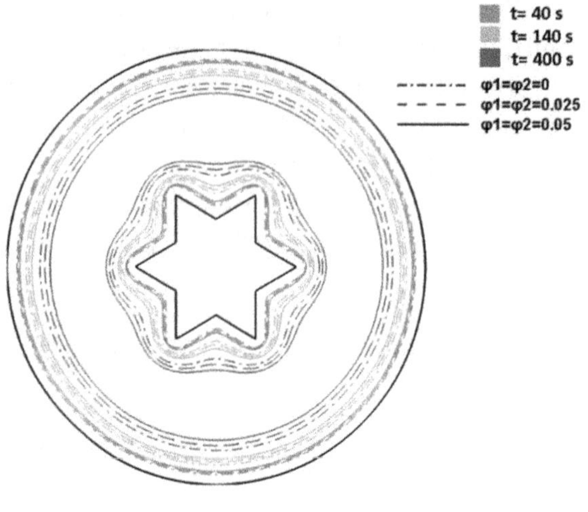

3.3.3 Assessment of HNEPCM With Fins

For illustrating the effect of nanoparticles and fins on the solidification process, Figure 34 presents isotherms distribution in the annulus over three time steps (60, 160, and 260 s) while the volume fraction of HNEPCM equals 0.025. During the early period, isotherms occupy approximately all the void between walls, and there is an insignificant gap in the annulus, but this initial condition starts to depart with time. During the second and third stages, a somewhat uniform isotherm establish near the walls while the middle ground is occupied by rotating cells implying the heat transfer enhancement produced by adding nanoparticles and fins.

Figure 34. The combination influence of HNEPCM and fin on temperature contours in three temporal stages during the solidification process of LHTESS with fin

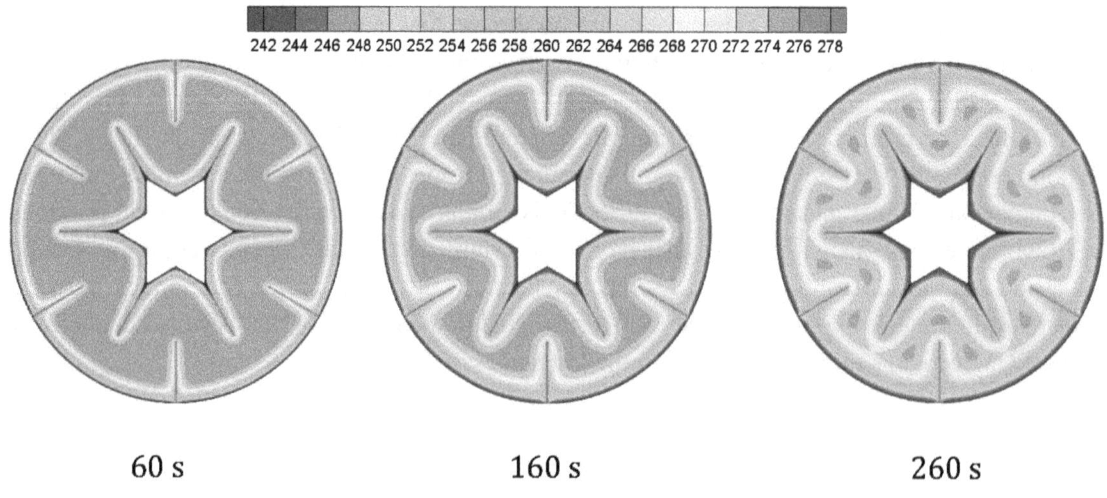

The solid fraction is depicted in Figure 35 in which periods are divided into three points, as Figure 35-a. Similarly, at 60 s, there is hardly any difference with no solidified PCM, except in the proximity of walls and fins. However, in the next periods, the fins and nano inclusion gradually stick out, especially at 260 s. It can be stated confidently that the whole amount of PCM turns into solid, except for small V-shaped areas enclosed between internal fins. Therefore, due to fins and nano utilization, the conduction phenomenon in this study is bolstered, leading to faster solidification rates, which is a crucial factor in LHTESS systems.

3.3.4 Average Temperature, Total Energy, and Solid Fraction Analysis

In this section, the analysis of different parameters is presented to elucidate further the process occurring in the annulus, and the particular effect of various techniques used in this paper. Figure 36 reveals the temperature (K) against time (s) for different cases. The pure PCM without fins and nanoparticles delivers the weakest temperature gradient over time since there is no improvement procedure incorporated in this case.

Figure 35. The combination influence of HNEPCM and fin on solid fraction contours in three temporal stages during the solidification process of LHTESS with fin

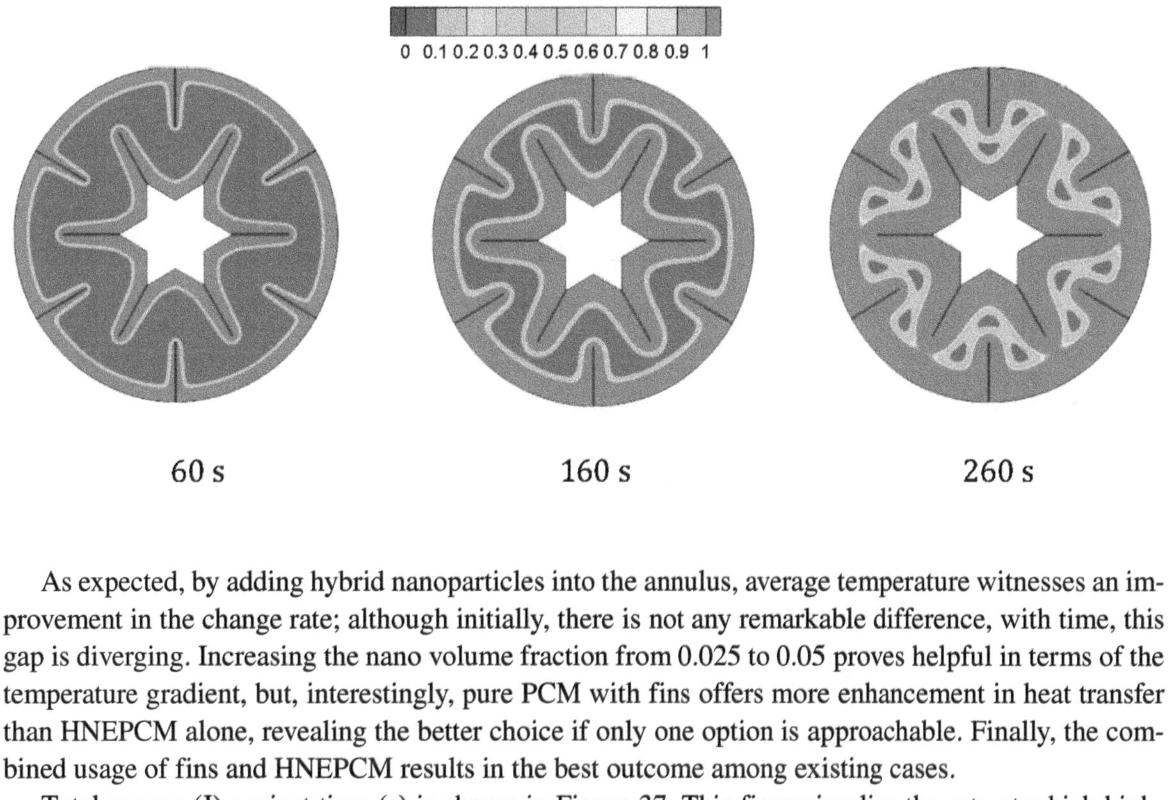

As expected, by adding hybrid nanoparticles into the annulus, average temperature witnesses an improvement in the change rate; although initially, there is not any remarkable difference, with time, this gap is diverging. Increasing the nano volume fraction from 0.025 to 0.05 proves helpful in terms of the temperature gradient, but, interestingly, pure PCM with fins offers more enhancement in heat transfer than HNEPCM alone, revealing the better choice if only one option is approachable. Finally, the combined usage of fins and HNEPCM results in the best outcome among existing cases.

Total energy (J) against time (s) is shown in Figure 37. This figure implies the rate at which high-energy liquid PCM transfers its energy to cold fluid flowing in the inside and outside tubes. Combined finned annulus with nanoparticles holds the first place regarding the energy damping rate as it is the case in the previous figure. However, the prevailing trend that existed in Fig. 36 cannot be seen at this plot, with the HNEPCM and fins case experiencing a sharp plunge from the start of the process to approximately 400 s and then following a linear path to the end. This pattern reveals the appreciable impact of rate promoters when using simultaneously, as for other cases presented in Figure 37, the trend pursues a similar approach, unlike the pronounced rate of total energy reduction in the HNEPCM and fins. The closest case to HNEPCM and fins is pure PCM with fins that still there is a massive gap between them since its' first phase of energy transfer is prolonged until somewhere around 900 s after which hardly any variation in its' total energy transpires.

Figure 36. Temperature plot for different modes

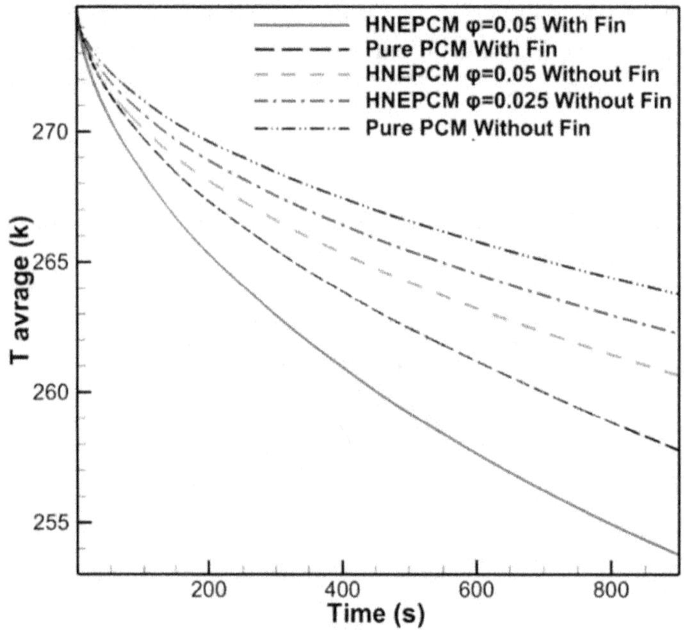

Figure 37. Total energy plot for different modes

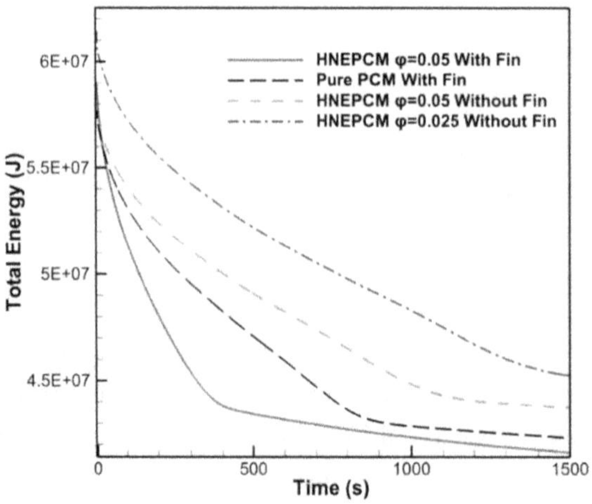

Figure 38. Solid fraction plot for different modes

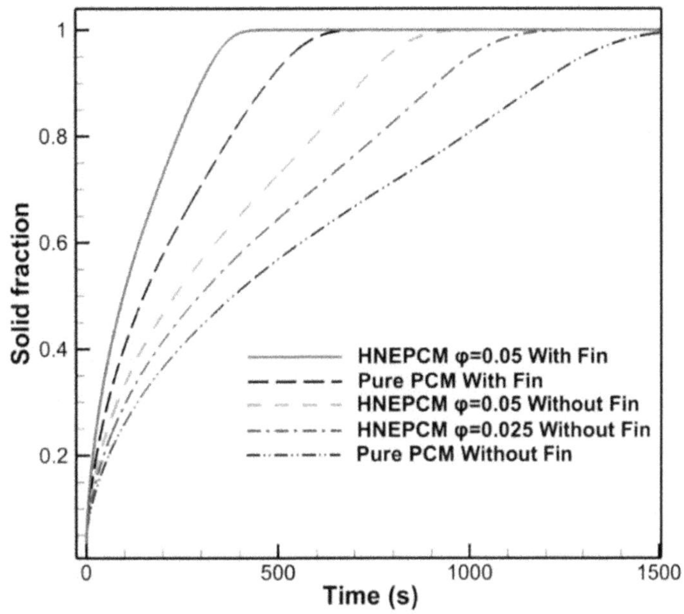

Figure 38 shows the solid fraction over time for each studied case. The results presented in this figure also confirm the points mentioned above, declaring the best condition for the solidification rate to be HNEPCM with fins, which solidified at roughly 300 s. This case is approximately five times as fast as pure PCM without fins, justifying the introduction of the fins and hybrid nanoparticles into the annulus. The next best case solidifying completely at about 600 s is pure PCM with fins which can be favorable procedure to employ if HNEPCM is not available.

4. EXAMPLE 4: EFFECT OF FIN AND HYBRID NANO-PARTICLES ON SOLID PROCESS IN HEXAGONAL TRIPLEX LATENT HEAT THERMAL ENERGY STORAGE SYSTEM

4.1 Problem Definition

The geometry and boundary condition are shown in Figure 39. The geometry demonstrates a triplex tube along with hexagonal fins made of copper. Dimensions of the geometry and fins are located in Table 10 and physical properties of HNPs and PCM are located in Table 11. Water with nanoparticles (TiO_2-GO) is used as HNEPCM. HNEPCM is liquid at 278 K and as it is shown in Figure 39 that coolant fluid with constant temperature of 240 K is used for the solidification procedure. Figure. 39-b indicates that:

'*a*' represents dimension of the hexagonal side and the fins length. '*b*' represents the inner diameter of the tube. '*c*' represents thickness of the fins and '*d*' represents the external diameter of the tube.

Solidification procedure is depended on time and is simulated at an unsteady state. In this research, time step is equal to $2\times10^5(s)$. The general equations governing the freezing process are reviewed in Chapter 6.

Figure 39. (a) Three-dimensional view of LHTESS with fin, (b) two-dimensional geometry and boundary conditions of present study

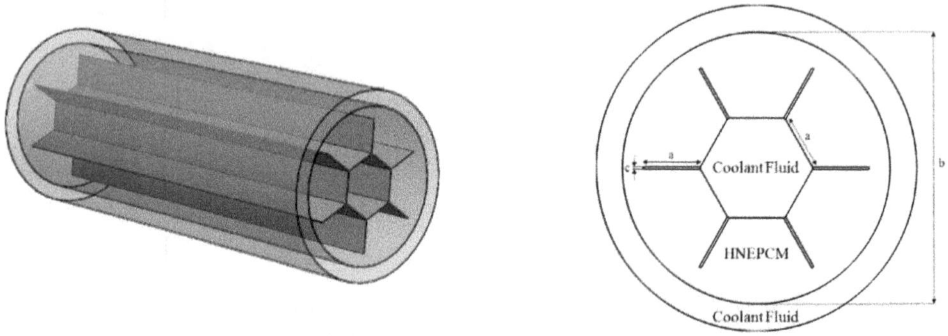

(a) (b)

Table 10. Geometry parameters of LHTESS

Parameters	Dimensions [cm]
a	4.25
b	10
c	0.17320508

Table 11. The physical properties of water as PCM, hybrid nanoparticles, and fin

Property	PCM	TiO_2	Go	Fin
$\rho\,(\frac{kg}{m^3})$	997 (Liquid phase)	4250	1800	8960
	916 (Solid phase)			
$C_p\,(\frac{J}{kgK})$	4184 (Liquid phase)	686.2	717	385
	2030 (Solid phase)			
$k\,(\frac{W}{mK})$	0.6 (Liquid phase)	8.9538	5000	400
	2.22 (Solid phase)			
$L_f\,(\frac{J}{kg})$	335,000	-	-	-

4.2 Numerical Simulation

The GFEM is employing PDE integrals to form discredited equations in mesh nodes. This method solves equations by cubic interpolation over triangles and nodal amounts are placed on the corners of cells. Definitions are replaced to form the Galerkin couple matrix and segregate the dependents from the variables. Integration is applied on cells and equations are formed, based on variables and Newton-Raphson method is used to solve the final equations and it employs backward difference techniques to solve the equations containing time-based integration.

Figure 40. Adaptive mesh refinement in two time steps including 20 and 100 (s)

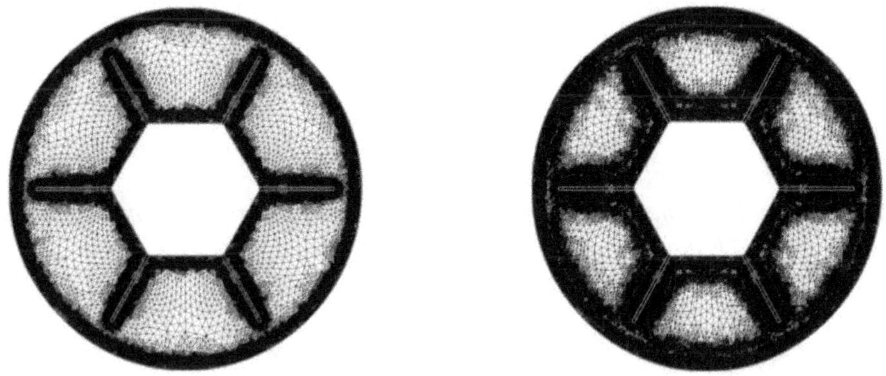

20 s 100 s

As it is demonstrated in Figure 40 the solidification procedure is simulated by adaptive grid refinement. At unsteady conditions, this method is utilized to refine meshes that had instant changes in their initial values.

In this study, the numerical method is utilized by comparing the equations based on GFEM with former empirical study published by Ismaeil et al. (2001). Figure 41 shows that this comparison contains maximum error of 5 percent. As a conclusion, assumption of discarding the convection terms is logical.

4.3 Impacts of Active Parameters

4.3.1 Comparison Between LHTESS With and Without Fins for Pure PCM

Temperature and solidification contours for with fins and without fins modes can be observed in Figure 42 for three different time steps (20, 100 and 300 (s)). There is equal amount of PCM in container in these two investigated modes. Results show that adding fins to the container causes an increment in heat penetration and heat exchange area which causes more decrement in temperature compared to without fins mode and also increases the solidification rate. Comparison of solidification rate between with fins and without fins container is shown in Figure 43 in three different time steps. It can be obtained from the results that when the fins are added, more volume of container is solidified and it is obvious that over time, solidification rate is increased.

Figure 41. Comparison of present study with experimental study published by Ismaeil et al. (2001)

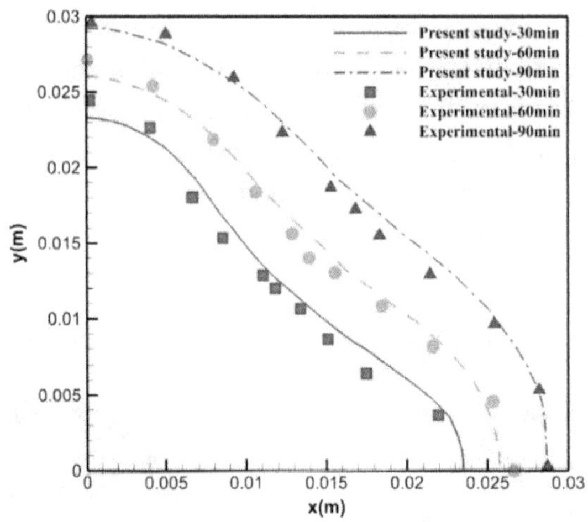

Figure 42. Solid fraction (left) and temperature (right) contour plots in three time steps during the solidification process of LHTESS without and with fin mode for pure PCM

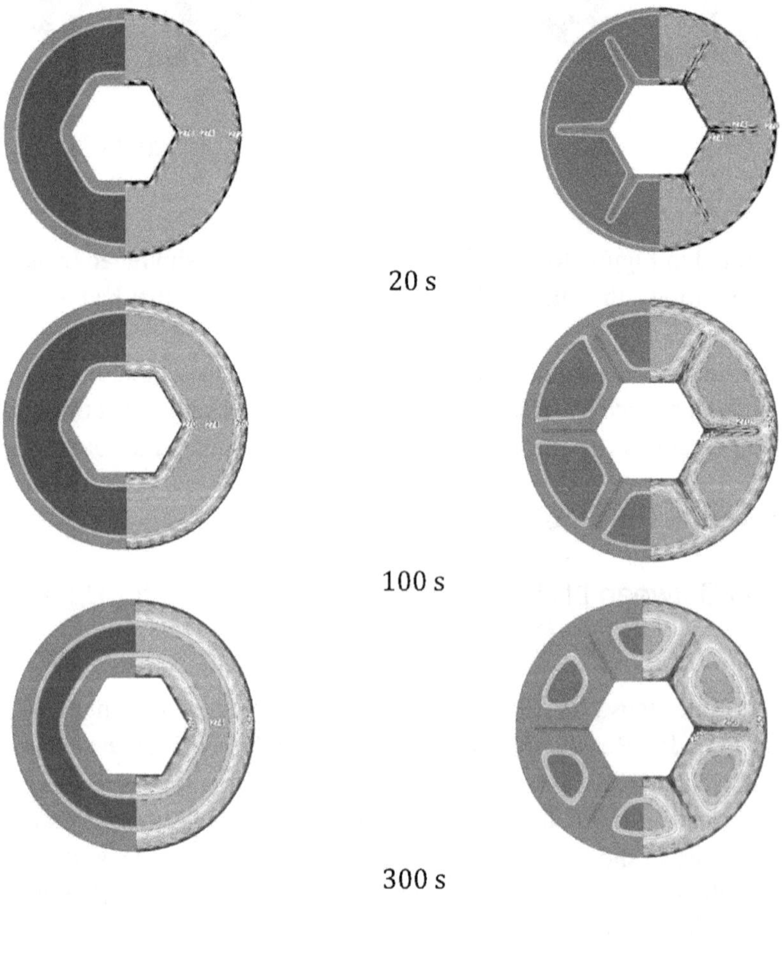

Figure 43. Comparison of solidification rate between LHTESS with and without fin mode

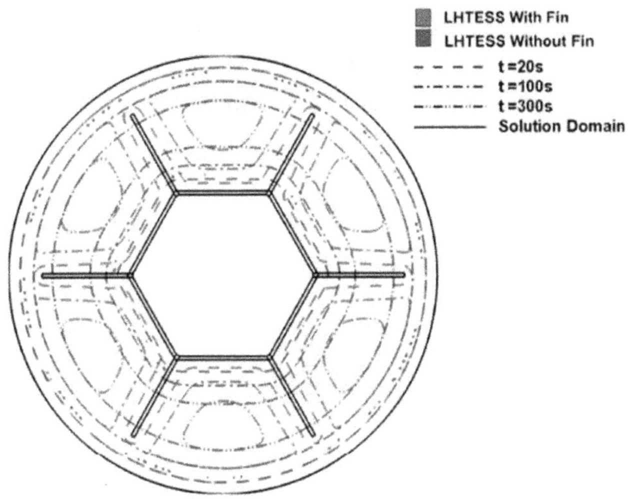

Figure 44. The effect of HNEPCM volume fractions on without fin LHTESS

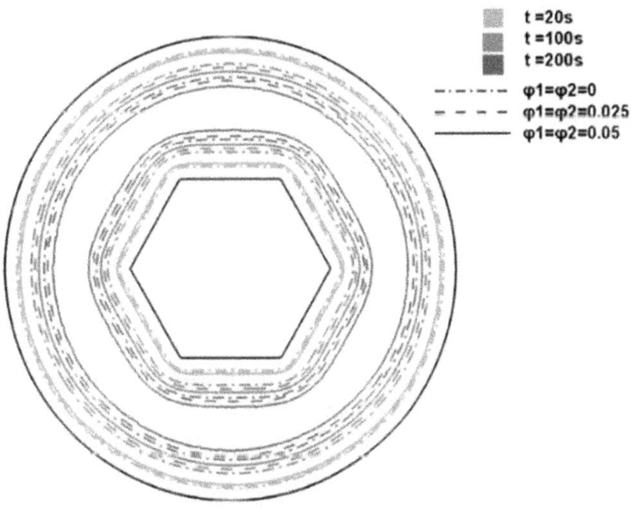

4.3.2 The Effect of HNEPCM Volume Fractions on LHTESS Without Fins

Figure 44 demonstrates the effect of different volume fractions (0, 0.025 and 0.05) on solidification process in 3 various time steps (20,100 and 300 (s)). $\varphi 1$ and $\varphi 2$ respectively represent TiO2 and Go volume fractions. As the results show, when HNEPCM volume fractions are increased, the solidification rate is simultaneously increased because of growth in heat transfer coefficient and also the solidification rate is increased over time.

4.3.3 The Simultaneous Effect of HNEPCM and Fin on LHTESS

Figure 45-a shows the simultaneous effect of HNEPCM with volume fraction amount of 0.025, with fin and nanoparticles (TiO_2-Go) as methods of increasing heat transfer in three different time steps on the temperature contour. The results illustrate that adding fins and nanoparticles to the container have a great effect on temperature reduction.

Figure 45a. The simultaneous effect of HNEPCM and fin on temperature contour plots in three time steps during the solidification process of LHTESS with fin mode

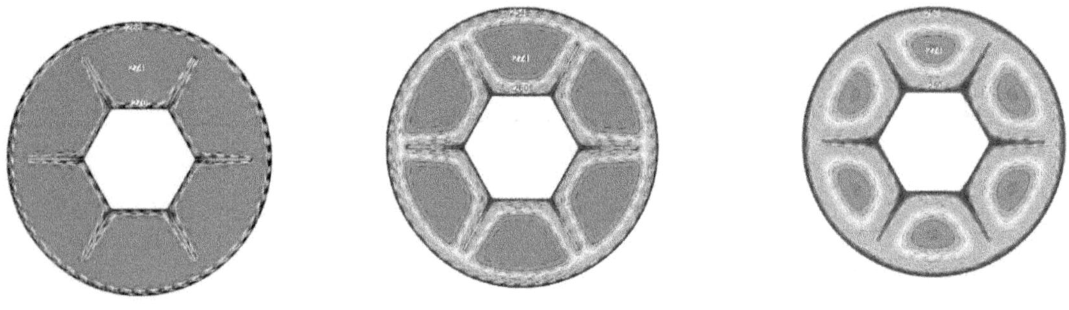

20 s 100 s 300 s

Also Figure 45-b shows the effect of HNPs and fins on solidification process. As results demonstrate when the fins and nanoparticles are added to the container solidification rate is strongly increased due to heat transfer coefficient and penetration increment, also heat exchange area amount is much higher with fins.

Figure 45b. The simultaneous effect of HNEPCM and fin on solid fraction contour plots in three time steps during the solidification process of LHTESS with fin mode

20 s 100 s 300 s

4.3.4 Comparison of Temperature and Total Energy of the Solidification Process Between With Fins and Without Fins LHTESS With HNEPCM

Figure 46 shows the temperature plot for different modes investigated and is drawn from 0 to 1500s. It can be observed from the results that HNEPCM with fins and nanoparticles volume fraction of 0.05 has the most temperature reduction. Also pure PCM without fins shows the least changes in temperature rate between the cases. Figure 47 depicts the total energy alterations in the container from 0 to 1500s. The results show that HNEPCM with fins and nanoparticles volume fraction of 0.05 has the most internal energy reduction which causes temperature reduction and an increment in the solidification rate. The alteration of the solid fraction process from 0 to 1500 (s) which is shown in Figure 48 results that HNEPCM with fins and nanoparticle volume fraction of 0.05 has the greatest solidification rate due to higher amount of heat transfer coefficient, Also full solidification of LHTESS is occurred in 580 (s). The lowest solidification rate is observed in pure PCM without fins and full solidification is occurred in 1093 (s). This issue represents the intense effect of HNPs and fins on LHTESS.

Figure 46. Temperature plot for different modes

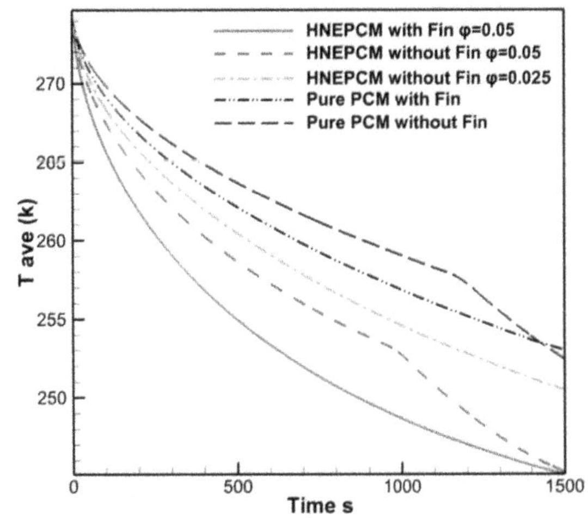

The effect of shape factor and volume fraction of nanoparticles on solidification time is shown in Figure 49 and as it is seen in the results when volume fraction of nanoparticles and shape factor is increased the solidification time is reduced.

Figure 47. Total energy plot for different modes

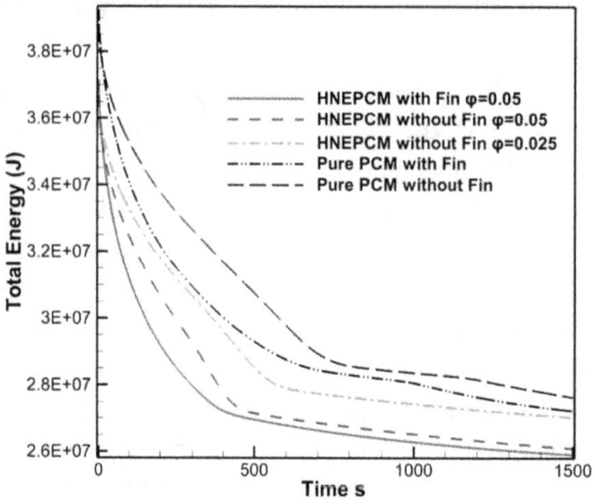

Figure 48. Solid fraction plot for different modes

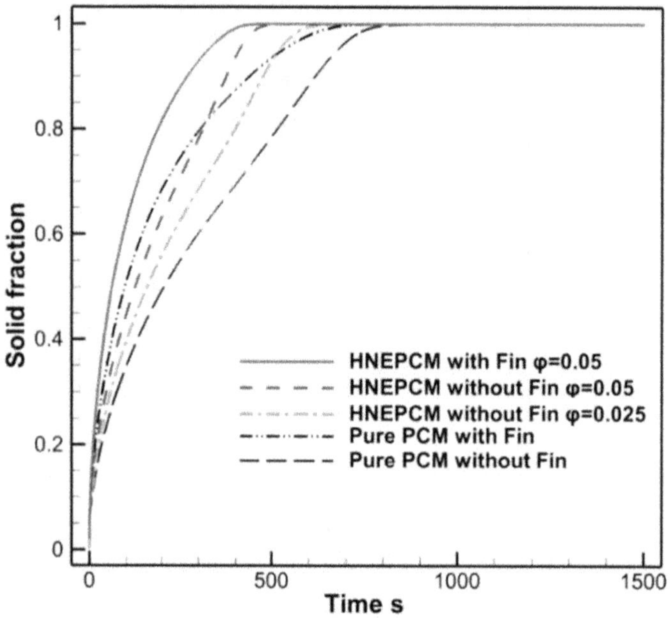

Figure 49. The effect of shape factor and nanoparticle volume fraction on solidification time

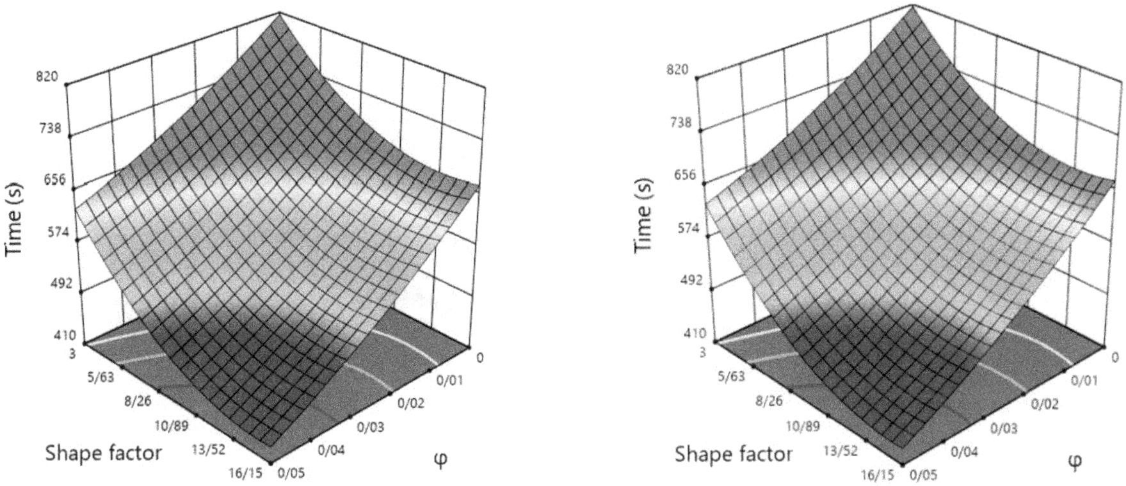

5. EXAMPLE 5: SOLIDIFICATION ACCELERATION IN A TRIPLEX-TUBE LATENT HEAT THERMAL ENERGY STORAGE SYSTEM USING V-SHAPED FIN AND NANO-ENHANCED PHASE CHANGE MATERIAL

5.1 Problem Definition

Figure 50 represents the two-dimensional schematic and boundary conditions. This figure shows the cross-section of V-shaped fin assisted triplex-tube LHTESS in the physical model. The Heat Transfer Fluid (HTF) fills the internal and external tubes of the triplex-tube LHTESS and the PCM is solidified in the middle tube of the system. The temperature of the inner and the middle tubes of triplex-tube LHTESS is assumed to be 240 K, which is constant during the process. Moreover, the pure PCM fills the middle tube with the primary temperature of 278 K.

The initial solid fraction of PCM area is assumed to be equal to 0.0, because the PCM is in liquid phase at the start of the process. Also, the solid fraction equal to 1 is enforced for the fin region. Moreover, V-shaped fin is added to the inner surface of the middle tube to augment the Thermal Penetration Depth (TPD) in regions containing PCM. The internal and external diameters of the triplex-tube LHTESS are 20 and 80 mm, respectively. Firstly, a two-dimensional simulation is performed using Finite Element Method to evaluate the performance of V-shaped fin within the triplex-tube LHTESS. Also, by using RSM, the geometric parameters of V-shaped fin are optimized to achieve the best fin arrangement. Secondly, SWCNT nanoparticles with three different volume fractions are dispersed into the pure PCM and the results are compared with the optimized V-shaped fin assisted LHTESS from the sight of solidification rate. The geometric parameters of V-shaped fin which are used in the optimization procedure are presented in Table 12 with their ranges (see Figure 51).

Figure 50. (a) Three-dimensional view of triplex-tube LHTESS, (b) two-dimensional view of triplex-tube LHTESS, and (c) computational domain and boundary conditions

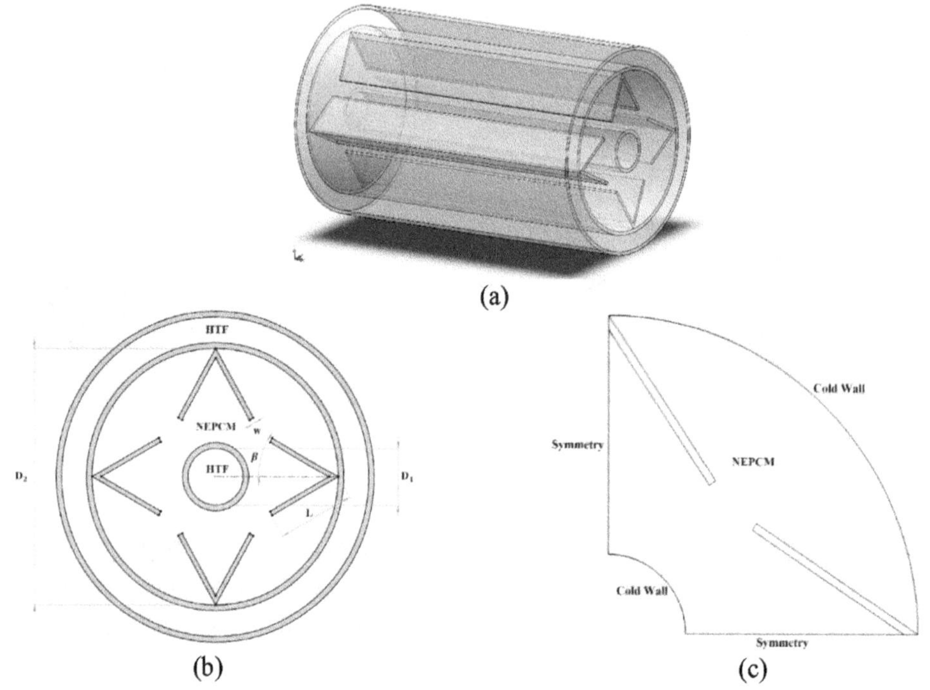

Figure 51. Adaptive mesh refinement in three time steps including 100, 200, and 300 (s)

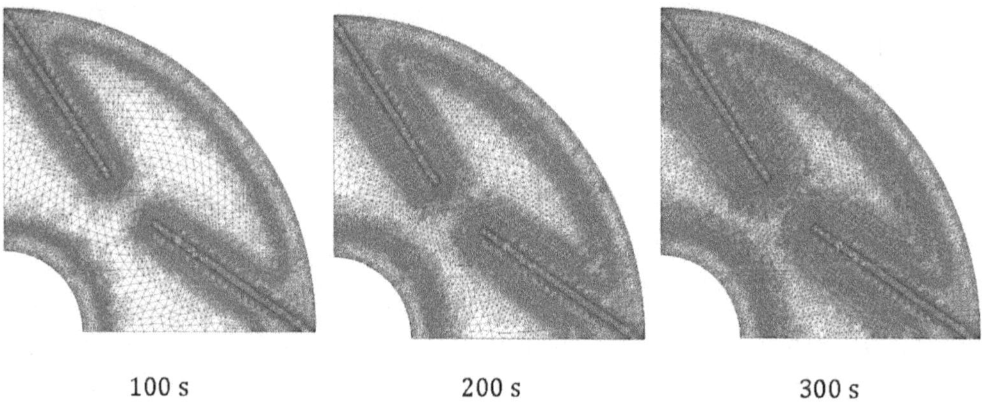

In Table 13, the PCM thermophysical properties, SWCNT nanoparticles and fin material are listed. As phase change occurs in the solidification procedure, the variation of thermophysical properties during the process is considered, in this simulation. Therefore, the thermophysical properties of PCM are presented for both phases. In this investigation, a time-step of $2\times10^{-5}(s)$ is considered due to the dependency of the solidification procedure to time in the triplex-tube LHTESS.

Table 12. Geometric parameters of V-shaped fin

Geometric Parameter	Range
β	[11.25°, 22.5°, 33.75°, 45°]
L	[15, 20, 25mm]
w	[0.5, 0.75, 1mm]

Table 13. The thermophysical properties of water as PCM, nanoparticles, and fin

Property	Water (PCM)	SWCNT (Nanoparticle)	Copper (Fin)
$\rho(\frac{\text{kg}}{\text{m}^3})$	997 (Liquid phase) 916 (Solid phase)	2600	8960
$C_p(\frac{\text{J}}{\text{kgK}})$	4184 (Liquid phase) 2030 (Solid phase)	425	385
$k(\frac{\text{W}}{\text{mK}})$	0.6 (Liquid phase) 2.22 (Solid phase)	6600	400
$L_f(\frac{\text{J}}{\text{kg}})$	335,000		

According to abovementioned points, following assumptions are assigned for simulating the solidification procedure in the V-shaped fin assisted triplex-tube LHTESS:

- A two-dimensional simulation in Cartesian coordinates is performed.
- An unsteady simulation is performed for the solidification procedure in the V-shaped fin assisted triplex-tube LHTESS.
- The influence of free convection on solidification process is ignored and the conduction is the predominant mechanism for the heat transfer.
- The variation of thermophysical properties of PCM is taken into account, because phase change is involved in this process.
- Constant thermophysical properties are considered for PCM and NEPCM in each phase.

According to these assumptions, the general governing equations are included in Chapter 6.

5.2 Numerical Simulation

In present study, the simulation of solidification process is carried out using SGFEM with the aid of FlexPDE 5.0.18 software. In order to solve the phase change procedure, cubic interpolation is applied over triangles. In the problems with triangle grid, the calculations are performed on the corner of cells which is called as node. In this method, the differential equations are made by replacing the descriptions, applying the dependencies to the variables, integration over cells and differentiating the final system regarding to variables. Then, these differential equations complete the Galerkin matrix. As the solidification is a time-dependent process, an implicit backward method is employed for time integration and solving the system. Moreover, in this simulation an adaptive grid refinement strategy is used to refine the grid when the variable alterations are considerable. The refinement procedure continues during the process until the required precision which is determined by the user is obtained.

Moreover, a comparison between present numerical results and previous investigation (Ismail et al., 2001) is done to approve the accuracy of this simulation, as depicted in Figure 52. This figure compares the solidification fronts in the system at three time steps and indicates an excellent agreement between the results, so that the maximum reported error is equal to 4.9%.

5.3 Impacts of Active Parameters

5.3.1 Optimization Procedure of Fin Arrangement

This section focuses on the optimization of geometric parameters of V-shaped fin by employing RSM. The parameters which are selected to be optimized include the angle between fin branches (β), the fin length (L) and the fin thickness (w). The FST is set as the response of the optimization process which is required to be minimized.

Figure 52. Comparison between present study and experimental study published by Ismaeil et al. (2001)

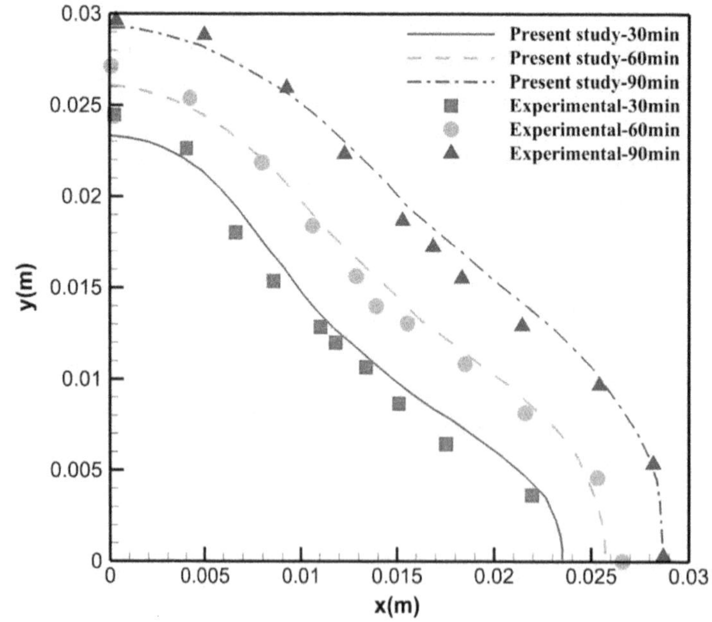

Figure 53 shows the influence of fin thickness on FST for three fin lengths. It can be seen that in all fin lengths, the change of FST is not considerable by augmenting the fin thickness and only a slight acceleration in solidification process is obtained. The reason of this slight acceleration is that increasing the fin thickness leads to augment the heat transfer area, while it does not increase the TPD into the PCM region and zone of fin impact. Consequently, the maximum fin thickness in the reviewed range is considered as its optimal value (w=1mm).

The influence of fin length on the solidification acceleration is illustrated in Figure 54. It can be seen that increasing the fin length expedites the solidification process significantly. Because, by augmenting the fin length, the TPD is enhanced to the PCM region which causes higher solidification rate. Moreover, due to the weak thermal conductivity of PCM, applying highly conductive fin in this region leads to a higher solidification rate. Consequently, the optimal fin length in the considered range is 25mm.

Figures53 and 54 also demonstrate the influence of the fin branches angle branches on solidification acceleration. It can be seen that for all fin lengths and thicknesses, by increasing this angle, first the FST drops and then increases. Because, in lower values of β, the V-shaped fin branches get closer to the axes and the PCM amount increases in the area between two fin branches in each quarter of the domain. Since the V-shaped fin does not affect the TPD in this region, the solidification experiences lower rate which lead to increase the FST.

Figure 53. Effect of fin thickness on FST when a) L=15mm, b) L=20mm, and c) L=25mm

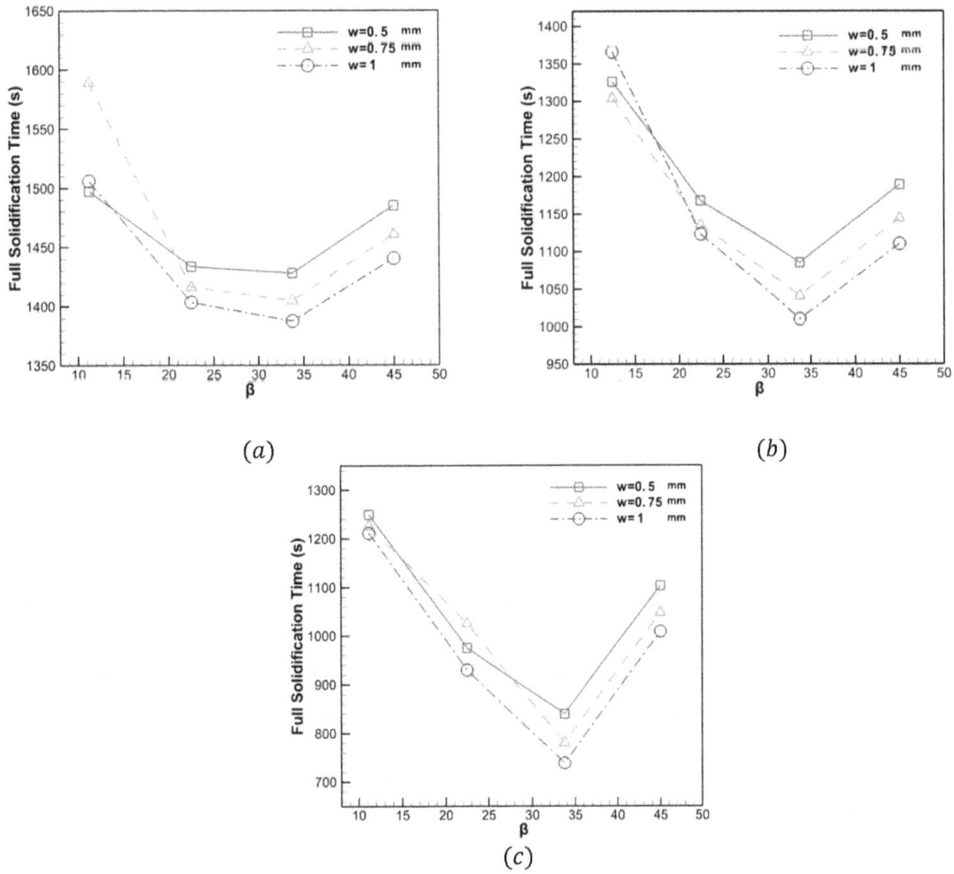

Also, in higher values of β, the fin branches get closer to each other which causes the TPD to be enhanced in the area between two branches, while in the region between the fin branches and the axes, the solidification procedure is still slow. Therefore, the optimal value of β is expected to be between the minimum value (β=11.25°) and maximum value (β=45°) of studied range. In current study, the optimal value of β is equal to 33.75° which leads to the fastest solidification process. Based on the this analysis, β=33.75°, L=25mm and w=1mm are considered as the best fin design from the sight of solidification acceleration. By considering this optimized fin arrangement in the triplex-tube LHTESS, the solidification is completed in 738.63 s.

Figure 54. Effect of fin length on FST when a) w=0.5mm, b) w=0.75mm, and c) w=1mm

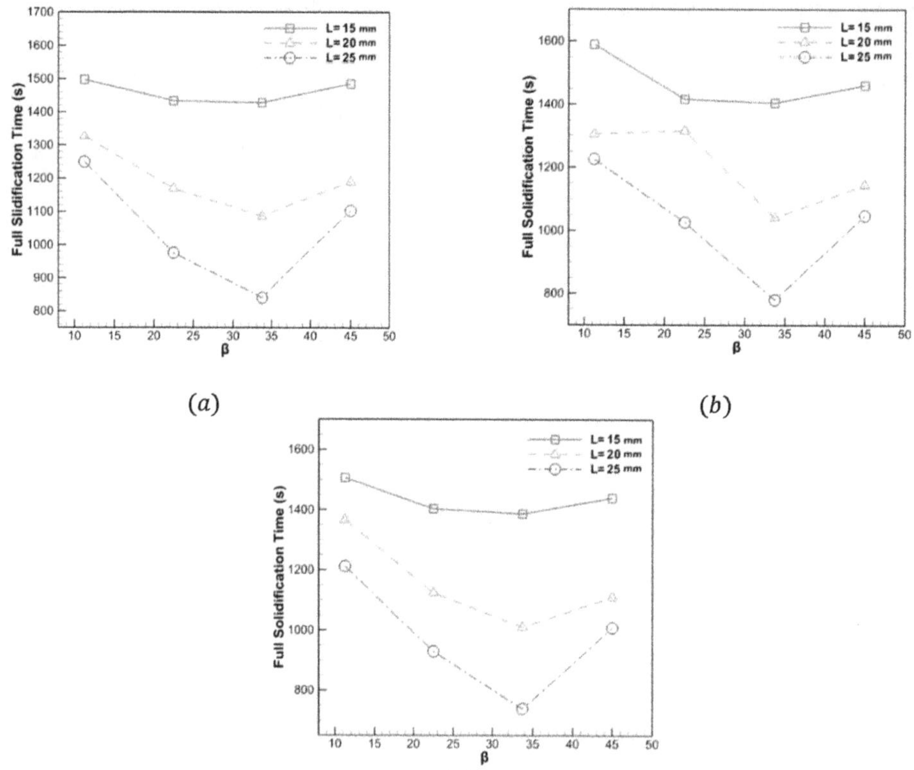

(a) (b) (c)

Figure 55 demonstrates the response surfaces of FST for several values of β, L and w. As mentioned before, it is obvious that the fin thickness does not change the FST considerably, while the branches angle and the fin length affect it remarkably.

According to RSM analysis, a correlation for FST is extracted as a function of branches angle (β), fin length (L) and fin thickness (w), which is presented as below:

Time = 3613.67 – 176.96(L) – 26.86(w) – 19.64(β) - 4.78(L)(w) – 0.36(L)(β) – 3.42(w)(β) - 3.81(L2) + 8^5.26(w2) + 0·46(β2) (3)

Figure 55. Response surfaces of FST for geometric parameters of V-shaped fin (β,L,w)

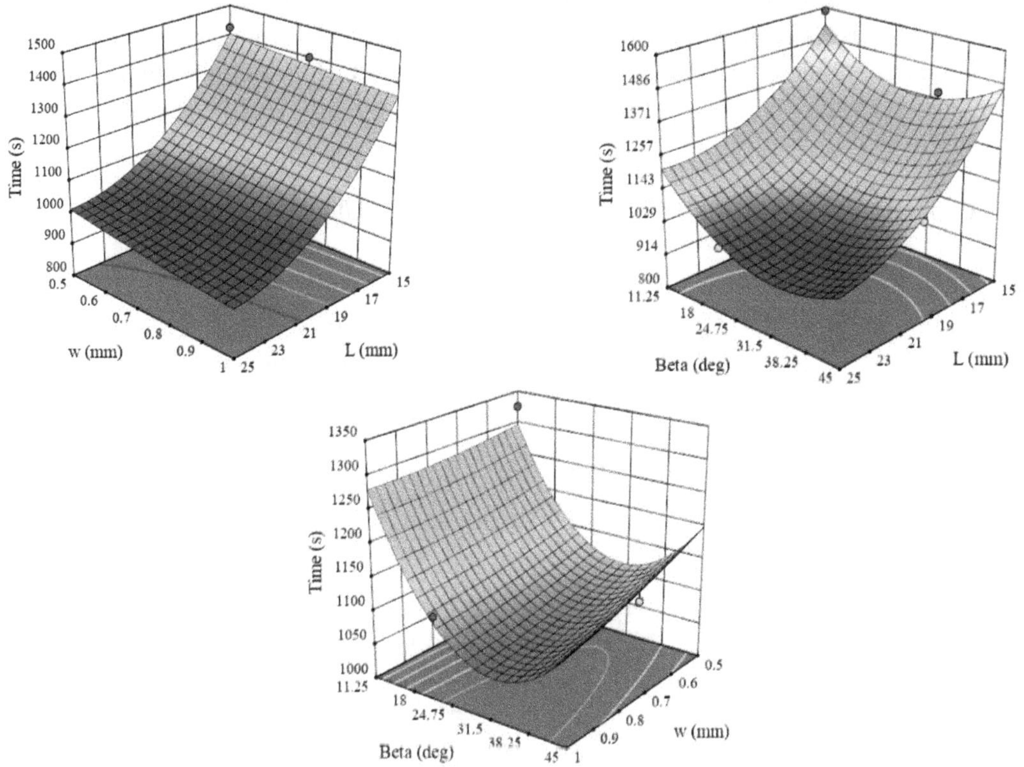

5.3.2 Comparing the LHTESS Performance in the Presence of Optimized V-Shaped Fin and Without Fin

When the PCM undergoes the phase change, the liberated thermal energy is stored by the triplex-tube LHTESS. Despite the high storage capacity of PCM, their low thermal conductivity undermines the solidification process. Present study employs the V-shaped fin on the middle tube of the triplex-tube LHTESS to fix this defect. In order to demonstrate the benefit of such fin arrangement in the system, the solidification fronts of the LHTESS containing optimized V-shaped and without fin are compared in three various time steps, as depicted in Figure 56. It can be seen that in the absence of fins, inception of solidification is only at the neighborhood of cold walls, while attaching V-shaped fin to the middle tube of the triplex-tube LHTESS causes the TPD to be increased into the PCM region.

Consequently, the solidification procedure starts near the cold walls and also in the vicinity of fin surface, which results in increasing the rate of heat transfer and expediting the process. It is noteworthy that the internal diameter of the LHTESS in the absence of fin is supposed to be greater than that of the LHTESS containing optimized fin, to satisfy the equality of PCM mass in current comparison, as depicted in Figure 56. Because, in the presence of V-shaped fin, less space is occupied by the PCM which reduces the PCM mass. The contour plots of temperature and solid fraction for both LHTESS containing optimized V-shaped and without fin are illustrated in Fig. 57a and 57b, in three various time steps. This figure shows that using optimized V-shaped fin in the triplex-tube LHTESS causes more uniform temperature distribution as compared to the LHTESS without fin, which leads to desired results in this study. In addition, as adding V-shaped fin enhances the TPD into the PCM region uniformly, the unsolidified PCM mass is reduced in the system. Consequently, the thermal resistance of the system decreases and the solidification procedure is completed faster than when no fin is employed in the system, as depicted in Figure 57(b).

Figure 56. Comparison between solidification front in LHTESS without fin and LHTESS with optimized V-shaped fin

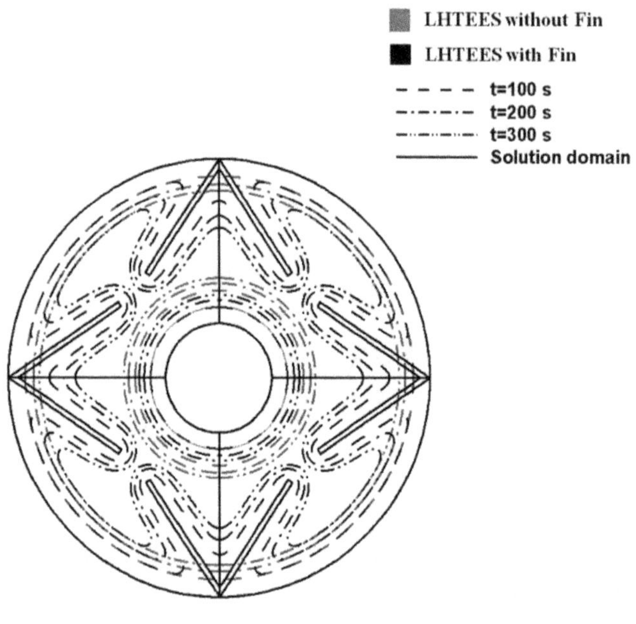

5.3.3 Dispersion of Nanoparticle Into the PCM

This section deals with the influence of dispersing nanoparticles into the PCM and making NEPCM. For this purpose, SWCNT nanoparticles with three different volume fractions equal to 0, 2.5, 5% are added to the PCM. In order to demonstrate the impact of nanoparticles on the efficiency of triplex-tube LHTESS, Figure 58 and Table 14 are presented. As shown in Table 14, making NEPCM in the LHTESS impressed the solidification process considerably.

It is remarkable that the impact of nanoparticles is more considerable when the conduction mechanism predominates the process. As conduction is the main heat transfer mechanism which governs the solidification process, dispersing nanoparticles with higher volume fractions causes higher solid fraction. For example, a triplex-tube LHTESS containing nanoparticles with volume fraction equal to 5% expedites the process by about 33.1%, as tabulated in Table 14.

Figure 57a. Temperature contour plots of LHTESS with optimized V-shaped fin (left) and LHTESS without fin (right) at three different time steps

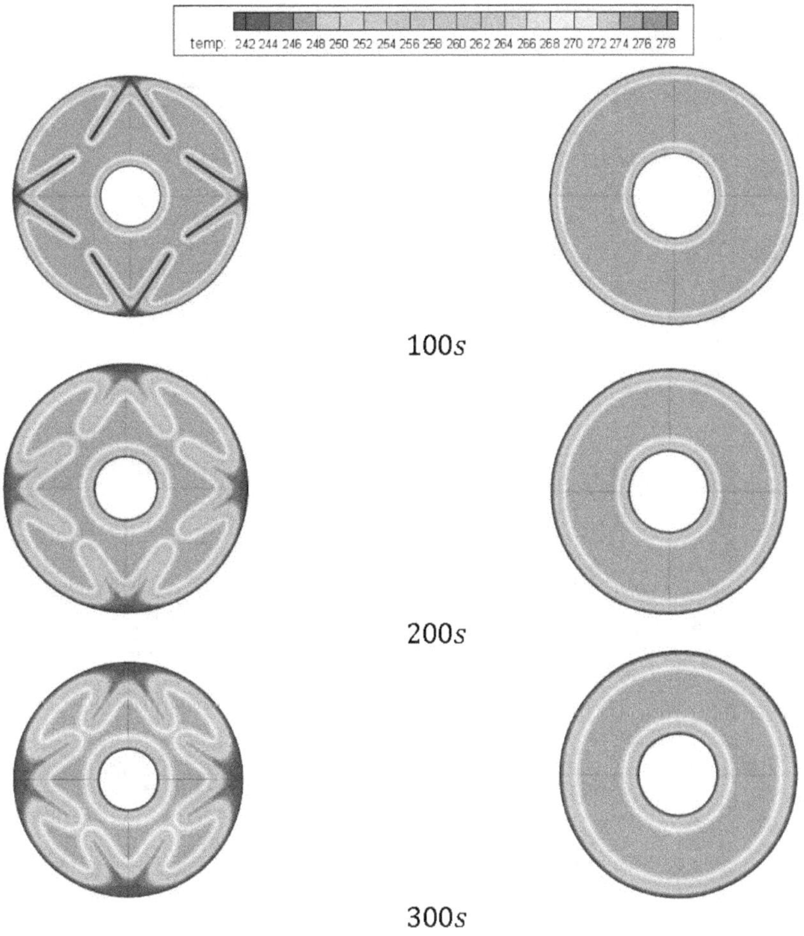

Figure 58 illustrates the influence of nanoparticle volume fraction on the solidification front at three various time steps. As shown, at the inception of the process, using nanoparticles with higher volume fractions does not increase the solid fraction noticeably. However, as time progresses, increasing the volume fraction of nanoparticles causes a considerable augmentation to the solid fraction. Because, at the inception of process, only near cold wall regions which have a high-temperature variation with unsolidified PCM experience the solidification process. Accordingly, at the inception of the procedure, the solidification is expedited due to this high temperature difference and there is no need to enhancement approaches. But, as the time progresses, since ice with poor thermal conductivity is in the region between the cold walls and unsolidified PCM, the solidification procedure is weakened. Consequently, the thermal resistance of the system increases and the solidification procedure is done too slow which requires an enhancement approach to be accelerated.

Figure 57b. Solid fraction contour plots of LHTESS with optimized V-shaped fin (left) and LHTESS without fin (right) at three different time steps

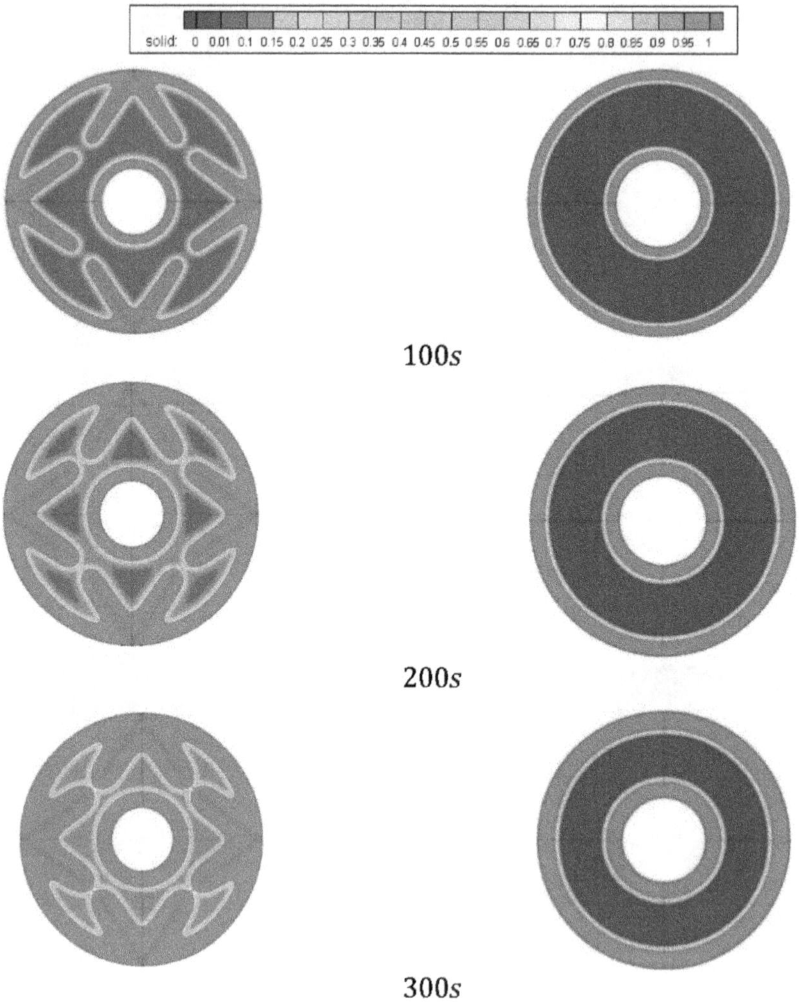

Table 14. Effect of nanoparticles volume fraction on FST and improvement

Nanoparticles Volume Fraction (φ)	Full Solidification Time (s)	Increment in Solidification Rate (%)
0	1438.68	-
0.025	1237.35	13.99
0.05	962.36	33.1

Figure 58. Effect of nanoparticles volume fraction on solidification front in the triplex-tube LHTESS without fin at three different time steps

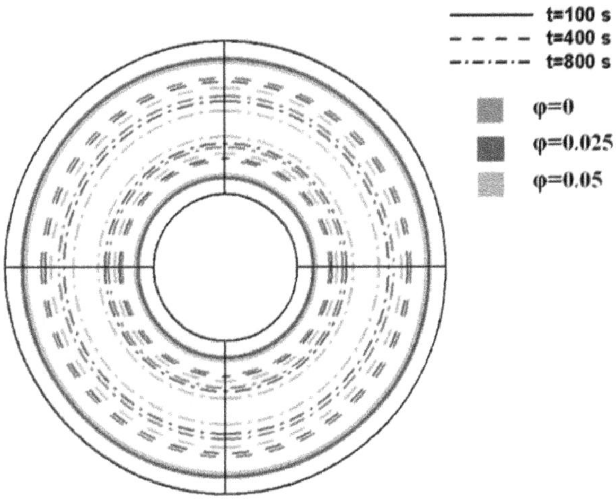

5.3.4 Comparison Between the Enhancement Approaches

In this study, two enhancement approaches including the V-shaped fin and SWCNT nanoparticles are employed to quicken the solidification procedure in the triplex-tube LHTESS. Figures 59 and 60 are displayed to compare these approaches from the sight of solidification rate and maximum capacity of energy storage. The average solid fraction for different cases during the process is depicted in Figure 59. It should be notices that solidification rate is expressed as the slope of this figure. The triplex-tube LHTESS equipped with optimized fin experiences greater solidification rate than that of nanoparticle dispersion, which means that lower time is required for the solidification to be completed, as illustrated in Figure 59. Results showed that the FST for the LHTESS containing optimized V-shaped fin is 1.94 times faster than that of the LHTESS without fin, and also 1.3 times faster than when SWCNT nanoparticles with 5% volume fraction are added to the PCM. The total energy during the solidification procedure is indicated in Fig. 60. The summation of sensible and latent heat is expressed as the total energy. As shown, the energy storage capacity experiences its maximum value at the inception of the process. Moreover, it is obvious that the maximum capacity of energy storage for all cases including the LHTESS without fin, LHTESS containing optimized fin and LHTESS with nanoparticles dispersion is approximately the same.

However, the maximum solidification rate is obtained in the LHTESS containing optimized fin. Therefore, as the maximum solidification rate and also maximum capacity of energy storage are obtained in optimized V-shaped fin assisted LHTESS, attaching V-shaped fin to the triplex-tube LHTESS is reported as a more effective enhancement approach as compared to the nanoparticles dispersion.

Figure 59. Comparison between solid fraction of triplex-tube LHTESS with pure PCM without fin, pure PCM with optimized V-shaped fin, NEPCM with 2.5 and 5% volume fractions during the solidification process

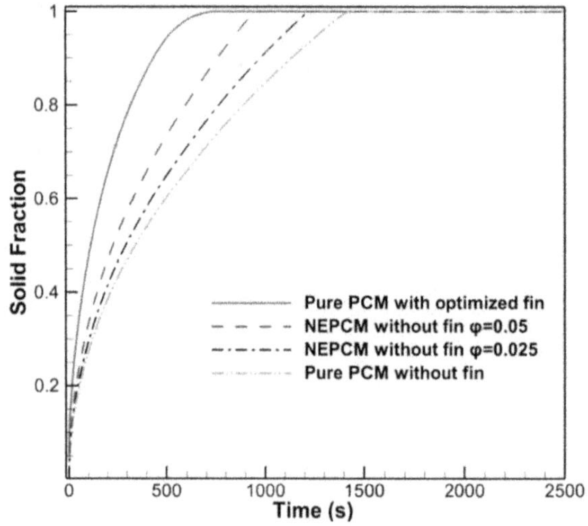

Figure 60. Comparison between total energy of triplex-tube LHTESS with pure PCM without fin, pure PCM with optimized V-shaped fin, NEPCM with 2.5 and 5% volume fractions during the solidification process

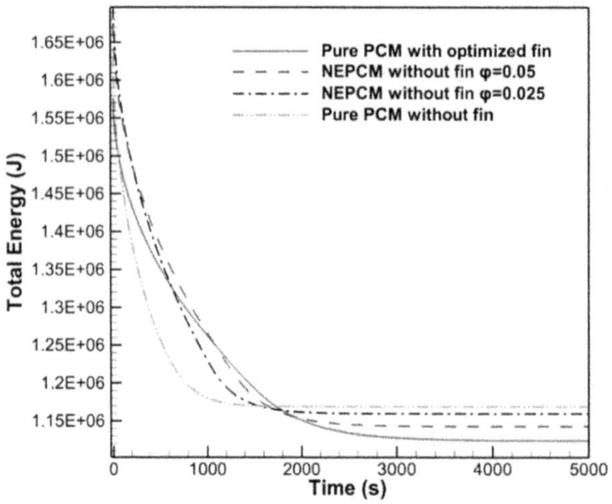

6. EXAMPLE 6: THE EFFECT OF TERNARY HYBRID NANOPARTICLES (GO-MgO-TiO_2) AND RADIATIVE HEAT TRANSFER ON THE SOLIDIFICATION PROCESS OF PCM INSIDE A TRIPLEX TUBE WITH HOLLOW FINS

6.1. System Description

A PCM-based TES system is consisting of an enclosure with an inner complex-shaped tube is taken into account. The proposed 2-D enclosure is filled with ternary hybrid nanoparticles enhanced phase change material (THNEPCM). For analysis of the solidification process, it is considered that Fo, MgO, and TiO_2 as ternary hybrid nanoparticles are homogeneously dispersed into the water as the base PCM. Water is considered the base PCM due to its several desirable properties, namely its reliability and nontoxicity, during numerous phase transition cycles. Thermo-physical features of the PCM in both solid and liquid phases and nanoparticles are illustrated in Table 15. Moreover, a heat transfer fluid (water) streams inside the tubes and exchanges thermal energy with the PCM so as to PCM convert to the solid phase. In spite of the fact that buoyancy influence can be considered at the beginning steps of the solidification, the prevailing conduction overcomes the convection mode of heat transfer. Resultingly, the convection term can be omitted from the governing equations.

Table 15. The physical properties of base PCM (water) and hybrid nanoparticles (TiO_2- MgO- Go)

Property	PCM	TiO_2	MgO	Go
$\rho\,(\frac{kg}{m^3})$	997 (Liquid phase) 916 (Solid phase)	4250	3560	1800
$C_p\,(\frac{J}{kgK})$	4184 (Liquid phase) 2030 (Solid phase)	686.2	955	717
$k\,(\frac{W}{mK})$	0.6 (Liquid phase) 2.22 (Solid phase)	18	45	5000
$L_f\,(\frac{J}{kg})$	335,000	-	-	-

6.1.1 Geometrical Arrangement and Boundary Condition

For the investigation of the influence of different geometrical configurations, three cases have been employed; the first case is constant temperature, in which the outer surface temperature of the container (shell) is set at 298 K. In the second case, the external surface of the shell container is adiabatically isolated. In both cases, the coolant fluid streams in the hollow fin with a complex-shaped structure, making the THNEPCM convert to solid in the enclosure. The isothermal coolant fluid with a temperature of 240 K is selected for the discharging process. The last case has a TTHX. In this arrangement, three tubes exist inside the heat exchanger in which the complex-shaped internal tube (hollow fin) and the external tube incorporate cold HTF and the middle space between them is filled with THNEPCM. This configuration supplies an appropriate condition for more rapid solidification because of double-sided heat transfer. Since the initial temperature of the PCM container is $T_{ini} = 275$ K at $t = 0$ s, which is greater than the solidification point, the PCM is taken into account to be fully liquid. In the current investigation, all the cases are simulated until full solidification of the PCM occurs. A schematic of the geometry, along with the boundary conditions for all considered cases, is presented in Figure 61.

Figure 61. Two-dimensional section and boundary conditions of present study configurations and computational domain

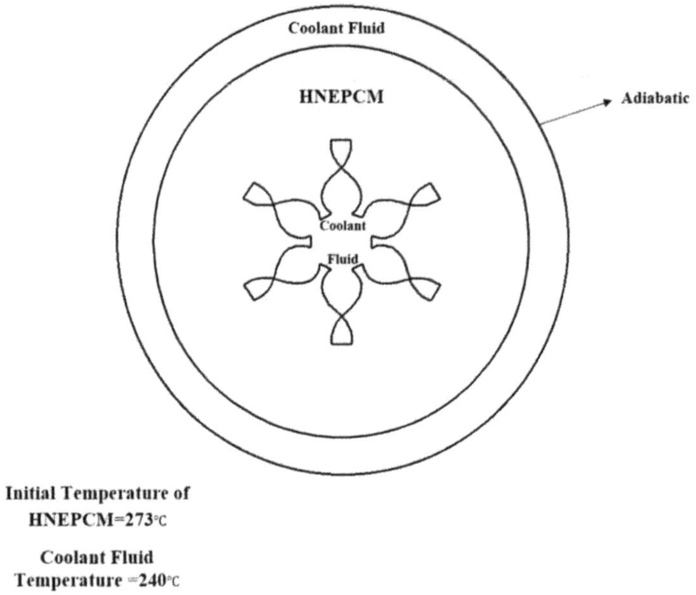

6.1.2 Governing Equations

In order to drive the governing equations, the following hypotheses are adopted.

- Although the thermo-physical specifications of the solid and liquid PCM are different, they are not dependent on the temperature.
- The PCM is homogenous and isotropic.
- The melt PCM is supposed to be an incompressible Newtonian fluid.
- Viscous dissipations are considered to be negligible.
- No-slip conditions for velocities at the boundaries
- Volume variation of PCM corresponding to the phase change is ignored.
- Flow is modeled as 2D due to the geometry of the unit and selected boundary conditions (i.e., there is not any temperature gradient in the z-direction).

Considering the above assumptions and the geometrical configuration of the system, the governing equations are given in Chapter 6.

6.1.3 Numerical Solution Procedure

FlexPDE, an open-source modeling software, was chosen to execute the numerical process of the solidification process. This software is based on Galerkin Finite Element Method (GFEM), which has been demonstrated to be a trustworthy technique for solving this problem. This methodology uses a residual-driven application and assigns a weighting function to each variable to trade off the importance of individual factors. Subsequently, the nonlinear equations transform to provide a system of linear equations that can be solved more easily. Since the GFEM is a numerical modeling of the original system, errors are unavoidable and should be minimized by dividing the domain into sufficient parts. Therefore, we realized dynamic mesh placement, as shown in Fig. 62. This compatibility paves the route for significantly more valid outcomes while reducing the areas above the error threshold (Dekhil et al., 2021). Experimental research by Ismail et al. (2001) is selected to be reproduced by GFEM so as to support the reliability of this study. For an acceptable comparability of outcomes computed via the presented model and literature data, the geometry is adopted to imitate the system considered by Ismail et al. (2001) as accurately as possible. Ismail et al. conducted computational and experimental research on finned tubes in thermal storage systems. The model is based on the enthalpy formulation technique, the control volume method, and the pure conduction mechanism of heat transfer. Figure 63.a. shows the ultimate comparison of the present technique and empirical study. From this figure, we can infer that the outcomes of both studies are in perfect suitability, with an error of 5% or less at any step of the procedure, illustrating that this simulation is reliable and authoritative. A mesh-independent analysis is performed to find the optimal magnitude for the mesh size. Regarding the geometry, an unstructured adaptive grid is applied in the current work, as seen in Figure 62. Also, it can be observed that the quality and density of the mesh are fulfilled at the edge points.

In this study, three mesh sizes are utilized to ensure mesh independency. Mesh 1, 2, and 3 have approximately 16312, 38278, and 95624 elements, respectively. Based on the numerical domain, average temperatures are calculated for all three grid sizes. As shown, the meshes for the three cases are slightly different. Figure 63.b. shows that the plot for mesh 3 is slightly off from two other meshes. However, the results obtained from mesh 1 and 2 agree with each other to a high degree. It is clear that the greater the number of grids, the higher the accuracy. However, as the number of grids increases, the runtime increases. Based on central processing unit (CPU) uptime, mesh two was chosen for simulation in this research. At last, time independency should be performed to ensure the numerical solution. Three different time steps are chosen for the computational domain based on Figure 63. c. As can be seen, the shorter the time step, the more computational cost and accuracy. Accordingly, the time step 0.005 s is considered for this study's simulations to reduce the computational cost.

Figure 62. Grid of studied domain

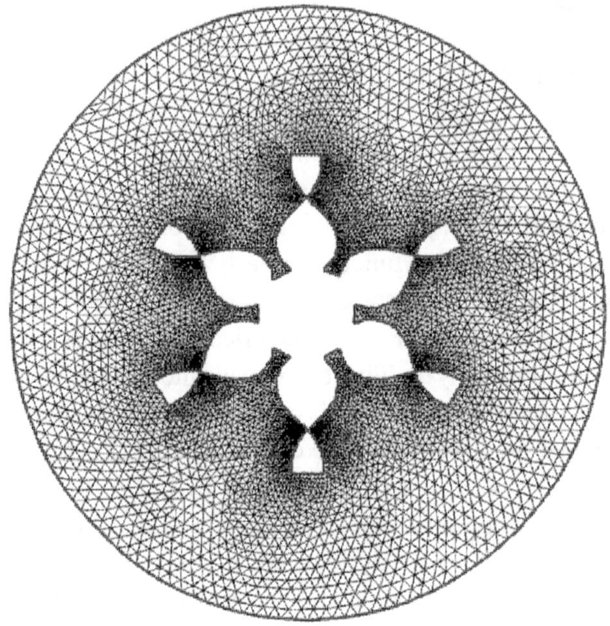

Figure 63a. Validation of present simulation with experimental study done by Ismaeil et al. (2001)

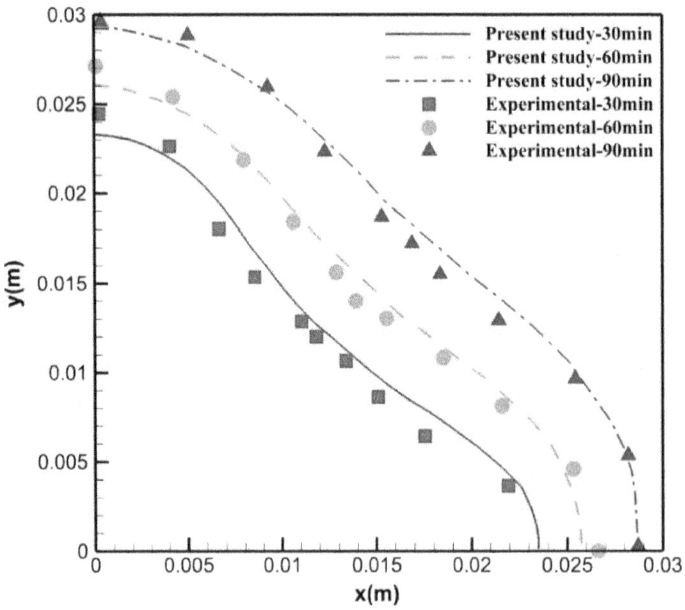

Figure 63b. Comparison of the average temperature at various grid sizes

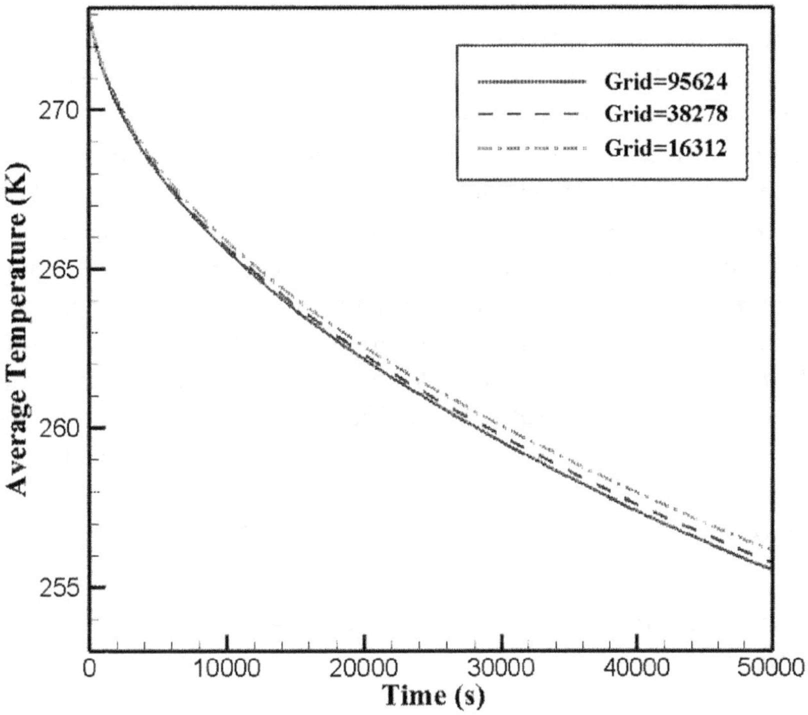

Figure 63c. Comparison of the average temperature at various time steps

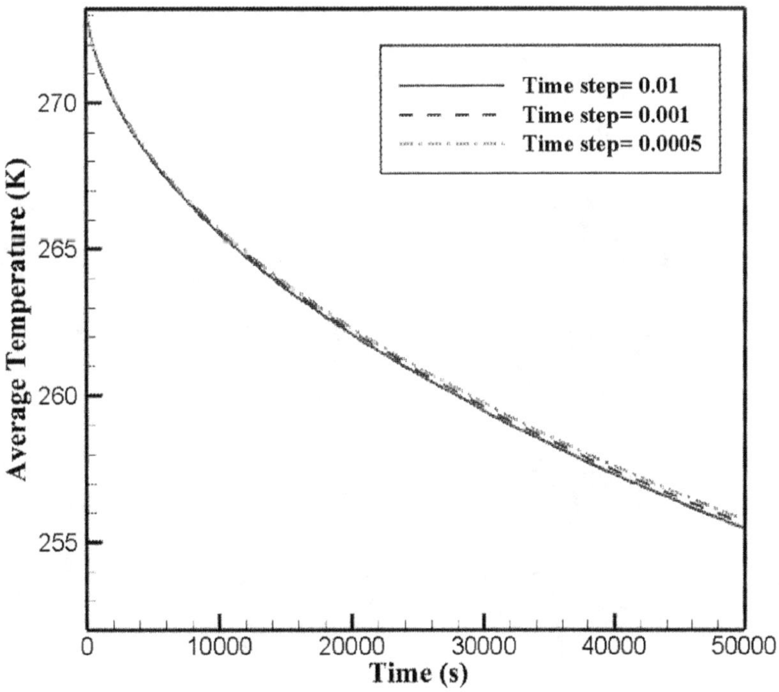

6.2 Results and Discussion

Fig. 64 shows the isothermal and solid fraction contours of the PCM at different temporal stages for the three cases studied: the adiabatic, the constant temperature, and the triplex tube arrangement. At the initial period of the solidification procedure, liquid PCM adjacent to the cold HTF streaming through the inner tube (pertinent to the adiabatic and the constant temperature cases) or the inner and outer tubes (pertinent to the triplex case) loses thermal energy by the conduction mechanism and the temperature of liquid PCM decreases. This period keeps going until the PCM temperature, which equals the initial value of 275 K at the onset, decreases to the value of the beginning temperature of the liquid-solid transition. When the temperature of PCM approaches the freezing point, a solid-liquid interface is formed, and a thin layer of the solid PCM is gently constructed to surround the tube surfaces. The solid layer continuously grows with time, and the solid fraction is effectively raised. A thicker sheet of the solid augments thermal resistance and decelerates the total PCM solidification procedure in the LTES system. Hereupon, the temperature of the PCM moderately diminishes until it gains a certain magnitude at the end of the discharging process. It can also be clearly observed that the triplex case has higher solidification kinetics than the other two cases, owing to its greater heat transfer surface area.

Figure 64. Contours of the solid-liquid interface (left) and isotherms (right) at different time frames in discharging mode for three considered cases

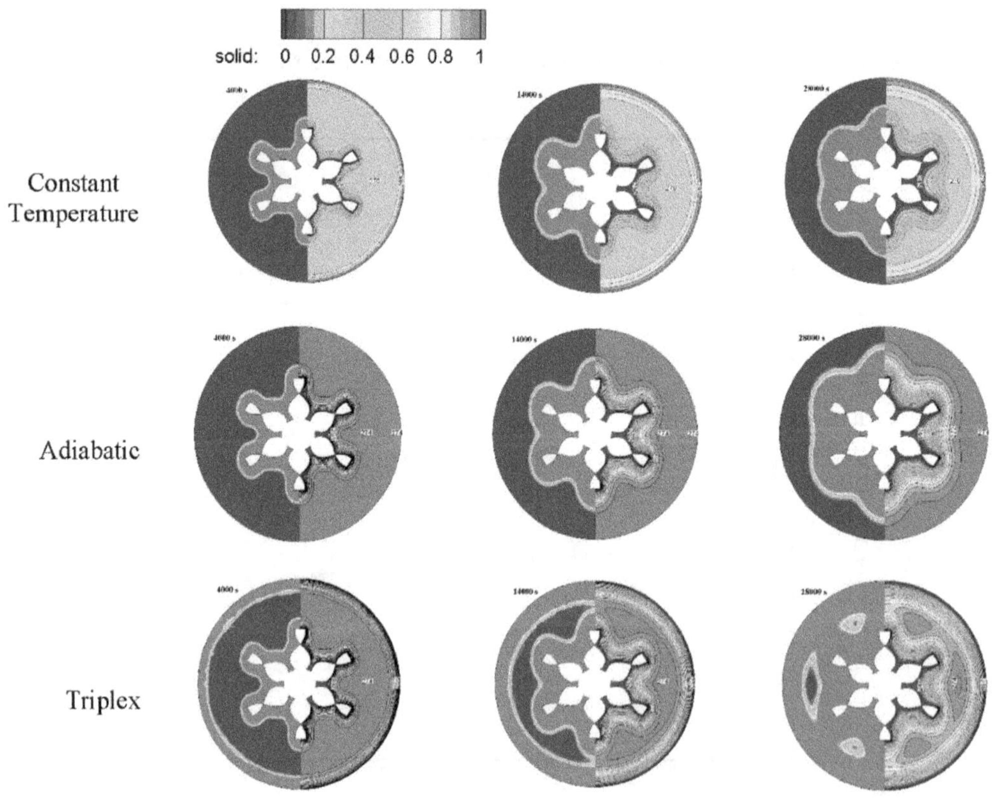

Figure 65 illustrates the average temperature of THNEPCM during the solidification procedure for the three investigated cases. According to this figure, it can be seen that the tilt of the average temperature curve in all the considered cases falls off sharply during the early time steps of the solidification process and subsequently lessens as time further elapses. This infers that a quicker discharging rate is obtained within the initial stage of the procedure. However, one can conclude that the average temperature of the solidification procedure for PCM material in the triplex case reduces at a greater rate than that in the two other considered cases due to the greater surface area in contact with the cold HTF. As time progresses, the differences in the temperature between the considered cases enhance.

Figure 65. Temporal evolution of the average temperature for different boundary conditions

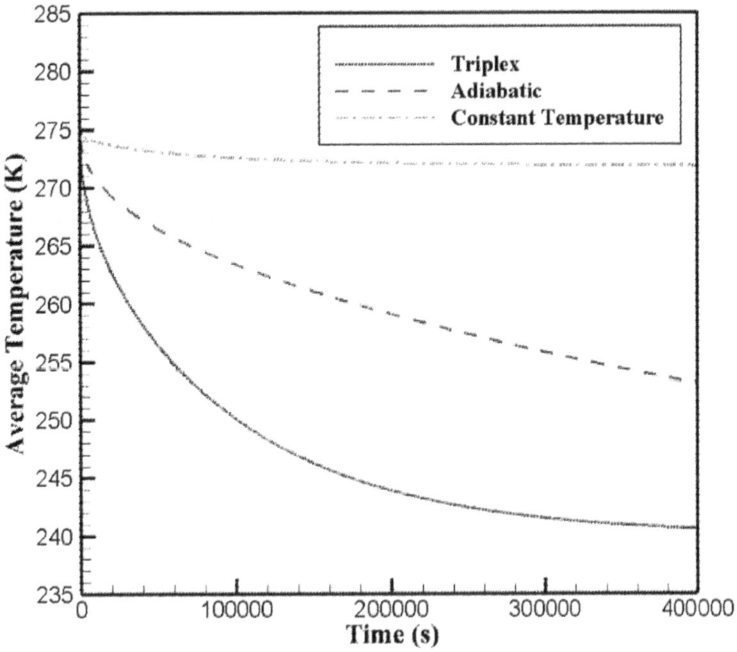

In Figure 66, the total energy released from the PCM during the solidification process for the three considered cases is compared. For all three cases, total heat exchanged is maximum at the very early stages of the discharging procedure, owing to the significantly high thermal gradients in the PCM. Afterward, the heat exchanged reduces with the diminution of temperature variations in the PCM. It is obvious that the heat exchanged is more remarkable for the triplex case compared to the other two cases at a given time. As a result, the triplex case is the most efficacious. At 400000 s, the triplex case improved total discharging energy by 23% compared to the constant temperature case and 4% compared to the adiabatic case. One can deduce that for the triplex case, the solidification phenomenon rate would virtually be doubled in comparison to the adiabatic case.

The solidification process of PCM embedded in the cylindrical enclosure is shown in Figure 67 for three different boundary conditions. The solidification procedure continues until the entire PCM solidifies. The result indicates that at a given time, the amount of the solid fraction for a triplex case is more significant than that for other cases, indicating a faster solidification process. The main reason for this phenomenon is that the surface area in contact with the HTF for the triplex case is the greatest among the investigated cases, which improves the heat conduction rate. At 100000s, the solid fraction is increased by 19% and 55% for triplex cases compared to adiabatic and constant temperature, respectively. Moreover, the constant temperature case cannot reach full solidification. Also, the full solidification times for adiabatic and triplex are 178500 and 27000, respectively. It is noteworthy to mention that the time requisite for complete discharging is a substantial factor in assessing the system's performance. Hence, the triplex tube heat exchanger is adopted in the following of this research.

Figure 66. Temporal evolution of the total energy for different boundary conditions

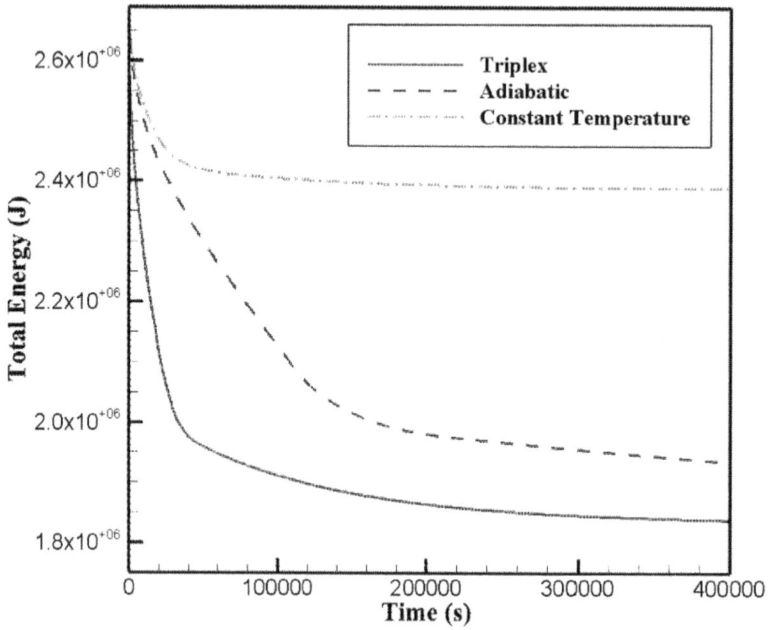

The solid fraction and temperature contours during the discharging process for different volumetric concentrations of nanoparticles at three different time steps are displayed in Fig. 68. At the initial time, there is direct contact between the liquid PCM and the cold surface of the inner and outer tubes. By decreasing the temperature of PCM, a thin layer of the solid PCM is created adjacent to the cold tube and thus becomes thicker as time progresses. The thicker layer of the solid compels a greater heat resistance; hence, the discharging rate decreases consistently. Furthermore, it can be clearly observed that the solid layers have the same thickness in all directions around the internal and external tubes.

Figure 67. Temporal evolution of the solid fraction for different boundary conditions

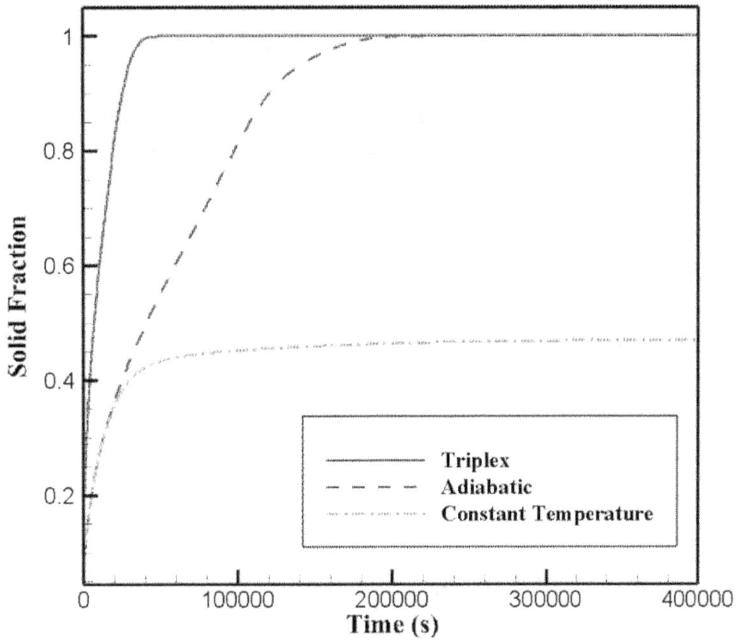

Figure 68. Contours of liquid-solid phase change (left) and temperature (right) at different time frames in discharging mode for three considered cases

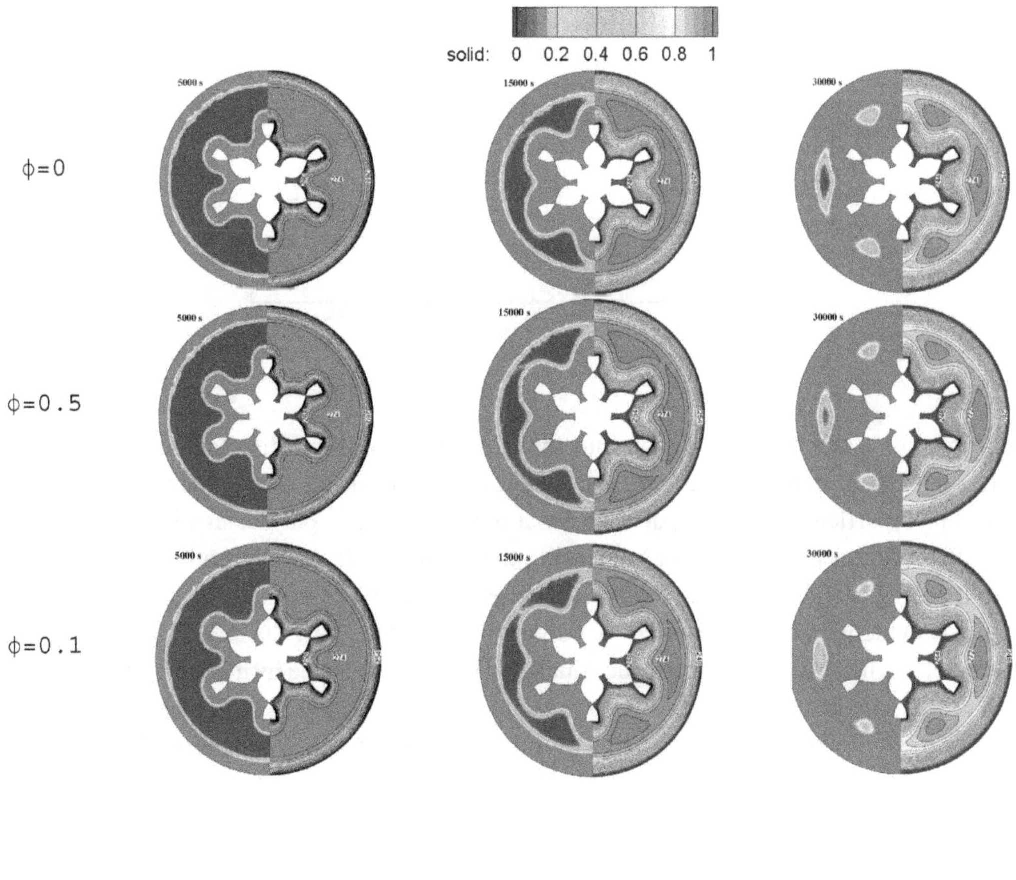

This phenomenon is because the conductive heat transfer is dominant during the discharging procedure. Compared with the case with pure PCM, the solidification process has a higher rate of solidification for the case with PCM containing THNP. At a given time, a lower temperature and more rapid evolution of solid front can be noticed for a higher volume fraction of particles. This is due to the fact that dispersing nanoparticles boosts the magnitudes of specific heat and heat conductivity of the THNEPCM. An increment in thermal conductivity leads to higher absorption of thermal energy from the THNEPCM, and the performance of the heat exchanger in view of the kinetics of solidification enhances. Consequently, dispersing nanomaterial into PCM is a beneficial method to achieve the quickest discharging. The behavior of average nano-PCM temperature in terms of discharging time is indicated in Figure 69 for various volume percentages of nanoparticles. A good enhancement in the discharging procedure is apparent when the volume percentage of nanoparticles increases from 0 to 10%. As a matter of fact, adding nanoparticles to the PCM improves the overall heat transfer process during discharging, owing to the high conductivity of nanoparticles. It can be observed that adding 5% and 10% of nanoparticles to the base PCM results in the reduction of the average temperature by 0.5 K and 1 K, orderly, at a time of 50000s.

Figure 69. Time-varying variation of the average temperature for different volumetric concentration of nanoparticles

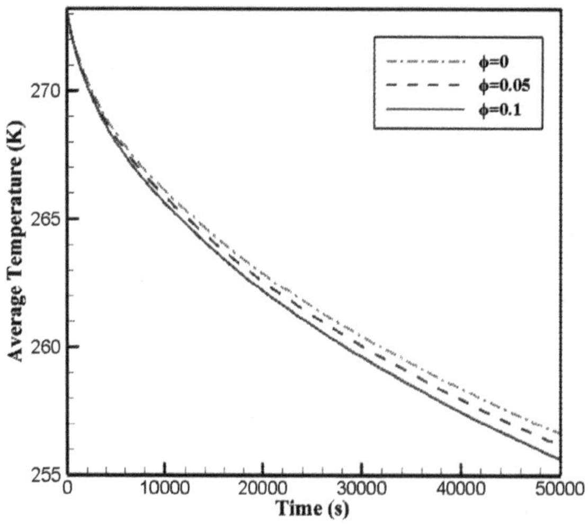

The impact of nanoparticle insertion into the base PCM on the instantaneous distribution of total energy has been demonstrated in Figure 70. As shown, the heat release rate diminishes with the embedment of nanoparticles into PCM, denoting that the nanoparticle plays a striking role in speeding up the PCM discharging. In point of fact, increasing the nanoparticles concentration ameliorates the heat conductivity of PCM. The role of ϕ becomes more effective by progressing the time of discharge. It is quantitatively evident that the total energy lessens by 4.2% and 8.5% by the incorporation of 5% and 10% of nanoparticles when the discharging time equals 50000s. To study the influence of the volume fraction on the solidification procedures, Figure 71 is presented. At the beginning of the discharging process, the liquid PCM cools down through the conduction mechanism. The thermal energy is transferred to the inner and outer tubes via conduction until its temperature attains the solidification point. At this stage, only sensible thermal energy exists, and the solid-liquid transition does not still appear. As soon as the temperature of PCM reaches the solidification point, the liquid PCM starts to solidify and emit energy as latent heat. This process continues until the whole PCM becomes solid. When the total PCM solidifies, the solid fraction reaches one, and no curve variations occur. Moreover, it can be deduced that the solidification rate increases as the volume fraction grows, attributed to the improvement in heat transfer.

Figure 70. Time-varying variation of total energy for different volumetric concentration of nanoparticles

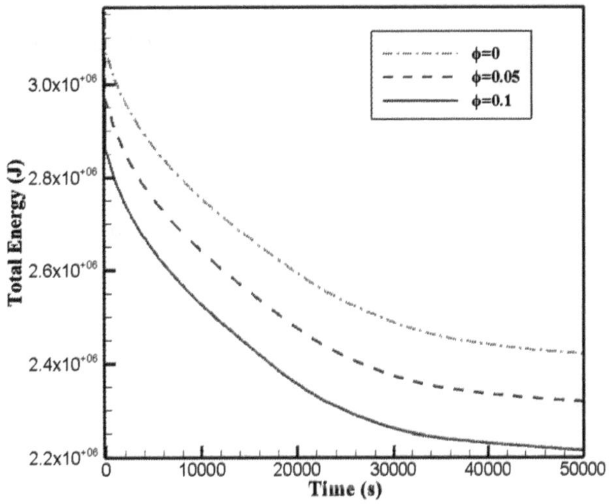

Figure 71. Time-varying variation of the solid fraction for different volumetric concentration of nanoparticles

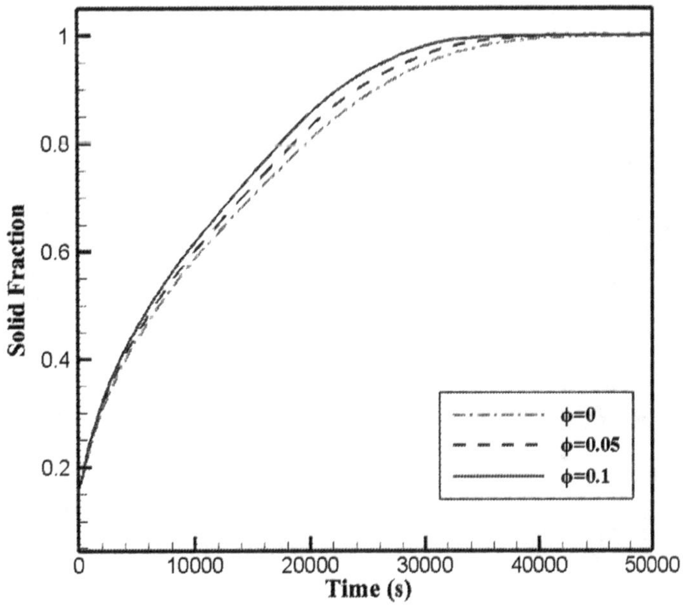

A significant difference between the solid fractions for different values of particles' volume fraction is observed at the middle stages of the discharging process. Indeed, the existence of nanoparticles augments thermal conduction because of the great heat conductivity. The full solidification times are 39000, 36000, and 33000 for 0%, 5%, and 10% concentration, respectively. From a quantitative point of view, the full solidification time reduces by about 5.7% and 13.1% by inserting the nanoparticles into the base PCM with a 5% and 10% concentration, respectively.

Figure 72 depicts the isothermal and solid fraction contours of nano-PCM within an enclosure over time during solidification for the different values of the radiation parameter. At the initial time of solidification, the liquid PCM loses its thermal energy to the Cold HTF, and subsequently, two solid layers are created near the inner and outer tubes. Since conductive heat transfer takes control of the discharging process, an approximately uniform solid-liquid interface is observed during the entire procedure. The formed layer consistently grows with elapsing time and impedes the thermal diffusion from the liquid portion of the PCM to the cold HTF. Hence, as time goes on, the solidification rate progressively declines. The process continues until the whole PCM completely turns solid. This figure illustrates that considering radiation heat transfer shortens the complete solidification time for the PCM due to the enhanced heat transfer rate. In other words, the greater the radiation heat transfer, the sooner the PCM converts into a solid state.

Figure 72. Solid fraction contours (left) and temperature fields (right) at different time frames in discharging mode for three considered cases

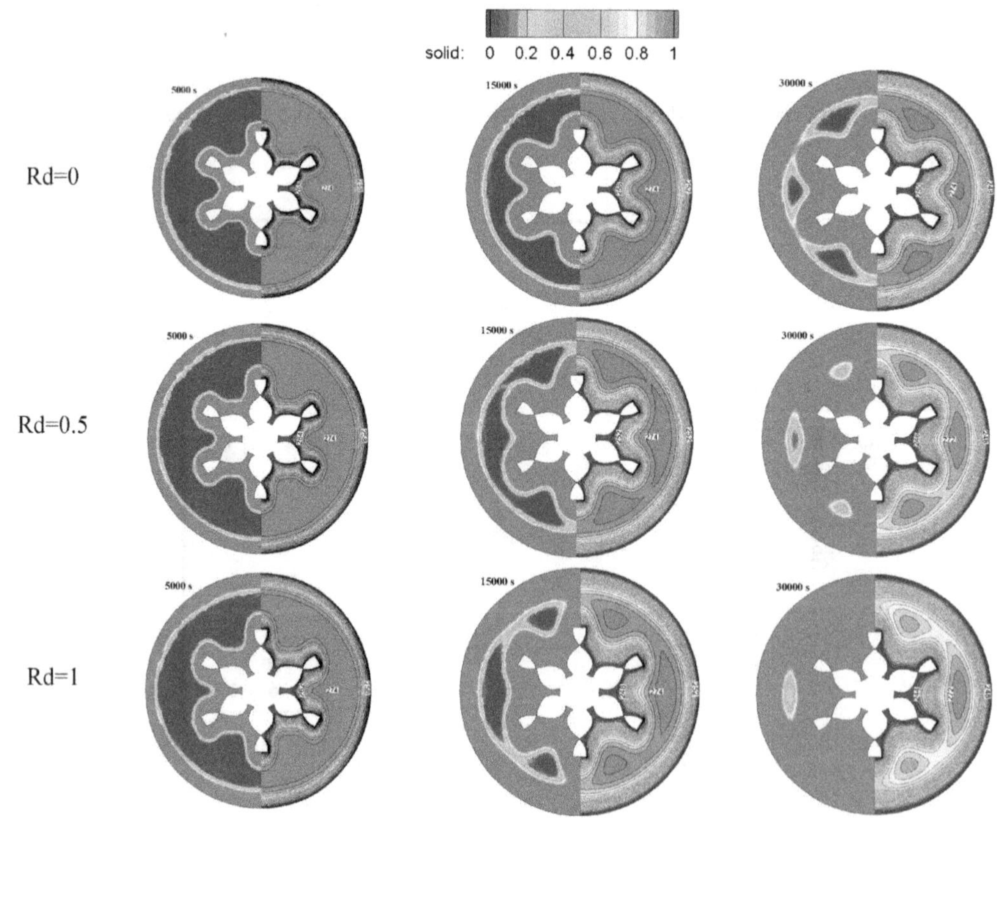

Figure 73 demonstrates the average PCM temperature versus time for various amounts of the thermal radiation parameter. As illustrated, the minimum temperature is lower for high values of the radiation parameter. Indeed, improving the thermal radiation leads to an increment in the thermal penetration depth and hence the solidification rate increase. Upon complete solidification of the PCM, the PCM temperature reaches the magnitude of 259 K, 256.1K, and 253.9K when the radiation parameter is 0, 0.5, and 1, respectively. Nevertheless, even after complete solidification is achieved, the PCM temperature needs additional time to reach the inlet HTF temperature. The influence of the radiation parameter on the transient profile of total energy is displayed in Fig. 74. At the early times of the discharging phenomenon, the heat flux reduces swiftly since the temperature gradient between the cold wall and liquid PCM declines quickly. By elapsing time, a thin layer of the solid PCM is constructed all over the inner and outer cold tubes, becoming thicker as time further progresses. The thicker layer of the solid prevents higher thermal energy transfer; hence, the discharging rate diminishes with further elapses in time. Hence, the rate of total energy reduction decreases temporally. It can also be seen that the greater the radiation heat transfer, the lower the total energy of PCM. For instance, at the fixed time of 50000s, decreasing the radiation parameter from 0 to 1, the total energy improves by about 2.1%.

The impact of the thermal radiation parameter on the temporal distribution of nano-PCM solid fraction during the solidification is plotted in Figure 75. As shown, at the inception of the procedure, the solidification rate is maximum due to the direct contact of liquid PCM and HTF streaming through the inner and outer tubes. By progressing the process, the solidification rate reduces gradually because of the growth of the solid layer between the liquid PCM and cold HTF. In the terminal moments of the process, the solidification rate decreases much more than before. Ultimately, no changes will occur in the solid fraction when completed. It can also be implied that the shortest full solidification time of the THNEPCM is related to the case that the radiation parameter is considered. In fact, the presence of radiative heat transfer enables the THNEPCM to obtain better thermal performance and exhibit a greater discharging rate. The full solidification times are 50000, 36000, and 32000 for 0, 0.5, and 1 radiation parameter, respectively. The total solidification times of the PCMs alleviate up to 36% by increasing the radiation parameter from 0 to 1.

Figure 73. Time-dependent average temperature profile for different values of radiation parameter

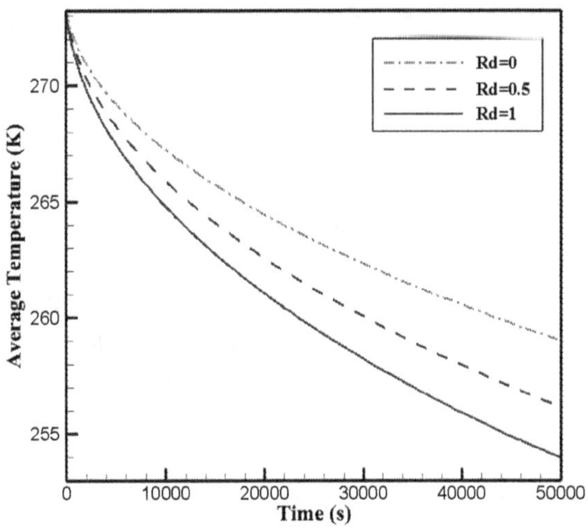

Figure 74. Time-dependent total energy profile for different values of radiation parameter

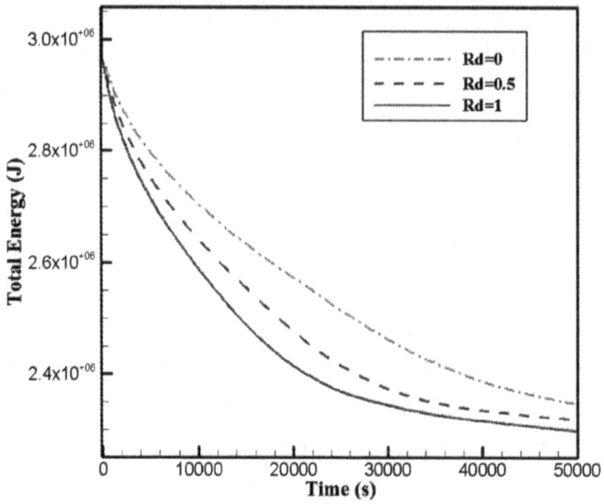

Figure 75. Time-dependent solid fraction profile for different values of radiation parameter

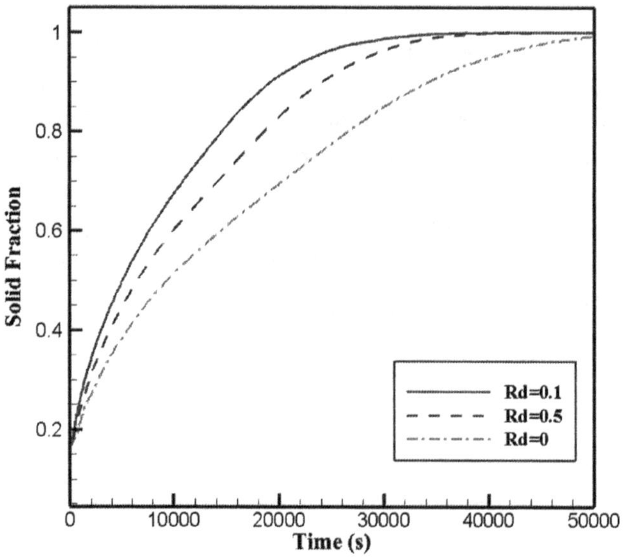

Figure 76 shows the impact of nanoparticles' shape factor on temperature fields and solid fraction contours at different time stages. It can be indicated that the PCM temperature constantly decreases with time increment until it reaches the respective minimum value at the end of the process. However, this growth is intense firstly but becomes more moderate with further progress in the storage's melting procedure. As observable, the formation of solid PCM is axisymmetric at any given time. The asymmetry of the solid extent is principally due to the domination of conductive heat transfer. Furthermore, a fast thickening of the solid is noticed in the PCM consisting of nanoparticles with a high shape factor. Figure 77 illustrates the influences of the shape factor of particles on the temporal behavior of average nano-PCM temperature throughout the solidification process. We can also observe that the average temperature changes remarkably with the shape factor of particles, especially at the end of the solidification process. In fact, the existence of nanoparticles with a higher shape factor leads to an increase in the thermal conductivity of THNEPCM and finally results in better enhancement of the solidification rate. Therefore, the shape factor of nanoparticles dispersing in the base PCM can be a significant parameter for a better solidification rate. When the time is 50000 min, the average temperature of nano-PCM is 256.6, 256.1, and 254.8 for cases containing nanoparticles with the shape factor whose values are 3, 5.7, and 16.1, respectively.

Figure 78 presents the time-varying profile of total energy for different values of shape factor. Initially, the value of total energy is found to be the same irrespective of the shape factor of the particles. However, a difference in the solidification rate for the three curves is observed by progressing the discharging procedure. Regarding the figure, the high shape factor contributes to obtaining full solidification in an approximately shorter time. For instance, 0.2–0.7% diminution in total energy has been achieved by using inserting nanoparticles with a shape factor of 16.1 compared to that with a shape factor of 5.7 and 3, respectively, when the solidification time is 50000s.

Figure 76. Solidification front (left) and temperature field (right) of nano-PCM at different time frames in discharging mode for three considered cases

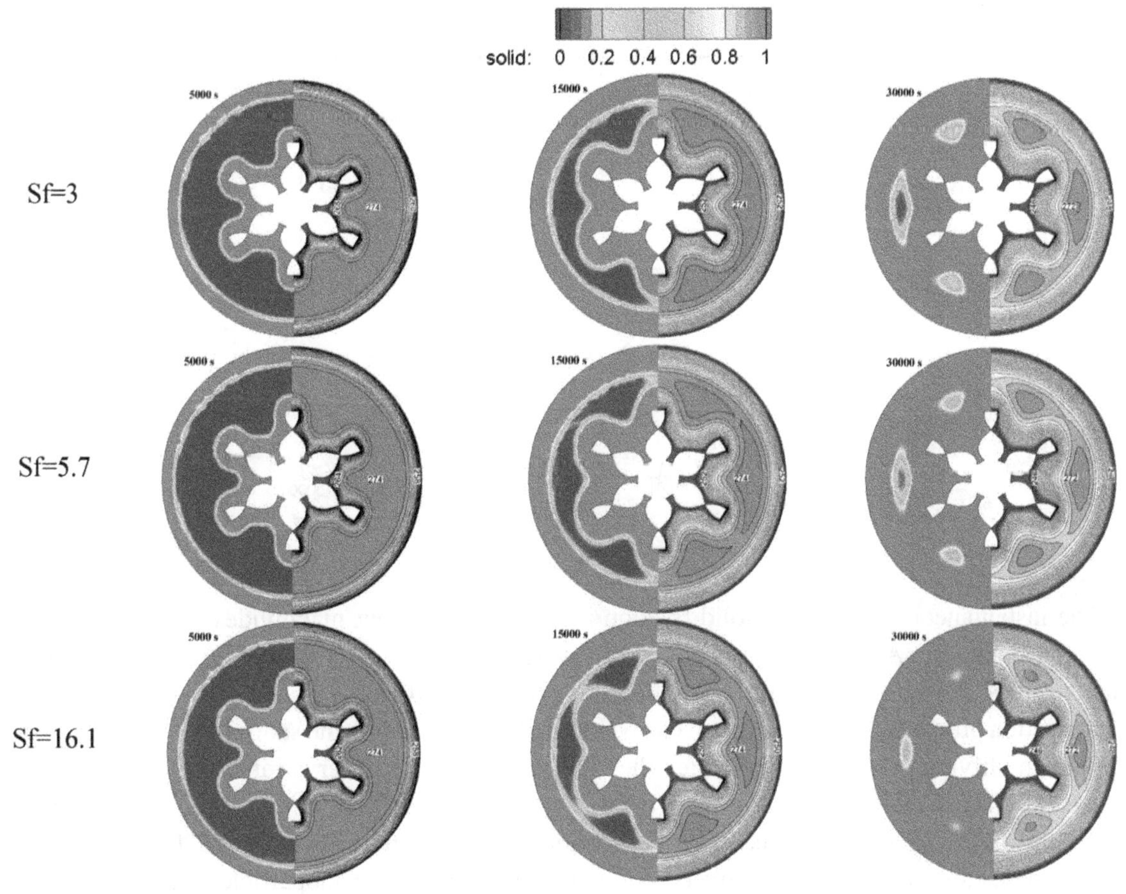

Figure 77. Transient evolution of the average temperature during the nano-PCM solidification for various values of shape factor

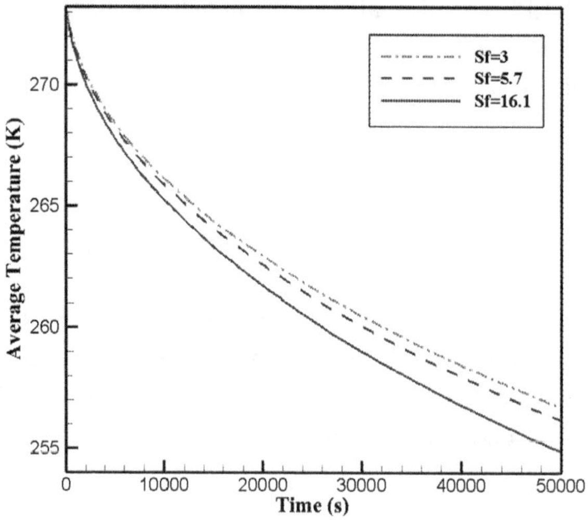

Figure 78. Transient evolution of the total energy during the nano-PCM solidification for various values of shape factor

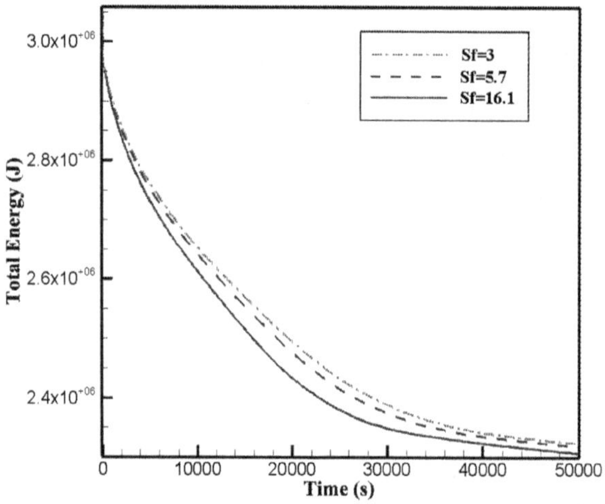

The instantaneous changes of solid fraction in PCM for different magnitudes of shape factors are shown in Figure 79. As time progresses, solidification curves provide a steeper trend at initial periods when the heat transfer rate is great. The solidification rate is significantly abated as time elapses until the solidification process completes. Also, it can be implied that increasing the shape factor of nanoparticles leads to amplifying the solidification rate. The main reason for this observation is that the thermal conductivity of THNEPCM and, in sequence, the depth of heat penetration in the unit grows by enlarging the nanoparticles' shape factor. The full solidification times are 28000, 26000, and 22000 for 3, 5.7, and 16.1 shape factor parameter, respectively. For example, upon using nanoparticles with a shape factor of 16.1 instead of those with a shape factor of 3, full solidification time decreases up to 21.9%. Figure

80 portrays the change of full solidification time (FST) with the crucial non-dimensional parameters including nanoparticles' volume fraction, nanoparticles' shape factor, and radiation parameter. As expected, increasing all the aforementioned factors would precipitate the discharging process and reduce solidification time because of the improved heat conductivity and thermal penetration depth.

Figure 79. Transient evolution of the solid fraction during the nano-PCM solidification for various values of shape factor

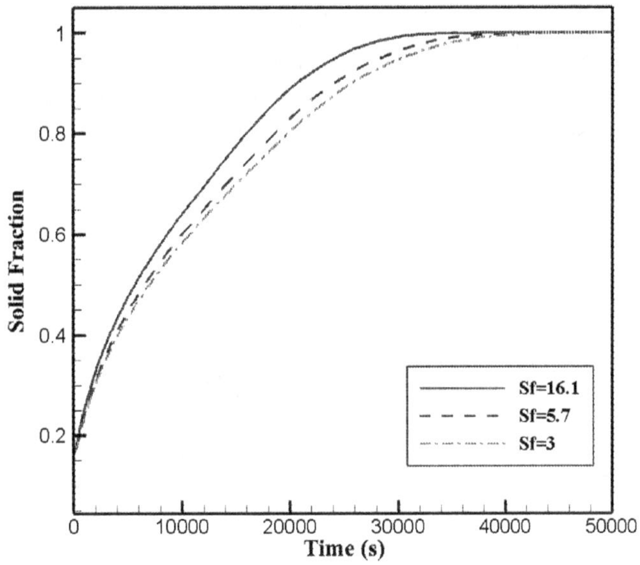

Figure 80. The effect of shape factor and nanoparticle volume fraction on solidification time

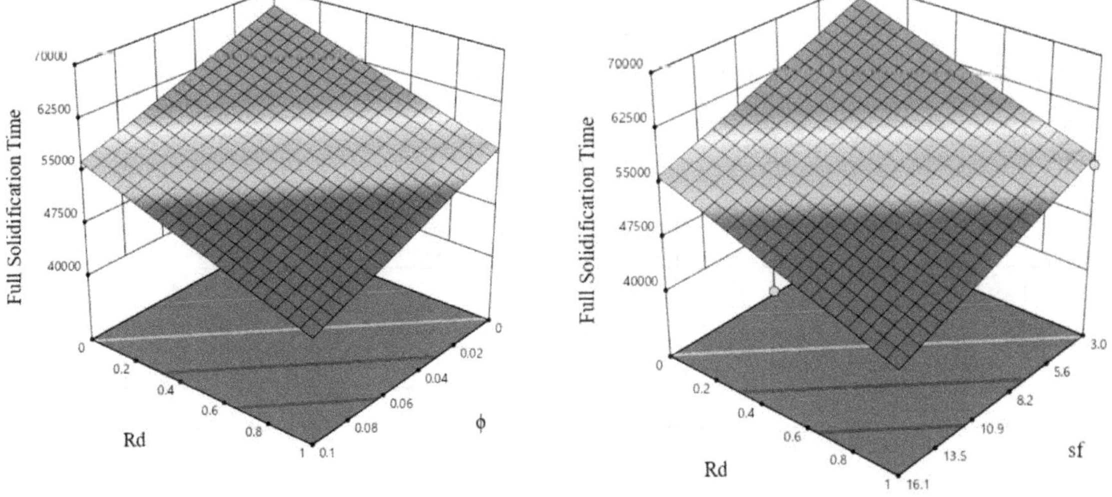

6.3 Optimization and Derivation of an FST Correlation

In order to reach the highest heat transfer rate by an optimal status, the optimization process is implemented. The Taguchi method as a powerful optimization technique is undertaken so as to optimize the engineering trials of the design process. The main advantage of this technique is diminishing the cost and time consumption by presenting an experimental layout for the analysis. The considered design parameters affecting FST consists of Rd, φ, and Sf. It is worth stating that although, a complete factorial analysis needs 81 tests for the aforementioned variables, using Taguchi analysis reduces the number of trials to nine. The data about the effective parameters in the nine considered samples are listed in Table 16. The response surface methodology (RSM) is utilized to find the highest amount of the FST. To do so, a formula is generated by RSM to evaluate the impact of design variables on the objective function. Equation 4 indicates the correlation for FST in terms of the adopted design parameters. The corresponding optimal values of the design variables in Eq. 4 are given in Table 17.

$$\text{Full Solidification Time} = 76036 - 110667(\phi) - 11400(Rd) - 892.8(sf) \quad (4)$$

Table 16. Designed tests by Taguchi method

Case Number	Volume Fraction	Radiation Parameter	Shape Factor
1	0.1	1	5.7
2	0	0.5	5.7
3	0	0	3
4	0.1	0.5	3
5	0.05	1	3
6	0.05	0.5	16.1
7	0.1	0	16.1
8	0	1	16.1
9	0.05	0	5.7

Table 17. The variance analysis (ANOVA)

Source	Sum of Squares	df	Mean Square	F-Value	P-Value	
Model	6.075E+08	3	2.025E+08	9.47	0.0167	significant
A-ϕ	1.837E+08	1	1.837E+08	8.59	0.0326	
B-Rd	1.949E+08	1	1.949E+08	9.11	0.0295	
C-sf	2.288E+08	1	2.288E+08	10.70	0.0222	
Residual	1.069E+08	5	2.139E+07			
R^2=0.8503						

6.4 Analysis of Variance (ANOVA)

In order to ascertain the comparative importance of the design parameters on the ultimate solution, analysis of variance (ANOVA) is applied to the outcomes of the Taguchi method. All the means are computed, and assumptions are formulized to perform the ANOVA technique. Thereafter, Degrees of Freedom (DF), the sum of squares (SS), the F-statistics, and the Mean Squares are computed. Eventually, based on the F-Static, a p-value can be obtained. The results acquired via ANOVA for FST are listed in Table 18, comprising the mean square, F ratio, the sum of squares (SS), degree of freedom (DF), and P ratio. It can be implied that the R^2 value is 0.8579, which affirms that there is appropriate adaptability between the calculations gained by the obtained formula and the numerical results under the specified optimum combination for the full freezing time.

Table 18. Outcome of the optimization procedure by RSM

Parameter	Optimum Value
ϕ	0.086
Rd	0.59
sf	15.3

Figure 81. Predicted vs. actual of the full solidification time

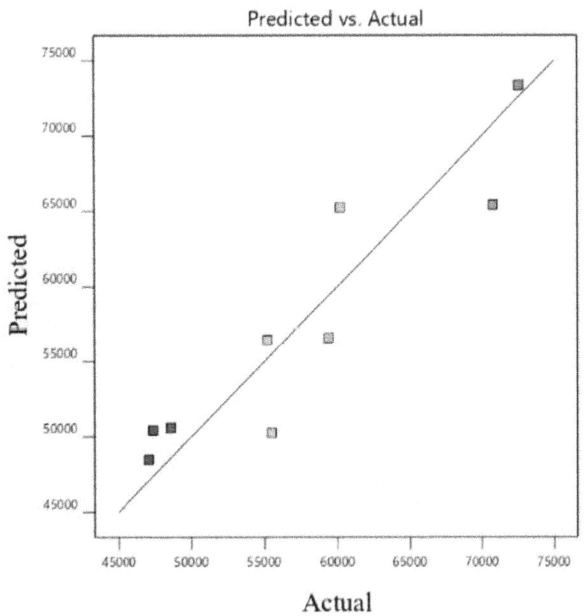

Further, Figure 81 compares estimated optimal data and actual results afforded by the numerical method. Closer inspection leads one to deduce that the accuracy between the estimated value and numerical simulation of the FST for the optimum combination is acceptable, with the highest difference of 11%. Figure 82 represents the alteration of desirability in terms of defined controlling factors. It is obvious that enhancing the nanoparticles' volume fraction, the radiation parameter, and the shape factor augments the desirability. Interval evaluations can be added to the optimization plot to illustrate the uncertainty in the point estimation. Besides, Interval estimates can help in understanding the influence of uncertainty on obtaining procedure objectives.

In Figure 83, the vivid yellow region implies that the whole range of all intervals satisfies the benchmark under consideration. However, the dark gold region relates to where the point approximation satisfies the benchmark demands, but the portion of an interval evaluation does not.

Figure 82. The effect of controlling parameters on the desirability

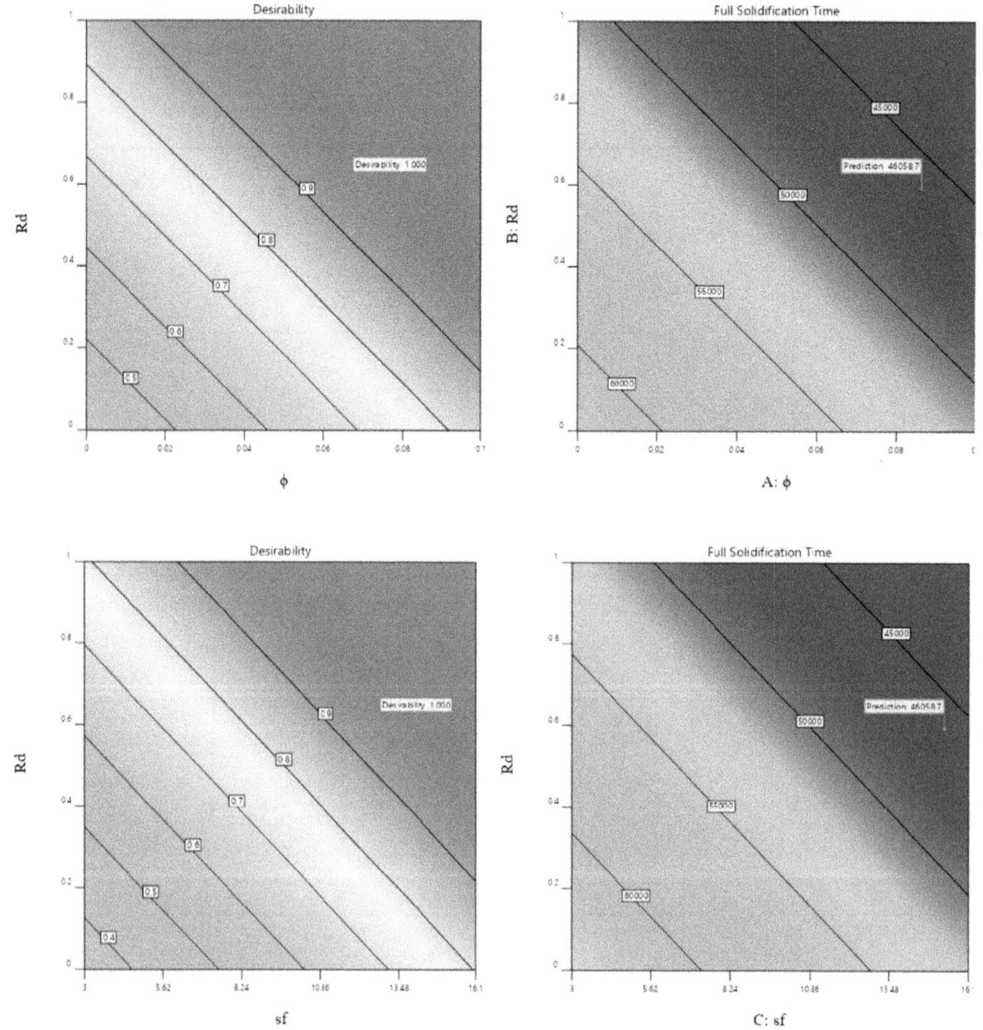

Figure 83. Overlay plots

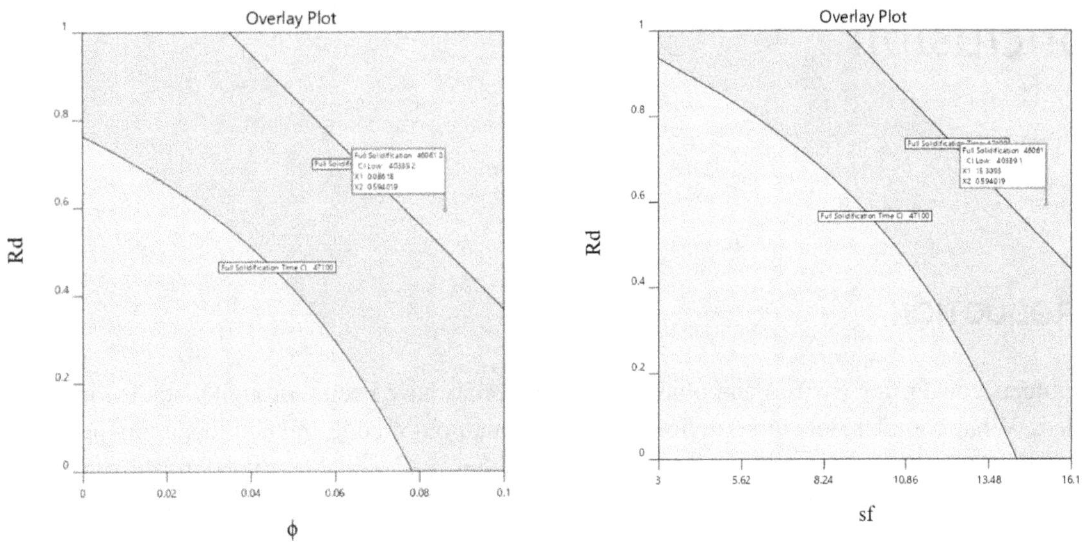

REFERENCES

Alizadeh, M., Hosseinzadeh, K., Mehrzadi, H., & Ganji, D. D. (2019). Investigation of LHTESS filled by Hybrid nano-enhanced PCM with Koch snowflake fractal cross section in the presence of thermal radiation. *Journal of Molecular Liquids*, *273*, 414–424. doi:10.1016/j.molliq.2018.10.049

Dekhil, M. A., Tala, J. V. S., Bulliard-Sauret, O., & Bougeard, D. (2021). Numerical analysis of the performance enhancement of a latent heat storage shell and tube unit using finned tubes during melting and solidification. *Applied Thermal Engineering*, *192*, 116866. doi:10.1016/j.applthermaleng.2021.116866

Hosseinzadeh, K., Alizadeh, M., Alipour, M. H., Jafari, B., & Ganji, D. D. (2019). Effect of nanoparticle shape factor and snowflake crystal structure on discharging acceleration LHTESS containing (Al2O3-GO) HNEPCM. *Journal of Molecular Liquids*, *289*, 111140. doi:10.1016/j.molliq.2019.111140

Hosseinzadeh, K., Alizadeh, M., Tavakoli, M. H., & Ganji, D. D. (2019). Investigation of phase change material solidification process in a LHTESS in the presence of fins with variable thickness and hybrid nanoparticles. *Applied Thermal Engineering*, *152*, 706–717. doi:10.1016/j.applthermaleng.2019.02.111

Ismail, K. A. R., Alves, C. L. F., & Modesto, M. S. (2001). Numerical and experimental study on the solidification of PCM around a vertical axially finned isothermal cylinder. *Applied Thermal Engineering*, *21*(1), 53–77. doi:10.1016/S1359-4311(00)00002-8

Venter, G., Haftka, R., & Starnes, J. Jr. (1996, August). Construction of response surfaces for design optimization applications. In *6th Symposium on Multidisciplinary Analysis and Optimization* (p. 4040). 10.2514/6.1996-4040

Conclusion

INTRODUCTION

All subjects concerning cavities and phase change materials have been thoroughly studied in previous chapters. What complements these reviews is the presentation of codes related to solving problems of holes and PCM in FlexPDE software. The publication of these codes enables the establishment of a foundation for further research in these disciplines. Coding in FlexPDE is such that problem data, equations, solution domain, boundary conditions, and commands related to output display should be written in it.

1. FlexPDE CODE FOR CAVITY

1.1 Sample 1

```
TITLE
'Cavity Sample Code 1'
SELECT
painted
smoothinit
threads=2
Spectral_colors
VARIABLES
psi
thetanf
!====================================================================
DEFINITIONS
epsilon=0.7
Rd=0.5
Ha=20
Ra=500
gama=pi/6
kf=0.6
knf=2.5
A2=1.028116649
```

```
A3=2.19555776456
A4=1.148382263
A5=1.292355436
A6=0.9702970356
 flux = -A4*grad(Thetanf)
!==========================================================================
EQUATIONS
psi:dxx(psi)+dyy(psi)=-Ha*(A6/A5)*(cos(gama)*cos(gama)*dxx(psi)+sin(gama)*sin(g
ama)*dyy(psi)+2*sin(gama)*cos(gama)*dx(dy(psi)))-(A2*A3/(A4*A5))*Ra*dx(thetanf)
thetanf: epsilon*((dxx(thetanf)+(1+(4/3*(kf/knf)*Rd))*dyy(thetanf)))=-dy(theta
nf)*dx(psi)+dy(psi)*dx(thetanf)
!==========================================================================
BOUNDARIES
 region 1
start 'inner'(0,0.07)
value(psi)=0   value(Thetanf)=1
ARC(CENTER=0,0) ANGLE=360 TO CLOSE
start 'outer' (-0.5,0.145028) value(psi)=0   value(Thetanf)=1
line to (-0.477688,0.152712) value(psi)=0   value(Thetanf)=1
line to (-0.46146,0.160395) value(psi)=0   value(Thetanf)=1
line to (-0.445233,0.166158) value(psi)=0   value(Thetanf)=1
line to (-0.426978,0.17) value(psi)=0   value(Thetanf)=1
line to (-0.400609,0.168079) value(psi)=0   value(Thetanf)=1
line to (-0.376268,0.160395) value(psi)=0   value(Thetanf)=1
line to (-0.355984,0.152712) value(psi)=0   value(Thetanf)=1
line to (-0.337728,0.145028) value(psi)=0   value(Thetanf)=1
line to (-0.315416,0.135424) value(psi)=0   value(Thetanf)=1
line to (-0.291075,0.125819) value(psi)=0   value(Thetanf)=1
line to (-0.266734,0.120056) value(psi)=0   value(Thetanf)=1
line to (-0.236308,0.118136) value(psi)=0   value(Thetanf)=1
line to (-0.207911,0.125819) value(psi)=0   value(Thetanf)=1
line to (-0.177485,0.137345) value(psi)=0   value(Thetanf)=1
line to (-0.151116,0.146949) value(psi)=0   value(Thetanf)=1
line to (-0.13286,0.156554) value(psi)=0   value(Thetanf)=1
line to (-0.112576,0.164237) value(psi)=0   value(Thetanf)=1
line to (-0.0922921,0.17) value(psi)=0   value(Thetanf)=1
line to (-0.06118661,0.166158) value(psi)=0   value(Thetanf)=1
line to (-0.0314402,0.158475) value(psi)=0   value(Thetanf)=1
line to (-0.00507099,0.14887) value(psi)=0   value(Thetanf)=1
line to (0.0212982,0.135424) value(psi)=0   value(Thetanf)=1
line to (0.0578093,0.123898) value(psi)=0   value(Thetanf)=1
line to (0.0922921,0.120056) value(psi)=0   value(Thetanf)=1
line to (0.124746,0.12774) value(psi)=0   value(Thetanf)=1
line to (0.151116,0.135424) value(psi)=0   value(Thetanf)=1
```

```
line to (0.1714,0.143107) value(psi)=0   value(Thetanf)=1
line to (0.197769,0.154633) value(psi)=0   value(Thetanf)=1
line to (0.216024,0.160395) value(psi)=0   value(Thetanf)=1
line to (0.236308,0.166158) value(psi)=0   value(Thetanf)=1
line to (0.252535,0.17) value(psi)=0   value(Thetanf)=1
line to (0.274848,0.166158) value(psi)=0   value(Thetanf)=1
line to (0.29716,0.160395) value(psi)=0   value(Thetanf)=1
line to (0.319473,0.150791) value(psi)=0   value(Thetanf)=1
line to (0.34787,0.139266) value(psi)=0   value(Thetanf)=1
line to (0.378296,0.125819) value(psi)=0   value(Thetanf)=1
line to (0.410751,0.120056) value(psi)=0   value(Thetanf)=1
line to (0.435091,0.120056) value(psi)=0   value(Thetanf)=1
line to (0.457404,0.12774) value(psi)=0   value(Thetanf)=1
line to (0.475659,0.133503) value(psi)=0   value(Thetanf)=1
line to (0.485801,0.139266) value(psi)=0   value(Thetanf)=1
line to (0.5,0.145028) value(psi)=0   value(Thetanf)=0
line to (0.15,-0.17) value(psi)=0   value(Thetanf)=0
line to (-0.15,-0.17) value(psi)=0   value(Thetanf)=0
line to close
!=========================================================================
MONITORS
PLOTS
tecplot(psi)
tecplot(Thetanf)
contour(psi)
contour(Thetanf)
elevation (normal (flux)) on "inner"as "inner-Nusselt" points = 100 export
format "#d#b#1" file="inner-Nusselt" !report(t_star)
elevation (normal (flux)) on "outer" as "outer-Nusselt" points = 100 export
format "#d#b#1" file="outer-Nusselt" !report(t_star)
elevation (normal (flux)) from  (-0.32,0) to  (0.32,0) as "Horizental Nusselt"
points = 100 export format "#d#b#1" file="Horizental Nusselt" !report(t_star)
elevation (Thetanf) from  (-0.32,0) to  (0.32,0) as "T"points = 100 export
format "#d#b#1" file="T"
End
```

1.2 Sample 2

```
TITLE
'Cavity Sample Code 2'
SELECT
painted
smoothinit
```

```
threads=2
Spectral_colors
VARIABLES
psi
Temp
S
!===========================================================================
DEFINITIONS
epsilon=0.6
Rd=0.4
Ha=30
Ra=1500
gama=pi/2
kf=0.6
knf=0.8
A2=0.96869091
A3=0.85737582
A4=1.97389886
A5=1.46917248
A6=0.93789474
 flux = -A4*grad(Temp)
!============================================================================
EQUATIONS
psi:dxx(psi)+dyy(psi)=-Ha*(A6/A5)*(cos(gama)*cos(gama)*dxx(psi)+sin(gama)*sin(
gama)*dyy(psi)+2*sin(gama)*cos(gama)*dx(dy(psi)))-(A2*A3/(A4*A5))*Ra*dx(Temp)
Temp: epsilon*((dxx(Temp)+(1+(4/3*(kf/knf)*Rd))*dyy(Temp)))=-
dy(Temp)*dx(psi)+dy(psi)*dx(Temp)
S:  S=0.8/0.6*((dx(Temp))^2+(dy(Temp))^2)+psi*2*dx(dy(psi))^2+2*(-
dx(dy(psi)))^2+(dyy(psi)-(dxx(psi))^2)+A6*Ha^2*psi*dy(psi)*sin(gama)+(dx(psi)*
cos(gama))^2
!============================================================================
BOUNDARIES
region 1
START 'inner' (-0.00433071,0.012619)
value (temp)=1  value(psi)=0  line to(-0.00695338,0.015)
value (temp)=1    value(psi)=0   line to (-0.00853018,0.0178571)
value (temp)=1   value(psi)=0   line to (-0.00905512,0.0204762)
value (temp)=1  value(psi)=0   line to (-0.00879265,0.022381)
value (temp)=1   value(psi)=0   line to (-0.00748031,0.0259524)
value (temp)=1   value(psi)=0   line to (-0.00433071,0.0295238)
value (temp)=1   value(psi)=0   line to (-0.00144357,0.0321429)
value (temp)=1   value(psi)=0   line to (0.0019685,0.0354762)
value (temp)=1  value(psi)=0   line to (0.00354331,0.0383333)
value (temp)=1   value(psi)=0   line to (0.00459318,0.0409524)
```

```
value (temp)=1   value(psi)=0   line to (0.00485564,0.0421429)
value (temp)=1   value(psi)=0   line to (0.00485564,0.0466667)
value (temp)=1   value(psi)=0   line to (-0.00433071,0.0466667)
value (temp)=1  value(psi)=0   line to (-0.00433071,0.042381)
value (temp)=1   value(psi)=0   line to (-0.00406824,0.0407143)
value (temp)=1   value(psi)=0   line to (-0.00328084,0.0388095)
value (temp)=1   value(psi)=0   line to (-0.00170604,0.0364286)
value (temp)=1   value(psi)=0   line to (-0.000131234,0.0345238)
value (temp)=1  value(psi)=0   line to (0.00223097,0.0321429)
value (temp)=1  value(psi)=0   line to (0.00406824,0.0307143)
value (temp)=1   value(psi)=0   line to (0.00590551,0.0288095)
value (temp)=1   value(psi)=0   line to (0.00748031,0.0269048)
value (temp)=1   value(psi)=0   line to (0.00905512,0.0242857)
value (temp)=1  value(psi)=0   line to (0.00958005,0.022619)
value (temp)=1  value(psi)=0   line to (0.00984252,0.0202381)
value (temp)=1   value(psi)=0   line to (0.00958005,0.0183333)
value (temp)=1   value(psi)=0   line to(0.00905512,0.0169048)
value (temp)=1   value(psi)=0   line to (0.0074672,0.0147229)
value (temp)=1   value(psi)=0   line to (0.00538058,0.012619)
value (temp)=1   value(psi)=0    line to (0.0103675,0.01)
value (temp)=1   value(psi)=0   line to (0.0108924,0.0121429)
value (temp)=1   value(psi)=0   line to (0.0119423,0.0145238)
value (temp)=1   value(psi)=0   line to (0.0135171,0.0164286)
value (temp)=1  value(psi)=0   line to (0.0158793,0.0180952)
value (temp)=1  value(psi)=0   line to (0.0185039,0.0188095)
value (temp)=1   value(psi)=0   line to (0.0216535,0.0188095)
value (temp)=1   value(psi)=0   line to (0.0248031,0.0188095)
value (temp)=1   value(psi)=0   line to (0.0279528,0.0180952)
value (temp)=1  value(psi)=0   line to (0.0311024,0.017381)
value (temp)=1  value(psi)=0   line to (0.0347769,0.0164286)
value (temp)=1   value(psi)=0   line to (0.0374016,0.0164286)
value (temp)=1   value(psi)=0   line to (0.041601,0.0169048)
value (temp)=1   value(psi)=0   line to (0.0439633,0.0178571)
value (temp)=1  value(psi)=0   line to (0.0476378,0.0195238)
value (temp)=1  value(psi)=0   line to (0.0429134,0.0269048)
value (temp)=1   value(psi)=0   line to (0.0389764,0.0247619)
value (temp)=1   value(psi)=0   line to (0.0366142,0.0228571)
value (temp)=1   value(psi)=0   line to (0.0347769,0.0207143)
value (temp)=1  value(psi)=0   line to (0.033727,0.0185714)
value (temp)=1  value(psi)=0   line to (0.0326772,0.0152381)
value (temp)=1   value(psi)=0   line to (0.0318898,0.012619)
value (temp)=1   value(psi)=0   line to (0.0308399,0.0097619)
value (temp)=1   value(psi)=0   line to (0.0295276,0.00714286)
value (temp)=1  value(psi)=0   line to (0.0276903,0.0047619)
```

```
value (temp)=1  value(psi)=0   line to (0.0253281,0.00285714)
value (temp)=1   value(psi)=0   line to (0.0216535,0.00166667)
value (temp)=1   value(psi)=0   line to (0.0187664,0.00166667)
value (temp)=1   value(psi)=0   line to (0.014791,0.00245715)
value (temp)=1   value(psi)=0    line to (0.0146787,-0.00309072)
value (temp)=1  value(psi)=0   line to (0.0177165,-0.00190476)
value (temp)=1   value(psi)=0   line to (0.0208661,-0.00190476)
value (temp)=1   value(psi)=0   line to  (0.0242782,-0.00261905)
value (temp)=1   value(psi)=0   line to (0.0274278,-0.0047619)
value (temp)=1  value(psi)=0   line to (0.02979,-0.00785714)
value (temp)=1  value(psi)=0   line to (0.0311024,-0.0107143)
value (temp)=1   value(psi)=0   line to (0.0321522,-0.0142857)
value (temp)=1   value(psi)=0   line to (0.0334646,-0.0188095)
value (temp)=1   value(psi)=0   line to (0.0347769,-0.0209524)
value (temp)=1   value(psi)=0   line to  (0.0371391,-0.0238095)
value (temp)=1  value(psi)=0   line to (0.0387139,-0.0252381)
value (temp)=1   value(psi)=0   line to (0.0429134,-0.027381)
value (temp)=1   value(psi)=0   line to (0.0476378,-0.02)
value (temp)=1   value(psi)=0   line to (0.0434383,-0.0178571)
value (temp)=1  value(psi)=0   line to (0.0405512,-0.0171429)
value (temp)=1  value(psi)=0   line to (0.037664,-0.0169048)
value (temp)=1   value(psi)=0   line to (0.0339895,-0.0169048)
value (temp)=1   value(psi)=0   line to (0.0308399,-0.017619)
value (temp)=1   value(psi)=0   line to (0.0274278,-0.0185714)
value (temp)=1  value(psi)=0   line to (0.0237533,-0.0192857)
value (temp)=1  value(psi)=0   line to (0.0203412,-0.0192857)
value (temp)=1   value(psi)=0   line to (0.0171916,-0.0188095)
value (temp)=1   value(psi)=0   line to (0.0143045,-0.017381)
value (temp)=1   value(psi)=0   line to (0.0124672,-0.0154762)
value (temp)=1  value(psi)=0   line to (0.0111549,-0.0130952)
value (temp)=1  value(psi)=0   line to (0.0103675,-0.0102381)
value (temp)=1   value(psi)=0    line to (0.00538058,-0.0128571)
value (temp)=1   value(psi)=0   line to (0.00774278,-0.0154762)
value (temp)=1   value(psi)=0   line to (0.00905512,-0.017381)
value (temp)=1  value(psi)=0   line to (0.00984252,-0.0197619)
value (temp)=1  value(psi)=0   line to (0.00958005,-0.022619)
value (temp)=1   value(psi)=0   line to (0.00879265,-0.0254762)
value (temp)=1   value(psi)=0   line to (0.00695538,-0.0278571)
value (temp)=1   value(psi)=0   line to (0.00511811,-0.03)
value (temp)=1  value(psi)=0   line to (0.00223097,-0.032619)
value (temp)=1  value(psi)=0   line to (0.00131234,-0.0345238)
value (temp)=1   value(psi)=0   line to (-0.0019685,-0.0369048)
value (temp)=1  value(psi)=0   line to (-0.00328084,-0.0397619)
value (temp)=1  value(psi)=0   line to (-0.00406824,-0.0419048)
```

```
value (temp)=1 value(psi)=0 line to (-0.00433071,-0.0430952)
value (temp)=1 value(psi)=0 line to (-0.00433071,-0.0471429)
value (temp)=1 value(psi)=0 line to (0.00511811,-0.0471429)
value (temp)=1 value(psi)=0 line to (0.00511811,-0.0430952)
value (temp)=1 value(psi)=0 line to (0.00459318,-0.0407143)
value (temp)=1 value(psi)=0 line to (0.00380577,-0.0390476)
value (temp)=1 value(psi)=0 line to (0.00249344,-0.0366667)
value (temp)=1 value(psi)=0 line to (-0.000131234,-0.0335714)
value (temp)=1 value(psi)=0 line to (-0.00275591,-0.0314286)
value (temp)=1 value(psi)=0 line to (-0.00564304,-0.0288095)
value (temp)=1 value(psi)=0 line to (-0.00721785,-0.0264286)
value (temp)=1 value(psi)=0 line to (-0.00853018,-0.0238095)
value (temp)=1 value(psi)=0 line to (-0.0090551,-0.0209524)
value (temp)=1 value(psi)=0 line to (-0.00879265,-0.0188095)
value (temp)=1 value(psi)=0 line to (-0.00774278,-0.0161905)
value (temp)=1 value(psi)=0 line to (-0.00614304,-0.0140476)
value (temp)=1 value(psi)=0 line to(-0.00433071,-0.0128571)
value (temp)=1 value(psi)=0 line to (-0.00984252,-0.0102381)
value (temp)=1 value(psi)=0 line to (-0.0103675,-0.0133333)
value (temp)=1 value(psi)=0 line to (-0.0119423,-0.0157143)
value (temp)=1 value(psi)=0 line to (-0.014042,-0.0178571)
value (temp)=1 value(psi)=0 line to (-0.0174541,-0.0190476)
value (temp)=1 value(psi)=0 line to (-0.0213911,-0.0192857)
value (temp)=1 value(psi)=0 line to (-0.025853,-0.0188095)
value (temp)=1 value(psi)=0 line to (-0.02979,-0.0178571)
value (temp)=1 value(psi)=0 line to (-0.0332021,-0.0169048)
value (temp)=1 value(psi)=0 line to (-0.0379265,-0.0169048)
value (temp)=1 value(psi)=0 line to (-0.0410761,-0.017381)
value (temp)=1 value(psi)=0 line to (-0.0429134,-0.0178571)
value (temp)=1 value(psi)=0 line to (-0.0465879,-0.02)
value (temp)=1 value(psi)=0 line to (-0.0418635,-0.027381)
value (temp)=1 value(psi)=0 line to (-0.038189,-0.0252381)
value (temp)=1 value(psi)=0 line to (-0.0355643,-0.0230952)
value (temp)=1 value(psi)=0 line to (-0.033727,-0.0204762)
value (temp)=1 value(psi)=0 line to (-0.0324147,-0.0178571)
value (temp)=1 value(psi)=0 line to (-0.0313648,-0.0142857)
value (temp)=1 value(psi)=0 line to (-0.0305774,-0.0114286)
value (temp)=1 value(psi)=0 line to (-0.0287402,-0.00761905)
value (temp)=1 value(psi)=0 line to (-0.0269029,-0.005)
value (temp)=1 value(psi)=0 line to (-0.0248031,-0.00333333)
value (temp)=1 value(psi)=0 line to (-0.0221785,-0.00238095)
value (temp)=1 value(psi)=0 line to (-0.0192913,-0.00214286)
value (temp)=1 value(psi)=0 line to (-0.0164042,-0.00214286)
value (temp)=1 value(psi)=0 line to (-0.0143045,-0.00285714)
```

```
value (temp)=1   value(psi)=0    line to (-0.0143045,0.00238095)
value (temp)=1  value(psi)=0   line to (-0.0174541,0.00166667)
value (temp)=1  value(psi)=0   line to (-0.0203412,0.00142857)
value (temp)=1   value(psi)=0   line to (-0.0232283,0.00238095)
value (temp)=1  value(psi)=0   line to (-0.0261155,0.00428571)
value (temp)=1   value(psi)=0   line to (-0.0284777,0.00666667)
value (temp)=1  value(psi)=0   line to (-0.030315,0.0102381)
value (temp)=1  value(psi)=0   line to (-0.0313648,0.0140476)
value (temp)=1   value(psi)=0   line to (-0.0326772,0.0178571)
value (temp)=1   value(psi)=0   line to (-0.0339895,0.0207143)
value (temp)=1   value(psi)=0   line to (-0.0360892,0.0230952)
value (temp)=1  value(psi)=0   line to (-0.0379265,0.0247619)
value (temp)=1  value(psi)=0   line to (-0.0418635,0.0269048)
value (temp)=1   value(psi)=0   line to (-0.0468504,0.0195238)
value (temp)=1   value(psi)=0   line to (-0.0426509,0.017381)
value (temp)=1   value(psi)=0   line to (-0.0395013,0.0164286)
value (temp)=1  value(psi)=0   line to (-0.0368766,0.0164286)
value (temp)=1  value(psi)=0   line to (-0.0329396,0.0166667)
value (temp)=1   value(psi)=0   line to (-0.0287402,0.0178571)
value (temp)=1   value(psi)=0   line to (-0.0237533,0.0188095)
value (temp)=1   value(psi)=0   line to (-0.0198163,0.0188095)
value (temp)=1  value(psi)=0   line to (-0.0166667,0.0183333)
value (temp)=1  value(psi)=0   line to (-0.014042,0.0171429)
value (temp)=1   value(psi)=0   line to (-0.0116798,0.0152381)
value (temp)=1  value(psi)=0   line to (-0.010105,0.012619)
value (temp)=1  value(psi)=0   line to (-0.00984252,0.0097619)
line to close
start 'outer' (0,0.09)
value(Temp)=0  value(psi)=0
ARC(CENTER=0,0) ANGLE=360 TO CLOSE

MONITORS
PLOTS
contour(Temp) as "Temp"painted
contour(psi) as "stream function" painted
contour(s) as "entropy generation" painted
tecplot(temp)
tecplot(psi)
tecplot (s)
elevation (temp) from  (-0.09,0) to  (0.09,0) as "T"points = 100 export format
"#d#b#1" file="T.txt"
elevation (normal (flux)) from    (-0.09,0) to  (0.09,0) as "Horizental Nus-
selt" points = 100 export format "#d#b#1" file="Horizental Nusselt" !report(t_
star)
```

```
elevation (normal (flux)) on "inner"as "inner-Nusselt" points = 100 export
format "#d#b#1" file="inner-Nusselt" !report(t_star)
elevation (normal (flux)) on "outer" as "outer-Nusselt" points = 100 export
format "#d#b#1" file="outer-Nusselt" !report(t_star)
End
```

1.3 Sample 3

```
TITLE
'Cavity Sample Code 3'
SELECT
painted
smoothinit
threads=2
Spectral_colors
VARIABLES
psi
thetanf
!=====================================================================
DEFINITIONS
epsilon=0.3
Rd=0.3
Ha=25
Ra=50
gama=pi/6
kf=0.6
knf=2.5
A2=1.028116649
A3=2.19555776456
A4=1.148382263
A5=1.292355436
A6=0.9702970356
 flux = -A4*grad(Thetanf)
!=====================================================================
EQUATIONS
psi:dxx(psi)+dyy(psi)=-Ha*(A6/A5)*(cos(gama)*cos(gama)*dxx(psi)+sin(gama)*sin(
gama)*dyy(psi)+2*sin(gama)*cos(gama)*dx(dy(psi)))-(A2*A3/
(A4*A5))*Ra*dx(thetanf)
thetanf: epsilon*((dxx(thetanf)+(1+(4/3*(kf/knf)*Rd))*dyy(thetanf)))=-dy(theta
nf)*dx(psi)+dy(psi)*dx(thetanf)
!=====================================================================
BOUNDARIES
 region 1
```

```
start 'inner'(0,-0.12)
value(psi)=0   value(Thetanf)=1
ARC(CENTER=0,0) ANGLE=360 TO CLOSE
START 'outer' (0.45,0)
value(psi)=0   value(Thetanf)=0   line to (0.4754289704,0.02075767713)
value(psi)=0   value(Thetanf)=0   line to (0.4980973490,0.04357787138)
value(psi)=0   value(Thetanf)=0   line to (0.5162559261,0.06796638205)
value(psi)=0   value(Thetanf)=0   line to (0.5284503420,0.09318005329)
value(psi)=0   value(Thetanf)=0   line to (0.5336361559,0.1183042875)
value(psi)=0   value(Thetanf)=0   line to (0.5312592044,0.1423504748)
value(psi)=0   value(Thetanf)=0   line to (0.5212946112,0.1643635596)
value(psi)=0   value(Thetanf)=0   line to (0.5042414476,0.1835288778)
value(psi)=0   value(Thetanf)=0   line to (0.4810739378,0.1992673496)
value(psi)=0   value(Thetanf)=0   line to (0.4531538936,0.2113091308)
value(psi)=0   value(Thetanf)=0   line to (0.4221124046,0.2197378094)
value(psi)=0   value(Thetanf)=0   line to (0.3897114317,0.2250000000)
value(psi)=0   value(Thetanf)=0   line to (0.3576975738,0.2278784866)
value(psi)=0   value(Thetanf)=0   line to (0.3276608177,0.2294305746)
value(psi)=0   value(Thetanf)=0   line to (0.3009104504,0.2308967096)
value(psi)=0   value(Thetanf)=0   line to (0.2783786046,0.2335873844)
value(psi)=0  value(Thetanf)=0   line to (0.2605592795,0.2387585905)
value(psi)=0   value(Thetanf)=0   line to (0.2474873734,0.2474873734)
value(psi)=0   value(Thetanf)=0   line to (0.2387585904,0.2605592795)
value(psi)=0   value(Thetanf)=0   line to (0.2335873845,0.2783786046)
value(psi)=0   value(Thetanf)=0   line to (0.2308967096,0.3009104505)
value(psi)=0   value(Thetanf)=0   line to (0.2294305745,0.3276608177)
value(psi)=0   value(Thetanf)=0   line to (0.2278784866,0.3576975739)
value(psi)=0   value(Thetanf)=0   line to (0.2250000000,0.3897114316)
value(psi)=0   value(Thetanf)=0   line to (0.2197378096,0.4221124042)
value(psi)=0   value(Thetanf)=0   line to (0.2113091310,0.4531538930)
value(psi)=0  value(Thetanf)=0   line to (0.1992673499,0.4810739372)
value(psi)=0   value(Thetanf)=0   line to (0.1835288783,0.5042414469)
value(psi)=0   value(Thetanf)=0   line to (0.1643635604,0.5212946107)
value(psi)=0   value(Thetanf)=0   line to (0.1423504759,0.5312592042)
value(psi)=0   value(Thetanf)=0   line to (0.1183042889,0.5336361560)
value(psi)=0   value(Thetanf)=0   line to (0.09318005476,0.5284503426)
value(psi)=0  value(Thetanf)=0   line to (0.06796638373,0.5162559272)
value(psi)=0   value(Thetanf)=0   line to (0.04357787311,0.4980973506)
value(psi)=0   value(Thetanf)=0   line to (0.02075767889,0.4754289723)
value(psi)=0   value(Thetanf)=0   line to (0,0.4500000023)
value(psi)=0   value(Thetanf)=0   line to (-0.01849976986,0.4237144315)
value(psi)=0   value(Thetanf)=0   line to (-0.03486229555,0.3984778817)
value(psi)=0   value(Thetanf)=0   line to (-0.04950718941,0.3760444511)
value(psi)=0   value(Thetanf)=0   line to (-0.06310330511,0.3578766375)
```

```
value(psi)=0   value(Thetanf)=0   line to (-0.07649136333,0.3450302516)
value(psi)=0   value(Thetanf)=0   line to (-0.09058666393,0.3380740397)
value(psi)=0   value(Thetanf)=0   line to (-0.1062716578,0.3370506443)
value(psi)=0   value(Thetanf)=0   line to (-0.1242892486,0.3414819105)
value(psi)=0   value(Thetanf)=0   line to (-0.1451477362,0.3504176397)
value(psi)=0   value(Thetanf)=0   line to (-0.1690473008,0.3625231129)
value(psi)=0   value(Thetanf)=0   line to (-0.1958359377,0.3761973427)
value(psi)=0   value(Thetanf)=0   line to (-0.2249999949,0.3897114296)
value(psi)=0   value(Thetanf)=0   line to (-0.2556911557,0.4013547260)
value(psi)=0   value(Thetanf)=0   line to (-0.2867882124,0.4095760209)
value(psi)=0   value(Thetanf)=0   line to (-0.3169885710,0.4131075557)
value(psi)=0  value(Thetanf)=0   line to (-0.3449214590,0.4110613950)
value(psi)=0   value(Thetanf)=0   line to (-0.3692725919,0.4029903259)
value(psi)=0   value(Thetanf)=0   line to (-0.3889087261,0.3889087332)
value(psi)=0   value(Thetanf)=0   line to (-0.4029903213,0.3692726013)
value(psi)=0   value(Thetanf)=0   line to (-0.4110613930,0.3449214701)
value(psi)=0   value(Thetanf)=0   line to (-0.4131075562,0.3169885835)
value(psi)=0   value(Thetanf)=0   line to (-0.4095760235,0.2867882255)
value(psi)=0   value(Thetanf)=0   line to (-0.4013547296,0.2556911683)
value(psi)=0   value(Thetanf)=0   line to (-0.3897114351,0.2250000077)
value(psi)=0   value(Thetanf)=0   line to (-0.3761973487,0.1958359496)
value(psi)=0   value(Thetanf)=0   line to (-0.3625231186,0.1690473116)
value(psi)=0   value(Thetanf)=0   line to (-0.3504176443,0.1451477457)
value(psi)=0  value(Thetanf)=0   line to (-0.34148191303,0.1242892566)
value(psi)=0   value(Thetanf)=0   line to (-0.3370506451,0.1062716648)
value(psi)=0   value(Thetanf)=0   line to (-0.3380740381,0.09058667012)
value(psi)=0   value(Thetanf)=0   line to (-0.3450302473,0.07649136902)
value(psi)=0   value(Thetanf)=0   line to (-0.3578766308,0.06310331074)
value(psi)=0  value(Thetanf)=0   line to (-0.3760444430,0.04950719537)
value(psi)=0   value(Thetanf)=0   line to (-0.3984778713,0.03486230203)
value(psi)=0   value(Thetanf)=0   line to (-0.4237144207,0.01849977720)
value(psi)=0   value(Thetanf)=0   line to (-0.4499999908,0)
value(psi)=0   value(Thetanf)=0   line to (-0.4754289621,-0.02075766968)
value(psi)=0   value(Thetanf)=0   line to (-0.4980973422,-0.04357786315)
value(psi)=0   value(Thetanf)=0   line to (-0.5162559205,-0.06796637321)
value(psi)=0   value(Thetanf)=0   line to (-0.5284503390,-0.09318004410)
value(psi)=0   value(Thetanf)=0   line to (-0.5336361553,-0.1183042784)
value(psi)=0   value(Thetanf)=0   line to (-0.5312592068,-0.1423504661)
value(psi)=0   value(Thetanf)=0  line to (-0.5212946163,-0.1643635516)
value(psi)=0  value(Thetanf)=0   line to (-0.5042414555,-0.1835288710)
value(psi)=0  value(Thetanf)=0   line to (-0.4810739480,-0.1992673440)
value(psi)=0   value(Thetanf)=0   line to (-0.4531539058,-0.2113091269)
value(psi)=0  value(Thetanf)=0   line to (-0.4221124177,-0.2197378067)
value(psi)=0  value(Thetanf)=0   line to (-0.3897114449,-0.2249999981)
```

```
value(psi)=0   value(Thetanf)=0   line to (-0.3576975872,-0.2278784859)
value(psi)=0   value(Thetanf)=0   line to (-0.3276608299,-0.2294305739)
value(psi)=0   value(Thetanf)=0   line to (-0.3009104616,-0.2308967090)
value(psi)=0   value(Thetanf)=0   line to (-0.2783786136,-0.2335873827)
value(psi)=0  value(Thetanf)=0   line to (-0.2605592863,-0.2387585873)
value(psi)=0   value(Thetanf)=0   line to (-0.2474873784,-0.2474873685)
value(psi)=0   value(Thetanf)=0   line to (-0.2387585937,-0.2605592724)
value(psi)=0  value(Thetanf)=0  line to (-0.2335873861,-0.2783785949)
value(psi)=0   value(Thetanf)=0   line to (-0.2308967105,-0.3009104386)
value(psi)=0   value(Thetanf)=0   line to (-0.2294305754,-0.3276608041)
value(psi)=0   value(Thetanf)=0   line to (-0.2278784875,-0.3576975582)
value(psi)=0   value(Thetanf)=0   line to (-0.2250000021,-0.3897114156)
value(psi)=0   value(Thetanf)=0   line to (-0.2197378127,-0.4221123881)
value(psi)=0   value(Thetanf)=0   line to (-0.2113091361,-0.4531538783)
value(psi)=0   value(Thetanf)=0   line to (-0.1992673569,-0.4810739247)
value(psi)=0   value(Thetanf)=0   line to (-0.1835288870,-0.5042414367)
value(psi)=0   value(Thetanf)=0   line to (-0.1643635707,-0.5212946040)
value(psi)=0   value(Thetanf)=0   line to (-0.1423504874,-0.5312592011)
value(psi)=0   value(Thetanf)=0   line to (-0.1183043011,-0.5336361567)
value(psi)=0   value(Thetanf)=0   line to (-0.09318006729,-0.5284503467)
value(psi)=0   value(Thetanf)=0  line to (-0.067966396087,-0.5162559348)
value(psi)=0   value(Thetanf)=0   line to (-0.04357788483,-0.4980973606)
value(psi)=0  value(Thetanf)=0   line to (-0.02075768971,-0.4754289847)
value(psi)=0   value(Thetanf)=0   line to (0,-0.4500000152)
value(psi)=0   value(Thetanf)=0   line to (0.01849976125,-0.4237144440)
value(psi)=0   value(Thetanf)=0   line to (0.03486228794,-0.3984778938)
value(psi)=0  value(Thetanf)=0   line to (0.04950718250,-0.3760444614)
value(psi)=0   value(Thetanf)=0   line to (0.06310329855,-0.3578766454)
value(psi)=0   value(Thetanf)=0   line to (0.07649135661,-0.3450302566)
value(psi)=0   value(Thetanf)=0   line to (0.09058665662,-0.3380740417)
value(psi)=0  value(Thetanf)=0   line to (0.1062716494,-0.3370506432)
value(psi)=0   value(Thetanf)=0   line to (0.1242892389,-0.3414819070)
value(psi)=0   value(Thetanf)=0   line to (0.1451477251,-0.3504176342)
value(psi)=0   value(Thetanf)=0   line to (0.1690472882,-0.3625231062)
value(psi)=0  value(Thetanf)=0   line to (0.1958359241,-0.3761973364)
value(psi)=0   value(Thetanf)=0   line to (0.2249999799,-0.3897114230)
value(psi)=0   value(Thetanf)=0   line to (0.2556911401,-0.4013547208)
value(psi)=0  value(Thetanf)=0   line to (0.2867881970,-0.4095760178)
value(psi)=0   value(Thetanf)=0   line to (0.3169885565,-0.4131075552)
value(psi)=0   value(Thetanf)=0   line to (0.3449214462,-0.4110613978)
value(psi)=0  value(Thetanf)=0   line to (0.3692725808,-0.4029903313)
value(psi)=0   value(Thetanf)=0   line to (0.3889087176,-0.3889087416)
value(psi)=0  value(Thetanf)=0   line to (0.4029903158,-0.3692726122)
value(psi)=0   value(Thetanf)=0   line to (0.4110613906,-0.3449214832)
```

```
value(psi)=0   value(Thetanf)=0   line to (0.4131075562,-0.3169885976)
value(psi)=0   value(Thetanf)=0   line to (0.4095760268,-0.2867882409)
value(psi)=0   value(Thetanf)=0   line to (0.4013547348,-0.2556911838)
value(psi)=0   value(Thetanf)=0   line to (0.3897114417,-0.2250000226)
value(psi)=0   value(Thetanf)=0   line to (0.3761973557,-0.1958359636)
value(psi)=0   value(Thetanf)=0   line to (0.3625231245,-0.1690473239)
value(psi)=0  value(Thetanf)=0   line to (0.3504176498,-0.1451477568)
value(psi)=0   value(Thetanf)=0   line to (0.3414819164,-0.1242892662)
value(psi)=0   value(Thetanf)=0   line to (0.3370506460,-0.1062716731)
value(psi)=0  value(Thetanf)=0   line to (0.3380740361,-0.09058667740)
value(psi)=0  value(Thetanf)=0   line to (0.3450302426,-0.07649137580)
value(psi)=0   value(Thetanf)=0   line to (0.3578766230,-0.06310331730)
value(psi)=0   value(Thetanf)=0   line to (0.3760444328,-0.04950720226)
value(psi)=0   value(Thetanf)=0   line to (0.3984778593,-0.03486230964)
value(psi)=0   value(Thetanf)=0   line to (0.4237144076,-0.01849978579)
value(psi)=0  value(Thetanf)=0  line to (0.4499999787,0)
value(psi)=0   value(Thetanf)=0   line to close
!========================================================================
MONITORS
PLOTS
tecplot(psi)
tecplot(Thetanf)
contour(psi)
contour(Thetanf)
elevation (normal (flux)) on "inner"as "inner-Nusselt" points = 100 export
format "#d#b#1" file="inner-Nusselt" !report(t_star)
elevation (normal (flux)) on "outer" as "outer-Nusselt" points = 100 export
format "#d#b#1" file="outer-Nusselt" !report(t_star)
End
```

1.4 Sample 4

```
TITLE
'Cavity Sample Code 4'
COORDINATES
Cartesian2
SELECT
!ngrid =10
regrid = on
errlim = 1e-6
VARIABLES
u(1)
v(1)
```

```
temp
psi(1)
DEFINITIONS
phi=0.03
rho_f=997.1
rho_p=3970
Cp_f=4179
Cp_p=765
Beta_f=21e-5
Beta_p=0.85e-5
kf=0.613
kp=40
rhoCp_nf=((1-phi)*(rho_f*Cp_f))+(phi*rho_p*Cp_p)
Gammaa = 10^(7)
Pr = 6.2
Ra=50
rho_nf=(1-phi)*rho_f+(phi*rho_p)
A0=rho_f/rho_nf
A1=(1/(1-phi)^2.5)
A2=((1-phi)*rho_f*Beta_f+phi*rho_p*Beta_p)/(rho_nf*Beta_f)
m=5.7
A3=(kp+(m-1)*kf-(m-1)*phi*(kf-kp))/(kp+(m-1)*kf+phi*(kf-kp))
Rd=0.4
Ha=30
flux = -(A3)*(1+(4/3)*(1/A3)*Rd)*grad(temp)
EQUATIONS
u:   u*dx(u)+v*dy(u)  = A0*Gammaa*dx(dx(u)+dy(v))+(Pr*A1*A0)*(div(grad(u)))
v:    u*dx(v)+v*dy(v)  =(A0)*Gammaa*dy(dx(u)+dy(v))+(Pr*A1)*(div(grad(v)))+(A2)
*Pr*Ra*temp-Pr*(Ha^2)*v
temp:  (u*dx(temp))+(v*dy(temp))=(A3)*(rho_f*Cp_f/rhoCp_nf)*(dxx(temp)+dyy(tem
p))+(4/3)*(Rd*rho_f*Cp_f/rhoCp_nf)*dyy(temp)
psi: div(grad(psi))=(dy(u)-dx(v))
BOUNDARIES
REGION 1
START 'inner' (0.45,0)
value(temp)=1 value(u)=0 value(v)=0 value(psi)=0  line to
(0.4754289704,0.02075767713)
value(temp)=1 value(u)=0 value(v)=0 value(psi)=0   line to
(0.4980973490,0.04357787138)
value(temp)=1 value(u)=0 value(v)=0 value(psi)=0   line to
(0.5162559261,0.06796638205)
value(temp)=1 value(u)=0 value(v)=0 value(psi)=0   line to
(0.5284503420,0.09318005329)
value(temp)=1 value(u)=0 value(v)=0 value(psi)=0    line to
```

```
(0.5336361559,0.1183042875)
value(temp)=1 value(u)=0 value(v)=0 value(psi)=0    line to
(0.5312592044,0.1423504748)
value(temp)=1 value(u)=0 value(v)=0 value(psi)=0    line to
(0.5212946112,0.1643635596)
value(temp)=1 value(u)=0 value(v)=0 value(psi)=0    line to
(0.5042414476,0.1835288778)
value(temp)=1 value(u)=0 value(v)=0 value(psi)=0    line to
(0.4810739378,0.1992673496)
value(temp)=1 value(u)=0 value(v)=0 value(psi)=0    line to
(0.4531538936,0.2113091308)
value(temp)=1 value(u)=0 value(v)=0 value(psi)=0    line to
(0.4221124046,0.2197378094)
value(temp)=1 value(u)=0 value(v)=0 value(psi)=0    line to
(0.3897114317,0.2250000000)
value(temp)=1 value(u)=0 value(v)=0 value(psi)=0    line to
(0.3576975738,0.2278784866)
value(temp)=1 value(u)=0 value(v)=0 value(psi)=0    line to
(0.3276608177,0.2294305746)
value(temp)=1 value(u)=0 value(v)=0 value(psi)=0    line to
(0.3009104504,0.2308967096)
value(temp)=1 value(u)=0 value(v)=0 value(psi)=0    line to
(0.2783786046,0.2335873844)
value(temp)=1 value(u)=0 value(v)=0 value(psi)=0    line to
(0.2605592795,0.2387585905)
value(temp)=1 value(u)=0 value(v)=0 value(psi)=0    line to
(0.2474873734,0.2474873734)
value(temp)=1 value(u)=0 value(v)=0 value(psi)=0    line to
(0.2387585904,0.2605592795)
value(temp)=1 value(u)=0 value(v)=0 value(psi)=0    line to
(0.2335873845,0.2783786046)
value(temp)=1 value(u)=0 value(v)=0 value(psi)=0    line to
(0.2308967096,0.3009104505)
value(temp)=1 value(u)=0 value(v)=0 value(psi)=0    line to
(0.2294305745,0.3276608177)
value(temp)=1 value(u)=0 value(v)=0 value(psi)=0    line to
(0.2278784866,0.3576975739)
value(temp)=1 value(u)=0 value(v)=0 value(psi)=0    line to
(0.2250000000,0.3897114316)
value(temp)=1 value(u)=0 value(v)=0 value(psi)=0    line to
(0.2197378096,0.4221124042)
value(temp)=1 value(u)=0 value(v)=0 value(psi)=0   line to
(0.2113091310,0.4531538930)
value(temp)=1 value(u)=0 value(v)=0 value(psi)=0   line to
```

```
(0.1992673499,0.4810739372)
value(temp)=1 value(u)=0 value(v)=0 value(psi)=0    line to
(0.1835288783,0.5042414469)
value(temp)=1 value(u)=0 value(v)=0 value(psi)=0    line to
(0.1643635604,0.5212946107)
value(temp)=1 value(u)=0 value(v)=0 value(psi)=0    line to
(0.1183042889,0.5336361560)
value(temp)=1 value(u)=0 value(v)=0 value(psi)=0    line to
(0.09318005476,0.5284503426)
value(temp)=1 value(u)=0 value(v)=0 value(psi)=0    line to
(0.06796638373,0.5162559272)
value(temp)=1 value(u)=0 value(v)=0 value(psi)=0    line to
(0.04357787311,0.4980973506)
value(temp)=1 value(u)=0 value(v)=0 value(psi)=0    line to
(0.02075767889,0.4754289723)
value(temp)=1 value(u)=0 value(v)=0 value(psi)=0    line to (0,0.4500000023)
value(temp)=1 value(u)=0 value(v)=0 value(psi)=0    line to
(-0.01849976986,0.4237144315)
value(temp)=1 value(u)=0 value(v)=0 value(psi)=0   line to
(-0.03486229555,0.3984778817)
value(temp)=1 value(u)=0 value(v)=0 value(psi)=0    line to
(-0.04950718941,0.3760444511)
value(temp)=1 value(u)=0 value(v)=0 value(psi)=0    line to
(-0.06310330511,0.3578766375)
value(temp)=1 value(u)=0 value(v)=0 value(psi)=0   line to
(-0.07649136333,0.3450302516)
value(temp)=1 value(u)=0 value(v)=0 value(psi)=0    line to
(-0.09058666393,0.3380740397)
value(temp)=1 value(u)=0 value(v)=0 value(psi)=0    line to
(-0.1062716578,0.3370506443)
value(temp)=1 value(u)=0 value(v)=0 value(psi)=0   line to
(-0.1242892486,0.3414819105)
value(temp)=1 value(u)=0 value(v)=0 value(psi)=0    line to
(-0.1451477362,0.3504176397)
value(temp)=1 value(u)=0 value(v)=0 value(psi)=0    line to
(-0.1690473008,0.3625231129)
value(temp)=1 value(u)=0 value(v)=0 value(psi)=0    line to
(-0.1958359377,0.3761973427)
value(temp)=1 value(u)=0 value(v)=0 value(psi)=0    line to
(-0.2249999949,0.3897114296)
value(temp)=1 value(u)=0 value(v)=0 value(psi)=0    line to
(-0.2556911557,0.4013547260)
value(temp)=1 value(u)=0 value(v)=0 value(psi)=0    line to
(-0.2867882124,0.4095760209)
```

```
value(temp)=1 value(u)=0 value(v)=0 value(psi)=0    line to
(-0.3169885710,0.4131075557)
value(temp)=1 value(u)=0 value(v)=0 value(psi)=0    line to
(-0.3449214590,0.4110613950)
value(temp)=1 value(u)=0 value(v)=0 value(psi)=0   line to
(-0.3692725919,0.4029903259)
value(temp)=1 value(u)=0 value(v)=0 value(psi)=0    line to
(-0.3889087261,0.3889087332)
value(temp)=1 value(u)=0 value(v)=0 value(psi)=0    line to
(-0.4029903213,0.3692726013)
value(temp)=1 value(u)=0 value(v)=0 value(psi)=0    line to
(-0.4110613930,0.3449214701)
value(temp)=1 value(u)=0 value(v)=0 value(psi)=0    line to
(-0.4131075562,0.3169885835)
value(temp)=1 value(u)=0 value(v)=0 value(psi)=0    line to
(-0.4095760235,0.2867882255)
value(temp)=1 value(u)=0 value(v)=0 value(psi)=0    line to
(-0.4013547296,0.2556911683)
value(temp)=1 value(u)=0 value(v)=0 value(psi)=0    line to
(-0.3897114351,0.2250000077)
value(temp)=1 value(u)=0 value(v)=0 value(psi)=0    line to
(-0.3761973487,0.1958359496)
value(temp)=1 value(u)=0 value(v)=0 value(psi)=0    line to
(-0.3625231186,0.1690473116)
value(temp)=1 value(u)=0 value(v)=0 value(psi)=0    line to
(-0.3504176443,0.1451477457)
value(temp)=1 value(u)=0 value(v)=0 value(psi)=0    line to
(-0.34148191303,0.1242892566)
value(temp)=1 value(u)=0 value(v)=0 value(psi)=0    line to
(-0.3370506451,0.1062716648)
value(temp)=1 value(u)=0 value(v)=0 value(psi)=0    line to
(-0.3380740381,0.09058667012)
value(temp)=1 value(u)=0 value(v)=0 value(psi)=0    line to
(-0.3450302473,0.07649136902)
value(temp)=1 value(u)=0 value(v)=0 value(psi)=0    line to
(-0.3578766308,0.06310331074)
value(temp)=1 value(u)=0 value(v)=0 value(psi)=0   line to
(-0.3760444430,0.04950719537)
value(temp)=1 value(u)=0 value(v)=0 value(psi)=0    line to
(-0.3984778713,0.03486230203)
value(temp)=1 value(u)=0 value(v)=0 value(psi)=0    line to
(-0.4237144207,0.01849977720)
value(temp)=1 value(u)=0 value(v)=0 value(psi)=0    line to (-0.4499999908,0)
value(temp)=1 value(u)=0 value(v)=0 value(psi)=0    line to (-0.4754289621,-
```

```
0.02075766968)
value(temp)=1 value(u)=0 value(v)=0 value(psi)=0    line to (-0.4980973422,-
0.04357786315)
value(temp)=1 value(u)=0 value(v)=0 value(psi)=0    line to (-0.5162559205,-
0.06796637321)
value(temp)=1 value(u)=0 value(v)=0 value(psi)=0    line to (-0.5284503390,-
0.09318004410)
value(temp)=1 value(u)=0 value(v)=0 value(psi)=0    line to (-0.5336361553,-
0.1183042784)
value(temp)=1 value(u)=0 value(v)=0 value(psi)=0    line to (-0.5312592068,-
0.1423504661)
value(temp)=1 value(u)=0 value(v)=0 value(psi)=0   line to (-0.5212946163,-
0.1643635516)
value(temp)=1 value(u)=0 value(v)=0 value(psi)=0    line to (-0.5042414555,-
0.1835288710)
value(temp)=1 value(u)=0 value(v)=0 value(psi)=0    line to (-0.4810739480,-
0.1992673440)
value(temp)=1 value(u)=0 value(v)=0 value(psi)=0   line to (-0.4531539058,-
0.2113091269)
value(temp)=1 value(u)=0 value(v)=0 value(psi)=0    line to (-0.4221124177,-
0.2197378067)
value(temp)=1 value(u)=0 value(v)=0 value(psi)=0    line to (-0.3897114449,-
0.2249999981)
value(temp)=1 value(u)=0 value(v)=0 value(psi)=0    line to (-0.3576975872,-
0.2278784859)
value(temp)=1 value(u)=0 value(v)=0 value(psi)=0    line to (-0.3276608299,-
0.2294305739)
value(temp)=1 value(u)=0 value(v)=0 value(psi)=0   line to (-0.3009104616,-
0.2308967090)
value(temp)=1 value(u)=0 value(v)=0 value(psi)=0    line to (-0.2783786136,-
0.2335873827)
value(tcmp)-1 value(u)=0 value(v)=0 value(psi)=0    line to (-0.2605592863,-
0.2387585873)
value(temp)=1 value(u)=0 value(v)=0 value(psi)=0    line to (-0.2474873784,-
0.2474873685)
value(temp)=1 value(u)=0 value(v)=0 value(psi)=0    line to (-0.2387585937,-
0.2605592724)
value(temp)=1 value(u)=0 value(v)=0 value(psi)=0   line to (-0.2335873861,-
0.2783785949)
value(temp)=1 value(u)=0 value(v)=0 value(psi)=0    line to (-0.2308967105,-
0.3009104386)
value(temp)=1 value(u)=0 value(v)=0 value(psi)=0    line to (-0.2294305754,-
0.3276608041)
value(temp)=1 value(u)=0 value(v)=0 value(psi)=0    line to (-0.2278784875,-
```

```
0.3576975582)
value(temp)=1 value(u)=0 value(v)=0 value(psi)=0    line to (-0.2250000021,-
0.3897114156)
value(temp)=1 value(u)=0 value(v)=0 value(psi)=0   line to (-0.2197378127,-
0.4221123881)
value(temp)=1 value(u)=0 value(v)=0 value(psi)=0    line to (-0.2113091361,-
0.4531538783)
value(temp)=1 value(u)=0 value(v)=0 value(psi)=0    line to (-0.1992673569,-
0.4810739247)
value(temp)=1 value(u)=0 value(v)=0 value(psi)=0    line to (-0.1835288870,-
0.5042414367)
value(temp)=1 value(u)=0 value(v)=0 value(psi)=0    line to (-0.1643635707,-
0.5212946040)
value(temp)=1 value(u)=0 value(v)=0 value(psi)=0    line to (-0.1423504874,-
0.5312592011)
value(temp)=1 value(u)=0 value(v)=0 value(psi)=0    line to (-0.1183043011,-
0.5336361567)
value(temp)=1 value(u)=0 value(v)=0 value(psi)=0    line to (-0.09318006729,-
0.5284503467)
value(temp)=1 value(u)=0 value(v)=0 value(psi)=0   line to (-0.067966396087,-
0.5162559348)
value(temp)=1 value(u)=0 value(v)=0 value(psi)=0    line to (-0.04357788483,-
0.4980973606)
value(temp)=1 value(u)=0 value(v)=0 value(psi)=0    line to (-0.02075768971,-
0.4754289847)
value(temp)=1 value(u)=0 value(v)=0 value(psi)=0    line to (0,-0.4500000152)
value(temp)=1 value(u)=0 value(v)=0 value(psi)=0    line to (0.01849976125,-
0.4237144440)
value(temp)=1 value(u)=0 value(v)=0 value(psi)=0    line to (0.03486228794,-
0.3984778938)
value(temp)=1 value(u)=0 value(v)=0 value(psi)=0    line to (0.04950718250,-
0.3760444614)
value(temp)=1 value(u)=0 value(v)=0 value(psi)=0    line to (0.06310329855,-
0.3578766454)
value(temp)=1 value(u)=0 value(v)=0 value(psi)=0    line to (0.07649135661,-
0.3450302566)
value(temp)=1 value(u)=0 value(v)=0 value(psi)=0    line to (0.09058665662,-
0.3380740417)
value(temp)=1 value(u)=0 value(v)=0 value(psi)=0    line to (0.1062716494,-
0.3370506432)
value(temp)=1 value(u)=0 value(v)=0 value(psi)=0    line to (0.1242892389,-
0.3414819070)
value(temp)=1 value(u)=0 value(v)=0 value(psi)=0    line to (0.1451477251,-
0.3504176342)
```

```
value(temp)=1 value(u)=0 value(v)=0 value(psi)=0   line to (0.1690472882,-
0.3625231062)
value(temp)=1 value(u)=0 value(v)=0 value(psi)=0    line to (0.1958359241,-
0.3761973364)
value(temp)=1 value(u)=0 value(v)=0 value(psi)=0    line to (0.2249999799,-
0.3897114230)
value(temp)=1 value(u)=0 value(v)=0 value(psi)=0   line to (0.2556911401,-
0.4013547208)
value(temp)=1 value(u)=0 value(v)=0 value(psi)=0    line to (0.2867881970,-
0.4095760178)
value(temp)=1 value(u)=0 value(v)=0 value(psi)=0    line to (0.3169885565,-
0.4131075552)
value(temp)=1 value(u)=0 value(v)=0 value(psi)=0    line to (0.3449214462,-
0.4110613978)
value(temp)=1 value(u)=0 value(v)=0 value(psi)=0    line to (0.3692725808,-
0.4029903313)
value(temp)=1 value(u)=0 value(v)=0 value(psi)=0    line to (0.3889087176,-
0.3889087416)
value(temp)=1 value(u)=0 value(v)=0 value(psi)=0    line to (0.4029903158,-
0.3692726122)
value(temp)=1 value(u)=0 value(v)=0 value(psi)=0    line to (0.4110613906,-
0.3449214832)
value(temp)=1 value(u)=0 value(v)=0 value(psi)=0    line to (0.4131075562,-
0.3169885976)
value(temp)=1 value(u)=0 value(v)=0 value(psi)=0    line to (0.4095760268,-
0.2867882409)
value(temp)=1 value(u)=0 value(v)=0 value(psi)=0    line to (0.4013547348,-
0.2556911838)
value(temp)=1 value(u)=0 value(v)=0 value(psi)=0    line to (0.3897114417,-
0.2250000226)
value(temp)=1 value(u)=0 value(v)=0 value(psi)=0    line to (0.3761973557,-
0.1958359636)
value(temp)=1 value(u)=0 value(v)=0 value(psi)=0    line to (0.3625231245,-
0.1690473239)
value(temp)=1 value(u)=0 value(v)=0 value(psi)=0    line to (0.3504176498,-
0.1451477568)
value(temp)=1 value(u)=0 value(v)=0 value(psi)=0    line to (0.3414819164,-
0.1242892662)
value(temp)=1 value(u)=0 value(v)=0 value(psi)=0    line to (0.3370506460,-
0.1062716731)
value(temp)=1 value(u)=0 value(v)=0 value(psi)=0    line to (0.3380740361,-
0.09058667740)
value(temp)=1 value(u)=0 value(v)=0 value(psi)=0    line to (0.3450302426,-
0.07649137580)
```

```
value(temp)=1 value(u)=0 value(v)=0 value(psi)=0    line to (0.3578766230,-0.06310331730)
value(temp)=1 value(u)=0 value(v)=0 value(psi)=0    line to (0.3760444328,-0.04950720226)
value(temp)=1 value(u)=0 value(v)=0 value(psi)=0    line to (0.3984778593,-0.03486230964)
value(temp)=1 value(u)=0 value(v)=0 value(psi)=0    line to (0.4237144076,-0.01849978579)
value(temp)=1 value(u)=0 value(v)=0 value(psi)=0   line to (0.4499999787,0)
value(temp)=1 value(u)=0 value(v)=0 value(psi)=0    line to close
start  'outer' (0,-1.2)
value(temp)=0 value(u)=0 value(v)=0 value(psi)=0
ARC(CENTER=0,0) ANGLE=360 TO CLOSE
PLOTS
grid(x,y)
contour(temp)
contour(temp) as "Temp"painted
contour(psi)
contour(psi) as "stream function" painted
contour(u) as "u" painted
contour(v) as "v" painted
tecplot(temp)
tecplot(u)
tecplot(v)
tecplot(psi)
elevation(u) from(0,-1.20334)to  (0,1.196) as "U"points = 100 export format "#d#b#1" file="U.txt"
elevation(v) from  (0,-1.20334)to  (0,1.196) as "V"points = 100 export format "#d#b#1" file="V.txt"
elevation (temp) from  (0,-1.20334)to  (0,1.196) as "T"points = 100 export format "#d#b#1" file="T.txt"
elevation (normal (flux)) on "inner"as "inner-Nusselt" points = 100 export format "#d#b#1" file="inner-Nusselt" !report(t_star)
elevation (normal (flux)) on "outer" as "outer-Nusselt" points = 100 export format "#d#b#1" file="outer-Nusselt" !report(t_star)
END
```

1.5 Sample 5

```
TITLE
'Cavity Sample Code 10'
SELECT
regrid=on
```

```
errlim=0.0001
painted
smoothinit
Spectral_colors
VARIABLES
psi
thetas
thetanf
!==========================================================================
DEFINITIONS
epsilon=0.5
NHs=1000
Ha=0
Ra=100
delta=0.01609862219
gama=pi/6
A2=1.005663230
A3=3.381165919
A4=1.119229944
A5=1.107420884
A6=0.9411764821
 flux = -A4*grad(Thetanf)
!==========================================================================
EQUATIONS
psi: dxx(psi)+dyy(psi)=-(A6/A5)*Ha*(dyy(psi)*sin(gama)*sin(gama)+dxx(psi)*cos(
gama)*cos(gama)+dx(dy(psi))*sin(2*gama))-((A3*A2)/(A4*A5))*Ra*dx(thetanf)
thetanf: epsilon*(div(grad(thetanf)))+NHs*(thetas-thetanf)=-dy(thetanf)*dx(psi
)+dy(psi)*dx(thetanf)
thetas:   epsilon*(div(grad(thetas)))+NHs*delta*(thetanf-thetas)=0
!==========================================================================
BOUNDARIES
region 1
start 'inner'(0,-1.35920656)
value(psi)=0 VALUE(Thetas)=1  value(Thetanf)=1   line to (-0.03492237,-
1.22718095)
value(psi)=0 VALUE(Thetas)=1  value(Thetanf)=1   line to (-0.03492237,-
1.14355409)
value(psi)=0 VALUE(Thetas)=1  value(Thetanf)=1   line to (-0.09617891,-
1.14355409)
value(psi)=0 VALUE(Thetas)=1  value(Thetanf)=1   line to (-0.13053571,-
1.0365436)
value(psi)=0 VALUE(Thetas)=1  value(Thetanf)=1   line to (-0.18276194,-
1.04622154)
value(psi)=0 VALUE(Thetas)=1  value(Thetanf)=1   line to (-0.04078158,-
```

```
0.91706466)
value(psi)=0 VALUE(Thetas)=1  value(Thetanf)=1   line to (-0.1448325,-
0.95619158)
value(psi)=0 VALUE(Thetas)=1  value(Thetanf)=1   line to (-0.04078158,-
0.8503311)
value(psi)=0 VALUE(Thetas)=1  value(Thetanf)=1   line to (-0.12561513,-
0.87101407)
value(psi)=0 VALUE(Thetas)=1  value(Thetanf)=1   line to (-0.05125659,-0.75807)
value(psi)=0 VALUE(Thetas)=1  value(Thetanf)=1   line to (-0.08843586,-
0.76713455)
value(psi)=0 VALUE(Thetas)=1  value(Thetanf)=1   line to (-0.09636985,-
0.7910336)
value(psi)=0 VALUE(Thetas)=1  value(Thetanf)=1   line to (-0.12996232,-
0.79422735)
value(psi)=0 VALUE(Thetas)=1  value(Thetanf)=1   line to (-0.12996232,-
0.81902953)
value(psi)=0 VALUE(Thetas)=1  value(Thetanf)=1   line to (-0.17165121,-
0.82919355)
value(psi)=0 VALUE(Thetas)=1  value(Thetanf)=1   line to (-0.18684765,-
0.90263043)
value(psi)=0 VALUE(Thetas)=1  value(Thetanf)=1   line to (-0.2165302,-
0.87345527)
value(psi)=0 VALUE(Thetas)=1  value(Thetanf)=1   line to (-0.24301343,-
0.97281425)
value(psi)=0 VALUE(Thetas)=1  value(Thetanf)=1   line to (-0.2746183,-
0.92870708)
value(psi)=0 VALUE(Thetas)=1  value(Thetanf)=1   line to (-0.30425434,-
0.94994269)
value(psi)=0 VALUE(Thetas)=1  value(Thetanf)=1   line to (-0.31843562,-
0.89177655)
value(psi)=0 VALUE(Thetas)=1  value(Thetanf)=1   line to (-0.36098375,-
0.92496424)
value(psi)=0 VALUE(Thetas)=1  value(Thetanf)=1   line to (-0.40246424,-
0.92496424)
value(psi)=0 VALUE(Thetas)=1  value(Thetanf)=1   line to (-0.40246424,-
0.8888193)
value(psi)=0 VALUE(Thetas)=1  value(Thetanf)=1   line to (-0.35122886,-
0.86735574)
value(psi)=0 VALUE(Thetas)=1  value(Thetanf)=1   line to (-0.38893737,-
0.81473044)
value(psi)=0 VALUE(Thetas)=1  value(Thetanf)=1   line to (-0.35122886,-
0.8247813)
value(psi)=0 VALUE(Thetas)=1  value(Thetanf)=1   line to (-0.35122886,-
0.79422735)
```

Conclusion

```
value(psi)=0 VALUE(Thetas)=1  value(Thetanf)=1   line to (-0.30425434,-
0.80394788)
value(psi)=0 VALUE(Thetas)=1  value(Thetanf)=1   line to (-0.31789696,-0.75807)
value(psi)=0 VALUE(Thetas)=1  value(Thetanf)=1   line to (-0.24879945,-
0.78100894)
value(psi)=0 VALUE(Thetas)=1  value(Thetanf)=1   line to (-0.29843063,-
0.73222614)
value(psi)=0 VALUE(Thetas)=1  value(Thetanf)=1   line to (-0.30979574,-
0.69400716)
value(psi)=0 VALUE(Thetas)=1  value(Thetanf)=1   line to (-0.27180152,-
0.70186938)
value(psi)=0 VALUE(Thetas)=1  value(Thetanf)=1   line to (-0.22451885,-
0.73299868)
value(psi)=0 VALUE(Thetas)=1  value(Thetanf)=1   line to (-0.16690895,-
0.71586733)
value(psi)=0 VALUE(Thetas)=1  value(Thetanf)=1   line to (-0.17199279,-
0.6987712)
value(psi)=0 VALUE(Thetas)=1  value(Thetanf)=1   line to (-0.2266176,-
0.66280809)
value(psi)=0 VALUE(Thetas)=1  value(Thetanf)=1   line to (-0.23643767,-
0.62978482)
value(psi)=0 VALUE(Thetas)=1  value(Thetanf)=1   line to (-0.18684765,-
0.62978482)
value(psi)=0 VALUE(Thetas)=1  value(Thetanf)=1   line to (-0.121243,-0.6667641)
value(psi)=0 VALUE(Thetas)=1  value(Thetanf)=1   line to (-0.121243,-
0.57811902)
value(psi)=0 VALUE(Thetas)=1  value(Thetanf)=1   line to (-0.17199279,-
0.50729353)
value(psi)=0 VALUE(Thetas)=1  value(Thetanf)=1   line to (-0.0593776,-
0.39271982)
value(psi)=0 VALUE(Thetas)=1  value(Thetanf)=1   line to (-0.27180152,-
0.50729353)
value(psi)=0 VALUE(Thetas)=1  value(Thetanf)=1   line to (-0.30342833,-
0.48903379)
value(psi)=0 VALUE(Thetas)=1  value(Thetanf)=1   line to (-0.31041655,-
0.24778242)
value(psi)=0 VALUE(Thetas)=1  value(Thetanf)=1   line to (-0.35333269,-
0.40259689)
value(psi)=0 VALUE(Thetas)=1  value(Thetanf)=1   line to (-0.44004426,-
0.39405902)
value(psi)=0 VALUE(Thetas)=1  value(Thetanf)=1   line to (-0.51681315,-
0.43838157)
value(psi)=0 VALUE(Thetas)=1  value(Thetanf)=1   line to (-0.45198583,-
0.47670722)
```

```
value(psi)=0 VALUE(Thetas)=1  value(Thetanf)=1   line to (-0.42719082,-
0.51965344)
value(psi)=0 VALUE(Thetas)=1  value(Thetanf)=1   line to (-0.46069984,-
0.52766064)
value(psi)=0 VALUE(Thetas)=1  value(Thetanf)=1   line to (-0.51915722,-
0.49833573)
value(psi)=0 VALUE(Thetas)=1  value(Thetanf)=1   line to (-0.53650482,-
0.50248106)
value(psi)=0 VALUE(Thetas)=1  value(Thetanf)=1   line to (-0.52253605,-
0.56093837)
value(psi)=0 VALUE(Thetas)=1  value(Thetanf)=1   line to (-0.47193596,-
0.58632171)
value(psi)=0 VALUE(Thetas)=1  value(Thetanf)=1   line to (-0.44612996,-
0.61529456)
value(psi)=0 VALUE(Thetas)=1  value(Thetanf)=1   line to (-0.48491113,-
0.62456158)
value(psi)=0 VALUE(Thetas)=1  value(Thetanf)=1   line to (-0.55197386,-
0.60597111)
value(psi)=0 VALUE(Thetas)=1  value(Thetanf)=1   line to (-0.4975594,-
0.65434184)
value(psi)=0 VALUE(Thetas)=1  value(Thetanf)=1   line to (-0.54411212,-
0.66546592)
value(psi)=0 VALUE(Thetas)=1  value(Thetanf)=1   line to (-0.51220663,-
0.70128678)
value(psi)=0 VALUE(Thetas)=1  value(Thetanf)=1   line to (-0.53866713,-
0.71656376)
value(psi)=0 VALUE(Thetas)=1  value(Thetanf)=1   line to (-0.51110857,-
0.74419486)
value(psi)=0 VALUE(Thetas)=1  value(Thetanf)=1   line to (-0.57553768,-
0.73785098)
value(psi)=0 VALUE(Thetas)=1  value(Thetanf)=1   line to (-0.56850797,-
0.7929539)
value(psi)=0 VALUE(Thetas)=1  value(Thetanf)=1   line to (-0.59981041,-
0.81102638)
value(psi)=0 VALUE(Thetas)=1  value(Thetanf)=1   line to (-0.62055066,-
0.77510322)
value(psi)=0 VALUE(Thetas)=1  value(Thetanf)=1   line to (-0.61308334,-
0.72166162)
value(psi)=0 VALUE(Thetas)=1  value(Thetanf)=1   line to (-0.67054734,-
0.73846333)
value(psi)=0 VALUE(Thetas)=1  value(Thetanf)=1   line to (-0.66697477,-
0.70217996)
value(psi)=0 VALUE(Thetas)=1  value(Thetanf)=1   line to (-0.72097514,-
0.69686293)
```

```
value(psi)=0 VALUE(Thetas)=1  value(Thetanf)=1   line to (-0.64816936,-
0.62424829)
value(psi)=0 VALUE(Thetas)=1  value(Thetanf)=1   line to (-0.68827706,-
0.61313003)
value(psi)=0 VALUE(Thetas)=1  value(Thetanf)=1   line to (-0.63227708,-
0.56325108)
value(psi)=0 VALUE(Thetas)=1  value(Thetanf)=1   line to (-0.64431922,-
0.52206543)
value(psi)=0 VALUE(Thetas)=1  value(Thetanf)=1   line to (-0.6228399,-
0.50966434)
value(psi)=0 VALUE(Thetas)=1  value(Thetanf)=1   line to (-0.63687027,-
0.47897554)
value(psi)=0 VALUE(Thetas)=1  value(Thetanf)=1   line to (-0.62014008,-
0.46015497)
value(psi)=0 VALUE(Thetas)=1  value(Thetanf)=1   line to (-0.63087958,-
0.42342451)
value(psi)=0 VALUE(Thetas)=1  value(Thetanf)=1   line to (-0.69151275,-
0.54429293)
value(psi)=0 VALUE(Thetas)=1  value(Thetanf)=1   line to (-0.71601754,-
0.46048344)
value(psi)=0 VALUE(Thetas)=1  value(Thetanf)=1   line to (-0.75566995,-
0.60352442)
value(psi)=0 VALUE(Thetas)=1  value(Thetanf)=1   line to (-0.7738105,-
0.49385022)
value(psi)=0 VALUE(Thetas)=1  value(Thetanf)=1   line to (-0.81467346,-
0.68138725)
value(psi)=0 VALUE(Thetas)=1  value(Thetanf)=1   line to (-0.83240523,-
0.63131904)
value(psi)=0 VALUE(Thetas)=1  value(Thetanf)=1   line to (-0.8819745,-
0.58725536)
value(psi)=0 VALUE(Thetas)=1  value(Thetanf)=1   line to (-0.94225744,-
0.65507042)
value(psi)=0 VALUE(Thetas)=1  value(Thetanf)=1   line to (-0.97288571,-
0.6020207)
value(psi)=0 VALUE(Thetas)=1  value(Thetanf)=1   line to (-0.99952409,-
0.65512265)
value(psi)=0 VALUE(Thetas)=1  value(Thetanf)=1   line to (-1.04105297,-
0.62735857)
value(psi)=0 VALUE(Thetas)=1  value(Thetanf)=1   line to (-1.12498956,-
0.64951302)
value(psi)=0 VALUE(Thetas)=1  value(Thetanf)=1   line to (-1.06383495,-
0.58789904)
value(psi)=0 VALUE(Thetas)=1  value(Thetanf)=1   line to (-1.0671149,-
0.53805193)
```

```
value(psi)=0 VALUE(Thetas)=1  value(Thetanf)=1   line to (-1.00780807,-
0.54153339)
value(psi)=0 VALUE(Thetas)=1  value(Thetanf)=1   line to (-1.03843634,-
0.48848367)
value(psi)=0 VALUE(Thetas)=1  value(Thetanf)=1   line to (-0.94956531,-
0.47018464)
value(psi)=0 VALUE(Thetas)=1  value(Thetanf)=1   line to (-0.96294094,-
0.40522456)
value(psi)=0 VALUE(Thetas)=1  value(Thetanf)=1   line to (-0.9974354,-
0.36483429)
value(psi)=0 VALUE(Thetas)=1  value(Thetanf)=1   line to (-0.81459209,-
0.42321444)
value(psi)=0 VALUE(Thetas)=1  value(Thetanf)=1   line to (-0.90050245,-
0.35266716)
value(psi)=0 VALUE(Thetas)=1  value(Thetanf)=1   line to (-0.75679912,-
0.38984766)
value(psi)=0 VALUE(Thetas)=1  value(Thetanf)=1   line to (-0.81712788,-
0.32672114)
value(psi)=0 VALUE(Thetas)=1  value(Thetanf)=1   line to (-0.68213617,-
0.33464549)
value(psi)=0 VALUE(Thetas)=1  value(Thetanf)=1   line to (-0.70857594,-
0.30697957)
value(psi)=0 VALUE(Thetas)=1  value(Thetanf)=1   line to (-0.73324012,-
0.31205806)
value(psi)=0 VALUE(Thetas)=1  value(Thetanf)=1   line to (-0.75280222,-
0.28456301)
value(psi)=0 VALUE(Thetas)=1  value(Thetanf)=1   line to (-0.77428154,-
0.2969641)
value(psi)=0 VALUE(Thetas)=1  value(Thetanf)=1   line to (-0.80392829,-
0.26594247)
value(psi)=0 VALUE(Thetas)=1  value(Thetanf)=1   line to (-0.87512471,-
0.28950041)
value(psi)=0 VALUE(Thetas)=1  value(Thetanf)=1   line to (-0.86469956,-
0.24920698)
value(psi)=0 VALUE(Thetas)=1  value(Thetanf)=1   line to (-0.96398857,-
0.27595132)
value(psi)=0 VALUE(Thetas)=1  value(Thetanf)=1   line to (-0.94159307,-
0.22652711)
value(psi)=0 VALUE(Thetas)=1  value(Thetanf)=1   line to (-0.97480167,-
0.21147936)
value(psi)=0 VALUE(Thetas)=1  value(Thetanf)=1   line to (-0.93151896,-
0.17011494)
value(psi)=0 VALUE(Thetas)=1  value(Thetanf)=1   line to (-0.98153441,-
0.14986102)
```

```
value(psi)=0 VALUE(Thetas)=1  value(Thetanf)=1   line to (-1.00227465,-
0.11393787)
value(psi)=0 VALUE(Thetas)=1  value(Thetanf)=1   line to (-0.97097221,-
0.09586539)
value(psi)=0 VALUE(Thetas)=1  value(Thetanf)=1   line to (-0.92676654,-
0.12950476)
value(psi)=0 VALUE(Thetas)=1  value(Thetanf)=1   line to (-0.90004594,-
0.07053558)
value(psi)=0 VALUE(Thetas)=1  value(Thetanf)=1   line to (-0.88989598,-
0.10821754)
value(psi)=0 VALUE(Thetas)=1  value(Thetanf)=1   line to (-0.86343549,-
0.09294056)
value(psi)=0 VALUE(Thetas)=1  value(Thetanf)=1   line to (-0.84836645,-
0.13848196)
value(psi)=0 VALUE(Thetas)=1  value(Thetanf)=1   line to (-0.81545635,-
0.10372816)
value(psi)=0 VALUE(Thetas)=1  value(Thetanf)=1   line to (-0.8007733,-
0.17503783)
value(psi)=0 VALUE(Thetas)=1  value(Thetanf)=1   line to (-0.78334176,-
0.10766456)
value(psi)=0 VALUE(Thetas)=1  value(Thetanf)=1   line to (-0.7559257,-
0.0787126)
value(psi)=0 VALUE(Thetas)=1  value(Thetanf)=1   line to (-0.74373748,-
0.11554767)
value(psi)=0 VALUE(Thetas)=1  value(Thetanf)=1   line to (-0.74705491,-
0.17206031)
value(psi)=0 VALUE(Thetas)=1  value(Thetanf)=1   line to (-0.70341377,-
0.21338628)
value(psi)=0 VALUE(Thetas)=1  value(Thetanf)=1   line to (-0.69115001,-
0.20043547)
value(psi)=0 VALUE(Thetas)=1  value(Thetanf)=1   line to (-0.68731744,-
0.13514745)
value(psi)=0 VALUE(Thetas)=1  value(Thetanf)=1   line to (-0.66362849,-
0.11013139)
value(psi)=0 VALUE(Thetas)=1  value(Thetanf)=1   line to (-0.63883348,-
0.1530776)
value(psi)=0 VALUE(Thetas)=1  value(Thetanf)=1   line to (-0.63805615,-
0.22838254)
value(psi)=0 VALUE(Thetas)=1  value(Thetanf)=1   line to (-0.56128726,-0.18406)
value(psi)=0 VALUE(Thetas)=1  value(Thetanf)=1   line to (-0.52532548,-
0.10469664)
value(psi)=0 VALUE(Thetas)=1  value(Thetanf)=1   line to (-0.36979414,-
0.14493741)
value(psi)=0 VALUE(Thetas)=1  value(Thetanf)=1   line to (-0.57522985,-
```

```
0.01825974)
value(psi)=0 VALUE(Thetas)=1  value(Thetanf)=1   line to
(-0.57522985,0.01825974)
value(psi)=0 VALUE(Thetas)=1  value(Thetanf)=1   line to
(-0.36979414,0.14493741)
value(psi)=0 VALUE(Thetas)=1  value(Thetanf)=1   line to
(-0.52532548,0.10469664)
value(psi)=0 VALUE(Thetas)=1  value(Thetanf)=1   line to (-0.56128726,0.18406)
value(psi)=0 VALUE(Thetas)=1  value(Thetanf)=1   line to
(-0.63805615,0.22838254)
value(psi)=0 VALUE(Thetas)=1  value(Thetanf)=1   line to
(-0.63883348,0.1530776)
value(psi)=0 VALUE(Thetas)=1  value(Thetanf)=1   line to
(-0.66362849,0.11013139)
value(psi)=0 VALUE(Thetas)=1  value(Thetanf)=1   line to
(-0.68731744,0.13514745)
value(psi)=0 VALUE(Thetas)=1  value(Thetanf)=1   line to
(-0.69115001,0.20043547)
value(psi)=0 VALUE(Thetas)=1  value(Thetanf)=1   line to
(-0.70341377,0.21338628)
value(psi)=0 VALUE(Thetas)=1  value(Thetanf)=1   line to
(-0.74705491,0.17206031)
value(psi)=0 VALUE(Thetas)=1  value(Thetanf)=1   line to
(-0.74373748,0.11554767)
value(psi)=0 VALUE(Thetas)=1  value(Thetanf)=1   line to (-0.7559257,0.0787126)
value(psi)=0 VALUE(Thetas)=1  value(Thetanf)=1   line to
(-0.78334176,0.10766456)
value(psi)=0 VALUE(Thetas)=1  value(Thetanf)=1   line to
(-0.8007733,0.17503783)
value(psi)=0 VALUE(Thetas)=1  value(Thetanf)=1   line to
(-0.81545635,0.10372816)
value(psi)=0 VALUE(Thetas)=1  value(Thetanf)=1   line to
(-0.84836645,0.13848196)
value(psi)=0 VALUE(Thetas)=1  value(Thetanf)=1   line to
(-0.86343549,0.09294056)
value(psi)=0 VALUE(Thetas)=1  value(Thetanf)=1   line to
(-0.88989598,0.10821754)
value(psi)=0 VALUE(Thetas)=1  value(Thetanf)=1   line to
(-0.90004594,0.07053558)
value(psi)=0 VALUE(Thetas)=1  value(Thetanf)=1   line to
(-0.92676654,0.12950476)
value(psi)=0 VALUE(Thetas)=1  value(Thetanf)=1   line to
(-0.97097221,0.09586539)
value(psi)=0 VALUE(Thetas)=1  value(Thetanf)=1   line to
```

```
(-1.00227465,0.11393787)
value(psi)=0 VALUE(Thetas)=1  value(Thetanf)=1   line to
(-0.98153441,0.14986102)
value(psi)=0 VALUE(Thetas)=1  value(Thetanf)=1   line to
(-0.93151896,0.17011494)
value(psi)=0 VALUE(Thetas)=1  value(Thetanf)=1   line to
(-0.97480167,0.21147936)
value(psi)=0 VALUE(Thetas)=1  value(Thetanf)=1   line to
(-0.94159307,0.22652711)
value(psi)=0 VALUE(Thetas)=1  value(Thetanf)=1   line to
(-0.96398857,0.27595132)
value(psi)=0 VALUE(Thetas)=1  value(Thetanf)=1   line to
(-0.86469956,0.24920698)
value(psi)=0 VALUE(Thetas)=1  value(Thetanf)=1   line to
(-0.87512471,0.28950041)
value(psi)=0 VALUE(Thetas)=1  value(Thetanf)=1   line to
(-0.80392829,0.26594247)
value(psi)=0 VALUE(Thetas)=1  value(Thetanf)=1   line to
(-0.77428154,0.2969641)
value(psi)=0 VALUE(Thetas)=1  value(Thetanf)=1   line to
(-0.75280222,0.28456301)
value(psi)=0 VALUE(Thetas)=1  value(Thetanf)=1   line to
(-0.73324012,0.31205806)
value(psi)=0 VALUE(Thetas)=1  value(Thetanf)=1   line to
(-0.70857594,0.30697957)
value(psi)=0 VALUE(Thetas)=1  value(Thetanf)=1   line to
(-0.68213617,0.33464549)
value(psi)=0 VALUE(Thetas)=1  value(Thetanf)=1   line to
(-0.81712788,0.32672114)
value(psi)=0 VALUE(Thetas)=1  value(Thetanf)=1   line to
(-0.75679912,0.38984766)
value(psi)=0 VALUE(Thetas)=1  value(Thetanf)=1   line to
(-0.90050245,0.35266716)
value(psi)=0 VALUE(Thetas)=1  value(Thetanf)=1   line to
(-0.81459209,0.42321444)
value(psi)=0 VALUE(Thetas)=1  value(Thetanf)=1   line to
(-0.9974354,0.36483429)
value(psi)=0 VALUE(Thetas)=1  value(Thetanf)=1   line to
(-0.96294094,0.40522456)
value(psi)=0 VALUE(Thetas)=1  value(Thetanf)=1   line to
(-0.94956531,0.47018464)
value(psi)=0 VALUE(Thetas)=1  value(Thetanf)=1   line to
(-1.03843634,0.48848367)
value(psi)=0 VALUE(Thetas)=1  value(Thetanf)=1   line to
```

```
(-1.00780807,0.54153339)
value(psi)=0 VALUE(Thetas)=1  value(Thetanf)=1   line to
(-1.0671149,0.53805193)
value(psi)=0 VALUE(Thetas)=1  value(Thetanf)=1   line to
(-1.06383495,0.58789904)
value(psi)=0 VALUE(Thetas)=1  value(Thetanf)=1   line to
(-1.12498956,0.64951302)
value(psi)=0 VALUE(Thetas)=1  value(Thetanf)=1   line to
(-1.04105297,0.62735857)
value(psi)=0 VALUE(Thetas)=1  value(Thetanf)=1   line to
(-0.99952409,0.65512265)
value(psi)=0 VALUE(Thetas)=1  value(Thetanf)=1   line to
(-0.97288571,0.6020207)
value(psi)=0 VALUE(Thetas)=1  value(Thetanf)=1   line to
(-0.94225744,0.65507042)
value(psi)=0 VALUE(Thetas)=1  value(Thetanf)=1   line to
(-0.8819745,0.58725536)
value(psi)=0 VALUE(Thetas)=1  value(Thetanf)=1   line to
(-0.83240523,0.63131904)
value(psi)=0 VALUE(Thetas)=1  value(Thetanf)=1   line to
(-0.81467346,0.68138725)
value(psi)=0 VALUE(Thetas)=1  value(Thetanf)=1   line to
(-0.7738105,0.49385022)
value(psi)=0 VALUE(Thetas)=1  value(Thetanf)=1   line to
(-0.75566995,0.60352442)
value(psi)=0 VALUE(Thetas)=1  value(Thetanf)=1   line to
(-0.71601754,0.46048344)
value(psi)=0 VALUE(Thetas)=1  value(Thetanf)=1   line to
(-0.69151275,0.54429293)
value(psi)=0 VALUE(Thetas)=1  value(Thetanf)=1   line to
(-0.63087958,0.42342451)
value(psi)=0 VALUE(Thetas)=1  value(Thetanf)=1   line to
(-0.62014008,0.46015497)
value(psi)=0 VALUE(Thetas)=1  value(Thetanf)=1   line to
(-0.63687027,0.47897554)
value(psi)=0 VALUE(Thetas)=1  value(Thetanf)=1   line to
(-0.6228399,0.50966434)
value(psi)=0 VALUE(Thetas)=1  value(Thetanf)=1   line to
(-0.64431922,0.52206543)
value(psi)=0 VALUE(Thetas)=1  value(Thetanf)=1   line to
(-0.63227708,0.56325108)
value(psi)=0 VALUE(Thetas)=1  value(Thetanf)=1   line to
(-0.68827706,0.61313003)
value(psi)=0 VALUE(Thetas)=1  value(Thetanf)=1   line to
```

```
(-0.64816936,0.62424829)
value(psi)=0 VALUE(Thetas)=1  value(Thetanf)=1   line to
(-0.72097514,0.69686293)
value(psi)=0 VALUE(Thetas)=1  value(Thetanf)=1   line to
(-0.66697477,0.70217996)
value(psi)=0 VALUE(Thetas)=1  value(Thetanf)=1   line to
(-0.67054734,0.73846333)
value(psi)=0 VALUE(Thetas)=1  value(Thetanf)=1   line to
(-0.61308334,0.72166162)
value(psi)=0 VALUE(Thetas)=1  value(Thetanf)=1   line to
(-0.62055066,0.77510322)
value(psi)=0 VALUE(Thetas)=1  value(Thetanf)=1   line to
(-0.59981041,0.81102638)
value(psi)=0 VALUE(Thetas)=1  value(Thetanf)=1   line to
(-0.56850797,0.7929539)
value(psi)=0 VALUE(Thetas)=1  value(Thetanf)=1   line to
(-0.57553768,0.73785098)
value(psi)=0 VALUE(Thetas)=1  value(Thetanf)=1   line to
(-0.51110857,0.74419486)
value(psi)=0 VALUE(Thetas)=1  value(Thetanf)=1   line to
(-0.53866713,0.71656376)
value(psi)=0 VALUE(Thetas)=1  value(Thetanf)=1   line to
(-0.51220663,0.70128679)
value(psi)=0 VALUE(Thetas)=1  value(Thetanf)=1   line to
(-0.54411212,0.66546592)
value(psi)=0 VALUE(Thetas)=1  value(Thetanf)=1   line to
(-0.4975594,0.65434184)
value(psi)=0 VALUE(Thetas)=1  value(Thetanf)=1   line to
(-0.55197386,0.60597111)
value(psi)=0 VALUE(Thetas)=1  value(Thetanf)=1   line to
(-0.48491113,0.62456158)
value(psi)=0 VALUE(Thetas)=1  value(Thetanf)=1   line to
(-0.44612996,0.61529456)
value(psi)=0 VALUE(Thetas)=1  value(Thetanf)=1   line to
(-0.47193596,0.58632171)
value(psi)=0 VALUE(Thetas)=1  value(Thetanf)=1   line to
(-0.52253605,0.56093837)
value(psi)=0 VALUE(Thetas)=1  value(Thetanf)=1   line to
(-0.53650482,0.50248106)
value(psi)=0 VALUE(Thetas)=1  value(Thetanf)=1   line to
(-0.51915722,0.49833573)
value(psi)=0 VALUE(Thetas)=1  value(Thetanf)=1   line to
(-0.46069984,0.52766064)
value(psi)=0 VALUE(Thetas)=1  value(Thetanf)=1   line to
```

```
(-0.42719082,0.51965344)
value(psi)=0 VALUE(Thetas)=1  value(Thetanf)=1   line to
(-0.45198583,0.47670722)
value(psi)=0 VALUE(Thetas)=1  value(Thetanf)=1   line to
(-0.51681315,0.43838157)
value(psi)=0 VALUE(Thetas)=1  value(Thetanf)=1   line to
(-0.44004426,0.39405902)
value(psi)=0 VALUE(Thetas)=1  value(Thetanf)=1   line to
(-0.35333269,0.40259689)
value(psi)=0 VALUE(Thetas)=1  value(Thetanf)=1   line to
(-0.31041655,0.24778242)
value(psi)=0 VALUE(Thetas)=1  value(Thetanf)=1   line to
(-0.30342833,0.48903379)
value(psi)=0 VALUE(Thetas)=1  value(Thetanf)=1   line to
(-0.27180152,0.50729353)
value(psi)=0 VALUE(Thetas)=1  value(Thetanf)=1   line to
(-0.0593776,0.39271982)
value(psi)=0 VALUE(Thetas)=1  value(Thetanf)=1   line to
(-0.17199279,0.50729353)
value(psi)=0 VALUE(Thetas)=1  value(Thetanf)=1   line to (-0.121243,0.57811902)
value(psi)=0 VALUE(Thetas)=1  value(Thetanf)=1   line to (-0.121243,0.6667641)
value(psi)=0 VALUE(Thetas)=1  value(Thetanf)=1   line to
(-0.18684765,0.62978482)
value(psi)=0 VALUE(Thetas)=1  value(Thetanf)=1   line to
(-0.23643767,0.62978482)
value(psi)=0 VALUE(Thetas)=1  value(Thetanf)=1   line to
(-0.2266176,0.66280809)
value(psi)=0 VALUE(Thetas)=1  value(Thetanf)=1   line to
(-0.17199279,0.6987712)
value(psi)=0 VALUE(Thetas)=1  value(Thetanf)=1   line to
(-0.16690895,0.71586733)
value(psi)=0 VALUE(Thetas)=1  value(Thetanf)=1   line to
(-0.22451885,0.73299868)
value(psi)=0 VALUE(Thetas)=1  value(Thetanf)=1   line to
(-0.27180152,0.70186938)
value(psi)=0 VALUE(Thetas)=1  value(Thetanf)=1   line to
(-0.30979574,0.69400716)
value(psi)=0 VALUE(Thetas)=1  value(Thetanf)=1   line to
(-0.29843063,0.73222615)
value(psi)=0 VALUE(Thetas)=1  value(Thetanf)=1   line to
(-0.24879945,0.78100894)
value(psi)=0 VALUE(Thetas)=1  value(Thetanf)=1   line to (-0.31789696,0.75807)
value(psi)=0 VALUE(Thetas)=1  value(Thetanf)=1   line to
(-0.30425434,0.80394788)
```

```
value(psi)=0 VALUE(Thetas)=1  value(Thetanf)=1   line to
(-0.35122886,0.79422735)
value(psi)=0 VALUE(Thetas)=1  value(Thetanf)=1   line to
(-0.35122886,0.8247813)
value(psi)=0 VALUE(Thetas)=1  value(Thetanf)=1   line to
(-0.38893737,0.81473044)
value(psi)=0 VALUE(Thetas)=1  value(Thetanf)=1   line to
(-0.35122886,0.86735575)
value(psi)=0 VALUE(Thetas)=1  value(Thetanf)=1   line to
(-0.40246424,0.8888193)
value(psi)=0 VALUE(Thetas)=1  value(Thetanf)=1   line to
(-0.40246424,0.92496424)
value(psi)=0 VALUE(Thetas)=1  value(Thetanf)=1   line to
(-0.36098375,0.92496424)
value(psi)=0 VALUE(Thetas)=1  value(Thetanf)=1   line to
(-0.31843562,0.89177655)
value(psi)=0 VALUE(Thetas)=1  value(Thetanf)=1   line to
(-0.30425434,0.94994269)
value(psi)=0 VALUE(Thetas)=1  value(Thetanf)=1   line to
(-0.2746183,0.92870708)
value(psi)=0 VALUE(Thetas)=1  value(Thetanf)=1   line to
(-0.24301343,0.97281425)
value(psi)=0 VALUE(Thetas)=1  value(Thetanf)=1   line to
(-0.2165302,0.87345527)
value(psi)=0 VALUE(Thetas)=1  value(Thetanf)=1   line to
(-0.18684765,0.90263043)
value(psi)=0 VALUE(Thetas)=1  value(Thetanf)=1   line to
(-0.17165121,0.82919356)
value(psi)=0 VALUE(Thetas)=1  value(Thetanf)=1   line to
(-0.12996232,0.81902953)
value(psi)=0 VALUE(Thetas)=1  value(Thetanf)=1   line to
(-0.12996232,0.79422735)
value(psi)=0 VALUE(Thetas)=1  value(Thetanf)=1   line to
(-0.09636985,0.7910336)
value(psi)=0 VALUE(Thetas)=1  value(Thetanf)=1   line to
(-0.08843586,0.76713455)
value(psi)=0 VALUE(Thetas)=1  value(Thetanf)=1   line to (-0.05125659,0.75807)
value(psi)=0 VALUE(Thetas)=1  value(Thetanf)=1   line to
(-0.12561513,0.87101407)
value(psi)=0 VALUE(Thetas)=1  value(Thetanf)=1   line to
(-0.04078158,0.8503311)
value(psi)=0 VALUE(Thetas)=1  value(Thetanf)=1   line to
(-0.1448325,0.95619158)
value(psi)=0 VALUE(Thetas)=1  value(Thetanf)=1   line to
```

```
(-0.04078158,0.91706466)
value(psi)=0 VALUE(Thetas)=1  value(Thetanf)=1   line to
(-0.18276194,1.04622154)
value(psi)=0 VALUE(Thetas)=1  value(Thetanf)=1   line to
(-0.13053571,1.0365436)
value(psi)=0 VALUE(Thetas)=1  value(Thetanf)=1   line to
(-0.09617891,1.14355409)
value(psi)=0 VALUE(Thetas)=1  value(Thetanf)=1   line to
(-0.03492237,1.14355409)
value(psi)=0 VALUE(Thetas)=1  value(Thetanf)=1   line to
(-0.03492237,1.22718095)
value(psi)=0 VALUE(Thetas)=1  value(Thetanf)=1   line to (0.00,1.35920656)
value(psi)=0 VALUE(Thetas)=1  value(Thetanf)=1   line to (0.00,1.35920656)
value(psi)=0 VALUE(Thetas)=1  value(Thetanf)=1   line to
(0.03492237,1.22718095)
value(psi)=0 VALUE(Thetas)=1  value(Thetanf)=1   line to
(0.03492237,1.14355409)
value(psi)=0 VALUE(Thetas)=1  value(Thetanf)=1   line to
(0.09617891,1.14355409)
value(psi)=0 VALUE(Thetas)=1  value(Thetanf)=1   line to (0.13053571,1.0365436)
value(psi)=0 VALUE(Thetas)=1  value(Thetanf)=1   line to
(0.18276194,1.04622154)
value(psi)=0 VALUE(Thetas)=1  value(Thetanf)=1   line to
(0.04078158,0.91706466)
value(psi)=0 VALUE(Thetas)=1  value(Thetanf)=1   line to (0.1448325,0.95619158)
value(psi)=0 VALUE(Thetas)=1  value(Thetanf)=1   line to (0.04078158,0.8503311)
value(psi)=0 VALUE(Thetas)=1  value(Thetanf)=1   line to
(0.12561513,0.87101407)
value(psi)=0 VALUE(Thetas)=1  value(Thetanf)=1   line to (0.05125659,0.75807)
value(psi)=0 VALUE(Thetas)=1  value(Thetanf)=1   line to
(0.08843586,0.76713455)
value(psi)=0 VALUE(Thetas)=1  value(Thetanf)=1   line to (0.09636985,0.7910336)
value(psi)=0 VALUE(Thetas)=1  value(Thetanf)=1   line to
(0.12996232,0.79422735)
value(psi)=0 VALUE(Thetas)=1  value(Thetanf)=1   line to
(0.12996232,0.81902953)
value(psi)=0 VALUE(Thetas)=1  value(Thetanf)=1   line to
(0.17165121,0.82919356)
value(psi)=0 VALUE(Thetas)=1  value(Thetanf)=1   line to
(0.18684765,0.90263043)
value(psi)=0 VALUE(Thetas)=1  value(Thetanf)=1   line to (0.2165302,0.87345527)
value(psi)=0 VALUE(Thetas)=1  value(Thetanf)=1   line to
(0.24301343,0.97281425)
value(psi)=0 VALUE(Thetas)=1  value(Thetanf)=1   line to (0.2746183,0.92870708)
```

```
value(psi)=0 VALUE(Thetas)=1  value(Thetanf)=1   line to
(0.30425434,0.94994269)
value(psi)=0 VALUE(Thetas)=1  value(Thetanf)=1   line to
(0.31843562,0.89177655)
value(psi)=0 VALUE(Thetas)=1  value(Thetanf)=1   line to
(0.36098375,0.92496424)
value(psi)=0 VALUE(Thetas)=1  value(Thetanf)=1   line to
(0.40246424,0.92496424)
value(psi)=0 VALUE(Thetas)=1  value(Thetanf)=1   line to (0.40246424,0.8888193)
value(psi)=0 VALUE(Thetas)=1  value(Thetanf)=1   line to
(0.35122886,0.86735575)
value(psi)=0 VALUE(Thetas)=1  value(Thetanf)=1   line to
(0.38893737,0.81473044)
value(psi)=0 VALUE(Thetas)=1  value(Thetanf)=1   line to (0.35122886,0.8247813)
value(psi)=0 VALUE(Thetas)=1  value(Thetanf)=1   line to
(0.35122886,0.79422735)
value(psi)=0 VALUE(Thetas)=1  value(Thetanf)=1   line to
(0.30425434,0.80394788)
value(psi)=0 VALUE(Thetas)=1  value(Thetanf)=1   line to (0.31789696,0.75807)
value(psi)=0 VALUE(Thetas)=1  value(Thetanf)=1   line to
(0.24879945,0.78100894)
value(psi)=0 VALUE(Thetas)=1  value(Thetanf)=1   line to
(0.29843063,0.73222615)
value(psi)=0 VALUE(Thetas)=1  value(Thetanf)=1   line to
(0.30979574,0.69400716)
value(psi)=0 VALUE(Thetas)=1  value(Thetanf)=1   line to
(0.27180152,0.70186938)
value(psi)=0 VALUE(Thetas)=1  value(Thetanf)=1   line to
(0.22451885,0.73299868)
value(psi)=0 VALUE(Thetas)=1  value(Thetanf)=1   line to
(0.16690895,0.71586733)
value(psi)=0 VALUE(Thetas)=1  value(Thetanf)=1   line to (0.17199279,0.6987712)
value(psi)=0 VALUE(Thetas)=1  value(Thetanf)=1   line to (0.2266176,0.66280809)
value(psi)=0 VALUE(Thetas)=1  value(Thetanf)=1   line to
(0.23643767,0.62978482)
value(psi)=0 VALUE(Thetas)=1  value(Thetanf)=1   line to
(0.18684765,0.62978482)
value(psi)=0 VALUE(Thetas)=1  value(Thetanf)=1   line to (0.121243,0.6667641)
value(psi)=0 VALUE(Thetas)=1  value(Thetanf)=1   line to (0.121243,0.57811902)
value(psi)=0 VALUE(Thetas)=1  value(Thetanf)=1   line to
(0.17199279,0.50729353)
value(psi)=0 VALUE(Thetas)=1  value(Thetanf)=1   line to (0.0593776,0.39271982)
value(psi)=0 VALUE(Thetas)=1  value(Thetanf)=1   line to
(0.27180152,0.50729353)
```

```
value(psi)=0 VALUE(Thetas)=1  value(Thetanf)=1   line to
(0.30342833,0.48903379)
value(psi)=0 VALUE(Thetas)=1  value(Thetanf)=1   line to
(0.31041655,0.24778242)
value(psi)=0 VALUE(Thetas)=1  value(Thetanf)=1   line to
(0.35333269,0.40259689)
value(psi)=0 VALUE(Thetas)=1  value(Thetanf)=1   line to
(0.44004426,0.39405902)
value(psi)=0 VALUE(Thetas)=1  value(Thetanf)=1   line to
(0.51681315,0.43838157)
value(psi)=0 VALUE(Thetas)=1  value(Thetanf)=1   line to
(0.45198583,0.47670722)
value(psi)=0 VALUE(Thetas)=1  value(Thetanf)=1   line to
(0.42719082,0.51965344)
value(psi)=0 VALUE(Thetas)=1  value(Thetanf)=1   line to
(0.46069984,0.52766064)
value(psi)=0 VALUE(Thetas)=1  value(Thetanf)=1   line to
(0.51915722,0.49833573)
value(psi)=0 VALUE(Thetas)=1  value(Thetanf)=1   line to
(0.53650482,0.50248106)
value(psi)=0 VALUE(Thetas)=1  value(Thetanf)=1   line to
(0.52253605,0.56093837)
value(psi)=0 VALUE(Thetas)=1  value(Thetanf)=1   line to
(0.47193596,0.58632171)
value(psi)=0 VALUE(Thetas)=1  value(Thetanf)=1   line to
(0.44612996,0.61529456)
value(psi)=0 VALUE(Thetas)=1  value(Thetanf)=1   line to
(0.48491113,0.62456158)
value(psi)=0 VALUE(Thetas)=1  value(Thetanf)=1   line to
(0.55197386,0.60597111)
value(psi)=0 VALUE(Thetas)=1  value(Thetanf)=1   line to (0.4975594,0.65434184)
value(psi)=0 VALUE(Thetas)=1  value(Thetanf)=1   line to
(0.54411212,0.66546592)
value(psi)=0 VALUE(Thetas)=1  value(Thetanf)=1   line to
(0.53866713,0.71656376)
value(psi)=0 VALUE(Thetas)=1  value(Thetanf)=1   line to
(0.51110857,0.74419486)
value(psi)=0 VALUE(Thetas)=1  value(Thetanf)=1   line to
(0.57553768,0.73785098)
value(psi)=0 VALUE(Thetas)=1  value(Thetanf)=1   line to (0.56850797,0.7929539)
value(psi)=0 VALUE(Thetas)=1  value(Thetanf)=1   line to
(0.59981041,0.81102638)
value(psi)=0 VALUE(Thetas)=1  value(Thetanf)=1   line to
(0.62055066,0.77510322)
```

```
value(psi)=0 VALUE(Thetas)=1  value(Thetanf)=1   line to
(0.61308334,0.72166162)
value(psi)=0 VALUE(Thetas)=1  value(Thetanf)=1   line to
(0.67054734,0.73846333)
value(psi)=0 VALUE(Thetas)=1  value(Thetanf)=1   line to
(0.66697477,0.70217996)
value(psi)=0 VALUE(Thetas)=1  value(Thetanf)=1   line to
(0.72097514,0.69686293)
value(psi)=0 VALUE(Thetas)=1  value(Thetanf)=1   line to
(0.64816936,0.62424829)
value(psi)=0 VALUE(Thetas)=1  value(Thetanf)=1   line to
(0.68827706,0.61313003)
value(psi)=0 VALUE(Thetas)=1  value(Thetanf)=1   line to
(0.63227708,0.56325108)
value(psi)=0 VALUE(Thetas)=1  value(Thetanf)=1   line to
(0.64431922,0.52206543)
value(psi)=0 VALUE(Thetas)=1  value(Thetanf)=1   line to (0.6228399,0.50966434)
value(psi)=0 VALUE(Thetas)=1  value(Thetanf)=1   line to
(0.63687027,0.47897554)
value(psi)=0 VALUE(Thetas)=1  value(Thetanf)=1   line to
(0.62014008,0.46015497)
value(psi)=0 VALUE(Thetas)=1  value(Thetanf)=1   line to
(0.63087958,0.42342451)
value(psi)=0 VALUE(Thetas)=1  value(Thetanf)=1   line to
(0.69151275,0.54429293)
value(psi)=0 VALUE(Thetas)=1  value(Thetanf)=1   line to
(0.71601754,0.46048344)
value(psi)=0 VALUE(Thetas)=1  value(Thetanf)=1   line to
(0.75566995,0.60352442)
value(psi)=0 VALUE(Thetas)=1  value(Thetanf)=1   line to (0.7738105,0.49385022)
value(psi)=0 VALUE(Thetas)=1  value(Thetanf)=1   line to
(0.81467346,0.68138725)
value(psi)=0 VALUE(Thetas)=1  value(Thetanf)=1   line to
(0.83240523,0.63131904)
value(psi)=0 VALUE(Thetas)=1  value(Thetanf)=1   line to (0.8819745,0.58725536)
value(psi)=0 VALUE(Thetas)=1  value(Thetanf)=1   line to
(0.94225744,0.65507042)
value(psi)=0 VALUE(Thetas)=1  value(Thetanf)=1   line to (0.97288571,0.6020207)
value(psi)=0 VALUE(Thetas)=1  value(Thetanf)=1   line to
(0.99952409,0.65512265)
value(psi)=0 VALUE(Thetas)=1  value(Thetanf)=1   line to
(1.04105297,0.62735857)
value(psi)=0 VALUE(Thetas)=1  value(Thetanf)=1   line to
(1.12498956,0.64951302)
```

```
value(psi)=0 VALUE(Thetas)=1  value(Thetanf)=1   line to
(1.06383495,0.58789904)
value(psi)=0 VALUE(Thetas)=1  value(Thetanf)=1   line to (1.0671149,0.53805193)
value(psi)=0 VALUE(Thetas)=1  value(Thetanf)=1   line to
(1.00780807,0.54153339)
value(psi)=0 VALUE(Thetas)=1  value(Thetanf)=1   line to
(1.03843634,0.48848367)
value(psi)=0 VALUE(Thetas)=1  value(Thetanf)=1   line to
(0.94956531,0.47018464)
value(psi)=0 VALUE(Thetas)=1  value(Thetanf)=1   line to
(0.96294094,0.40522456)
value(psi)=0 VALUE(Thetas)=1  value(Thetanf)=1   line to (0.9974354,0.36483429)
value(psi)=0 VALUE(Thetas)=1  value(Thetanf)=1   line to
(0.81459209,0.42321444)
value(psi)=0 VALUE(Thetas)=1  value(Thetanf)=1   line to
(0.90050245,0.35266716)
value(psi)=0 VALUE(Thetas)=1  value(Thetanf)=1   line to
(0.75679912,0.38984766)
value(psi)=0 VALUE(Thetas)=1  value(Thetanf)=1   line to
(0.81712788,0.32672114)
value(psi)=0 VALUE(Thetas)=1  value(Thetanf)=1   line to
(0.68213617,0.33464549)
value(psi)=0 VALUE(Thetas)=1  value(Thetanf)=1   line to
(0.70857594,0.30697957)
value(psi)=0 VALUE(Thetas)=1  value(Thetanf)=1   line to
(0.73324012,0.31205806)
value(psi)=0 VALUE(Thetas)=1  value(Thetanf)=1   line to
(0.75280222,0.28456301)
value(psi)=0 VALUE(Thetas)=1  value(Thetanf)=1   line to (0.77428154,0.2969641)
value(psi)=0 VALUE(Thetas)=1  value(Thetanf)=1   line to
(0.80392829,0.26594247)
value(psi)=0 VALUE(Thetas)=1  value(Thetanf)=1   line to
(0.87512471,0.28950041)
value(psi)=0 VALUE(Thetas)=1  value(Thetanf)=1   line to
(0.86469956,0.24920698)
value(psi)=0 VALUE(Thetas)=1  value(Thetanf)=1   line to
(0.96398857,0.27595132)
value(psi)=0 VALUE(Thetas)=1  value(Thetanf)=1   line to
(0.94159307,0.22652711)
value(psi)=0 VALUE(Thetas)=1  value(Thetanf)=1   line to
(0.97480167,0.21147936)
value(psi)=0 VALUE(Thetas)=1  value(Thetanf)=1   line to
(0.93151896,0.17011494)
value(psi)=0 VALUE(Thetas)=1  value(Thetanf)=1   line to
```

```
(0.98153441,0.14986102)
value(psi)=0 VALUE(Thetas)=1  value(Thetanf)=1   line to
(1.00227465,0.11393787)
value(psi)=0 VALUE(Thetas)=1  value(Thetanf)=1   line to
(0.97097221,0.09586539)
value(psi)=0 VALUE(Thetas)=1  value(Thetanf)=1   line to
(0.92676654,0.12950476)
value(psi)=0 VALUE(Thetas)=1  value(Thetanf)=1   line to
(0.90004594,0.07053558)
value(psi)=0 VALUE(Thetas)=1  value(Thetanf)=1   line to
(0.88989598,0.10821754)
value(psi)=0 VALUE(Thetas)=1  value(Thetanf)=1   line to
(0.86343549,0.09294056)
value(psi)=0 VALUE(Thetas)=1  value(Thetanf)=1   line to
(0.84836645,0.13848196)
value(psi)=0 VALUE(Thetas)=1  value(Thetanf)=1   line to
(0.81545635,0.10372816)
value(psi)=0 VALUE(Thetas)=1  value(Thetanf)=1   line to (0.8007733,0.17503783)
value(psi)=0 VALUE(Thetas)=1  value(Thetanf)=1   line to
(0.78334176,0.10766456)
value(psi)=0 VALUE(Thetas)=1  value(Thetanf)=1   line to (0.7559257,0.0787126)
value(psi)=0 VALUE(Thetas)=1  value(Thetanf)=1   line to
(0.74373748,0.11554767)
value(psi)=0 VALUE(Thetas)=1  value(Thetanf)=1   line to
(0.74705491,0.17206031)
value(psi)=0 VALUE(Thetas)=1  value(Thetanf)=1   line to
(0.70341377,0.21338628)
value(psi)=0 VALUE(Thetas)=1  value(Thetanf)=1   line to
(0.69115001,0.20043547)
value(psi)=0 VALUE(Thetas)=1  value(Thetanf)=1   line to
(0.68731744,0.13514745)
value(psi)=0 VALUE(Thetas)=1  value(Thetanf)=1   line to
(0.66362849,0.11013139)
value(psi)=0 VALUE(Thetas)=1  value(Thetanf)=1   line to (0.63883348,0.1530776)
value(psi)=0 VALUE(Thetas)=1  value(Thetanf)=1   line to
(0.63805615,0.22838254)
value(psi)=0 VALUE(Thetas)=1  value(Thetanf)=1   line to (0.56128726,0.18406)
value(psi)=0 VALUE(Thetas)=1  value(Thetanf)=1   line to
(0.52532548,0.10469664)
value(psi)=0 VALUE(Thetas)=1  value(Thetanf)=1   line to
(0.36979414,0.14493741)
value(psi)=0 VALUE(Thetas)=1  value(Thetanf)=1   line to
(0.57522985,0.01825974)
value(psi)=0 VALUE(Thetas)=1  value(Thetanf)=1   line to (0.57522985,-
```

```
0.01825974)
value(psi)=0 VALUE(Thetas)=1  value(Thetanf)=1   line to (0.36979414,-
0.14493741)
value(psi)=0 VALUE(Thetas)=1  value(Thetanf)=1   line to (0.52532548,-
0.10469664)
value(psi)=0 VALUE(Thetas)=1  value(Thetanf)=1   line to (0.56128726,-0.18406)
value(psi)=0 VALUE(Thetas)=1  value(Thetanf)=1   line to (0.63805615,-
0.22838254)
value(psi)=0 VALUE(Thetas)=1  value(Thetanf)=1   line to (0.63883348,-
0.1530776)
value(psi)=0 VALUE(Thetas)=1  value(Thetanf)=1   line to (0.66362849,-
0.11013139)
value(psi)=0 VALUE(Thetas)=1  value(Thetanf)=1   line to (0.68731744,-
0.13514745)
value(psi)=0 VALUE(Thetas)=1  value(Thetanf)=1   line to (0.69115001,-
0.20043547)
value(psi)=0 VALUE(Thetas)=1  value(Thetanf)=1   line to (0.70341377,-
0.21338628)
value(psi)=0 VALUE(Thetas)=1  value(Thetanf)=1   line to (0.74705491,-
0.17206031)
value(psi)=0 VALUE(Thetas)=1  value(Thetanf)=1   line to (0.74373748,-
0.11554767)
value(psi)=0 VALUE(Thetas)=1  value(Thetanf)=1   line to (0.7559257,-0.0787126)
value(psi)=0 VALUE(Thetas)=1  value(Thetanf)=1   line to (0.78334176,-
0.10766456)
value(psi)=0 VALUE(Thetas)=1  value(Thetanf)=1   line to (0.8007733,-
0.17503783)
value(psi)=0 VALUE(Thetas)=1  value(Thetanf)=1   line to (0.81545635,-
0.10372816)
value(psi)=0 VALUE(Thetas)=1  value(Thetanf)=1   line to (0.84836645,-
0.13848196)
value(psi)=0 VALUE(Thetas)=1  value(Thetanf)=1   line to (0.86343549,-
0.09294056)
value(psi)=0 VALUE(Thetas)=1  value(Thetanf)=1   line to (0.88989598,-
0.10821754)
value(psi)=0 VALUE(Thetas)=1  value(Thetanf)=1   line to (0.90004594,-
0.07053558)
value(psi)=0 VALUE(Thetas)=1  value(Thetanf)=1   line to (0.92676654,-
0.12950476)
value(psi)=0 VALUE(Thetas)=1  value(Thetanf)=1   line to (0.97097221,-
0.09586539)
value(psi)=0 VALUE(Thetas)=1  value(Thetanf)=1   line to (1.00227465,-
0.11393787)
value(psi)=0 VALUE(Thetas)=1  value(Thetanf)=1   line to (0.98153441,-
```

```
0.14986102)
value(psi)=0 VALUE(Thetas)=1  value(Thetanf)=1   line to (0.93151896,-
0.17011494)
value(psi)=0 VALUE(Thetas)=1  value(Thetanf)=1   line to (0.97480167,-
0.21147936)
value(psi)=0 VALUE(Thetas)=1  value(Thetanf)=1   line to (0.94159307,-
0.22652711)
value(psi)=0 VALUE(Thetas)=1  value(Thetanf)=1   line to (0.96398857,-
0.27595132)
value(psi)=0 VALUE(Thetas)=1  value(Thetanf)=1   line to (0.86469956,-
0.24920698)
value(psi)=0 VALUE(Thetas)=1  value(Thetanf)=1   line to (0.87512471,-
0.28950041)
value(psi)=0 VALUE(Thetas)=1  value(Thetanf)=1   line to (0.80392829,-
0.26594247)
value(psi)=0 VALUE(Thetas)=1  value(Thetanf)=1   line to (0.77428154,-
0.2969641)
value(psi)=0 VALUE(Thetas)=1  value(Thetanf)=1   line to (0.75280222,-
0.28456301)
value(psi)=0 VALUE(Thetas)=1  value(Thetanf)=1   line to (0.73324012,-
0.31205806)
value(psi)=0 VALUE(Thetas)=1  value(Thetanf)=1   line to (0.70857594,-
0.30697957)
value(psi)=0 VALUE(Thetas)=1  value(Thetanf)=1   line to (0.68213617,-
0.33464549)
value(psi)=0 VALUE(Thetas)=1  value(Thetanf)=1   line to (0.81712788,-
0.32672114)
value(psi)=0 VALUE(Thetas)=1  value(Thetanf)=1   line to (0.75679912,-
0.38984766)
value(psi)=0 VALUE(Thetas)=1  value(Thetanf)=1   line to (0.90050245,-
0.35266716)
value(psi)=0 VALUE(Thetas)=1  value(Thetanf)=1   line to (0.81459209,-
0.42321444)
value(psi)=0 VALUE(Thetas)=1  value(Thetanf)=1   line to (0.9974354,-
0.36483429)
value(psi)=0 VALUE(Thetas)=1  value(Thetanf)=1   line to (0.96294094,-
0.40522456)
value(psi)=0 VALUE(Thetas)=1  value(Thetanf)=1   line to (0.94956531,-
0.47018464)
value(psi)=0 VALUE(Thetas)=1  value(Thetanf)=1   line to (1.03843634,-
0.48848367)
value(psi)=0 VALUE(Thetas)=1  value(Thetanf)=1   line to (1.00780807,-
0.54153339)
value(psi)=0 VALUE(Thetas)=1  value(Thetanf)=1   line to (1.0671149,-
```

```
0.53805193)
value(psi)=0 VALUE(Thetas)=1  value(Thetanf)=1   line to (1.06383495,-
0.58789904)
value(psi)=0 VALUE(Thetas)=1  value(Thetanf)=1   line to (1.12498956,-
0.64951302)
value(psi)=0 VALUE(Thetas)=1  value(Thetanf)=1   line to (1.04105297,-
0.62735857)
value(psi)=0 VALUE(Thetas)=1  value(Thetanf)=1   line to (0.99952409,-
0.65512265)
value(psi)=0 VALUE(Thetas)=1  value(Thetanf)=1   line to (0.97288571,-
0.6020207)
value(psi)=0 VALUE(Thetas)=1  value(Thetanf)=1   line to (0.94225744,-
0.65507042)
value(psi)=0 VALUE(Thetas)=1  value(Thetanf)=1   line to (0.8819745,-
0.58725536)
value(psi)=0 VALUE(Thetas)=1  value(Thetanf)=1   line to (0.83240523,-
0.63131904)
value(psi)=0 VALUE(Thetas)=1  value(Thetanf)=1   line to (0.81467346,-
0.68138725)
value(psi)=0 VALUE(Thetas)=1  value(Thetanf)=1   line to (0.7738105,-
0.49385022)
value(psi)=0 VALUE(Thetas)=1  value(Thetanf)=1   line to (0.75566995,-
0.60352442)
value(psi)=0 VALUE(Thetas)=1  value(Thetanf)=1   line to (0.71601754,-
0.46048344)
value(psi)=0 VALUE(Thetas)=1  value(Thetanf)=1   line to (0.69151275,-
0.54429293)
value(psi)=0 VALUE(Thetas)=1  value(Thetanf)=1   line to (0.63087958,-
0.42342451)
value(psi)=0 VALUE(Thetas)=1  value(Thetanf)=1   line to (0.62014008,-
0.46015497)
value(psi)=0 VALUE(Thetas)=1  value(Thetanf)=1   line to (0.63687027,-
0.47897554)
value(psi)=0 VALUE(Thetas)=1  value(Thetanf)=1   line to (0.6228399,-
0.50966434)
value(psi)=0 VALUE(Thetas)=1  value(Thetanf)=1   line to (0.64431922,-
0.52206543)
value(psi)=0 VALUE(Thetas)=1  value(Thetanf)=1   line to (0.63227708,-
0.56325108)
value(psi)=0 VALUE(Thetas)=1  value(Thetanf)=1   line to (0.68827706,-
0.61313003)
value(psi)=0 VALUE(Thetas)=1  value(Thetanf)=1   line to (0.64816936,-
0.62424829)
value(psi)=0 VALUE(Thetas)=1  value(Thetanf)=1   line to (0.72097514,-
```

```
0.69686293)
value(psi)=0 VALUE(Thetas)=1  value(Thetanf)=1   line to (0.66697477,-
0.70217996)
value(psi)=0 VALUE(Thetas)=1  value(Thetanf)=1   line to (0.67054734,-
0.73846333)
value(psi)=0 VALUE(Thetas)=1  value(Thetanf)=1   line to (0.61308334,-
0.72166162)
value(psi)=0 VALUE(Thetas)=1  value(Thetanf)=1   line to (0.62055066,-
0.77510322)
value(psi)=0 VALUE(Thetas)=1  value(Thetanf)=1   line to (0.59981041,-
0.81102638)
value(psi)=0 VALUE(Thetas)=1  value(Thetanf)=1   line to (0.56850797,-
0.7929539)
value(psi)=0 VALUE(Thetas)=1  value(Thetanf)=1   line to (0.57553768,-
0.73785098)
value(psi)=0 VALUE(Thetas)=1  value(Thetanf)=1   line to (0.51110857,-
0.74419486)
value(psi)=0 VALUE(Thetas)=1  value(Thetanf)=1   line to (0.53866713,-
0.71656376)
value(psi)=0 VALUE(Thetas)=1  value(Thetanf)=1   line to (0.51220663,-
0.70128678)
value(psi)=0 VALUE(Thetas)=1  value(Thetanf)=1   line to (0.54411212,-
0.66546592)
value(psi)=0 VALUE(Thetas)=1  value(Thetanf)=1   line to (0.4975594,-
0.65434184)
value(psi)=0 VALUE(Thetas)=1  value(Thetanf)=1   line to (0.55197386,-
0.60597111)
value(psi)=0 VALUE(Thetas)=1  value(Thetanf)=1   line to (0.48491113,-
0.62456158)
value(psi)=0 VALUE(Thetas)=1  value(Thetanf)=1   line to (0.44612996,-
0.61529456)
value(psi)=0 VALUE(Thetas)=1  value(Thetanf)=1   line to (0.47193596,-
0.58632171)
value(psi)=0 VALUE(Thetas)=1  value(Thetanf)=1   line to (0.52253605,-
0.56093837)
value(psi)=0 VALUE(Thetas)=1  value(Thetanf)=1   line to (0.53650482,-
0.50248106)
value(psi)=0 VALUE(Thetas)=1  value(Thetanf)=1   line to (0.51915722,-
0.49833573)
value(psi)=0 VALUE(Thetas)=1  value(Thetanf)=1   line to (0.46069984,-
0.52766064)
value(psi)=0 VALUE(Thetas)=1  value(Thetanf)=1   line to (0.42719082,-
0.51965344)
value(psi)=0 VALUE(Thetas)=1  value(Thetanf)=1   line to (0.45198583,-
```

```
0.47670722)
value(psi)=0 VALUE(Thetas)=1  value(Thetanf)=1   line to (0.51681315,-
0.43838157)
value(psi)=0 VALUE(Thetas)=1  value(Thetanf)=1   line to (0.44004426,-
0.39405902)
value(psi)=0 VALUE(Thetas)=1  value(Thetanf)=1   line to (0.35333269,-
0.40259689)
value(psi)=0 VALUE(Thetas)=1  value(Thetanf)=1   line to (0.31041655,-
0.24778242)
value(psi)=0 VALUE(Thetas)=1  value(Thetanf)=1   line to (0.30342833,-
0.48903379)
value(psi)=0 VALUE(Thetas)=1  value(Thetanf)=1   line to (0.27180152,-
0.50729353)
value(psi)=0 VALUE(Thetas)=1  value(Thetanf)=1   line to (0.0593776,-
0.39271982)
value(psi)=0 VALUE(Thetas)=1  value(Thetanf)=1   line to (0.17199279,-
0.50729353)
value(psi)=0 VALUE(Thetas)=1  value(Thetanf)=1   line to (0.121243,-0.57811902)
value(psi)=0 VALUE(Thetas)=1  value(Thetanf)=1   line to (0.121243,-0.6667641)
value(psi)=0 VALUE(Thetas)=1  value(Thetanf)=1   line to (0.18684765,-
0.62978482)
value(psi)=0 VALUE(Thetas)=1  value(Thetanf)=1   line to (0.23643767,-
0.62978482)
value(psi)=0 VALUE(Thetas)=1  value(Thetanf)=1   line to (0.2266176,-
0.66280809)
value(psi)=0 VALUE(Thetas)=1  value(Thetanf)=1   line to (0.17199279,-
0.6987712)
value(psi)=0 VALUE(Thetas)=1  value(Thetanf)=1   line to (0.16690895,-
0.71586733)
value(psi)=0 VALUE(Thetas)=1  value(Thetanf)=1   line to (0.22451885,-
0.73299868)
value(psi)=0 VALUE(Thetas)=1  value(Thetanf)=1   line to (0.27180152,-
0.70186938)
value(psi)=0 VALUE(Thetas)=1  value(Thetanf)=1   line to (0.30979574,-
0.69400716)
value(psi)=0 VALUE(Thetas)=1  value(Thetanf)=1   line to (0.29843063,-
0.73222614)
value(psi)=0 VALUE(Thetas)=1  value(Thetanf)=1   line to (0.24879945,-
0.78100894)
value(psi)=0 VALUE(Thetas)=1  value(Thetanf)=1   line to (0.31789696,-0.75807)
value(psi)=0 VALUE(Thetas)=1  value(Thetanf)=1   line to (0.30425434,-
0.80394788)
value(psi)=0 VALUE(Thetas)=1  value(Thetanf)=1   line to (0.35122886,-
0.79422735)
```

```
value(psi)=0 VALUE(Thetas)=1  value(Thetanf)=1   line to (0.35122886,-
0.8247813)
value(psi)=0 VALUE(Thetas)=1  value(Thetanf)=1   line to (0.38893737,-
0.81473044)
value(psi)=0 VALUE(Thetas)=1  value(Thetanf)=1   line to (0.35122886,-
0.86735574)
value(psi)=0 VALUE(Thetas)=1  value(Thetanf)=1   line to (0.40246424,-
0.8888193)
value(psi)=0 VALUE(Thetas)=1  value(Thetanf)=1   line to (0.40246424,-
0.92496424)
value(psi)=0 VALUE(Thetas)=1  value(Thetanf)=1   line to (0.36098375,-
0.92496424)
value(psi)=0 VALUE(Thetas)=1  value(Thetanf)=1   line to (0.31843562,-
0.89177655)
value(psi)=0 VALUE(Thetas)=1  value(Thetanf)=1   line to (0.30425434,-
0.94994269)
value(psi)=0 VALUE(Thetas)=1  value(Thetanf)=1   line to (0.2746183,-
0.92870708)
value(psi)=0 VALUE(Thetas)=1  value(Thetanf)=1   line to (0.24301343,-
0.97281425)
value(psi)=0 VALUE(Thetas)=1  value(Thetanf)=1   line to (0.2165302,-
0.87345527)
value(psi)=0 VALUE(Thetas)=1  value(Thetanf)=1   line to (0.18684765,-
0.90263043)
value(psi)=0 VALUE(Thetas)=1  value(Thetanf)=1   line to (0.17165121,-
0.82919355)
value(psi)=0 VALUE(Thetas)=1  value(Thetanf)=1   line to (0.12996232,-
0.81902953)
value(psi)=0 VALUE(Thetas)=1  value(Thetanf)=1   line to (0.12996232,-
0.79422735)
value(psi)=0 VALUE(Thetas)=1  value(Thetanf)=1   line to (0.09636985,-
0.7910336)
value(psi)=0 VALUE(Thetas)=1  value(Thetanf)=1   line to (0.08843586,-
0.76713455)
value(psi)=0 VALUE(Thetas)=1  value(Thetanf)=1   line to (0.05125659,-0.75807)
value(psi)=0 VALUE(Thetas)=1  value(Thetanf)=1   line to (0.12561513,-
0.87101407)
value(psi)=0 VALUE(Thetas)=1  value(Thetanf)=1   line to (0.04078158,-
0.8503311)
value(psi)=0 VALUE(Thetas)=1  value(Thetanf)=1   line to (0.1448325,-
0.95619158)
value(psi)=0 VALUE(Thetas)=1  value(Thetanf)=1   line to (0.04078158,-
0.91706466)
value(psi)=0 VALUE(Thetas)=1  value(Thetanf)=1   line to (0.18276194,-
```

```
1.04622154)
value(psi)=0 VALUE(Thetas)=1  value(Thetanf)=1   line to (0.13053571,-
1.0365436)
value(psi)=0 VALUE(Thetas)=1  value(Thetanf)=1   line to (0.09617891,-
1.14355409)
value(psi)=0 VALUE(Thetas)=1  value(Thetanf)=1   line to (0.03492237,-
1.14355409)
value(psi)=0 VALUE(Thetas)=1  value(Thetanf)=1   line to (0.03492237,-
1.22718095)
value(psi)=0 VALUE(Thetas)=1  value(Thetanf)=1   line to close
START 'outer' (0,-2)
value(psi)=0 VALUE(Thetas)=0  value(Thetanf)=0
ARC(CENTER=0,0) ANGLE=360 TO CLOSE
MONITORS
PLOTS
tecplot(psi)
tecplot(Thetas)
tecplot(Thetanf)
contour(psi)
contour(Thetas)
contour(Thetanf)
elevation (normal (flux)) on "inner"as "inner-Nusselt" points = 100 export
format "#d#b#1" file="inner-Nusselt" !report(t_star)
elevation (normal (flux)) on "outer" as "outer-Nusselt" points = 100 export
format "#d#b#1" file="outer-Nusselt" !report(t_sta
End
```

2. FlexPDE CODE FOR PCM

2.1 Sample 1

```
TITLE
'PCM Sample Code 1'
SELECT
painted
smoothinit
threads=2
Spectral_colors
VARIABLES
temp(threshold=1)
solid(threshold=0.01)
DEFINITIONS
Lf
```

Conclusion

```
Tm
T0= 1
temp_hot=273+2
temp_cold=240
Tinit
sinit
rho { Density kg/m3 }
cp { heat capacity }
r2=0.0452
r1=0.22
k
L_1=0.007
L_2=0.01
L=4.2*r1
X1=0.1*L
X2=0.6*L
L1=L/2
L2=L/4
beta1=pi/4
theta=0.050745
 t_fin=0.3e-3
gamma=Arctan(t_fin/(2*r1))
eps=1e-6
response = 0.1
s0 = 0.5*erfc((temp-Tm)/T0)
rho_f=997
rho_s1=3970
rho_s2=1800
cp_f=4179
cp_s1=765
cp_s2=717
ks1=40
ks2=5000
s=8.6
phi1=0.05
phi2=0.05
Rd=0.5
Kf=if solid=0 then 2.24 else 0.75
rho_nf=1145
Cp_nf=4008
Lf_nf=2.6e5
INITIAL VALUES
temp=Tinit
solid = 0.5*erfc((tinit-Tm)/T0)
```

```
EQUATIONS
temp: rho*cp*dt(temp) - div(k*grad(temp)) = Lf*dt(solid)+4/3*Rd*dyy(temp)
solid:   dt(solid) - eps*div(grad(solid)) =  (S0-solid)/response
BOUNDARIES
region 1 'melting Plate'
cp = cp_nf
rho=rho_nf
k=Kf
Tm=273
Lf= Lf_nf
Tinit = temp_hot
sinit = 0
START  (-0.0433071,0.12619)
value(temp)=temp_cold  line to(-0.0695338,0.15)
value(temp)=temp_cold   line to (-0.0853018,0.178571)
value(temp)=temp_cold   line to (-0.0905512,0.204762)
value(temp)=temp_cold   line to (-0.0879265,0.22381)
value(temp)=temp_cold   line to (-0.0748031,0.259524)
value(temp)=temp_cold   line to (-0.0433071,0.295238)
value(temp)=temp_cold   line to (-0.0144357,0.321429)
value(temp)=temp_cold   line to (0.019685,0.354762)
value(temp)=temp_cold   line to (0.0354331,0.383333)
value(temp)=temp_cold   line to (0.0459318,0.409524)
value(temp)=temp_cold   line to (0.0485564,0.421429)
value(temp)=temp_cold   line to (0.0485564,0.466667)
value(temp)=temp_cold   line to (-0.0433071,0.466667)
value(temp)=temp_cold   line to (-0.0433071,0.42381)
value(temp)=temp_cold   line to (-0.0406824,0.407143)
value(temp)=temp_cold   line to (-0.0328084,0.388095)
value(temp)=temp_cold   line to (-0.0170604,0.364286)
value(temp)=temp_cold   line to (-0.00131234,0.345238)
value(temp)=temp_cold   line to (0.0223097,0.321429)
value(temp)=temp_cold   line to (0.0406824,0.307143)
value(temp)=temp_cold   line to (0.0590551,0.288095)
value(temp)=temp_cold   line to (0.0748031,0.269048)
value(temp)=temp_cold   line to (0.0905512,0.242857)
value(temp)=temp_cold   line to (0.0958005,0.22619)
value(temp)=temp_cold   line to (0.0984252,0.202381)
value(temp)=temp_cold   line to (0.0958005,0.183333)
value(temp)=temp_cold   line to(0.0905512,0.169048)
value(temp)=temp_cold   line to (0.074672,0.147229)
value(temp)=temp_cold   line to (0.0538058,0.12619)
value(temp)=temp_cold    line to (0.10355,0.108524)
value(temp)=temp_cold   line to (0.108924,0.121429)
```

```
value(temp)=temp_cold   line to (0.119423,0.145238)
value(temp)=temp_cold   line to (0.135171,0.164286)
value(temp)=temp_cold   line to (0.158793,0.180952)
value(temp)=temp_cold   line to (0.185039,0.188095)
value(temp)=temp_cold   line to (0.216535,0.188095)
value(temp)=temp_cold   line to (0.248031,0.188095)
value(temp)=temp_cold   line to (0.279528,0.180952)
value(temp)=temp_cold   line to (0.311024,0.17381)
value(temp)=temp_cold   line to (0.347769,0.164286)
value(temp)=temp_cold   line to (0.374016,0.164286)
value(temp)=temp_cold   line to (0.41601,0.169048)
value(temp)=temp_cold   line to (0.439633,0.178571)
value(temp)=temp_cold   line to (0.476378,0.195238)
value(temp)=temp_cold   line to (0.429134,0.269048)
value(temp)=temp_cold   line to (0.389764,0.247619)
value(temp)=temp_cold   line to (0.366142,0.228571)
value(temp)=temp_cold   line to (0.347769,0.207143)
value(temp)=temp_cold   line to (0.33727,0.185714)
value(temp)=temp_cold   line to (0.326772,0.152381)
value(temp)=temp_cold   line to (0.318898,0.12619)
value(temp)=temp_cold   line to (0.308399,0.097619)
value(temp)=temp_cold   line to (0.295276,0.0714286)
value(temp)=temp_cold   line to (0.276903,0.047619)
value(temp)=temp_cold   line to (0.253281,0.0285714)
value(temp)=temp_cold   line to (0.216535,0.0166667)
value(temp)=temp_cold   line to (0.187664,0.0166667)
value(temp)=temp_cold   line to (0.14791,0.0245715)
value(temp)=temp_cold    line to (0.146787,-0.0309072)
value(temp)=temp_cold   line to (0.177165,-0.0190476)
value(temp)=temp_cold   line to (0.208661,-0.0190476)
value(temp)=temp_cold   line to  (0.242782,-0.0261905)
value(temp)=temp_cold   line to (0.274278,-0.047619)
value(temp)=temp_cold   line to (0.2979,-0.0785714)
value(temp)=temp_cold   line to (0.311024,-0.107143)
value(temp)=temp_cold   line to (0.321522,-0.142857)
value(temp)=temp_cold   line to (0.334646,-0.188095)
value(temp)=temp_cold   line to (0.347769,-0.209524)
value(temp)=temp_cold   line to  (0.371391,-0.238095)
value(temp)=temp_cold   line to (0.387139,-0.252381)
value(temp)=temp_cold   line to (0.429134,-0.27381)
value(temp)=temp_cold   line to (0.476378,-0.2)
value(temp)=temp_cold   line to (0.434383,-0.178571)
value(temp)=temp_cold   line to (0.405512,-0.171429)
value(temp)=temp_cold   line to (0.37664,-0.169048)
```

```
value(temp)=temp_cold   line to (0.339895,-0.169048)
value(temp)=temp_cold   line to (0.308399,-0.17619)
value(temp)=temp_cold   line to (0.274278,-0.185714)
value(temp)=temp_cold   line to (0.237533,-0.192857)
value(temp)=temp_cold   line to (0.203412,-0.192857)
value(temp)=temp_cold   line to (0.171916,-0.188095)
value(temp)=temp_cold   line to (0.143045,-0.17381)
value(temp)=temp_cold   line to (0.124672,-0.154762)
value(temp)=temp_cold   line to (0.111549,-0.130952)
value(temp)=temp_cold   line to (0.103675,-0.102381)
value(temp)=temp_cold    line to (0.0538058,-0.128571)
value(temp)=temp_cold   line to (0.0774278,-0.154762)
value(temp)=temp_cold   line to (0.0905512,-0.17381)
value(temp)=temp_cold   line to (0.0984252,-0.197619)
value(temp)=temp_cold   line to (0.0958005,-0.22619)
value(temp)=temp_cold   line to (0.0879265,-0.254762)
value(temp)=temp_cold   line to (0.0695538,-0.278571)
value(temp)=temp_cold   line to (0.0511811,-0.3)
value(temp)=temp_cold   line to (0.0223097,-0.32619)
value(temp)=temp_cold   line to (0.00131234,-0.345238)
value(temp)=temp_cold   line to (-0.019685,-0.369048)
value(temp)=temp_cold   line to (-0.0328084,-0.397619)
value(temp)=temp_cold   line to (-0.0406824,-0.419048)
value(temp)=temp_cold   line to (-0.0433071,-0.430952)
value(temp)=temp_cold   line to (-0.0433071,-0.471429)
value(temp)=temp_cold   line to (0.0511811,-0.471429)
value(temp)=temp_cold   line to (0.0511811,-0.430952)
value(temp)=temp_cold   line to (0.0459318,-0.407143)
value(temp)=temp_cold   line to (0.0380577,-0.390476)
value(temp)=temp_cold   line to (0.0249344,-0.366667)
value(temp)=temp_cold   line to (-0.00131234,-0.335714)
value(temp)=temp_cold   line to (-0.0275591,-0.314286)
value(temp)=temp_cold   line to (-0.0564304,-0.288095)
value(temp)=temp_cold   line to (-0.0721785,-0.264286)
value(temp)=temp_cold   line to (-0.0853018,-0.238095)
value(temp)=temp_cold   line to (-0.090551,-0.209524)
value(temp)=temp_cold   line to (-0.0879265,-0.188095)
value(temp)=temp_cold   line to (-0.0774278,-0.161905)
value(temp)=temp_cold   line to (-0.0614304,-0.140476)
value(temp)=temp_cold   line to(-0.0433071,-0.128571)
value(temp)=temp_cold    line to  (-0.0984252,-0.102381)
value(temp)=temp_cold   line to (-0.103675,-0.133333)
value(temp)=temp_cold   line to (-0.119423,-0.157143)
value(temp)=temp_cold   line to (-0.14042,-0.178571)
```

```
value(temp)=temp_cold   line to (-0.174541,-0.190476)
value(temp)=temp_cold   line to (-0.213911,-0.192857)
value(temp)=temp_cold   line to (-0.25853,-0.188095)
value(temp)=temp_cold   line to (-0.2979,-0.178571)
value(temp)=temp_cold   line to (-0.332021,-0.169048)
value(temp)=temp_cold   line to (-0.379265,-0.169048)
value(temp)=temp_cold   line to (-0.410761,-0.17381)
value(temp)=temp_cold   line to (-0.429134,-0.178571)
value(temp)=temp_cold   line to (-0.465879,-0.2)
value(temp)=temp_cold   line to (-0.418635,-0.27381)
value(temp)=temp_cold   line to (-0.38189,-0.252381)
value(temp)=temp_cold   line to (-0.355643,-0.230952)
value(temp)=temp_cold   line to (-0.33727,-0.204762)
value(temp)=temp_cold   line to (-0.324147,-0.178571)
value(temp)=temp_cold   line to (-0.313648,-0.142857)
value(temp)=temp_cold   line to (-0.305774,-0.114286)
value(temp)=temp_cold   line to (-0.287402,-0.0761905)
value(temp)=temp_cold   line to (-0.269029,-0.05)
value(temp)=temp_cold   line to (-0.248031,-0.0333333)
value(temp)=temp_cold   line to (-0.221785,-0.0238095)
value(temp)=temp_cold   line to (-0.192913,-0.0214286)
value(temp)=temp_cold   line to (-0.164042,-0.0214286)
value(temp)=temp_cold   line to (-0.143045,-0.0285714)
value(temp)=temp_cold    line to (-0.143045,0.0238095)
value(temp)=temp_cold   line to (-0.174541,0.0166667)
value(temp)=temp_cold   line to (-0.203412,0.0142857)
value(temp)=temp_cold   line to (-0.232283,0.0238095)
value(temp)=temp_cold   line to (-0.261155,0.0428571)
value(temp)=temp_cold   line to (-0.284777,0.0666667)
value(temp)=temp_cold   line to (-0.30315,0.102381)
value(temp)=temp_cold   line to (-0.313648,0.140476)
value(temp)=temp_cold   line to (-0.326772,0.178571)
value(temp)=temp_cold   line to (-0.339895,0.207143)
value(temp)=temp_cold   line to (-0.360892,0.230952)
value(temp)=temp_cold   line to (-0.379265,0.247619)
value(temp)=temp_cold   line to (-0.418635,0.269048)
value(temp)=temp_cold   line to (-0.468504,0.195238)
value(temp)=temp_cold   line to (-0.426509,0.17381)
value(temp)=temp_cold   line to (-0.395013,0.164286)
value(temp)=temp_cold   line to (-0.368766,0.164286)
value(temp)=temp_cold   line to (-0.329396,0.166667)
value(temp)=temp_cold   line to (-0.287402,0.178571)
value(temp)=temp_cold   line to (-0.237533,0.188095)
value(temp)=temp_cold   line to (-0.198163,0.188095)
```

```
value(temp)=temp_cold   line to (-0.166667,0.183333)
value(temp)=temp_cold   line to (-0.14042,0.171429)
value(temp)=temp_cold   line to (-0.116798,0.152381)
value(temp)=temp_cold   line to (-0.10105,0.12619)
value(temp)=temp_cold   line to (-0.0984252,0.097619)
line to close
start 'Outer' (0,0.9)
value(temp)=temp_cold
ARC(CENTER=0,0) ANGLE=360 TO CLOSE
TIME 0 by 5e-4 to 50000
MONITORS
for cycle=10
grid(x,y)
PLOTS
for t= 0  by 1000 to endtime
tecplot(temp)
tecplot(solid)
contour(temp)
contour(solid) range=(0,1)
Summary
HISTORIES
history(integral(cp*(temp)+Lf*(1-solid))) as "Total Energy"points = 100 export
format "#d#b#1" file="total energy parameter.txt" !report(t_star)
history(integral(rho*cp*(temp)+rho*Lf*(1-solid))) as "Total Energy"points =
100 export format "#d#b#1" file="total energy.txt" !report(t_star)
history((integral(1-solid))/(integral(1))) as "Liq_ave_main"points = 100 ex-
port format "#d#b#1" file="Liq_ave_main.txt"
history((integral(solid))/(integral(1))) as "solid_ave_main"points = 100 ex-
port format "#d#b#1" file="Solid_ave_main.txt"
history((integral(temp))/(integral(1))) as "T_ave"points = 100 export format
"#d#b#1" file="T_ave.txt"
End
```

2.2 Sample 2

```
TITLE
'PCM Sample Code 2'
SELECT
painted
smoothinit
threads=2
Spectral_colors
VARIABLES
```

```
temp(threshold=1)
solid(threshold=0.01)
DEFINITIONS
Lf
Tm
T0= 1
temp_hot=273+2
temp_cold=240
T_infinity=298
h=40
Tinit
sinit
rho { Density kg/m3 }
cp { heat capacity }
r2=0.0452
r1=0.22
k
L_1=0.007
L_2=0.01
L=4.2*r1
X1=0.1*L
X2=0.6*L
L1=L/2
L2=L/4
beta1=pi/4
theta=0.050745
 t_fin=0.3e-3
gamma=Arctan(t_fin/(2*r1))
eps=1e-6
response = 0.1
s0 = 0.5*erfc((temp-Tm)/T0)
rho_f=997
rho_s1=3970
rho_s2=1800
cp_f=4179
cp_s1=765
cp_s2=717
ks1=40
ks2=5000
s=8.6
phi1=0.05
phi2=0.05
Rd=0.5
Kf=if solid=0 then 1 else 0.764
```

```
rho_nf=1200.2
Cp_nf=3384.2
Lf_nf=2.6e5
INITIAL VALUES
temp=Tinit
solid = 0.5*erfc((tinit-Tm)/T0)
EQUATIONS
temp: rho*cp*dt(temp) - div(k*grad(temp)) = Lf*dt(solid)+4/3*Rd*dyy(temp)
solid:   dt(solid) - eps*div(grad(solid)) =  (S0-solid)/response
BOUNDARIES
region 1 'melting Plate'
cp = cp_nf
rho=rho_nf
k=Kf
Tm=273
Lf= Lf_nf
Tinit = temp_hot
sinit = 0
start 'inner'(0.06,0.06)
value(temp)=temp_cold
ARC(CENTER=0.05,0.05) ANGLE=360 TO CLOSE
START 'outer' (0,0.1)
 natural (temp)=0  line to (0.0994505,0.1)
 natural (temp)=h*(temp-T_infinity)   line to (0.0989011,0.0972527)
 natural (temp)=h*(temp-T_infinity)  line to (0.0978022,0.0956044)
 natural (temp)=h*(temp-T_infinity)   line to (0.0956044,0.093956)
 natural (temp)=h*(temp-T_infinity)  line to (0.0928571,0.0923077)
 natural (temp)=h*(temp-T_infinity)   line to (0.0901099,0.0906593)
 natural (temp)=h*(temp-T_infinity)   line to (0.0873626,0.089011)
 natural (temp)=h*(temp-T_infinity)   line to (0.0851648,0.0879121)
 natural (temp)=h*(temp-T_infinity)   line to (0.082967,0.0868132)
 natural (temp)=h*(temp-T_infinity)   line to (0.0813187,0.0851648)
 natural (temp)=h*(temp-T_infinity)   line to (0.0802198,0.0840659)
 natural (temp)=h*(temp-T_infinity)   line to (0.0796703,0.0824176)
 natural (temp)=h*(temp-T_infinity)   line to (0.0796703,0.0813187)
 natural (temp)=h*(temp-T_infinity)   line to (0.0802198,0.0802198)
 natural (temp)=h*(temp-T_infinity)   line to (0.0813187,0.0785714)
 natural (temp)=h*(temp-T_infinity)   line to (0.0824176,0.0774725)
 natural (temp)=h*(temp-T_infinity)   line to (0.0840659,0.0763736)
 natural (temp)=h*(temp-T_infinity)   line to (0.0862637,0.0752747)
 natural (temp)=h*(temp-T_infinity)   line to (0.0879121,0.0741758)
 natural (temp)=h*(temp-T_infinity)  line to (0.0895604,0.0730769)
 natural (temp)=h*(temp-T_infinity)  line to (0.0917582,0.071978)
 natural (temp)=h*(temp-T_infinity)   line to (0.093956,0.0708791)
```

```
natural (temp)=h*(temp-T_infinity)   line to (0.0956044,0.0697802)
natural (temp)=h*(temp-T_infinity)   line to (0.0978022,0.0681319)
natural (temp)=h*(temp-T_infinity)   line to (0.0989011,0.0664835)
natural (temp)=h*(temp-T_infinity)   line to (0.0989011,0.0642857)
natural (temp)=h*(temp-T_infinity)   line to (0.0983516,0.0626374)
natural (temp)=h*(temp-T_infinity)   line to (0.0967033,0.0615385)
natural (temp)=h*(temp-T_infinity)   line to (0.093956,0.0598901)
natural (temp)=h*(temp-T_infinity)   line to (0.0917582,0.0582418)
natural (temp)=h*(temp-T_infinity)   line to (0.089011,0.0571429)
natural (temp)=h*(temp-T_infinity)  line to (0.0868132,0.056044)
natural (temp)=h*(temp-T_infinity)   line to (0.0846154,0.0549451)
natural (temp)=h*(temp-T_infinity)     line to (0.0818681,0.0527473)
natural (temp)=h*(temp-T_infinity)   line to (0.0802198,0.0510989)
natural (temp)=h*(temp-T_infinity)  line to (0.0796703,0.05)
natural (temp)=h*(temp-T_infinity)   line to (0.0796703,0.0489011)
natural (temp)=h*(temp-T_infinity)   line to (0.0796703,0.0478022)
natural (temp)=h*(temp-T_infinity)   line to (0.0807692,0.0461538)
natural (temp)=h*(temp-T_infinity)   line to (0.0818681,0.0450549)
natural (temp)=h*(temp-T_infinity)  line to (0.0835165,0.043956)
natural (temp)=h*(temp-T_infinity)   line to (0.0851648,0.0428571)
natural (temp)=h*(temp-T_infinity)  line to (0.0873626,0.0417582)
natural (temp)=h*(temp-T_infinity)   line to (0.0895604,0.0406593)
natural (temp)=h*(temp-T_infinity)  line to (0.0923077,0.039011)
natural (temp)=h*(temp-T_infinity) line to (0.0945055,0.0373626)
natural (temp)=h*(temp-T_infinity)   line to (0.0972527,0.0357143)
natural (temp)=h*(temp-T_infinity)   line to (0.0983516,0.0340659)
natural (temp)=h*(temp-T_infinity)   line to (0.0989011,0.0324176)
natural (temp)=h*(temp-T_infinity)   line to (0.0989011,0.0313187)
natural (temp)=h*(temp-T_infinity)  line to (0.0983516,0.0302198)
natural (temp)=h*(temp-T_infinity) line to (0.0967033,0.0285714)
natural (temp)=h*(temp-T_infinity)  line to (0.0945055,0.0269231)
natural (temp)=h*(temp-T_infinity)  line to (0.0928571,0.0258242)
natural (temp)=h*(temp-T_infinity)  line to (0.0906593,0.0241758)
natural (temp)=h*(temp-T_infinity)   line to (0.089011,0.0236264)
natural (temp)=h*(temp-T_infinity)   line to (0.0868132,0.0225275)
natural (temp)=h*(temp-T_infinity)  line to (0.0851648,0.0214286)
natural (temp)=h*(temp-T_infinity)   line to (0.0835165,0.0203297)
natural (temp)=h*(temp-T_infinity) line to (0.0818681,0.0192308)
natural (temp)=h*(temp-T_infinity)  line to (0.0802198,0.0175824)
natural (temp)=h*(temp-T_infinity)   line to (0.0796703,0.0159341)
natural (temp)=h*(temp-T_infinity)  line to (0.0796703,0.0142857)
natural (temp)=h*(temp-T_infinity)   line to (0.0813187,0.0126374)
natural (temp)=h*(temp-T_infinity)  line to (0.082967,0.010989)
natural (temp)=h*(temp-T_infinity)  line to (0.0851648,0.00934066)
```

```
natural (temp)=h*(temp-T_infinity)  line to (0.0879121,0.00824176)
natural (temp)=h*(temp-T_infinity)  line to (0.0901099,0.00714286)
natural (temp)=h*(temp-T_infinity)   line to (0.0917582,0.00604396)
natural (temp)=h*(temp-T_infinity)  line to (0.093956,0.00494505)
natural (temp)=h*(temp-T_infinity)   line to (0.0956044,0.00384615)
natural (temp)=h*(temp-T_infinity)   line to (0.0972527,0.00274725)
natural (temp)=h*(temp-T_infinity)  line to (0.0983516,0.00164835)
natural (temp)=h*(temp-T_infinity)  line to (0.0994505,0.000549451)
natural (temp)=h*(temp-T_infinity)  line to (0.1,0)
natural (temp)=0   line to (0,0)
natural (temp)=h*(temp-T_infinity)   line to (0.00164835,0.0021978)
natural (temp)=h*(temp-T_infinity)   line to (0.0032967,0.00384615)
natural (temp)=h*(temp-T_infinity)  line to (0.00549451,0.00549451)
natural (temp)=h*(temp-T_infinity)  line to (0.00879121,0.00714286)
natural (temp)=h*(temp-T_infinity)  line to (0.0115385,0.00879121)
natural (temp)=h*(temp-T_infinity)   line to (0.0142857,0.0104396)
natural (temp)=h*(temp-T_infinity)   line to (0.017033,0.0126374)
natural (temp)=h*(temp-T_infinity)  line to (0.0186813,0.0137363)
natural (temp)=h*(temp-T_infinity)   line to (0.0197802,0.0153846)
natural (temp)=h*(temp-T_infinity)   line to (0.0197802,0.017033)
natural (temp)=h*(temp-T_infinity)   line to (0.0192308,0.0186813)
natural (temp)=h*(temp-T_infinity)  line to (0.0175824,0.0197802)
natural (temp)=h*(temp-T_infinity)   line to (0.0148352,0.0214286)
natural (temp)=h*(temp-T_infinity)  line to (0.0126374,0.0225275)
natural (temp)=h*(temp-T_infinity)   line to (0.0104396,0.0241758)
natural (temp)=h*(temp-T_infinity)  line to (0.00769231,0.0252747)
natural (temp)=h*(temp-T_infinity)   line to (0.00549451,0.0269231)
natural (temp)=h*(temp-T_infinity)  line to (0.00384615,0.028022)
natural (temp)=h*(temp-T_infinity)   line to (0.00164835,0.0296703)
natural (temp)=h*(temp-T_infinity)   line to (0.000549451,0.0307692)
natural (temp)=h*(temp-T_infinity)   line to (0,0.0324176)
natural (temp)=h*(temp-T_infinity)   line to (0.000549451,0.0346154)
natural (temp)=h*(temp-T_infinity)   line to (0.0021978,0.0362637)
natural (temp)=h*(temp-T_infinity)line to (0.00494505,0.0379121)
natural (temp)=h*(temp-T_infinity)   line to (0.00824176,0.0395604)
natural (temp)=h*(temp-T_infinity)   line to (0.0126374,0.0423077)
natural (temp)=h*(temp-T_infinity)  line to (0.0153846,0.043956)
natural (temp)=h*(temp-T_infinity)   line to (0.0181319,0.0456044)
natural (temp)=h*(temp-T_infinity)  line to (0.0192308,0.0472527)
natural (temp)=h*(temp-T_infinity)  line to (0.0197802,0.0494505)
natural (temp)=h*(temp-T_infinity)   line to (0.0186813,0.0516484)
natural (temp)=h*(temp-T_infinity)   line to (0.017033,0.0532967)
natural (temp)=h*(temp-T_infinity)  line to (0.0142857,0.0549451)
natural (temp)=h*(temp-T_infinity)   line to (0.0115385,0.0565934)
```

```
 natural (temp)=h*(temp-T_infinity)   line to (0.00879121,0.0582418)
 natural (temp)=h*(temp-T_infinity)   line to (0.00549451,0.0598901)
 natural (temp)=h*(temp-T_infinity)   line to (0.0032967,0.060989)
 natural (temp)=h*(temp-T_infinity)  line to (0.0010989,0.0626374)
 natural (temp)=h*(temp-T_infinity)   line to (0.000549451,0.0637363)
 natural (temp)=h*(temp-T_infinity)   line to (0,0.0648352)
 natural (temp)=h*(temp-T_infinity)   line to (0.000549451,0.0664835)
 natural (temp)=h*(temp-T_infinity)   line to (0.00164835,0.0681319)
 natural (temp)=h*(temp-T_infinity)   line to (0.0032967,0.0697802)
 natural (temp)=h*(temp-T_infinity)  line to (0.00604396,0.0714286)
 natural (temp)=h*(temp-T_infinity)   line to (0.00824176,0.0725275)
 natural (temp)=h*(temp-T_infinity)   line to (0.0104396,0.0741758)
 natural (temp)=h*(temp-T_infinity)   line to (0.0131868,0.0752747)
 natural (temp)=h*(temp-T_infinity)   line to (0.0159341,0.0769231)
 natural (temp)=h*(temp-T_infinity)   line to (0.0186813,0.0791209)
 natural (temp)=h*(temp-T_infinity)   line to (0.0197802,0.0807692)
 natural (temp)=h*(temp-T_infinity)   line to (0.0197802,0.0824176)
 natural (temp)=h*(temp-T_infinity)   line to (0.0181319,0.0851648)
 natural (temp)=h*(temp-T_infinity)  line to (0.0159341,0.0868132)
 natural (temp)=h*(temp-T_infinity)   line to (0.0131868,0.0884615)
 natural (temp)=h*(temp-T_infinity)   line to (0.010989,0.0895604)
 natural (temp)=h*(temp-T_infinity)  line to (0.00824176,0.0912088)
 natural (temp)=h*(temp-T_infinity)   line to (0.00494505,0.0928571)
 natural (temp)=h*(temp-T_infinity)   line to (0.00274725,0.0945055)
 natural (temp)=h*(temp-T_infinity)   line to (0.00164835,0.0956044)
 natural (temp)=h*(temp-T_infinity)  line to (0.000549451,0.0972527)
 natural (temp)=0   line to close
TIME 0 by 5e-4 to 9000
MONITORS
for cycle=10
grid(x,y)
PLOTS
for t= 0  by 50 to endtime
tecplot(temp)
tecplot(solid)
contour(temp)
contour(solid) range=(0,1)
Summary
HISTORIES
history(integral(cp*(temp)+Lf*(1-solid))) as "Total Energy"points = 100 export
format "#d#b#1" file="total energy parameter.txt" !report(t_star)
history(integral(rho*cp*(temp)+rho*Lf*(1-solid))) as "Total Energy"points =
100 export format "#d#b#1" file="total energy.txt" !report(t_star)
history((integral(1-solid))/(integral(1))) as "Liq_ave_main"points = 100 ex-
```

```
port format "#d#b#1" file="Liq_ave_main.txt"
history((integral(solid))/(integral(1))) as "solid_ave_main"points = 100 ex-
port format "#d#b#1" file="Solid_ave_main.txt"
history((integral(temp))/(integral(1))) as "T_ave"points = 100 export format
"#d#b#1" file="T_ave.txt"
End
```

2.3 Sample 3

```
TITLE
'PCM Sample Code 3'
SELECT
painted
smoothinit
threads=2
Spectral_colors
VARIABLES
temp(threshold=1)
solid(threshold=0.01)
DEFINITIONS
Lf
Tm
T0= 1
temp_hot=273+2
temp_cold=240
Tinit
sinit
rho { Density kg/m3 }
cp { heat capacity }
r2=0.0452
r1=0.22
k
L_1=0.007
L_2=0.01
L=4.2*r1
X1=0.1*L
X2=0.6*L
L1=L/2
L2=L/4
beta1=pi/4
theta=0.050745
 t_fin=0.3e-3
gamma=Arctan(t_fin/(2*r1))
```

```
eps=1e-6
response = 0.1
s0 = 0.5*erfc((temp-Tm)/T0)
rho_f=997
rho_s1=3970
rho_s2=1800
cp_f=4179
cp_s1=765
cp_s2=717
ks1=40
ks2=5000
s=8.6
phi1=0.05
phi2=0.05
Rd=0.5
Kf=if solid=0 then 1 else 0.764
rho_nf=1200.2
Cp_nf=3384.2
Lf_nf=2.6e5
INITIAL VALUES
temp=Tinit
solid = 0.5*erfc((tinit-Tm)/T0)
EQUATIONS
temp: rho*cp*dt(temp) - div(k*grad(temp)) = Lf*dt(solid)+4/3*Rd*dyy(temp)
solid:   dt(solid) - eps*div(grad(solid)) =  (S0-solid)/response
BOUNDARIES
region 1 'melting Plate'
cp = cp_nf
rho=rho_nf
k=Kf
Tm=273
Lf= Lf_nf
Tinit = temp_hot
sinit = 0

  start 'inner' (0.5,0.0677) value(temp)=temp_cold  {bottom wall}
  line to (0.6609,0.1846)  value(temp)=temp_cold   {right wall}
  line to (0.5995,0.3738)  value(temp)=temp_cold   {right wall up}
  line to (0.4005,0.3738)  value(temp)=temp_cold   {left up wall}
  line to (0.3391,0.1846)  value(temp)=temp_cold  {left wall}
  line to close
  start 'outer' (0,0)   natural (temp)=0  {Bottom wall}
  line to (1,0)  natural (temp)=0  {right wall}
  line to (0.5,0.85)   natural (temp)=0  {left wall}
```

```
  line to close
TIME 0 by 5e-4 to 100000
MONITORS
for cycle=10
grid(x,y)
PLOTS
for t= 0  by 50 to endtime
tecplot(temp)
tecplot(solid)
contour(temp)
contour(solid) range=(0,1)
Summary
HISTORIES
history(integral(cp*(temp)+Lf*(1-solid))) as "Total Energy"points = 100 export
format "#d#b#1" file="total energy parameter.txt" !report(t_star)
history(integral(rho*cp*(temp)+rho*Lf*(1-solid))) as "Total Energy"points =
100 export format "#d#b#1" file="total energy.txt" !report(t_star)
history((integral(1-solid))/(integral(1))) as "Liq_ave_main"points = 100 ex-
port format "#d#b#1" file="Liq_ave_main.txt"
history((integral(solid))/(integral(1))) as "solid_ave_main"points = 100 ex-
port format "#d#b#1" file="Solid_ave_main.txt"
history((integral(temp))/(integral(1))) as "T_ave"points = 100 export format
"#d#b#1" file="T_ave.txt"
End
```

2.4 Sample 4

```
TITLE
'PCM Sample Code 4'
SELECT
painted
smoothinit
threads=2
Spectral_colors
VARIABLES
temp(threshold=1)
solid(threshold=0.01)
DEFINITIONS
Lf
Tm
T0= 1
temp_hot=273+2
temp_cold=240
```

Conclusion

```
Tinit
sinit
rho { Density kg/m3 }
cp { heat capacity }
r2=0.0452
r1=0.22
k
L_1=0.007
L_2=0.01
L=4.2*r1
X1=0.1*L
X2=0.6*L
L1=L/2
L2=L/4
beta1=pi/4
theta=0.050745
 t_fin=0.3e-3
gamma=Arctan(t_fin/(2*r1))
eps=1e-6
response = 0.1
s0 = 0.5*erfc((temp-Tm)/T0)
rho_f=997
rho_s1=3970
rho_s2=1800
cp_f=4179
cp_s1=765
cp_s2=717
ks1=40
ks2=5000
s=8.6
phi1=0.05
phi2=0.05
Rd=0.5
Kf=if solid=0 then 2.24 else 0.764
rho_nf=1200.2
Cp_nf=3384.2
Lf_nf=2.6e5
INITIAL VALUES
temp=Tinit
solid = 0.5*erfc((tinit-Tm)/T0)
EQUATIONS
temp: rho*cp*dt(temp) - div(k*grad(temp)) = Lf*dt(solid)+4/3*Rd*dyy(temp)
solid:   dt(solid) - eps*div(grad(solid)) =  (S0-solid)/response
BOUNDARIES
```

```
region 1 'melting Plate'
cp = cp_nf
rho=rho_nf
k=Kf
Tm=273
Lf= Lf_nf
Tinit = temp_hot
sinit = 0
start 'inner' (0,0.01252)
value(temp)=temp_cold
ARC(CENTER=0,0) ANGLE=360 TO CLOSE
start 'Outer' (0,0.04)
value(temp)=temp_cold
ARC(CENTER=0,0) ANGLE=360 TO CLOSE
region 2 'Fin1'
rho=7900
cp =560
Lf= 205e3
Tm=273+1085
Tinit = temp_hot
sinit = 1
K=20.22
start  'Fin' (-0.00384,        0.011896487)
value(solid)=1
line to (-0.00384,        0.02244606)
line to (-0.00344,        0.02244606)
line to (-0.00344,        0.012018253)
line to close
start  'Fin' (-0.000239975,        0.012498576)
value(solid)=1
line to (-0.00024,        0.028499856)
line to (0.00016,        0.028499856)
line to (0.00016,        0.012499856)
line to close
start  'Fin' (0.00336,        0.012040864)
value(solid)=1
line to (0.00336,        0.02244606)
line to (0.00376,        0.02244606)
line to (0.00376,        0.011922013)
line to close
region 3 'Fin2'
rho=2720
cp =1050
Lf= 205e3
```

Conclusion

```
Tm=273+1085
Tinit = temp_hot
sinit = 1
K=229.05
start  'Fin' (0.012204766,        -0.002704751)
value(solid)=1
line to (0.021318858,        -0.007966774)
line to (0.021118858,        -0.008313184)
line to (0.012107694,        -0.003110587)
line to close
start  'Fin' (0.010944072,        -0.006041463)
value(solid)=1
line to (0.024801101,        -0.014041823)
line to (0.024561126,        -0.014457472)
line to (0.010704097,        -0.006457113)
line to close
start  'Fin' (0.008747694,        -0.008930277)
value(solid)=1
line to (0.017758858,        -0.014132875)
line to (0.017558858,        -0.014479285)
line to (0.008444766,        -0.009217262)
line to close
region 4 'Fin3'
rho=8960
cp =385
Lf= 205e3
Tm=273+1085
Tinit = temp_hot
sinit = 1
K=400
start  'Fin'(-0.008444766,        -0.009217262)
value(solid)=1
line to (-0.017558858,        -0.014479285)
line to (-0.017758858,        -0.014132875)
line to (-0.008747694,        -0.008930277)
line to close
start  'Fin' (-0.010704097,        -0.006457113)
value(solid)=1
line to (-0.024561126,        -0.014457472)
line to (-0.024801101,        -0.014041823)
line to (-0.010944072,        -0.006041463)
line to close
start  'Fin' (-0.012107694,        -0.003110587)
value(solid)=1
```

```
line to (-0.021118858,        -0.008313184)
line to (-0.021318858,        -0.007966774)
line to (-0.012204766,        -0.002704751)
line to close
TIME 0 by 5e-4 to 500
MONITORS
for cycle=10
grid(x,y)
PLOTS
for t= 0  by 5 to endtime
tecplot(temp)
tecplot(solid)
contour(temp)
contour(solid) range=(0,1)
Summary
HISTORIES
history(integral(cp*(temp)+Lf*(1-solid))) as "Total Energy"points = 100 export
format "#d#b#1" file="total energy parameter.txt" !report(t_star)
history(integral(rho*cp*(temp)+rho*Lf*(1-solid))) as "Total Energy"points =
100 export format "#d#b#1" file="total energy.txt" !report(t_star)
history((integral(1-solid))/(integral(1))) as "Liq_ave_main"points = 100 ex-
port format "#d#b#1" file="Liq_ave_main.txt"
history((integral(solid))/(integral(1))) as "solid_ave_main"points = 100 ex-
port format "#d#b#1" file="Solid_ave_main.txt"
history((integral(temp))/(integral(1))) as "T_ave"points = 100 export format
"#d#b#1" file="T_ave.txt"
End
```

2.5 sample 5

```
TITLE
'PCM Sample Code 5'
SELECT
painted
smoothinit
threads=2
Spectral_colors
VARIABLES
temp(threshold=1)
solid(threshold=0.01)
DEFINITIONS
Lf
Tm
```

Conclusion

```
TO= 1
temp_hot=273+2
temp_cold=240
Tinit
sinit
rho { Density kg/m3 }
cp { heat capacity }
r2=0.0452
r1=0.22
k
L_1=0.007
L_2=0.01
L=4.2*r1
X1=0.1*L
X2=0.6*L
L1=L/2
L2=L/4
beta1=pi/4
theta=0.050745
t_fin=0.3e-3
gamma=Arctan(t_fin/(2*r1))
eps=1e-6
response = 0.1
s0 = 0.5*erfc((temp-Tm)/T0)
rho_f=997
rho_s1=3970
rho_s2=1800
cp_f=4179
cp_s1=765
cp_s2=717
ks1=40
ks2=5000
s-8.6
phi1=0.05
phi2=0.05
Rd=0.5
Kf=if solid=0 then 2.24 else 0.764
rho_nf=1200.2
Cp_nf=3384.2
Lf_nf=2.6e5
INITIAL VALUES
temp=Tinit
solid = 0.5*erfc((tinit-Tm)/T0)
EQUATIONS
```

```
temp: rho*cp*dt(temp) - div(k*grad(temp)) = Lf*dt(solid)+4/3*Rd*dyy(temp)
solid:   dt(solid) - eps*div(grad(solid)) =  (S0-solid)/response
BOUNDARIES
region 1 'melting Plate'
cp = cp_nf
rho=rho_nf
k=Kf
Tm=273
Lf= Lf_nf
Tinit = temp_hot
sinit = 0
start 'inner' (0,0.149041158)
value(temp)=temp_cold
ARC(CENTER=0,0) ANGLE=360 TO CLOSE
start 'Outer' (0,0.55)
value(temp)=temp_cold
ARC(CENTER=0,0) ANGLE=360 TO CLOSE
region 2 'Fin1'
rho=7900
cp =560
Lf= 205e3
Tm=273+1085
Tinit = temp_hot
sinit = 1
K=398
start  'Fin' (0.0072,         0.148867145)
value(solid)=1
line to (0.0072,        0.182180989)
line to (-0.0312,        0.204351239)
line to (0.06,        0.257005584)
line to(-0.0312,        0.309659928)
line to(0.06,        0.362314273)
line to(0.0072,        0.392798367)
line to(0.0072,        0.409484523)
line to(-0.0072,        0.409484523)
line to(-0.0072,        0.384484523)
line to(0.0312,        0.362314273)
line to(-0.06,        0.309659928)
line to(0.0312,        0.257005584)
line to(-0.06,        0.204351239)
line to(-0.0072,        0.173867145)
line to(-0.0072,        0.148867145)
line to close
start  'Fin' (0.132522729,        -0.06819819)
```

```
value(solid)=1
line to(0.161373364,        -0.084855111)
line to(0.161373364,        -0.129195612)
line to(0.252573364,        -0.076541268)
line to(0.252573364,        -0.181849957)
line to(0.343773364,        -0.129195612)
line to(0.343773364,        -0.190163801)
line to(0.358223999,        -0.198506879)
line to(0.351023999,        -0.210977644)
line to(0.329373364,        -0.198477644)
line to(0.329373364,        -0.154137144)
line to(0.238173364,        -0.206791488)
line to(0.238173364,        -0.101482799)
line to(0.146973364,        -0.154137144)
line to(0.146973364,        -0.093168955)
line to(0.125322729,        -0.080668955)
line to close
start  'Fin' (-0.125322729,        -0.080668955)
value(solid)=1
line to(-0.154173364,        -0.097325877)
line to(-0.192573364,        -0.075155627)
line to(-0.192573364,        -0.180464316)
line to(-0.283773364,        -0.127809971)
line to(-0.283773364,        -0.233118661)
line to(-0.336573364,        -0.202634566)
line to(-0.351023999,        -0.210977644)
line to(-0.358223999,        -0.198506879)
line to(-0.336573364,        -0.186006879)
line to(-0.298173364,        -0.208177129)
line to(-0.298173364,        -0.10286844)
line to(-0.206973364,        -0.155522784)
line to(-0.206973364,        -0.050214095)
line to(-0.154173364,        -0.08069819)
line to(-0.132522729,        -0.06819819)
line to close
TIME 0 by 5e-4 to 50000
MONITORS
for cycle=10
grid(x,y)
PLOTS
for t= 0  by 5 to endtime
tecplot(temp)
tecplot(solid)
contour(temp)
```

```
contour(solid) range=(0,1)
Summary
HISTORIES
history(integral(cp*(temp)+Lf*(1-solid))) as "Total Energy"points = 100 export
format "#d#b#1" file="total energy parameter.txt" !report(t_star)
history(integral(rho*cp*(temp)+rho*Lf*(1-solid))) as "Total Energy"points =
100 export format "#d#b#1" file="total energy.txt" !report(t_star)
history((integral(1-solid))/(integral(1))) as "Liq_ave_main"points = 100 ex-
port format "#d#b#1" file="Liq_ave_main.txt"
history((integral(solid))/(integral(1))) as "solid_ave_main"points = 100 ex-
port format "#d#b#1" file="Solid_ave_main.txt"
history((integral(temp))/(integral(1))) as "T_ave"points = 100 export format
"#d#b#1" file="T_ave.txt"
End
```

Compilation of References

Abbas, N., Rehman, K. U., Shatanawi, W., & Malik, M. Y. (2022). Numerical study of heat transfer in hybrid nanofluid flow over permeable nonlinear stretching curved surface with thermal slip. *International Communications in Heat and Mass Transfer, 135*, 106107. doi:10.1016/j.icheatmasstransfer.2022.106107

Abderrahmane, A., Manoongam, A., Alizadeh, A. A., Younis, O., Zekri, H., Isa, S. S. P. M., Baghaei, S., Jamshed, W., & Guedri, K. (2023). Investigation of the free convection of nanofluid flow in a wavy porous enclosure subjected to a magnetic field using the Galerkin finite element method. *Journal of Magnetism and Magnetic Materials, 569*, 170446. doi:10.1016/j.jmmm.2023.170446

Acharya, N., Maity, S., & Kundu, P. K. (2022). Entropy generation optimization of unsteady radiative hybrid nanofluid flow over a slippery spinning disk. *Proceedings of the Institution of Mechanical Engineers. Part C, Journal of Mechanical Engineering Science, 236*(11), 6007–6024. doi:10.1177/09544062211065384

Afridi, M. I., Alkanhal, T. A., Qasim, M., & Tlili, I. (2019). Entropy generation in Cu-Al2O3-H2O hybrid nanofluid flow over a curved surface with thermal dissipation. *Entropy (Basel, Switzerland), 21*(10), 941. doi:10.3390/e21100941

Ahmad, W., Syed, K. S., Ishaq, M., Hassan, A., & Iqbal, Z. (2017). Numerical study of conjugate heat transfer in a double-pipe with exponential fins using DGFEM. *Applied Thermal Engineering, 111*, 1184–1201. doi:10.1016/j.applthermaleng.2016.09.171

Alharbi, S. O., Nawaz, M., & Nazir, U. (2019). Thermal analysis for hybrid nanofluid past a cylinder exposed to magnetic field. *AIP Advances, 9*(11), 115022. doi:10.1063/1.5127327

Alipour, N., Jafari, B., & Hosseinzadeh, K. (2023). Optimization of wavy trapezoidal porous cavity containing mixture hybrid nanofluid (water/ethylene glycol Go–Al2O3) by response surface method. *Scientific Reports, 13*(1), 1635. doi:10.103841598-023-28916-2 PMID:36717648

Alizadeh, M., Hosseinzadeh, K., & Ganji, D. D. (2019). Investigating the effects of hybrid nanoparticles on solid-liquid phase change process in a Y-shaped fin-assisted LHTESS by means of FEM. *Journal of Molecular Liquids, 287*, 110931. doi:10.1016/j.molliq.2019.110931

Alizadeh, M., Hosseinzadeh, K., Mehrzadi, H., & Ganji, D. D. (2019). Investigation of LHTESS filled by Hybrid nano-enhanced PCM with Koch snowflake fractal cross section in the presence of thermal radiation. *Journal of Molecular Liquids, 273*, 414–424. doi:10.1016/j.molliq.2018.10.049

Alkanhal, T. A., Sheikholeslami, M., Usman, M., Haq, R. U., Shafee, A., Al-Ahmadi, A. S., & Tlili, I. (2019). Thermal management of MHD nanofluid within the porous medium enclosed in a wavy shaped cavity with square obstacle in the presence of radiation heat source. *International Journal of Heat and Mass Transfer, 139*, 87–94. doi:10.1016/j.ijheatmasstransfer.2019.05.006

Awan, A. U., Ali, B., Shah, S. A. A., Oreijah, M., Guedri, K., & Eldin, S. M. (2023). Numerical analysis of heat transfer in Ellis hybrid nanofluid flow subject to a stretching cylinder. *Case Studies in Thermal Engineering*, *49*, 103222. doi:10.1016/j.csite.2023.103222

Barman, P., & Rao, P. (2021). Effect of aspect ratio on natural convection in a wavy porous cavity submitted to a partial heat source. *International Communications in Heat and Mass Transfer*, *126*, 105453. doi:10.1016/j.icheatmasstransfer.2021.105453

Barnoon, P., Toghraie, D., Dehkordi, R. B., & Afrand, M. (2019). Two phase natural convection and thermal radiation of Non-Newtonian nanofluid in a porous cavity considering inclined cavity and size of inside cylinders. *International Communications in Heat and Mass Transfer*, *108*, 104285. doi:10.1016/j.icheatmasstransfer.2019.104285

Barreneche, C., Navarro, H., Serrano, S., Cabeza, L. F., & Fernández, A. I. (2014). New database on phase change materials for thermal energy storage in buildings to help PCM selection. *Energy Procedia*, *57*, 2408–2415. doi:10.1016/j.egypro.2014.10.249

Bejan, A. (2013). *Convection heat transfer*. John Wiley & Sons. doi:10.1002/9781118671627

Bhagat, K., Prabhakar, M., & Saha, S. K. (2018). Estimation of thermal performance and design optimization of finned multitube latent heat thermal energy storage. *Journal of Energy Storage*, *19*, 135–144. doi:10.1016/j.est.2018.06.014

Biswas, N., Mondal, M. K., Mandal, D. K., Manna, N. K., Gorla, R. S. R., & Chamkha, A. J. (2022). A narrative loom of hybrid nanofluid-filled wavy walled tilted porous enclosure imposing a partially active magnetic field. *International Journal of Mechanical Sciences*, *217*, 107028. doi:10.1016/j.ijmecsci.2021.107028

Buongiorno, J. (2006). *Convective transport in nanofluids*. Academic Press.

Cakmak, N. K., Said, Z., Sundar, L. S., Ali, Z. M., & Tiwari, A. K. (2020). Preparation, characterization, stability, and thermal conductivity of rGO-Fe3O4-TiO2 hybrid nanofluid: An experimental study. *Powder Technology*, *372*, 235–245. doi:10.1016/j.powtec.2020.06.012

Can, A., Selimefendigil, F., & Öztop, H. F. (2022). A review on soft computing and nanofluid applications for battery thermal management. *Journal of Energy Storage*, *53*, 105214. doi:10.1016/j.est.2022.105214

Chamkha, A. J., Doostanidezfuli, A., Izadpanahi, E., & Ghalambaz, M. J. A. P. T. (2017). Phase-change heat transfer of single/hybrid nanoparticles-enhanced phase-change materials over a heated horizontal cylinder confined in a square cavity. *Advanced Powder Technology*, *28*(2), 385–397. doi:10.1016/j.apt.2016.10.009

Chamkha, A., Ismael, M., Kasaeipoor, A., & Armaghani, T. (2016). Entropy generation and natural convection of CuO-water nanofluid in C-shaped cavity under magnetic field. *Entropy (Basel, Switzerland)*, *18*(2), 50. doi:10.3390/e18020050

David Müzel, S., Bonhin, E. P., Guimarães, N. M., & Guidi, E. S. (2020). Application of the finite element method in the analysis of composite materials: A review. *Polymers*, *12*(4), 818. doi:10.3390/polym12040818 PMID:32260389

Dekhil, M. A., Tala, J. V. S., Bulliard-Sauret, O., & Bougeard, D. (2021). Numerical analysis of the performance enhancement of a latent heat storage shell and tube unit using finned tubes during melting and solidification. *Applied Thermal Engineering*, *192*, 116866. doi:10.1016/j.applthermaleng.2021.116866

Devarajan, M., Parasumanna Krishnamurthy, N., Balasubramanian, M., Ramani, B., Wongwises, S., Abd El-Naby, K., & Sathyamurthy, R. (2018). Thermophysical properties of CNT and CNT/Al2O3 hybrid nanofluid. *Micro & Nano Letters*, *13*(5), 617–621. doi:10.1049/mnl.2017.0029

Dhatt, G., Lefrançois, E., & Touzot, G. (2012). *Finite element method*. John Wiley & Sons. doi:10.1002/9781118569764

Ebrahimi, D., Yousefzadeh, S., Akbari, O. A., Montazerifar, F., Rozati, S. A., Nakhjavani, S., & Safaei, M. R. (2021). Mixed convection heat transfer of a nanofluid in a closed elbow-shaped cavity (CESC). *Journal of Thermal Analysis and Calorimetry*, *144*(6), 2295–2316. doi:10.100710973-021-10548-1

Elbahjaoui, R., & El Qarnia, H. (2017). Thermal analysis of nanoparticle-enhanced phase change material solidification in a rectangular latent heat storage unit including natural convection. *Energy and Building*, *153*, 1–17. doi:10.1016/j.enbuild.2017.08.003

Esfahani, N. N., Toghraie, D., & Afrand, M. (2018). A new correlation for predicting the thermal conductivity of ZnO–Ag (50%–50%)/water hybrid nanofluid: An experimental study. *Powder Technology*, *323*, 367–373. doi:10.1016/j.powtec.2017.10.025

Fallah Najafabadi, M., Talebi Rostami, H., Hosseinzadeh, K., & Ganji, D. D. (2022). Hydrothermal study of nanofluid flow in channel by RBF method with exponential boundary conditions. *Proceedings of the Institution of Mechanical Engineers, Part E: Journal of Process Mechanical Engineering*.

Fallahi, A., Guldentops, G., Tao, M., Granados-Focil, S., & Van Dessel, S. (2017). Review on solid-solid phase change materials for thermal energy storage: Molecular structure and thermal properties. *Applied Thermal Engineering*, *127*, 1427–1441. doi:10.1016/j.applthermaleng.2017.08.161

Faraj, K., Khaled, M., Faraj, J., Hachem, F., & Castelain, C. (2020). Phase change material thermal energy storage systems for cooling applications in buildings: A review. *Renewable & Sustainable Energy Reviews*, *119*, 109579. doi:10.1016/j.rser.2019.109579

Farid, M. M., & Khalaf, A. N. (1994). Performance of direct contact latent heat storage units with two hydrated salts. *Solar Energy*, *52*(2), 179–189. doi:10.1016/0038-092X(94)90067-1

Ferrer, G., Solé, A., Barreneche, C., Martorell, I., & Cabeza, L. F. (2015). Review on the methodology used in thermal stability characterization of phase change materials. *Renewable & Sustainable Energy Reviews*, *50*, 665–685. doi:10.1016/j.rser.2015.04.187

Ghadikolaei, S. S., Hosseinzadeh, K., & Ganji, D. D. (2018). Investigation on ethylene glycol-water mixture fluid suspend by hybrid nanoparticles (TiO2-CuO) over rotating cone with considering nanoparticles shape factor. *Journal of Molecular Liquids*, *272*, 226–236. doi:10.1016/j.molliq.2018.09.084

Ghasemi, B., & Aminossadati, S. M. (2009). Natural convection heat transfer in an inclined enclosure filled with a water-CuO nanofluid. *Numerical Heat Transfer Part A*, *55*(8), 807–823. doi:10.1080/10407780902864623

Goharshadi, E. K., Ahmadzadeh, H., Samiee, S., & Hadadian, M. (2013). *Nanofluids for heat transfer enhancement-a review*. Academic Press.

Gorzin, M., Hosseini, M. J., Rahimi, M., & Bahrampoury, R. (2019). Nano-enhancement of phase change material in a shell and multi-PCM-tube heat exchanger. *Journal of Energy Storage*, *22*, 88–97. doi:10.1016/j.est.2018.12.023

Hamilton, R. L., & Crosser, O. (1962). Thermal conductivity of heterogeneous two-component systems. *Industrial & Engineering Chemistry Fundamentals*, *1*(3), 187–191. doi:10.1021/i160003a005

Haq, R. U., Soomro, F. A., Hammouch, Z., & Rehman, S. U. (2018). Heat exchange within the partially heated C-shape cavity filled with the water based SWCNTs. *International Journal of Heat and Mass Transfer*, *127*, 506–514. doi:10.1016/j.ijheatmasstransfer.2018.07.101

Hatami, M., Zhou, J., Geng, J., & Jing, D. J. J. O. M. (2018). Variable magnetic field (VMF) effect on the heat transfer of a half-annulus cavity filled by Fe3O4-water nanofluid under constant heat flux. *Journal of Magnetism and Magnetic Materials*, *451*, 173–182. doi:10.1016/j.jmmm.2017.10.110

Hayat, T., & Nadeem, S. (2017). Heat transfer enhancement with Ag–CuO/water hybrid nanofluid. *Results in Physics*, *7*, 2317–2324. doi:10.1016/j.rinp.2017.06.034

Hémery, C.-V., Pra, F., Robin, J.-F., & Marty, P. (2014). Experimental performances of a battery thermal management system using a phase change material. *Journal of Power Sources*, *270*, 349–358. doi:10.1016/j.jpowsour.2014.07.147

Heris, S. Z., Pour, M. B., Mahian, O., & Wongwises, S. (2014). A comparative experimental study on the natural convection heat transfer of different metal oxide nanopowders suspended in turbine oil inside an inclined cavity. *International Journal of Heat and Mass Transfer*, *73*, 231–238. doi:10.1016/j.ijheatmasstransfer.2014.01.071

Hosseinzadeh, K., Mardani, M. R., Paikar, M., Hasibi, A., Tavangar, T., Nimafar, M., ... & Shafii, M. B. (2022). Investigation of second grade viscoelastic non-Newtonian nanofluid flow on the curve stretching surface in presence of MHD. *Results in Engineering*, *17*.

Hosseinzadeh, K., Akbari, S., Faghiri, S., & Shafii, M. B. (2023). Optimization approach for MoS2-water ethylene glycol mixture nanofluid flow in a wavy enclosure. *International Journal of Thermofluids*, *18*, 100337. doi:10.1016/j.ijft.2023.100337

Hosseinzadeh, K., Alizadeh, M., Alipour, M. H., Jafari, B., & Ganji, D. D. (2019). Effect of nanoparticle shape factor and snowflake crystal structure on discharging acceleration LHTESS containing (Al2O3-GO) HNEPCM. *Journal of Molecular Liquids*, *289*, 111140. doi:10.1016/j.molliq.2019.111140

Hosseinzadeh, K., Alizadeh, M., & Ganji, D. D. (2019). Solidification process of hybrid nano-enhanced phase change material in a LHTESS with tree-like branching fin in the presence of thermal radiation. *Journal of Molecular Liquids*, *275*, 909–925. doi:10.1016/j.molliq.2018.11.109

Hosseinzadeh, K., Alizadeh, M., Tavakoli, M. H., & Ganji, D. D. (2019). Investigation of phase change material solidification process in a LHTESS in the presence of fins with variable thickness and hybrid nanoparticles. *Applied Thermal Engineering*, *152*, 706–717. doi:10.1016/j.applthermaleng.2019.02.111

Hosseinzadeh, K., Asadi, A., Mogharrebi, A., Ermia Azari, M., & Ganji, D. (2021). Investigation of mixture fluid suspended by hybrid nanoparticles over vertical cylinder by considering shape factor effect. *Journal of Thermal Analysis and Calorimetry*, *143*(2), 1081–1095. doi:10.100710973-020-09347-x

Hosseinzadeh, K., Faghiri, S., Akbari, S., Kermani, J. R., Jafari, B., & Shafii, M. B. (2023). Effect of ternary hybrid nanoparticles (GO-MgO-TiO2) and radiative heat transfer on the solidification process of PCM inside a triplex tube with hollow fins. *International Journal of Thermofluids*, *20*, 100443. doi:10.1016/j.ijft.2023.100443

Hosseinzadeh, K., Mardani, M. R., Salehi, S., Paikar, M., & Ganji, D. D. (2021). Investigation of micropolar hybrid nanofluid (iron oxide–molybdenum disulfide) flow across a sinusoidal cylinder in presence of magnetic field. *International Journal of Applied and Computational Mathematics*, *7*(5), 1–17. doi:10.100740819-021-01148-6

Hosseinzadeh, K., Moghaddam, M. E., Asadi, A., Mogharrebi, A. R., & Ganji, D. D. (2020). Effect of internal fins along with hybrid nano-particles on solid process in star shape triplex latent heat thermal energy storage system by numerical simulation. *Renewable Energy*, *154*, 497–507. doi:10.1016/j.renene.2020.03.054

Hosseinzadeh, K., Moghaddam, M. E., Asadi, A., Mogharrebi, A. R., Jafari, B., Hasani, M. R., & Ganji, D. D. (2021). Effect of two different fins (longitudinal-tree like) and hybrid nano-particles (MoS2-TiO2) on solidification process in triplex latent heat thermal energy storage system. *Alexandria Engineering Journal*, *60*(1), 1967–1979. doi:10.1016/j.aej.2020.12.001

Hosseinzadeh, K., Moghaddam, M. E., Nateghi, S., Shafii, M. B., & Ganji, D. D. (2023). Radiation and convection heat transfer optimization with MHD analysis of a hybrid nanofluid within a wavy porous enclosure. *Journal of Magnetism and Magnetic Materials*, *566*, 170328. doi:10.1016/j.jmmm.2022.170328

Hosseinzadeh, K., Mogharrebi, A. R., Asadi, A., Paikar, M., & Ganji, D. D. (2020). Effect of fin and hybrid nano-particles on solid process in hexagonal triplex latent heat thermal energy storage system. *Journal of Molecular Liquids*, *300*, 112347. doi:10.1016/j.molliq.2019.112347

Hosseinzadeh, K., Montazer, E., Shafii, M. B., & Ganji, A. R. D. (2021). Solidification enhancement in triplex thermal energy storage system via triplets fins configuration and hybrid nanoparticles. *Journal of Energy Storage*, *34*, 102177. doi:10.1016/j.est.2020.102177

Hosseinzadeh, K., Montazer, E., Shafii, M. B., & Ganji, D. D. (2021). Heat transfer hybrid nanofluid (1-Butanol/MoS2–Fe3O4) through a wavy porous cavity and its optimization. *International Journal of Numerical Methods for Heat & Fluid Flow*, *31*(5), 1547–1567. doi:10.1108/HFF-07-2020-0442

Hosseinzadeh, K., Roghani, S., Mogharrebi, A. R., Asadi, A., & Ganji, D. D. (2021). Optimization of hybrid nanoparticles with mixture fluid flow in an octagonal porous medium by effect of radiation and magnetic field. *Journal of Thermal Analysis and Calorimetry*, *143*(2), 1413–1424. doi:10.100710973-020-10376-9

Hosseinzadeh, K., Roshani, M., Attar, M. A., Ganji, D. D., & Shafii, M. B. (2023). Heat transfer study and optimization of nanofluid triangular cavity with a pentagonal barrier by finite element approach and RSM. *Heliyon*, *9*(9), e20193. doi:10.1016/j.heliyon.2023.e20193 PMID:37809404

Huebner, K. H., Dewhirst, D. L., Smith, D. E., & Byrom, T. G. (2001). *The finite element method for engineers*. John Wiley & Sons.

Hutton, D. V. (2004). *Fundamentals of finite element analysis*. The McGraw Hill Companies.

Hwang, K. S., Lee, J. H., & Jang, S. P. (2007). Buoyancy-driven heat transfer of water-based Al2O3 nanofluids in a rectangular cavity. *International Journal of Heat and Mass Transfer*, *50*(19-20), 4003–4010. doi:10.1016/j.ijheatmasstransfer.2007.01.037

Ismail, K. A. R., Alves, C. L. F., & Modesto, M. S. (2001). Numerical and experimental study on the solidification of PCM around a vertical axially finned isothermal cylinder. *Applied Thermal Engineering*, *21*(1), 53–77. doi:10.1016/S1359-4311(00)00002-8

Jelle, B., & Kalnæs, S. (2017). Phase change materials for application in energy-efficient buildings. *Cost-effective energy efficient building retrofitting*, 57-118.

Jouhara, H., Żabnieńska-Góra, A., Khordehgah, N., Ahmad, D., & Lipinski, T. (2020). Latent thermal energy storage technologies and applications: A review. *International Journal of Thermofluids*, *5*, 100039. doi:10.1016/j.ijft.2020.100039

Kaluri, R. S., & Basak, T. (2011). Role of entropy generation on thermal management during natural convection in porous square cavities with distributed heat sources. *Chemical Engineering Science*, *66*(10), 2124–2140. doi:10.1016/j.ces.2011.02.009

Kant, K., Shukla, A., Sharma, A., & Biwole, P. H. (2017). Heat transfer study of phase change materials with graphene nano particle for thermal energy storage. *Solar Energy*, *146*, 453–463. doi:10.1016/j.solener.2017.03.013

Khanafer, K., Vafai, K., & Lightstone, M. (2003). Buoyancy-driven heat transfer enhancement in a two-dimensional enclosure utilizing nanofluids. *International Journal of Heat and Mass Transfer*, *46*(19), 3639–3653. doi:10.1016/S0017-9310(03)00156-X

Khan, S. A., Hayat, T., & Alsaedi, A. (2022). Thermal conductivity performance for ternary hybrid nanomaterial subject to entropy generation. *Energy Reports*, *8*, 9997–10005. doi:10.1016/j.egyr.2022.07.149

Koçak, B., Fernandez, A. I., & Paksoy, H. (2020). Review on sensible thermal energy storage for industrial solar applications and sustainability aspects. *Solar Energy*, *209*, 135–169. doi:10.1016/j.solener.2020.08.081

Kuznetsov, A. V., & Nield, D. (2010). The onset of double-diffusive nanofluid convection in a layer of a saturated porous medium. *Transport in Porous Media*, *85*(3), 941–951. doi:10.100711242-010-9600-1

Kwok, K. W., Leung, K. M., Flahaut, E., Cheng, J., & Cheng, S. H. (2010). Chronic toxicity of double-walled carbon nanotubes to three marine organisms: Influence of different dispersion methods. *Nanomedicine (London)*, *5*(6), 951–961. doi:10.2217/nnm.10.59 PMID:20735229

Lane, G. A. (1992). Phase change materials for energy storage nucleation to prevent supercooling. *Solar Energy Materials and Solar Cells*, *27*(2), 135–160. doi:10.1016/0927-0248(92)90116-7

Li, D., Wu, Y., Wang, B., Liu, C., & Arıcı, M. (2020). Optical and thermal performance of glazing units containing PCM in buildings: A review. *Construction & Building Materials*, *233*, 117327. doi:10.1016/j.conbuildmat.2019.117327

Li, H. (2022). *Graded Finite Element Methods for Elliptic Problems in Nonsmooth Domains* (Vol. 10). Springer Nature. doi:10.1007/978-3-031-05821-9

Liu, Z., & Liu, Z. (2018). Numerical Solution of PDEs. *Multiphysics in Porous Materials*, 79-89.

Liu, S., Peng, H., Hu, Z., Ling, X., & Huang, J. (2019). Solidification performance of a latent heat storage unit with innovative longitudinal triangular fins. *International Journal of Heat and Mass Transfer*, *138*, 667–676. doi:10.1016/j.ijheatmasstransfer.2019.04.121

Li, Y., & Liu, S. (2014). Experimental study on thermal performance of a solar chimney combined with PCM. *Applied Energy*, *114*, 172–178. doi:10.1016/j.apenergy.2013.09.022

Lohrasbi, S., Sheikholeslami, M., & Ganji, D. D. (2016). Discharging process expedition of NEPCM in fin-assisted latent heat thermal energy storage system. *Journal of Molecular Liquids*, *221*, 833–841. doi:10.1016/j.molliq.2016.06.044

Luo, X. H., Li, B. W., & Hu, Z. M. (2016). Effects of thermal radiation on MHD flow and heat transfer in a cubic cavity. *International Journal of Heat and Mass Transfer*, *92*, 449–466. doi:10.1016/j.ijheatmasstransfer.2015.08.104

Maccarini, A., Hultmark, G., Bergsøe, N. C., & Afshari, A. (2018). Free cooling potential of a PCM-based heat exchanger coupled with a novel HVAC system for simultaneous heating and cooling of buildings. *Sustainable Cities and Society*, *42*, 384–395. doi:10.1016/j.scs.2018.06.016

Madhesh, D., Parameshwaran, R., & Kalaiselvam, S. (2014). Experimental investigation on convective heat transfer and rheological characteristics of Cu–TiO2 hybrid nanofluids. *Experimental Thermal and Fluid Science*, *52*, 104–115. doi:10.1016/j.expthermflusci.2013.08.026

Manjunatha, S., Puneeth, V., Gireesha, B., & Chamkha, A. (2022). Theoretical study of convective heat transfer in ternary nanofluid flowing past a stretching sheet. *Journal of Applied and Computational Mechanics*, *8*(4), 1279–1286.

Mehling, H., & Cabeza, L. F. (2007). Phase change materials and their basic properties. In *Thermal energy storage for sustainable energy consumption: fundamentals, case studies and design* (pp. 257–277). Springer. doi:10.1007/978-1-4020-5290-3_17

Mehling, H., & Cabeza, L. F. (2008). Heat and cold storage with PCM. *Heat and Mass Transfer*, 11–55.

Miroshnichenko, I. V., Sheremet, M. A., Oztop, H. F., & Al-Salem, K. (2016). MHD natural convection in a partially open trapezoidal cavity filled with a nanofluid. *International Journal of Mechanical Sciences*, *119*, 294–302. doi:10.1016/j.ijmecsci.2016.11.001

Mondal, P., & Mahapatra, T. R. (2021). MHD double-diffusive mixed convection and entropy generation of nanofluid in a trapezoidal cavity. *International Journal of Mechanical Sciences*, *208*, 106665. doi:10.1016/j.ijmecsci.2021.106665

Moradi, M. R., Hosseinzadeh, K., Hasibi, A., & Domiri Ganji, D. (2023). Hydrothermal study on nano-bioconvective fluid flow over a vertical plate under the effect of magnetic field. *Numerical Heat Transfer Part B*, 1–15. doi:10.1080/10407790.2023.2241632

Najafabadi, M. F., TalebiRostami, H., Hosseinzadeh, K., & Ganji, D. D. (2022). Investigation of nanofluid flow in a vertical channel considering polynomial boundary conditions by Akbari-Ganji's method. *Theoretical & Applied Mechanics Letters*, *12*(4), 100356. doi:10.1016/j.taml.2022.100356

Namburu, P. K., Kulkarni, D. P., Misra, D., & Das, D. K. (2007). Viscosity of copper oxide nanoparticles dispersed in ethylene glycol and water mixture. *Experimental Thermal and Fluid Science*, *32*(2), 397–402. doi:10.1016/j.expthermflusci.2007.05.001

Nield, D. A., & Kuznetsov, A. V. (2011). The Cheng–Minkowycz problem for the double-diffusive natural convective boundary layer flow in a porous medium saturated by a nanofluid. *International Journal of Heat and Mass Transfer*, *54*(1-3), 374–378. doi:10.1016/j.ijheatmasstransfer.2010.09.034

Nikishkov, G. P. (2004). Introduction to the finite element method. University of Aizu.

Parekh, K. (2014). Thermo-magnetic properties of ternary polydispersed Mn0. 5Zn0. 5Fe2O4 ferrite magnetic fluid. *Solid State Communications*, *187*, 33–37. doi:10.1016/j.ssc.2014.02.005

Pepper, D. W., & Heinrich, J. C. (2005). *The finite element method: basic concepts and applications*. Taylor & Francis. doi:10.1201/9780203942352

Pimenov, D. Y., Mia, M., Gupta, M. K., Machado, Á. R., Pintaude, G., Unune, D. R., Khanna, N., Khan, A. M., Tomaz, Í., Wojciechowski, S., & Kuntoğlu, M. (2022). Resource saving by optimization and machining environments for sustainable manufacturing: A review and future prospects. *Renewable & Sustainable Energy Reviews*, *166*, 112660. doi:10.1016/j.rser.2022.112660

Pustovalova, O., & Karyakin, M. (2017). Usage FlexPDE Package in the Courses of Mathematical Modeling and Mechanics. *Conference Proceedings. New Perspectives in Science Education 2017.*

Qi, C., Liang, L., & Rao, Z. (2016). Study on the flow and heat transfer of liquid metal based nanofluid with different nanoparticle radiuses using two-phase lattice Boltzmann method. *International Journal of Heat and Mass Transfer*, *94*, 316–326. doi:10.1016/j.ijheatmasstransfer.2015.11.068

Qi, C., Yang, L., & Wang, G. (2017). Numerical study on convective heat transfer enhancement in horizontal rectangle enclosures filled with Ag-Ga nanofluid. *Nanoscale Research Letters*, *12*(1), 1–10. doi:10.118611671-017-2095-8 PMID:28476081

Qi, C., Zhao, N., Cui, X., Chen, T., & Hu, J. (2018). Effects of half spherical bulges on heat transfer characteristics of CPU cooled by TiO2-water nanofluids. *International Journal of Heat and Mass Transfer*, *123*, 320–330. doi:10.1016/j.ijheatmasstransfer.2018.02.086

Rahman, M. M., Alam, M. S., Al-Salti, N., & Eltayeb, I. A. (2016). Hydromagnetic natural convective heat transfer flow in an isosceles triangular cavity filled with nanofluid using two-component nonhomogeneous model. *International Journal of Thermal Sciences*, *107*, 272–288. doi:10.1016/j.ijthermalsci.2016.04.009

Raj, V. A. A., & Velraj, R. (2010). Review on free cooling of buildings using phase change materials. *Renewable & Sustainable Energy Reviews*, *14*(9), 2819–2829. doi:10.1016/j.rser.2010.07.004

Reddy, J. N. (2021b). Introduction to the finite element method. Cambridge University Press.

Rostami, A. K., Hosseinzadeh, K., & Ganji, D. D. (2022). Hydrothermal analysis of ethylene glycol nanofluid in a porous enclosure with complex snowflake shaped inner wall. *Waves in Random and Complex Media*, *32*(1), 1–18. doi:10.1080/17455030.2020.1758358

Sabat, L., & Kundu, C. K. (2020). History of finite element method: A review. *Recent Developments in Sustainable Infrastructure: Select Proceedings of ICRDSI 2019*, 395-404.

Şahan, N., & Paksoy, H. (2018). Novel shapeable phase change material (PCM) composites for thermal energy storage (TES) applications. *Solar Energy Materials and Solar Cells*, *174*, 380–387. doi:10.1016/j.solmat.2017.09.022

Sahu, S. K., Shaw, S., Thatoi, D. N., & Nayak, M. K. (2022). A thermal management of Darcy-Forchheimer SWCNT–MWCNT Cross hybrid nanofluid flow due to vertical stretched cylinder with and without inertia effects. *Waves in Random and Complex Media*, 1–27. doi:10.1080/17455030.2022.2088889

Said, Z., Le, D. T. N., Sharma, P., Dang, V. H., Le, H. S., Nguyen, D. T., Bui, T. A. E., & Nguyen, V. G. (2022). Optimization of combustion, performance, and emission characteristics of a dual-fuel diesel engine powered with microalgae-based biodiesel/diesel blends and oxyhydrogen. *Fuel*, *326*, 124987. doi:10.1016/j.fuel.2022.124987

Said, Z., Nguyen, T. H., Sharma, P., Li, C., Ali, H. M., Ahmed, S. F., ... Truong, T. H. (2022). Multi-attribute optimization of sustainable aviation fuel production-process from microalgae source. *Fuel*, *324*, 124759. doi:10.1016/j.fuel.2022.124759

Saleh, H., Roslan, R., & Hashim, I. (2011). Natural convection heat transfer in a nanofluid-filled trapezoidal enclosure. *International Journal of Heat and Mass Transfer*, *54*(1-3), 194–201. doi:10.1016/j.ijheatmasstransfer.2010.09.053

Sarbu, I., & Sebarchievici, C. (2018). A comprehensive review of thermal energy storage. *Sustainability (Basel)*, *10*(1), 191. doi:10.3390u10010191

Sathiyamoorthy, M., & Chamkha, A. (2010). Effect of magnetic field on natural convection flow in a liquid gallium filled square cavity for linearly heated side wall (s). *International Journal of Thermal Sciences*, *49*(9), 1856–1865. doi:10.1016/j.ijthermalsci.2010.04.014

Selimefendigil, F., & Öztop, H. F. (2019). Fluid-solid interaction of elastic-step type corrugation effects on the mixed convection of nanofluid in a vented cavity with magnetic field. *International Journal of Mechanical Sciences*, *152*, 185–197. doi:10.1016/j.ijmecsci.2018.12.044

Shabani, B., & Biju, M. (2015). Theoretical modelling methods for thermal management of batteries. *Energies*, *8*(9), 10153–10177. doi:10.3390/en80910153

Sheikholeslami, M. (2019). Numerical approach for MHD Al2O3-water nanofluid transportation inside a permeable medium using innovative computer method. *Computer Methods in Applied Mechanics and Engineering*, *344*, 306–318. doi:10.1016/j.cma.2018.09.042

Sheikholeslami, M., Mehryan, S. A. M., Shafee, A., & Sheremet, M. A. (2019). Variable magnetic forces impact on magnetizable hybrid nanofluid heat transfer through a circular cavity. *Journal of Molecular Liquids*, *277*, 388–396. doi:10.1016/j.molliq.2018.12.104

Shekaramiz, M., Fathi, S., Ataabadi, H. A., Kazemi-Varnamkhasti, H., & Toghraie, D. (2021). MHD nanofluid free convection inside the wavy triangular cavity considering periodic temperature boundary condition and velocity slip mechanisms. *International Journal of Thermal Sciences*, *170*, 107179. doi:10.1016/j.ijthermalsci.2021.107179

Sheremet, M. A., Cimpean, D. S., & Pop, I. (2017). Free convection in a partially heated wavy porous cavity filled with a nanofluid under the effects of Brownian diffusion and thermophoresis. *Applied Thermal Engineering*, *113*, 413–418. doi:10.1016/j.applthermaleng.2016.11.033

Sheremet, M. A., Oztop, H. F., & Pop, I. (2016). MHD natural convection in an inclined wavy cavity with corner heater filled with a nanofluid. *Journal of Magnetism and Magnetic Materials*, *416*, 37–47. doi:10.1016/j.jmmm.2016.04.061

Sheremet, M. A., & Pop, I. (2014). Conjugate natural convection in a square porous cavity filled by a nanofluid using Buongiorno's mathematical model. *International Journal of Heat and Mass Transfer*, *79*, 137–145. doi:10.1016/j.ijheatmasstransfer.2014.07.092

Sheremet, M. A., Trîmbiţaş, R., Groşan, T., & Pop, I. (2018). Natural convection of an alumina-water nanofluid inside an inclined wavy-walled cavity with a non-uniform heating using Tiwari and Das' nanofluid model. *Applied Mathematics and Mechanics*, *39*(10), 1425–1436. doi:10.100710483-018-2377-7

Shulepova, E. V., Sheremet, M. A., Oztop, H. F., & Abu-Hamdeh, N. (2020). Mixed convection of Al2O3–H2O nanoliquid in a square chamber with complicated fin. *International Journal of Mechanical Sciences*, *165*, 105192. doi:10.1016/j.ijmecsci.2019.105192

Suresh, S., Venkitaraj, K., & Selvakumar, P. (2011). Synthesis, characterisation of Al2O3-Cu nano composite powder and water based nanofluids. *Advanced Materials Research*, *328-330*, 1560–1567. doi:10.4028/www.scientific.net/AMR.328-330.1560

Taguchi, G. (1986). *Introduction to quality engineering: designing quality into products and processes*. Academic Press.

Taguchi, G. (1987). *Quality Resources*. Taguchi Techniques for Quality Engineering.

Taguchi, G. (1990). *Introduction to Quality Engineering*. Asian Productivity Organization.

Thantong, P., & Chantawong, P. (2017). Experimental study of a solar wall collector with PCM towards the natural ventilation of model house. *Energy Procedia*, *138*, 32–37. doi:10.1016/j.egypro.2017.10.041

Tiari, S., Qiu, S., & Mahdavi, M. (2016). Discharging process of a finned heat pipe–assisted thermal energy storage system with high temperature phase change material. *Energy Conversion and Management*, *118*, 426–437. doi:10.1016/j.enconman.2016.04.025

Velraj, R., Seeniraj, R., Hafner, B., Faber, C., & Schwarzer, K. (1999). Heat transfer enhancement in a latent heat storage system. *Solar Energy*, *65*(3), 171–180. doi:10.1016/S0038-092X(98)00128-5

Venter, G., Haftka, R., & Starnes, J. Jr. (1996, August). Construction of response surfaces for design optimization applications. In *6th Symposium on Multidisciplinary Analysis and Optimization* (p. 4040). doi:10.2514/6.1996-4040

Vieira, D. R., Calmon, J. L., & Coelho, F. Z. (2016). Life cycle assessment (LCA) applied to the manufacturing of common and ecological concrete: A review. *Construction & Building Materials*, *124*, 656–666. doi:10.1016/j.conbuildmat.2016.07.125

Ward, W. (n.d.). *Propellant Preheat System Optimization*. Academic Press.

Wolff, F., Beckermann, C., & Viskanta, R. (1988). Natural convection of liquid metals in vertical cavities. *Experimental Thermal and Fluid Science*, *1*(1), 83–91. doi:10.1016/0894-1777(88)90051-9

Xuan, Y., & Roetzel, W. (2000). Conceptions for heat transfer correlation of nanofluids. *International Journal of Heat and Mass Transfer*, *43*(19), 3701–3707. doi:10.1016/S0017-9310(99)00369-5

Zalba, B., Marín, J. M., Cabeza, L. F., & Mehling, H. (2003). Review on thermal energy storage with phase change: Materials, heat transfer analysis and applications. *Applied Thermal Engineering*, *23*(3), 251–283. doi:10.1016/S1359-4311(02)00192-8

Zangooee, M. R., Hosseinzadeh, K., & Ganj, D. D. (2022a). Investigation of three-dimensional hybrid nanofluid flow affected by nonuniform MHD over exponential stretching/shrinking plate. *Nonlinear Engineering*, *11*(1), 143–155. doi:10.1515/nleng-2022-0019

Zangooee, M. R., Hosseinzadeh, K., & Ganji, D. D. (2022b). Hydrothermal analysis of hybrid nanofluid flow on a vertical plate by considering slip condition. *Theoretical & Applied Mechanics Letters*, *12*(5), 100357. doi:10.1016/j.taml.2022.100357

Zangooee, M. R., Hosseinzadeh, K., & Ganji, D. D. (2023). Hydrothermal analysis of Ag and CuO hybrid NPs suspended in mixture of water 20%+ EG 80% between two concentric cylinders. *Case Studies in Thermal Engineering*, *50*, 103398. doi:10.1016/j.csite.2023.103398

Zhai, X., Wang, X., Wang, T., & Wang, R. (2013). A review on phase change cold storage in air-conditioning system: Materials and applications. *Renewable & Sustainable Energy Reviews*, *22*, 108–120. doi:10.1016/j.rser.2013.02.013

Zienkiewicz, O. C., Taylor, R. L., & Zhu, J. Z. (2005). *The finite element method: Its basis and fundamentals*. Elsevier.

About the Authors

Khashayar Hosseinzadeh is a distinguished scholar and researcher in the field of Mechanical Engineering-Energy Conversion, holding a Post-Doctoral Fellowship at Sharif University of Technology. He earned his Ph.D. in Mechanical Engineering-Energy Conversion from Babol Noshirvani University of Technology, Iran. Dr. Hosseinzadeh's expertise spans a wide range of topics within the realm of mechanical engineering and energy conversion. His primary research interests include Energy Storage Systems, Phase Change Materials, Photovoltaic Thermal Hybrid Solar Collectors, Energy Management, Mini Channel Heat Sinks, Micro Mixers, Heat Exchangers, Multi-phase Flow, Thermosyphon, Nanofluids, Non-Newtonian Fluids, Porous Media, HVAC (Heating, Ventilation, and Air Conditioning), Analytical Methods, and Computational Fluid Dynamics (CFD). Dr. Hosseinzadeh has received significant recognition for his contributions to the field, being acknowledged as one of the World's Top 1% Scientists by both the Web of Science (WOS) and Essential Science Indicators (ESI) in 2022 and 2023. Additionally, he has been consistently recognized in the top 2% of scientists worldwide by Stanford University in the years 2020, 2021, 2022, and 2023.

Mehdi Mahboobtosi graduated from Noshirvani University of Technology in Babol with a master's degree in mechanical engineering, majoring in energy conversion. He was among the top students in his studies. He is a researcher in energy conversion mechanical engineering. His research interests include phase change materials, energy storage systems, nanofluids, hydrogen liquefaction, energy management, heat exchangers, nonlinear methods in solving engineering problems.

Davood Domiri Ganji is a full professor in Mechanical Engineering at Noshirvani Babol University of Technology, Iran. He obtained my Bachelor's degree in Heat and Fluids from the University of Tehran, followed by a Master's and PhD in Energy Conversion at Tarbiat Modares University. He published numerous books and papers, furthering knowledge in energy conversion, heat and fluids, and nonlinear dynamics. As editor-in-chief of the International Journal of Nonlinear Dynamics in Engineering and Sciences (IJNDES), he collaborates with other scholars in the field. He is passionate about teaching and mentoring students, sharing my expertise to shape future engineers. Outside of academia, He enjoys traveling and experiencing new cultures.

Index

Ensure Quality Research is Introduced to the Academic Community

Become an Reviewer for IGI Global Authored Book Projects

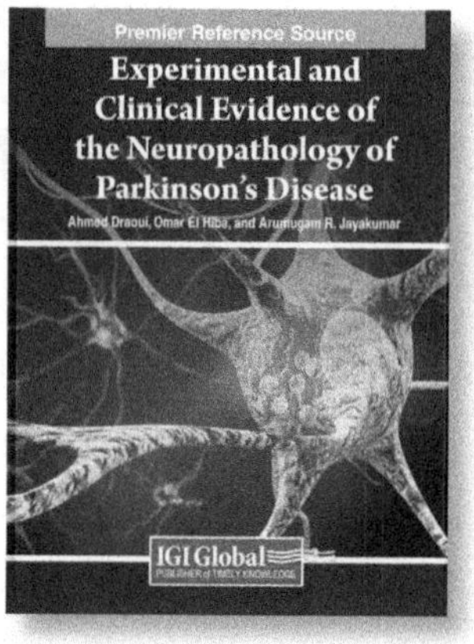

The overall success of an authored book project is dependent on quality and timely manuscript evaluations.

Applications and Inquiries may be sent to:
development@igi-global.com

Applicants must have a doctorate (or equivalent degree) as well as publishing, research, and reviewing experience. Authored Book Evaluators are appointed for one-year terms and are expected to complete at least three evaluations per term. Upon successful completion of this term, evaluators can be considered for an additional term.

If you have a colleague that may be interested in this opportunity, we encourage you to share this information with them.

www.ingramcontent.com/pod-product-compliance
Lightning Source LLC
LaVergne TN
LVHW062120300926
849161LV00010B/2501
* 9 7 9 8 3 6 9 3 0 9 3 2 2 *